PHOTOREFRACTIVE EFFECTS AND MATERIALS

The Kluwer International Series in Engineering and Computer Science

ELECTRONIC MATERIALS: SCIENCE AND TECHNOLOGY

Series Editor
Harry L. Tuller
Massachusetts Institute of Technology

Other books in the series:

Diamond: Electronic Properties and Applications, Lawrence S. Pan and Don R. Kania, editors
Ion Implantation: Basics to Device Fabrication, Emanuele Rimini, author
Solid State Batteries: Materials Design and Optimization, C. Julien and G. Nazri, authors
Sol-Gel Optics: Processing and Applications, L.C. Klein, editor

---

The Series **ELECTRONIC MATERIALS: Science and Technology** will address the following goals

* Bridge the gap between theory and application.

* Foster and facilitate communication among the materials scientists, electrical engineers, physicists and chemists.

* Provide publication with an interdisciplinary approach in the following topic areas:

- Sensors and Actuators
- Electrically Active Ceramics and Polymers
- Structure-Property-Processing -Performance Correlations in Electronic Materials
- Electrically Active Interfaces
- High T_c Superconducting Materials
- Optoelectronic Materials
- Composite Materials
- Defect Engineering
- Solid State Ionics
- Electronic Materials in Energy Conversion-Solar Cells, High Energy Density Microbatteries, Solid State Fuel Cells, etc.

With the dynamic growth evident in this field and the need to communicate findings between disciplines, this book series will provide a forum for information targeted toward

• Materials Scientists • Electrical Engineers • Physicists • Chemists

PHOTOREFRACTIVE EFFECTS AND MATERIALS

edited by

David D. Nolte

Purdue University

KLUWER ACADEMIC PUBLISHERS
Boston / Dordrecht / London

Distributors for North America:
Kluwer Academic Publishers
101 Philip Drive
Assinippi Park
Norwell, Massachusetts 02061 USA

Distributors for all other countries:
Kluwer Academic Publishers Group
Distribution Centre
Post Office Box 322
3300 AH Dordrecht, THE NETHERLANDS

Library of Congress Cataloging-in-Publication Data

A C.I.P. Catalogue record for this book is available
from the Library of Congress.

Printed on acid-free paper.

Printed in the United States of America

TABLE OF CONTENTS

Chapter 2

THE PHOTOREFRACTIVE EFFECT IN FERROELECTRIC OXIDES

M. H. Garrett, G. D. Fogarty, G. D. Bacher, R. N. Schwartz and B. A. Wechsler

Chapter 3

PERMANENTLY FIXED VOLUME PHASE GRATINGS IN FERROELECTRICS

A. S. Kewitsch, A. Yariv and M. Segev

Chapter 4

PHOTOREFRACTIVE SPATIAL SOLITONS

M. Segev, B. Crosignani, G. Salamo, G. Duree, P. Di Porto and A. Yariv

Chapter 5

PHOTOREFRACTIVE POLYMERS

S. M. Silence, D. M. Burland and W. E. Moerner

Chapter 6

NEAR-RESONANT PHOTOREFRACTIVE EFFECTS IN BULK SEMICONDUCTORS

J. E. Millerd, E. Garmire and A. Partovi

Chapter 7

PHOTOREFRACTIVE QUANTUM WELLS AND THIN FILMS

D. D. Nolte and M. R. Melloch

Chapter 8

PHOTOREFRACTIVE FIBERS

L. Hesselink

PREFACE

The photorefractive effect is now firmly established as one of the highest-sensitivity nonlinear optical effects, making it an attractive choice for use in many optical holographic processing applications. As with all technologies based on advanced materials, the rate of progress in the development of photorefractive applications has been principally limited by the rate at which breakthroughs in materials science have supplied better photorefractive materials. The last ten years have seen an upsurge of interest in photorefractive applications because of several advances in the synthesis and growth of new and sensitive materials.

This book is a collection of many of the most important recent developments in photorefractive effects and materials. The introductory chapter, which provides the necessary tools for understanding a wide variety of photorefractive phenomena, is followed by seven contributed chapters that offer views of the state-of-the-art in several different material systems. The second chapter represents the most detailed study to date on the growth and photorefractive performance of $BaTiO_3$, one of the most important photorefractive ferroelectrics. The third chapter describes the process of permanently fixing holographic gratings in ferroelectrics, important for volumetric data storage with ultra-high data densities. The fourth chapter describes the discovery and theory of photorefractive spatial solitons. Photorefractive polymers are an exciting new class of photorefractive materials, described in the fifth chapter. Polymers have many advantages, primarily related to fabrication, that could promise a breakthrough to the marketplace because of ease and low-cost of manufacturing. The sixth and seventh chapters describe the newest advances in photorefractive semiconductors that are particularly attractive because they can be naturally and easily integrated with existing semiconductor electronic and photonic technologies. The final chapter presents the fabrication and application of photorefractive fibers.

This book has been written to be accessible to material scientists who are not experts in photorefractive materials, while at the same time providing the most up-to-date information for photorefractive specialists. Students should find it to be a clear introduction and reference, and an excellent resource for researchers interested in learning about the most advanced discoveries of photorefractive effects, and applying that knowledge in the laboratory. References to over 400 original papers represent the personal selection of the contributing authors, and identify sources of additional information which will help guide the reader to the interesting and extensive literature on this subject.

I am grateful to all my colleagues for their assistance and contributions, and to Alastair Glass for introducing me to the exciting field of photorefractive research. Special thanks go to my wife and colleague Prof. Laura J. Pyrak-Nolte at the University of Notre Dame for her unique talent for making difficult tasks easy.

Purdue University, West Lafayette

David D. Nolte
December 1994

1.

PHOTOREFRACTIVE TRANSPORT AND MULTIWAVE MIXING

D. D. Nolte
Department of Physics
Purdue University
West Lafayette, IN 47907

1. THE PHOTOREFRACTIVE EFFECT

The photorefractive effect [1] can loosely be defined as a light-induced change in the optical properties of a material when the incident light is spatially nonuniform. The spatial nonuniformity is a key feature that distinguishes the photorefractive effect from other common nonlinear optical effects that occur under spatially uniform intensities [2, 3]. In other words, the optical changes are driven by the gradient of the intensity, rather than by the intensity itself.

This subtle distinction between optical nonlinearities that respond to intensity and the photorefractive effect that responds to intensity gradients has important consequences for optical applications that require low light levels, such as optical computing. The response to gradients corresponds to a nonlocal response of the material in both time and space; the maximum optical changes induced in the material do not occur instantaneously nor necessarily locally where the light stimulus is a maximum. The nonlocal response occurs because charges move and are stored inside the photorefractive material, making it possible to integrate and store the optical stimulus over time. The time integration significantly reduces the incident light intensity needed to drive the optical changes in the material.

Ultra-low intensity performance is one of the chief advantages of the photorefractive effect. Most nonlinear optical materials require large irradiances (often greater than kW/cm^2) to generate significant refractive index changes. Photorefractive materials, in contrast, operate using irradiances of only mW/cm^2 or $\mu W/cm^2$. Achieving such low irradiances is a prerequisite for low power consumption in optical image processing and optical computing. The price paid for the low-light operation is diminished speed. But the many different photorefractive materials and operating conditions provide a range of operating speeds that can span many orders of magnitude from nanoseconds and picoseconds [4-8] to hours and longer. Fast materials are suited for real-time processing of optical information, while slow materials are useful for optical memory applications [9-11]. The broad range of photorefractive effects, materials and applications have been described in several good reviews [12-18].

A response to gradients can have important applications, such as edge detection in images. However, the primary usefulness of photorefractive materials is found in holography [19]. Holography converts an image into sinusoidal intensity patterns which have well-defined spatial derivatives. Therefore, the photorefractive effect, which responds specifically to spatial derivatives of light intensity, is a perfect material for dynamic holography and its many applications. Dynamic holography involves the formation of light-induced gratings in a nonlinear optical material. Beams diffracted from the grating act to regenerate images without lenses, and under special circumstances (called phase conjugation) can remove aberrations in images.

The photorefractive process for steady-state conditions and for transport of a single photocarrier type (electrons) is shown in Fig. 1. It begins with the photogeneration of charge carriers that are free to move. The charge generation usually is associated with photoionization of defects within the material, although advanced photorefractive materials also rely on photogeneration of charge from special organic molecules [see Chap. 5] incorporated in photorefractive polymers, or from interband transitions [see Chaps. 6 and 7]. The photogeneration rate is locally a maximum at the positions of maximum intensity. The nonequilibrium photocarrier densities can be driven by diffusion or drift or both. When no external electric field is applied, the carriers diffuse away symmetrically from the intensity maxima. When an external electric field is applied, then drift dominates the transport. For any of these transport processes charge is separated only under spatially nonuniform photoexcitation.

The second stage of the photorefractive process after photogeneration and transport is the trapping of the charge carriers, forming a semi-permanent space-charge density inside the material. The trapping occurs at defect sites that are available (empty) to trap charge carriers. The immediate consequence of separated and trapped space-charge is the formation of space-charge electric fields. When drift dominates the transport under large applied electric fields, the electric field is maximum in the dark fringes, but is screened in the bright fringes. The electric field is therefore coincident with the maxima and minima of the intensity. On the other hand, when diffusion dominates the transport, the electric field maxima are shifted by a quarter fringe spacing. The location of the electric field inside the material, relative to the location of the interference fringes, plays a fundamental role in photorefractive nonlinear optical mixing [20].

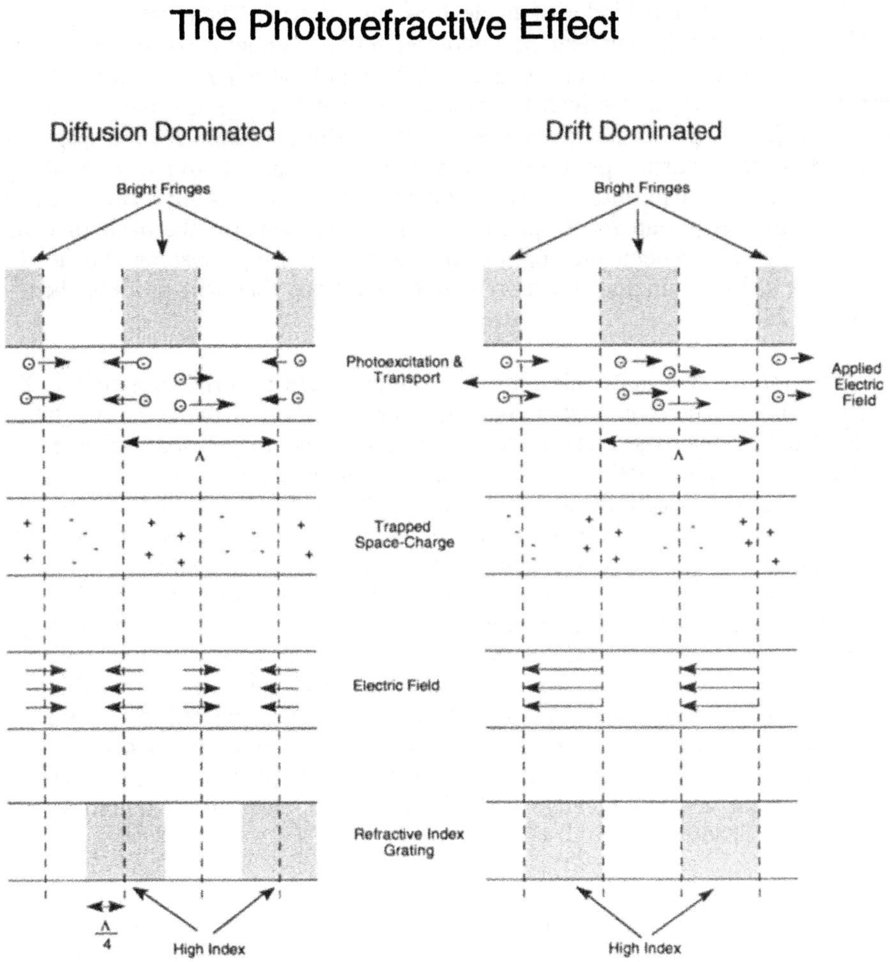

Fig. 1 The photorefractive process for diffusion-dominated and drift dominated transport. The interference fringe pattern is imprinted in the photorefractive material as a refractive index grating. A spatial shift of the grating can occur relative to the intensity pattern, with important consequences for photorefractive applications.

The final stage of the photorefractive grating formation process is the conversion of the internal space-charge electric field into a spatial modulation of the refractive index of the material. This conversion is made possible by the electro-optic effect; electric fields change the refractive index of an electro-optic material. Therefore, the spatially modulated light intensity incident on a photorefractive material is converted through the photorefractive process into a spatial modulation of the refractive index with the same spatial frequency as the intensity pattern. A quarter fringe spacing shift of the refractive index relative to the optical stimulus (for diffusion-dominated transport) has special properties during photorefractive wave mixing. It allows one of the interfering beams to gain intensity at the expense of the other beam. This is called nonreciprocal energy transfer, or photorefractive gain, which is the basis of many photorefractive phenomena and applications. A quarter-fringe-spacing shift is also possible for drift-dominated transport under conditions that will be described in section 4.1.2.

Useful photorefractive materials must have appropriate material properties. For instance photorefractive materials must be insulators or semi-insulating semiconductors, otherwise excess free carriers screen the trapped space-charge. On the other hand, photorefractive materials must have appreciable photoconductivity to allow the charge to separate and form space-charge fields. Defect states with sufficient concentrations are essential to the photorefractive process because they provide the sites to trap the space-charge. When insufficient trap sites are available, the space-charge fields and the optical gratings are limited in magnitude. Electro-optic effects ultimately determine the magnitude of the light-induced gratings. Ferroelectric oxides, such as $BaTiO_3$, $KNbO_3$ or $SrBaNbO_3$ have large linear electro-optic effects that make them attractive for photorefractive applications (see Chaps. 2 and 3). The large electro-optic effects in these materials are related to ferroelectric lattice instabilities and phase transitions. Other large electro-optic effects occur for resonant or near-resonant excitation in semiconductors for photon energies at or near the fundamental absorption edge of bulk semiconductors or in quantum-well structures (see Chaps. 6 and 7).

The eventual usefulness of photorefractive materials will depend on the many different requirements that optical system designs will demand of optical materials. At this early stage in the evolution of optical components and systems, it is difficult to predict whether photorefractive materials will find a niche in future applications. However, the rich variety of photorefractive materials and photorefractive behavior, with the increasing ability to tailor and design novel materials to fulfill specific functions, make research on photorefractive materials an exciting and fruitful area of study.

2. COHERENT INTERFERENCE

The optical stimulus that starts the photorefractive grating formation process is the coherent interference between two laser beams. Two mutually coherent light fields interfere to form interference fringes. For two linearly polarized coherent plane waves with wavevectors $\mathbf{k_1}$ and $\mathbf{k_2}$ inside a crystal, the plane waves are

$$\mathbf{E_1} = \hat{\mathbf{e}}_1 A_1 \exp\left[i\left(\mathbf{k}_1 \cdot \mathbf{r} - \omega t\right)\right]$$

$$\mathbf{E_2} = \hat{\mathbf{e}}_2 A_2 \exp\left[i\left(\mathbf{k}_2 \cdot \mathbf{r} - \omega t + \delta\right)\right] \tag{1}$$

where $\mathbf{e}_{1,2}$ are unit polarization vectors and the phase δ is the relative phase difference between the beams, which can be a function of time. The intensity for the interference is proportional to the square of the modulus of the field

$$|\mathbf{E}|^2 = |\mathbf{E}_1 + \mathbf{E}_2|^2$$

$$= \left|\left(\hat{\mathbf{e}}_1 A_1 \exp\left(i\,\mathbf{k}_1 \cdot \mathbf{r}\right) + \hat{\mathbf{e}}_2 A_2 \exp\left(i\left(\mathbf{k}_2 \cdot \mathbf{r} + \delta\right)\right)\right)\right|^2$$

$$= A_1{}^2 + A_2{}^2 + \left(\hat{\mathbf{e}}_1^* \cdot \hat{\mathbf{e}}_2\right) A_1^* A_2 \exp\left(i\left(\mathbf{K} \cdot \mathbf{r} + \delta\right)\right)$$
$$+ \left(\hat{\mathbf{e}}_1 \cdot \hat{\mathbf{e}}_2^*\right) A_1 A_2^* \exp\left(-i\left(\mathbf{K} \cdot \mathbf{r} + \delta\right)\right) \tag{2}$$

where the grating vector $\mathbf{K}$ is given by the difference of the beam k-vectors

$$\mathbf{K} = \mathbf{k}_2 - \mathbf{k}_1 . \tag{3}$$

The grating vector is related to the fringe spacing Λ through $K = 2\pi/\Lambda$. The intensity of the interference fringe pattern is

$$I = I_1 + I_2 + \left(\hat{\mathbf{e}}_1^* \cdot \hat{\mathbf{e}}_2\right) A_1^* A_2 \exp\left(i\left(\mathbf{K} \cdot \mathbf{r} + \delta\right)\right)$$
$$+ \left(\hat{\mathbf{e}}_1 \cdot \hat{\mathbf{e}}_2^*\right) A_1 A_2^* \exp\left(-i\left(\mathbf{K} \cdot \mathbf{r} + \delta\right)\right) \tag{4}$$

which can also be expressed as

$$I = I_0\left(1 + \frac{m}{2}\exp\left(i\left(\mathbf{K}\cdot\mathbf{r} - \delta\right)\right) + \frac{m^*}{2}\exp\left(-i\left(\mathbf{K}\cdot\mathbf{r} - \delta\right)\right)\right) \tag{5}$$

where the total intensity is

$$I_0 = I_1 + I_2 \tag{6}$$

The complex modulation index is defined as

$$m = 2\frac{\hat{e}_1^*\cdot\hat{e}_2}{I_1 + I_2}A_1^* A_2 \,. \tag{7}$$

with a magnitude

$$|m| = 2\frac{\sqrt{I_1 I_2}}{I_1 + I_2}\left|\hat{e}_1\cdot\hat{e}_2\right| \tag{8}$$

where the fundamental constants have been dropped, allowing us to write $I_1 = A_1{}^*A_1$. The modulation index is often expressed in terms of the ratio of the initial intensities of the two beams

$$\beta = \frac{I_1}{I_2} \tag{9}$$

where I_1 is conventionally the stronger of the two beams. The magnitude of the modulation index is therefore

$$|m| = \left|\hat{e}_1\cdot\hat{e}_2\right|\frac{2\sqrt{\beta}}{1+\beta} \tag{10}$$

The modulation index m plays several important roles in photorefractive phenomena and helps in the numerical modeling of those effects. When m is small, it serves as an expansion parameter that allows the photorefractive transport equations to be linearized. When m approaches unity for nearly equal intensities $I_1 \approx I_2$, the transport equations become strongly nonlinear, leading to complex behavior that can be difficult to interpret. The modulation ratio is also the source of phase conjugate beams.

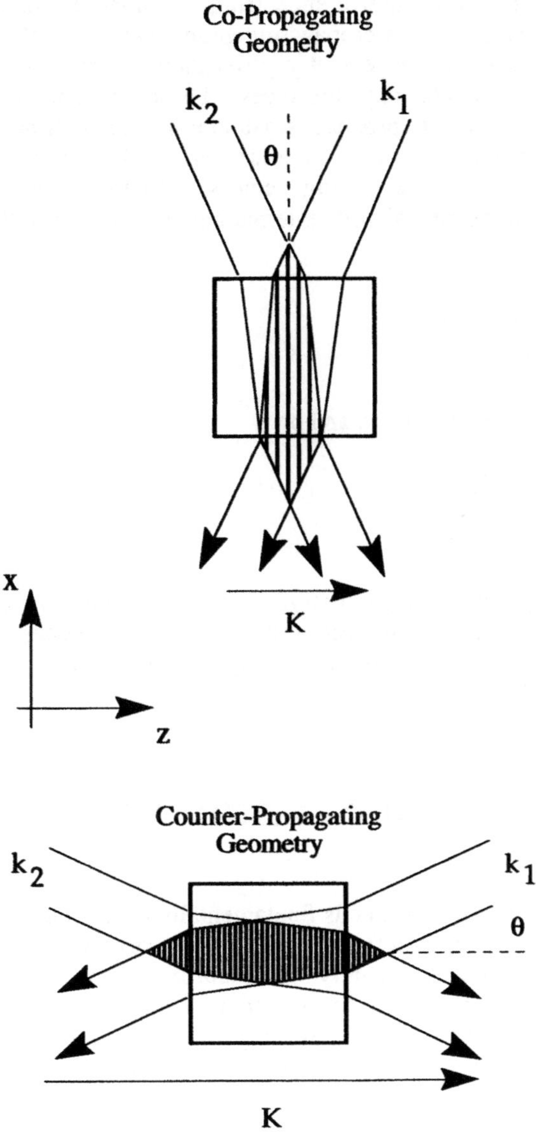

Fig. 2 The fringe spacings for the transmission geometry can be many times larger than the free-space wavelength, while the fringe spacings for the reflection geometry can be several times shorter than the free-space wavelength.

There are two common interference fringe geometries for K: the transmission geometry (also called the co-propagating geometry), in which both beams enter the same face; and the reflection geometry (also called the counter-propagating geometry), in which the beams enter opposite faces. These geometries are shown in Fig. 2. The transmission geometry produces small grating vectors K because refraction at the surface significantly reduces the internal angles. The reflection geometry, on the other hand, leads to large grating vectors. When the two beams are incident symmetrically on the crystal with external angle of incidence θ, the fringe spacings are given by

$$\Lambda = \frac{\lambda}{2 \sin\theta} \tag{11}$$

in the co-propagating geometry, and by

$$\Lambda = \frac{\lambda}{2\sqrt{n^2 - \sin^2\theta}} \tag{12}$$

in the counter-propagating geometry. In the co-propagating geometry, the fringe spacing is independent of the refractive index and depends only on the external incidence angle. In the counter-propagating geometry, the fringe spacing is approximately $\Lambda \approx \lambda/2n$.

3. DEFECT PHOTOIONIZATION AND CARRIER TRAPPING

The photorefractive effect depends fundamentally on crystal defects, and no detailed understanding of the photorefractive effect is possible without an understanding of defect physics. Once a spatially nonuniform light intensity is incident on the photorefractive materials, photocarriers are ionized out of point defects, they drift or diffuse, and are trapped at defects, forming trapped space-charge that is the source of the space-charge electric fields that mirror the input light intensity.

Many of the aspects of photorefractive performance can be traced back to properties of the point defects. Aberrations from "typical" photorefractive response, leading to pathological behavior degrading photorefractive device performance, can usually be attributed to the influence of multiple defects levels in the insulating or semi-insulating photorefractive materials. Point defects are both cause and culprit for ideal or deviant photorefractive behavior. Control of defects means control of many photorefractive properties. Unfortunately, it has been extremely difficult to control defects in many of the traditional photorefractive materials, especially the ferroelectric oxides. Even in semiconductors, which are relatively pure compared to ferroelectrics, multiple defect levels can lead to unexpected results [21].

When the translational symmetry of the crystal is broken by a localized flaw in the lattice, energy levels arise inside the forbidden energy gap of a crystal . The flaw can be a substitutional impurity, an interstitial or a vacancy. Such "point" defects can also form complexes of multiple point defects. Each of these different types of defects produce widely different energy levels within the bandgap. The vast number of possible point defects and complexes has made it difficult to identify the defect structures that are related to energy levels in crystals.

A defect "level" is defined by a change in the charge state of the point defect. A defect "energy" is therefore associated with a transition between two charge states. As an example, for a donor emitting an electron to the conduction band, the trap energy is given by

$$E(N-1) + e_{CB} - E(N) = E_T \tag{13}$$

where E(N) is the total energy of the Nth charge state, e_{CB} is the energy of an electron at the bottom of the conduction band and E_T is the trap energy measured with respect to the conduction band. A defect with multiple charge states is shown in Fig. 3a. An understanding of eq. (13) removes common misconceptions concerning defects. First, a defect energy level E_T represents a pair of charge states. Second, the energy level corresponds to differences in total energy, including elastic energy of the distorted lattice around the defect. Changes in charge state can be accompanied by changes in the lattice elastic energy. This elastic energy must be included when optical transition energies are compared against thermal activation energies, because thermal activation energies and optical transition energies are not equal [22, 23].

The first prerequisite for a photorefractive material is high resistivity, caused when the Fermi level is pinned near a defect level deep in the bandgap. The high resistivity is necessary to reduce dielectric relaxation, which is an erasure process in the photorefractive effect. If the dielectric relaxation rate is too large, then a large space-charge density cannot be supported by the usually weak illumination used in photorefractive applications.

In the case of a single deep acceptor with charge states $N^{-/0}$ and a transition E_A, the equilibrium defect occupancies are

$$\begin{aligned} N_A^- &= \frac{N_A}{1+\frac{1}{g}\exp\left[\left(E_c - E_A - E_F\right)\right]} \\ N_A^0 &= \frac{N_A}{1+\frac{1}{g}\exp\left[\left(E_F - E_v - E_A\right)\right]} \end{aligned} \tag{14}$$

where E_F is the Fermi level and g is a degeneracy factor, usually assumed to be g = 2. The Fermi level is pinned by the defect when the deep defect concentration exceeds the shallow donor concentration N_{sd}. The location of the Fermi level is given by the solution to

$$0 = N_{sd} - N_c \exp\left[\left(E_F - E_c\right)/k_BT\right] - \frac{N_A}{1 + \frac{1}{g}\exp\left[\left(E_c - E_A - E_F\right)/k_BT\right]} \qquad (15)$$

where N_c is the effective density of states of the conduction band. The Fermi level is usually pinned within several k_BT of the defect level.

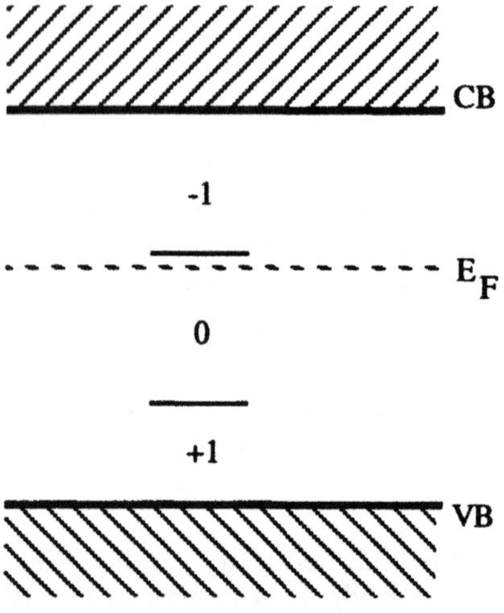

Fig. 3a Multiple levels associated with a single point defect. The two defect levels correspond to charge state transitions among three different charge states.

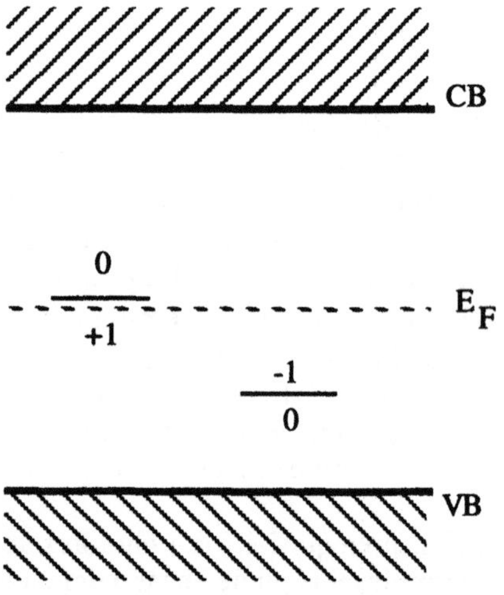

Fig. 3b Defect levels from two different point defects. Compensating shallow defects are assumed to be present but are not shown.

Optical transition energies and cross-sections determine photo-generation rates, which are a key part of the photorefractive effect. The extrinsic absorption from a deep acceptor is given by

$$\alpha(h\nu) = s_e(h\nu)\, N_A^- + s_h(h\nu)\, N_A^0 \qquad (16)$$

where $s_e(h\nu)$ and $s_h(h\nu)$ are the electron and hole optical cross-sections, and N^- and N^0 are the occupancies of the defects.

Optical cross-sections are characterized by two parameters: a threshold energy and a peak magnitude. The functional dependence of the cross-section on photon energy is known approximately for transitions to parabolic bands from highly localized defect states. The optical cross section from a simple model of a defect wavefunction bound by a delta-function potential is [24]

$$s(h\nu) = \frac{8}{3}\frac{e^2\hbar}{m^* c}\frac{E_I^{1/2}\left(h\nu - E_I\right)^{3/2}}{(h\nu)^3} \tag{17}$$

(in cgs units) where E_I is the optical ionization energy and m* is the effective mass for the parabolic band. The maximum occurs at $h\nu = 2E_I$, and decreases as $(h\nu)^{-3/2}$ for large ν. As an example of the cross-section magnitude, a defect with $E_I = 0.5$ eV and a photon with $h\nu = 1$ eV in a transition to a conduction band with $m^* = 0.067\ m_e$ has a peak cross-section $s \approx 10^{-16}\ cm^2$. The cross-section in eq. (17) agrees qualitatively with experimental data for deep levels in semiconductors [25]. The quantitative values for optical cross sections in practical systems vary widely and depend on the details of the defect wavefunction and the band states. For photorefractive applications the important aspects concerning optical cross-sections are that they can have relatively strong photon energy dependencies, and can have significantly different transition probabilities for electron or hole transitions. The energy dependence makes it difficult to compare photorefractive experiments performed using laser sources with different photon energies. Capture cross sections are even less well-behaved than optical cross-sections and cannot be described even qualitatively by a general formula.

When light is incident on the crystal, the defect occupancies are determined by a balance between photoexcitation and carrier trapping expressed by

$$\frac{\partial n}{\partial t} - \nabla\frac{j_e}{e} = \left(I\, s_e + \beta_e\right) N_A^- - \gamma_e\, n\, N_A^0 \tag{18}$$

where

$$\beta_e = \sigma_e\, v_{th}\, N_c \exp\left(-(E_{cb} - E_A)/k_B T\right) \tag{19}$$

is the thermal generation rate, N_c is the effective density of states for the conduction band, E_{cb} is the conduction band energy, and

$$\gamma_e = \sigma_e\, v_{th} \tag{20}$$

is the capture coefficient. The steady-state occupancy under illumination I is given by

$$\frac{N_A^-}{N_A^0} = \frac{N_{sd} - N_{sa} - n_0}{N_A - N_{sd} + N_{sa} + n_0} = \frac{\gamma_e n_0}{\left(I s_e + \beta_e\right)} \tag{21}$$

where N_{sd} and N_{sa} are the shallow donor and acceptor levels.

4. ONE-DIMENSIONAL TRANSPORT EQUATIONS AND SPACE-CHARGE GRATINGS

Photocarrier transport and the formation of photo-induced space-charge electric fields are the foundation of the photorefractive effect. This problem is much more general than only for photorefractive materials, and arises any time an electric field is applied to high-resistivity material [26]. One-dimensional transport equations are the starting point for an understanding of space-charge gratings.

The photorefractive effect is a photo-induced space-charge effect that balances the formation of space-charge trapped at defects against dielectric relaxation as free carriers move to neutralize the space charge. There are several transport models of increasing complexity that can be invoked to describe many of the basic features of photorefractive behavior. Some of these models are band transport models in which photocarriers are excited into a continuous energy band, where they are free to transport before trapping at defect sites [27-30]. An alternative transport model in ferroelectrics is the hopping transport model in which photogenerated charge hops from one defect site to another [31- 33].

The simplest model is a single-defect single-band model in which a single defect communicates with only one of the bands. While this model has the simplest solution, and is often used for heuristic purposes, it usually fails to quantitatively describe most photorefractive materials The next level of complexity, called the single-defect two-band model, includes communication with a second band [34-38]. This model is necessary to describe the effects of electron-hole competition, which can be pronounced. However, even this model often fails, especially at low temperatures or in nominally undoped crystals. Most materials, even high-quality ones, have many defect levels. The interplay of trapping and space-charge among multiple defect levels can lead to unexpected and complicated behavior in photorefractive crystals during photorefractive mixing [39-42]. The two-defect two-band model can be used to model experimentally observed effects [21]. Models that include large numbers of defects become prohibitively complicated, but may be invoked when necessary, and some phenomenological models can be invoked to handle the effects of multiple defects [39]. In this section, we describe the analytic results for the one-defect one-band, the one-defect two-band, and the two-defect two-band models, pointing out the important features that each successive model provides. These models can all be solved analytically in a small-signal analysis.

4.1 One-Defect One-Band Model

In the one-defect one-band model, a single defect level communicates with one bandedge. The transitions are shown in Fig. 4. The deep defect is assumed to be an acceptor with density N_A that is compensated by shallow donors with density N_{sd}. The deep acceptor communicates only with the conduction band in this model.

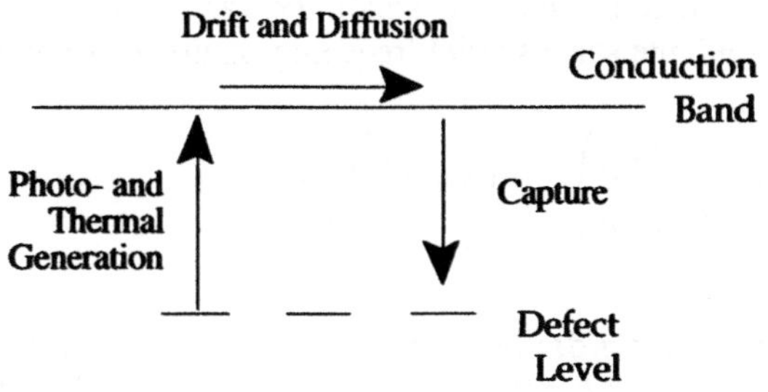

Fig. 4 One-defect one-band model for photorefractive transport showing the transport and transition processes. The generation of trapped space-charge is balanced against erasure by dielectric relaxation.

The transport and space-charge equations for electrons are

detailed balance: $$\frac{\partial n}{\partial t} - \nabla \frac{j_e}{e} = \left(I\, s_{eA} + \beta_{eA}\right) N_A^- - \gamma_{eA}\, n\, N_A^0 \qquad (22)$$

current density: $$j_e = e\, \mu_e\, n\, E + k_B\, T\, \mu_e\, \nabla n \qquad (23)$$

continuity: $$\frac{\partial}{\partial t}\left(n + N_A^-\right) = \frac{\nabla}{e}\, j_e \qquad (24)$$

ion density: $$N_A = N_A^0 + N_A^- \qquad (25)$$

Poisson's Eq'n: $$\nabla E = -\left(\frac{e}{\varepsilon\, \varepsilon_0}\right)\left(n + N_A^- - N_{sd}\right) \qquad (26)$$

Poisson's equation plays a key role by connecting the spatially modulated space-charge (represented by $N_A^- - N_{sd}$) with the electric field. The photovoltaic effect has been omitted from these transport equations, but can play a dominant role in many ferroelectrics. For a discussion of the important contributions from the photovoltaic effect see Refs [43-54].

The transport and space-charge equations are nonlinear differential equations in several nonlinear variables. A full nonlinear solution can be obtained numerically.

However, analytic solutions can be obtained using small signal analysis. The light pattern is assumed to have a space and time dependence given by

$$I(x,t) = I_0\left(1 + m\, e^{i(Kx - \Omega t)}\right) \tag{27}$$

where the modulation index m is given in eq. (8). This expression represents a moving fringe pattern with a fixed grating vector K moving at a velocity $v = \Omega/K$. The fringe pattern is stationary when $\Omega = 0$ and steady-state solutions result.

When the modulation index m is small, the variables are assumed to have sinusoidal components with the same spatial frequency as the interference pattern,

$$\begin{pmatrix} n(x) \\ E(x) \end{pmatrix} = \begin{pmatrix} n_0 \\ E_0 \end{pmatrix} + e^{i(Kx - \Omega t)}\begin{pmatrix} n_1 \\ E_1 \end{pmatrix} \tag{28}$$

Furthermore, higher-order harmonic components are neglected. It is convenient to work with solutions to the space-charge density rather than the space-charge field E_1. The modulated space-charge density is

$$N_1 = \frac{iK\varepsilon\varepsilon_0}{e} E_1 \tag{29}$$

The zero-order solution to the free electron density is given by the solution to

$$n_0 = \frac{\left(s_{eA} I_0 + \beta_{eA}\right)}{\gamma_{eA}}\left(\frac{N_{sd} - n_0}{N_A - N_{sd} + n_0}\right) \tag{30}$$

which requires an iterative solution. When the photocarrier density n_0 is much smaller than the defect densities, then the n_0 terms on the right hand side of the equation can be neglected.

Several transition rates are important when considering the balance among the transition processes in eqs. (22-26). These transition rates are defined in Table I [36, 55].

Table I. Electron Transport and Transition Rates:	
Dielectric relation rate:	$\Gamma_{die} = e\mu_e n_0/\varepsilon\,\varepsilon_0$
Transition rate:	$\Gamma_{Ie} = s_{eA} I_0 + \beta_{eA} + \gamma_{eA} n_0$
Ion Recombination rate:	$\Gamma_{Re} = \gamma_{eA}(N_A + n_0 - N_{sd})$
Drift rate:	$\Gamma_{Ee} = K\,\mu_e E_0$
Diffusion rate:	$\Gamma_{De} = K^2 k_B T\,\mu_e/e$

Using these transition rates, the linearized equations for the first-order solution are

$$\begin{pmatrix} -i\Omega - i\Gamma_{Ee} + \Gamma_{De} + \Gamma_{Ie} + \Gamma_{Re} & \Gamma_{Ie} - \Gamma_{die} \\ -i\Omega + \Gamma_{Ie} + \Gamma_{Re} & -i\Omega + \Gamma_{Ie} \end{pmatrix} \begin{pmatrix} n_1 \\ N_1 \end{pmatrix} = I_1 \begin{pmatrix} 1 \\ 1 \end{pmatrix} s_e\left(N_{sd} - n_0\right) \tag{31}$$

with the solution for the space-charge density given by

$$N_1 = m\, s_e I_0 \frac{\left(-i\Gamma_{Ee} + \Gamma_{De}\right)\left(N_{sd} - n_0\right)}{\left(\Gamma_{Ie} - i\Omega\right)\left(-i\Gamma_{Ee} + \Gamma_{De} + \Gamma_{die}\right) + \Gamma_{Re}\left(\Gamma_{die} - i\Omega\right) - \Omega^2} \tag{32}$$

It is also helpful to identify characteristic fields and transport lengths for charge transport. These are defined in Table II.

Table II. Electron Transport Lengths and Fields

Lengths:	
Drift Length:	$L_{Ee} = E_0 \mu\tau = \frac{1}{K}\frac{\Gamma_{Ee}}{\Gamma_{Re}}$
Diffusion Length:	$L_D^2 = \frac{\mu\tau k_B T}{e} = \frac{1}{K^2}\frac{\Gamma_{De}}{\Gamma_{Re}}$
Space-charge Debye Length:	$L_{sc}^2 = \frac{\varepsilon\varepsilon_0 k_B T}{N_T e^2} = \frac{1}{K^2}\frac{\Gamma_{De}}{\Gamma_{die}}\frac{n_0}{N_T}$
Fields:	
Drift Field:	$E_0 = \frac{\Gamma_{Ee}}{K\mu}$
Diffusion Field:	$E_D = \frac{K k_B T}{e} = \frac{\Gamma_{De}}{K\mu}$
Trap-Limited Field:	$E_{max} = \frac{e}{K\varepsilon\varepsilon_0}\frac{N_A^0 N_A^-}{N_A} = \frac{\Gamma_{die}}{K\mu}\frac{\Gamma_{Re}}{\Gamma_{Ie}}$
Space-Charge Field:	$E_1 = \frac{iK\varepsilon\varepsilon_0}{e} N_1$

4.1.1 Steady-State Solutions

The photorefractive space-charge electric field is limited in magnitude by a finite density of trap sites. The largest space-charge field is obtained when there is a full modulation of the charge density on the deep trap. This maximum space-charge field is

$$E_{max} = \frac{e}{K\varepsilon\varepsilon_0}\frac{N_A^- N_A^0}{N_A} = \frac{e}{K\varepsilon\varepsilon_0}\frac{\left(N_{sd} - n_0\right)\left(N_A - N_{sd} + n_0\right)}{N_A} \tag{33}$$

The steady-state solution to the space-charge field when $\Omega = 0$ can be expressed in terms of E_{max} by

$$E_1 = \frac{-m E_{max}}{\left(1 + \frac{\beta_{eA}}{s_e I_0}\right)}\left[\frac{E_0 + iE_D}{-iE_0 + E_D + \frac{e n_0}{K\varepsilon\varepsilon_0} + E_{max}}\right] \tag{34}$$

where

$$E_D = K \frac{k_B T}{e} \tag{35}$$

is the diffusion field and E_0 is the applied field.

When the carrier density n_0 is small, under usually weak illumination conditions, the solution becomes

$$E_1 = - m E_{max} \frac{E_0 + iE_D}{\left(-iE_0 + E_D + E_{max}\right)} \frac{\sigma_p}{\left(\sigma_p + \sigma_d\right)} \tag{36}$$

where σ_d is the dark conductivity and σ_p is the photoconductivity. The space-charge field is linearly proportional to intensity when the photoconductivity is smaller than the dark conductivity, and saturates to a fixed value when the photoconductivity exceeds the dark conductivity.

One of the most important features of the photorefractive effect is the operation at very low light intensities. Most nonlinear optical processes require large light intensities to drive the material into its range of nonlinear response. The advantage of the photorefractive effect is its reliance on the integration of photon flux, which is stored as space-charge at traps. Fully developed space-charge gratings require the photoconductivity to exceed the dark conductivity. This is because the dark conductivity represents an erasure mechanism. Equivalently, from eq.(34), the light intensity needed to fully develop the gratings is balanced against thermal emission of carriers from the traps. The saturation intensity is

$$\begin{aligned} I_{sat} &= \frac{n_d}{\alpha \tau} \\ &= \frac{n_d \gamma_{eA} \left(N_A - N_{sd}\right)}{s_{eA} N_{sd}} \end{aligned} \tag{37}$$

where n_d is the dark carrier density, α is the defect absorption coefficient and τ is the carrier lifetime.

The magnitude of the space-charge field is shown in Fig. 5 as a function of the photon flux. Typical intensities to saturate the gratings in GaAs are approximately I_{sat} = 1 mW/cm^2. The material parameters used to simulate the photorefractive response of GaAs are given in Table III. These values are representative, but should not be considered as specific to particular GaAs photorefractive performance. Different dopants will have different cross-sections.

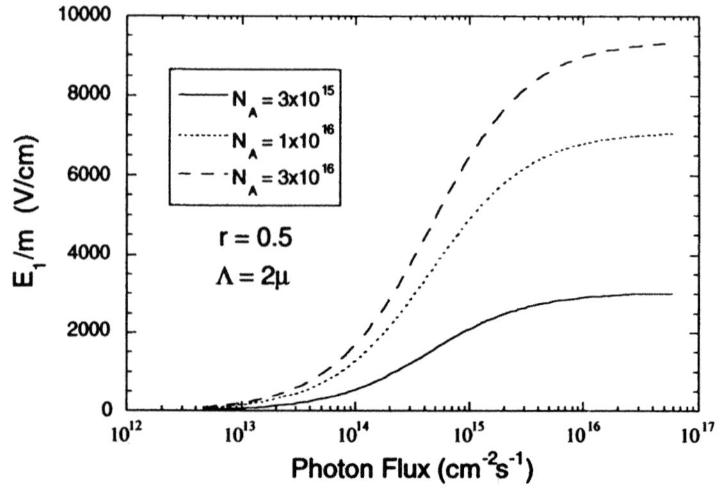

Fig. 5 Space-charge field magnitude as a function of photon flux for $E_0 = 10^4$ V/cm. When the photoconductivity exceeds the dark conductivity, the gratings become fully developed.

Table III Simulation Parameters for GaAs

$s_{eA} = 1\text{x}10^{-16}$ cm^{-2}
$s_{hA} = 1\text{x}10^{-15}$ cm^{-2}
$\sigma_{eA} = 1\text{x}10^{-14}$ cm^{-2}
$\sigma_{hA} = 1\text{x}10^{-15}$ cm^{-2}
$\alpha = 0$ cm^{-1}

In the absence of an applied field the space-charge field is driven by diffusion. At large fringe spacings, the space-charge field is

$$E_1 \approx -imE_D \tag{38}$$

At short fringe spacings the diffusion field can exceed the maximum field E_{max}, limiting the field. Trap-limited performance occurs when the drift field E_0 or diffusion field E_D become comparable to or larger than E_{max}. In this limit the magnitude of the space-charge field saturates to E_{max}. In addition, the space-charge field takes on the maximum phase shift. The photorefractive phase shift, which expresses the spatial shift of the space-charge field relative to the intensity, is given by

$$\phi = \pi + \arctan\left[\frac{E_D(E_D + E_{max}) + E_0^2}{E_0(E_D + E_{max}) - E_0E_D}\right] \tag{39}$$

In the limit as the applied field $E_0 >> E_{max}$ the phase shift simplifies to

$$\phi = \pi + \arctan\left[\frac{E_0}{E_{max}}\right] \tag{40}$$

and ϕ approaches $-\pi/2$ as the space-charge field saturates to E_{max}. The steady-state space-charge field is shown in Fig. 6 as a function of fringe spacing for several defect densities.

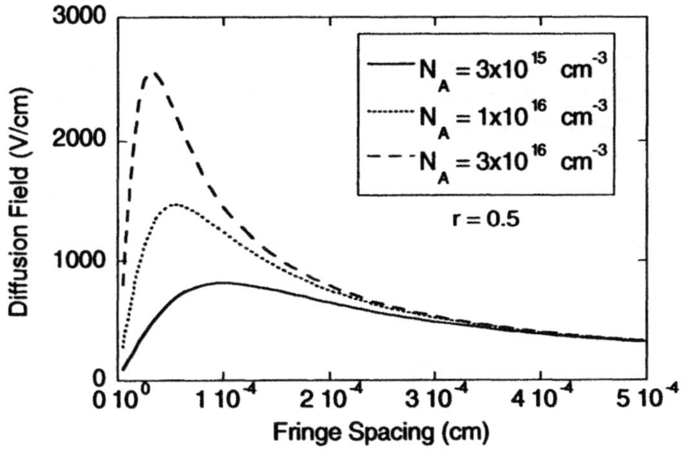

Fig. 6 Diffusion field as a function of fringe spacing for different defect densities. Shorter fringe spacings favor a larger diffusion field when sufficient defects are present to support the space-charge.

The magnitude of the space-charge field is shown in Fig. 7 as a function of applied electric field for the three different trap concentrations used in Fig. 6. In each case the compensation ratio is $r = 0.5$, for $N_{sd} = 0.5\ N_A$. The diffusion field is approximately 700 V/cm at room temperature for a fringe spacing of 2 μm. Trap-limited space-charge fields occur especially for $N_A = 3x10^{15}\ cm^{-3}$. In this case a maximum field of only 3000 V/cm can be supported by the trap density.

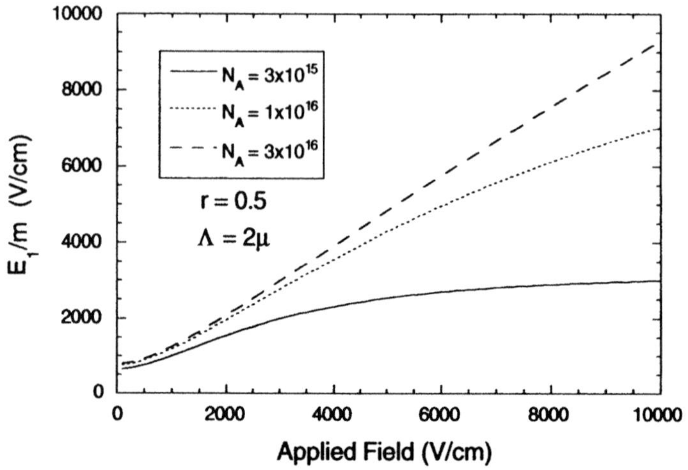

Fig. 7 Magnitude of the steady-state space-charge electric field as functions of applied field for different deep acceptor concentrations with a fixed compensation ratio of 0.5.

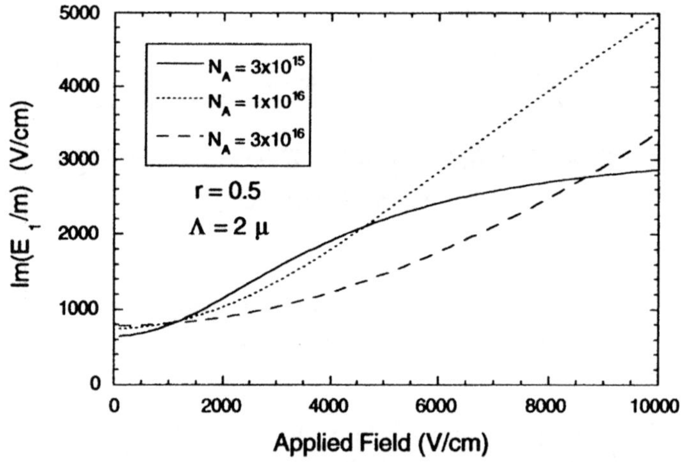

Fig. 8 Imaginary part of the space-charge field from Fig. 7. The imaginary part determines the magnitude of the photorefractive gain.

Photorefractive energy exchange depends on the imaginary part of the space-charge field. This is shown in Fig. 8 for the same conditions as Fig. 7. The behavior is complicated by the interplay between the magnitude of the space-charge field set by the trap density, and the phase shift from eq. (39). The largest imaginary part of the field is obtained in the intermediate case of $N_A = 1x10^{16}$ cm^{-3}. For large trap densities, the phase shift remains small, while for small trap densities, the phase shift approaches $\pi/2$ (Eq. 40), but the magnitude of the space-charge field is limited.

Large imaginary components of the space-charge field are important for large photorefractive gain. However when a large electric field is applied to increase the magnitude of the space-charge field, the phase shift approaches zero. The simplest approach to boost the phase shift while maintaining large field magnitudes simply uses the fringe-spacing dependence of the trap limited field in eq.(33). The space-charge field magnitude is shown in Fig. 9 for an applied field of $1x10^4$ V/cm as a function of fringe spacing. At small fringe spacings, the space-charge field is trap-limited. Only in the case of $N_A = 3x10^{16}$ for fringe spacings larger than $\Lambda = 5$ µ does the trap density fully support enough charge to screen the applied field. A maximum imaginary component of the space-charge field occurs for a specific fringe spacing for a given trap density, shown in Fig. 10. In all cases the maximum value for the imaginary component of the field is half of the applied electric field.

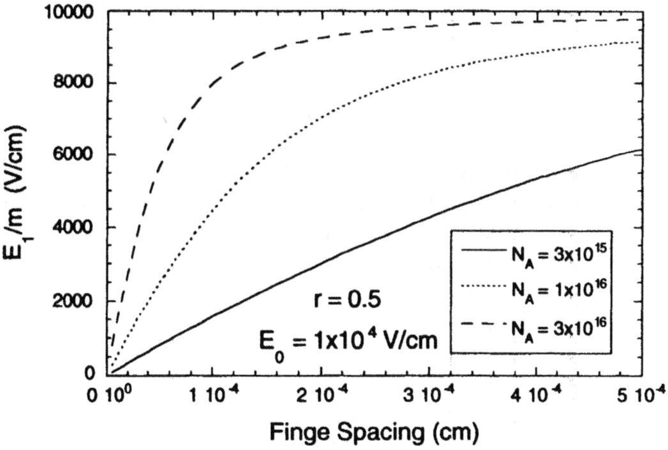

Fig. 9 Magnitude of the steady-state space-charge field as a function of fringe spacing for several trap densities with a fixed applied field of 10^4 V/cm.

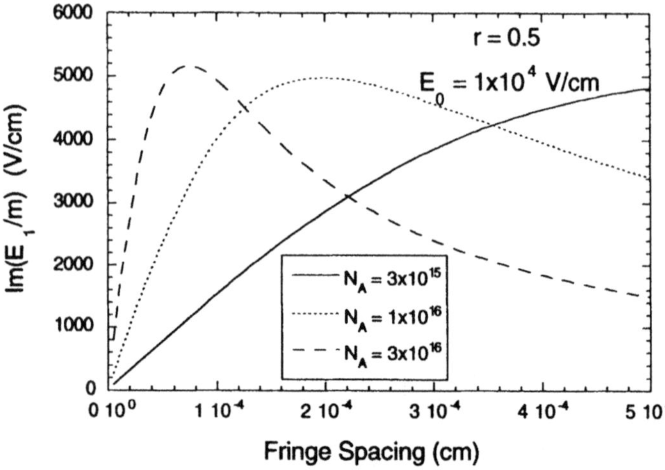

Fig. 10 The steady-state space-charge field becomes trap limited at small fringe spacings and small defect concentrations. The maximum value for the imaginary component as attained at a specific fringe spacing for a specific defect concentration.

4.1.2 Transient and Moving Gratings

The price paid for photorefractive operation at low light powers is relatively low speed. However many holographic image processing applications can be supported by devices with speeds compatible with video rates, which are in the kHz range. In addition, the relatively slow speed of the photorefractive effect can be used to advantage in non-steady-state conditions, such as with moving gratings [56-58] or with alternating electric fields [59, 60].

Many aspects of band transport enter into the response time of photorefractive grating formation under steady illumination [28, 30, 61], and under high-intensity pulses [62]. The basic response time for grating formation can be obtained by taking the determinant of eq. (31) and solving for $-i\Omega = 1/\tau_{PR}$, where τ_{PR} is the photorefractive response time. When the photogeneration rate is not too large, then the solution to the determinant is

$$\frac{1}{\tau_{PR}} = \frac{\Gamma_{die}\,\Gamma_{Re}}{\left(\Gamma_{De} + \Gamma_{Re}\right)^2 + \Gamma_{Ee}^2}\left(\Gamma_{De} + \Gamma_{Re} + i\Gamma_{Ee}\right) \qquad (41)$$

which has a real and imaginary component when an external electric field is applied. The device response time is given by the real part of eq. (41)

$$\tau_0 = \tau_{die}\left[1 + \frac{\Gamma_{De}}{\Gamma_{De}} + \frac{\Gamma_{Ee}^2}{\Gamma_{Re}^2 + \Gamma_{Re}\Gamma_{De}}\right] \tag{42}$$

which can also be expressed in terms of transport lengths as

$$\tau_0 = \tau_{die}\left[1 + K^2 L_D^2 + \frac{K^2 L_E^2}{1 + K^2 L_D^2}\right] \tag{43}$$

The important aspect of eq. (43) is the increase in the response time with increasing applied electric field and decreasing fringe spacing. The low-field long-fringe-spacing limit of the response time is the dielectric relaxation time. This corresponds to the maximum speed that a photorefractive material can achieve. When electric fields are applied to the material to increase the refractive index grating amplitudes, or when shorter fringe spacings are used, then the device speed is reduced.

A heuristic interpretation can be made to explain the diminishing speed with increasing applied field [63]. For a given field, enough time is required to allow the full charge transfer to screen the field. The larger the field, the more charge is needed to screen it, and the more time is needed. The charge integration time t_{Integ} is set by the charge transfer rate dN/dt and by the charge density N_E needed to screen an applied field E

$$\frac{dN}{dt} t_{Integ} = N_E = \frac{2\pi \varepsilon \varepsilon_0 E}{e \Lambda} \tag{44}$$

The charge transfer rate dN/dt is determined by the production rate of free carriers, and by the probability that the carriers are trapped within the first fringe spacing. This capture probability is the ratio of the transit time to the carrier lifetime, given by t_{trans}/τ ,where the transit time is $t_{trans} = 1/K\mu E$ and the trapping lifetime is τ. The charge transfer rate is therefore

$$\begin{aligned} \frac{dN}{dt} &= \frac{I\alpha}{h\nu}\frac{t_{trans}}{\tau} \\ &= \frac{I\alpha}{h\nu}\frac{\Lambda}{2\pi\mu\tau E} \end{aligned} \tag{45}$$

and the charge integration time is given by

$$t_{\text{Integ}} = \tau_{\text{die}} \left(\frac{2\,\pi\,\mu\,\tau\,E}{\Lambda} \right)^2 \tag{46}$$

This integration time is added to the zero-field response time to yield the high-field device response time

$$\tau_0 = \tau_{\text{die}} \left(1 + \frac{t_{\text{Integ}}}{\tau_{\text{die}}} \right) \tag{47}$$

which is identical to eq. (43) when diffusion has been neglected. The dependence of the device relaxation time on the square of the electric field decreases the speed of the device. This is especially significant for the high fields that are necessary to achieve reasonable diffraction efficiencies in semiconductors and sillenites.

An important feature of the transient response of photorefractive gratings is the lag between a moving fringe pattern and the space-charge field. When the gratings are stationary, an applied electric field produces a space-charge field that is in phase with the intensity pattern. Therefore, when the system is not trap-limited, an applied electric field does not significantly improve the two-wave mixing photorefractive gain, which requires a large phase shift between the space-charge field and the intensity pattern.

However, when the grating is moving, the space-charge will lag behind the intensity pattern. Under optimal grating velocities, a π-phase shift can be achieved and the imaginary part of the space-charge field can equal the magnitude of the applied electric field. This condition of high applied fields and moving gratings is therefore one method to achieve a maximum photorefractive gain. The space-charge density as a function of angular frequency of a moving grating was given in eq. (32).

Simulations of the space-charge field are shown in Fig. 11 and Fig. 12 as functions of angular frequency. An angular frequency of 100 rad/sec and a fringe spacing of 5 μm produces a fringe velocity of 80 μm/sec, or 16 fringe spacings per second. The real part of the space-charge field goes to zero and the imaginary part takes on a maximum value at the resonant frequency. It is important to note that this resonance may drive the system out of the regime where the small amplitude approximation is valid. A modulation index as small as $m = 0.01$ may be needed to keep the system in the small amplitude regime.

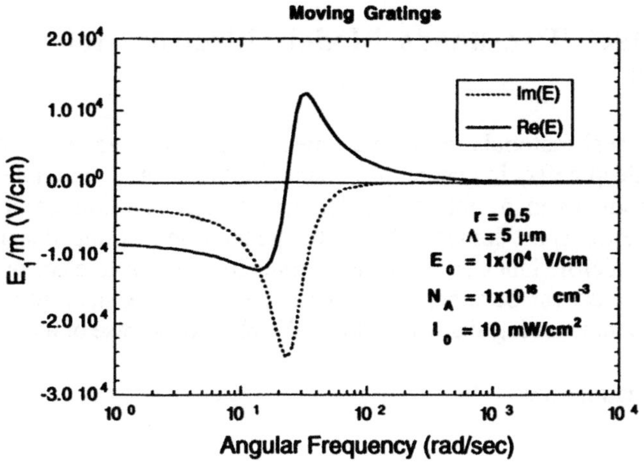

Fig. 11 Ratio of the space-charge field to the modulation index as a function of angular frequency of the moving grating for an applied electric field. At the resonant frequency, the imaginary part of the space-charge field acquires a maximum value, providing maximum photorefractive gain.

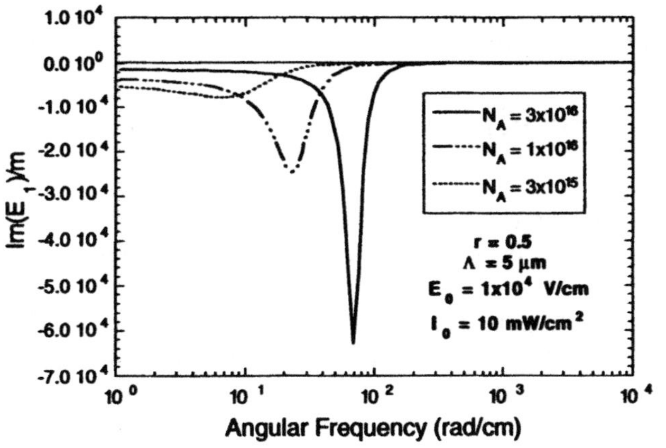

Fig. 12 Imaginary part of the space-charge field ratioed to the modulation index for three different trap densities. The resonance weakens when the material becomes trap limited. For high trap densities, the small-amplitude approximation breaks down.

4.2 One-Defect Two-Bands Model: Electron-Hole Competition

It is unusual, in real materials, to have a defect that communicates with only one band. A single deep defect in general has nonnegligible electron and hole ionization rates. This introduces many interesting features of electron-hole competition, as electron and hole transport compete during the generation of trapped space-charge. The two-band model includes hole transition rates in addition to electron transition rates. The deep defect can be either a deep donor compensating shallow acceptors, or a deep acceptor compensating shallow donors. We choose the case of a deep donor.

The transport equations are

$$\frac{\partial n}{\partial t} - \nabla \frac{j_e}{e} = I\alpha + \left(Is_e + \beta_e\right)N_D^0 - \gamma_e\, n\, N_D^+ - \gamma_{eh}\, np \qquad (48)$$

$$\frac{\partial p}{\partial t} + \nabla \frac{j_h}{e} = I\alpha + \left(Is_h + \beta_h\right)N_D^+ - \gamma_h\, p\, N_D^0 - \gamma_{eh}\, np \qquad (49)$$

$$j_e = e\,\mu_e\, n\, E + k_B T\,\mu_e\, \nabla n \qquad (50)$$

$$j_h = e\,\mu_h\, p\, E - k_B T\,\mu_h\, \nabla p \qquad (51)$$

$$N_D = N_D^0 + N_D^+ \qquad (52)$$

$$\frac{\partial}{\partial t}\left(n + N_{sa} - p - N_D^+\right) = \frac{\nabla}{e}\left(j_e + j_h\right) \qquad (53)$$

$$\nabla E = -\left(\frac{e}{\varepsilon\varepsilon_0}\right)\left(n + N_{sa} - p - N_D^+\right) \qquad (54)$$

In these equations, e and h refer to electron and hole properties, respectively, α is the interband absorption and γ_{eh} is the electron-hole recombination cross-section.

The role of electron-hole pair generation across the bandgap, as well as direct electron-hole recombination, have been included in the rate equations, shown in Fig. 13. The interband absorption coefficient α can vary from zero up to 1×10^5 cm^{-1} as the photon energy is increased. The limit $\alpha = 0$ is the case for nonresonant excitation. As α becomes larger than the extrinsic defect absorption, new regimes of behavior are observed as the photon energy becomes near-resonant and finally resonant, which will be discussed in Chap. 6 and Chap. 7.

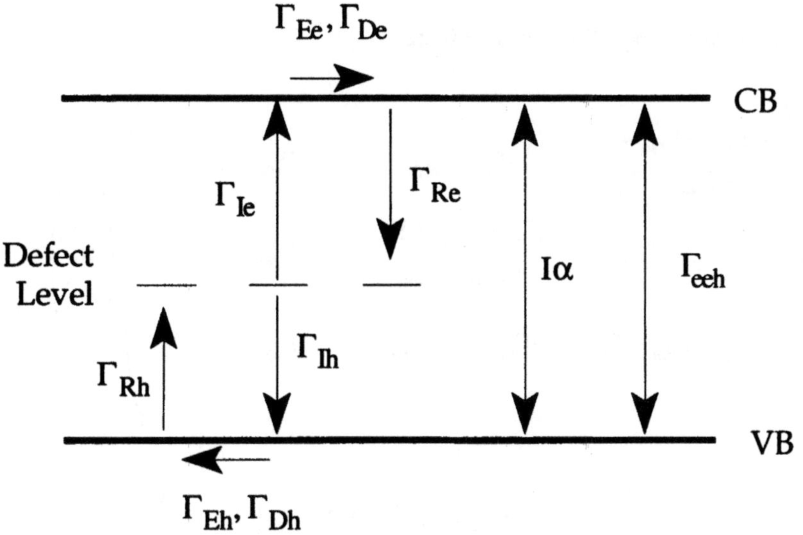

Fig. 13 Transition rates for the one-defect two-band model. Electron-hole competition can significantly alter the photorefractive behavior from the one-band model.

The zero-order solutions for n_0 and p_0 in eqs. (48) - (54) are solved by assuming homogeneous illumination I_0 and no interband absorption. These solutions are

$$n_0 = \frac{I_0 \alpha + \left(I_0 s_e + \beta_e\right)\left(N_D - N_{sa} - n_0 + p_0\right)}{\gamma_e\left(N_{sa} + n_0 - p_0\right) + \gamma_{eh}\, p_0} \tag{55}$$

$$p_0 = \frac{I_0 \alpha + \left(I_0 s_h + \beta_h\right)\left(N_{sa} + n_0 - p_0\right)}{\gamma_h\left(N_D - N_{sa} - n_0 + p_0\right) + \gamma_{eh}\, n_0} \tag{56}$$

The first-order solutions of eqs. (48) - (54) are obtained in the small modulation amplitude approximation. The solution is expressed in terms of the hole transition rates in Table IV (in addition to the electron transition rates given in Table I.)

The linearized equations for $\Omega = 0$ are

$$\begin{pmatrix} -i\Gamma_{Ee}+\Gamma_{De}+\Gamma_{Ie} + \Gamma_{Re}+\Gamma_{eeh} & -\Gamma_{Ie}+\Gamma_{ehh} & -\Gamma_{die}+\Gamma_{Ie} \\ -\Gamma_{Ih}+\Gamma_{eeh} & i\Gamma_{Eh}+\Gamma_{Dh}+\Gamma_{Ih} + \Gamma_{Rh}+\Gamma_{ehh} & \Gamma_{dih}-\Gamma_{Ih} \\ \Gamma_{Ie}+\Gamma_{Re}+\Gamma_{Ih} & -\Gamma_{Ie}-\Gamma_{Ih}-\Gamma_{Rh} & \Gamma_{Ie}+\Gamma_{Ih} \end{pmatrix} \begin{pmatrix} n_1 \\ p_1 \\ A_1 \end{pmatrix}$$

$$= I_1 \begin{pmatrix} \alpha + s_e\left(N_D - N_A + p_0 - n_0\right) \\ \alpha + s_h\left(N_A - p_0 + n_0\right) \\ s_e\left(N_D - N_A + p_0 - n_0\right) - s_h\left(N_A - p_0 + n_0\right) \end{pmatrix} \qquad (57)$$

Electron-hole competition has important consequences for photorefractive effects. In the two-band model it can lead to either reduced or enhanced space-charge fields depending on details of electron and hole thermal and optical emission. Reduced fields result from competition between the electrons and holes during the formation of the trapped space-charge [21, 36, 37, 39, 64, 65]. Enhanced space-charge fields result from a temperature-intensity resonance when the thermal emission of one carrier is balanced by optical emission by the opposite carrier [66].

Electron-hole competition is most pronounced for zero applied electric field. In this situation, the electron and hole diffusion currents cancel, leading to significant reductions in the diffusion field. When an electric field is applied, on the other hand, the electron and hole currents both contribute to the formation of trapped space-charge because the carriers transport in opposite directions.

The temperature-intensity resonance [66] relies on electron-hole competition to enhance space-charge fields when an external electric field is applied. The resonance is important for photorefractive gain because nearly a $\pi/2$ phase shift can be produced with a DC electric field without relying on moving grating or AC electric field techniques. The resonance occurs when the thermal generation of carriers to one band is balanced by photoionization to the other band. A photorefractive phase shift approaching $\pi/2$ is induced under these conditions, accompanied by a significant increase in the space-charge field. The temperature-intensity resonance therefore provides an important process to enhance photorefractive gain in selected photorefractive crystals. The drawback of this technique is that it is not entirely general, and relies on special conditions for defect cross sections. One material in which this technique has been successful is InP:Fe [66].

Table IV. Hole Transport and Transition Rates:

Dielectric relation rate:	$\Gamma_{dih} = e\,\mu_h\,p_0/\varepsilon\,\varepsilon_0$
Transition rate:	$\Gamma_{Ih} = s_h\,I_0 + \beta_e + \gamma_h\,n_0$
Ion Recombination rate:	$\Gamma_{Rh} = \gamma_h(N_{sd} - n_0)$
Drift rate:	$\Gamma_{Eh} = K\,\mu_h\,E_0$
Diffusion rate:	$\Gamma_{Dh} = K^2\,k_BT\,\mu_h/e$
Direct Recombination rate:	$\Gamma_{ehh} = \gamma_{eh}\,p_0$

4.3 Multiple Defect Levels

All materials have multiple defects as a simple consequence of thermodynamics. At the elevated temperatures at which optical materials are grown, the free energy is lowered by incorporating intrinsic and extrinsic defects into the crystal. Even in materials that are carefully controlled, with carefully introduced dopants, multiple defect levels will be present, often in significant concentrations that can affect photorefractive performance. One of the chief difficulties facing the eventual usefulness of photorefractive effects and devices is material control. It may be argued that all of the idiosyncratic properties of individual photorefractive crystals may be traced to the role of multiple defect levels. Multiple defect levels provide avenues for complicated electron-hole competition as well as additional modes for screening of internal and external electric fields. Multiple defect levels can also cause the dominant photocarrier to change sign, changing the direction of energy transfer in two-wave mixing.

In thermal equilibrium, only a single dominant defect level will be partially occupied, pinning the Fermi level. All other defect levels will be entirely occupied by electrons or holes. However, under steady-state photoexcitation, each defect level will have a nonequilibrium occupancy which can act as sources for thermally or photogenerated carriers, leading to electron-hole competition. In this section, we consider how photorefractive transport from a dominant deep acceptor level is modified by an additional deep donor level. The donor level can be either above or below the deep acceptor, with appropriate compensating shallow defects.

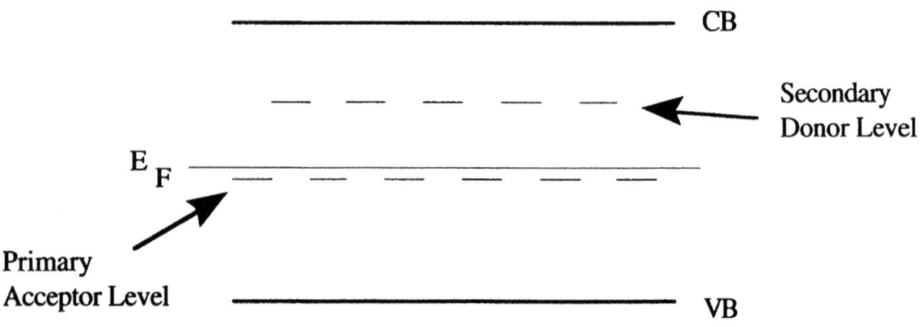

Fig.14 Multiple defect model for photorefractive transport. Additional deep defects may occur above or below a dominant deep level that pins the Fermi level. The additional defects can be either donors or acceptors, providing that there is appropriate shallow compensating defects to ensure that the Fermi level is pinned near one of the deep defect levels.

The transport and space-charge equations for the two-defect model with a deep acceptor and a deep donor are

$$\frac{\partial n}{\partial t} - \nabla \frac{j_e}{e} = \left(s_{eD} I + \beta_e\right) N_D^0 + \left(s_{eA} I + \beta_e\right) N_A^- - \gamma_{eD}\, n\, N_D^+ - \gamma_{eA}\, n\, N_A^0 + I\alpha - \gamma_{eh}\, np \tag{58}$$

$$\frac{\partial p}{\partial t} + \nabla \frac{j_h}{e} = \left(s_{hD} I + \beta_{hD}\right) N_D^+ + \left(s_{hA} I + \beta_{hA}\right) N_A^0 - \gamma_{hD}\, p\, N_D^0 - \gamma_{hA}\, p\, N_A^- + I\alpha - \gamma_{eh} np \tag{59}$$

$$j_e = e\, \mu_e\, n\, E + k_B\, T\, \mu_e\, \nabla n \tag{60}$$

$$j_h = e\, \mu_h\, p\, E - k_B\, T\, \mu_h\, \nabla p \tag{61}$$

$$\frac{\partial}{\partial t}\left(n + N_A^- - p - N_D^+\right) = \frac{\nabla}{e}\left(j_e + j_h\right) \tag{62}$$

$$\nabla E = -\left(\frac{e}{\varepsilon \varepsilon_0}\right)\left(n + N_A^- + N_{sa} - p - N_D^+ - N_{sd}\right) \tag{63}$$

An additional condition is needed to define the time dependence of the deep donor charge state

$$\begin{aligned}\frac{dN_{D0}^{+}}{dt} &= \left(s_{eD}I_0+\beta_{eD}\right)\left(N_D-N_{D0}^{+}\right)-\gamma_{eD}n_0N_D^{+}\\ &-\left(s_{hD}I_0+\beta_{hD}\right)N_{D0}^{+}+\gamma_{hD}p_0\left(N_D-N_{D0}^{+}\right)\end{aligned} \tag{64}$$

The zero-order solutions are

$$n_0=\frac{\left(s_{eD}I_0+\beta_{eD}\right)\left(N_D-N_{D0}^{+}\right)+\left(s_{eA}I_0+\beta_{eA}\right)N_{A0}^{-}+I_0\,\alpha}{\gamma_{eD}N_{D0}^{+}+\gamma_{eA}\left(N_A-N_{A0}^{-}\right)+\gamma_{eh}\,p_0} \tag{65}$$

$$p_0=\frac{\left(s_{hD}I_0+\beta_{hD}\right)N_{D0}^{+}+\left(s_{hA}I_0+\beta_{hA}\right)\left(N_A-N_{A0}^{-}\right)+I_0\,\alpha}{\gamma_{hD}\left(N_D-N_{D0}^{+}\right)+\gamma_{hA}N_{A0}^{-}+\gamma_{eh}\,n_0} \tag{66}$$

$$N_{D0}^{+}=\frac{\left(s_{eD}I_0+\beta_{eD}\right)N_D+\gamma_{hD}p_0N_D}{\left(s_{eD}I_0+\beta_{eD}\right)+\gamma_{eD}n_0+\left(s_{hD}I_0+\beta_{hD}\right)+\gamma_{hD}\,p_0} \tag{67}$$

$$N_{A0}^{-}=\frac{\left(s_{hA}I_0+\beta_{hA}\right)N_A+\gamma_{eA}p_0N_A}{\left(s_{eA}I_0+\beta_{eA}\right)+\gamma_{eA}n_0+\left(s_{hA}I_0+\beta_{hA}\right)+\gamma_{hA}\,p_0} \tag{68}$$

The first-order solutions are obtained in the small signal analysis and can be expressed in matrix form, including the spatial modulation of N_D^+ as a fourth variable

$$\begin{pmatrix}A_{11}&A_{12}&A_{13}&A_{14}\\A_{21}&A_{22}&A_{23}&A_{24}\\A_{31}&A_{32}&A_{33}&A_{34}\\A_{41}&A_{42}&A_{43}&A_{44}\end{pmatrix}\begin{pmatrix}n_1\\p_1\\A_1\\N_{D1}^{+}\end{pmatrix}=\begin{pmatrix}B_1\\B_2\\B_3\\B_4\end{pmatrix} \tag{69}$$

where the matrix elements are given in Table V.

Table V Matrix Elements for the Two-Defect Two-Band Model

$A_{11} = -i\Gamma_{Ee} + \Gamma_{De} + \gamma_{eD}N^{+}_{D0} + \Gamma_{IeA} + \gamma_{eA}\left(N_A - N^{-}_{A0}\right) + \gamma_{eh}p_0$
$A_{12} = -\Gamma_{IeA} + \gamma_{eh}n_0$
$A_{13} = -\Gamma_{die} + \Gamma_{IeA}$
$A_{14} = \Gamma_{IeD} - \Gamma_{IeA}$
$A_{21} = -\Gamma_{IhA} + \gamma_{eh}p_0$
$A_{22} = i\Gamma_{Eh} + \Gamma_{Dh} + \gamma_{hD}\left(N_D - N^{+}_{D0}\right) + \Gamma_{IhA} + \gamma_{hA}N^{-}_{A0} + \gamma_{eh}n_0$
$A_{23} = \Gamma_{dih} - \Gamma_{IhA}$
$A_{24} = -\Gamma_{IhD} + \Gamma_{IhA}$
$A_{31} = \Gamma_{IhA} + \gamma_{eA}\left(N_A - N^{-}_{A0}\right) + \Gamma_{IeA} + \gamma_{eD}N^{+}_{D0}$
$A_{32} = -\gamma_{hD}\left(N_D - N^{+}_{A0}\right) - \gamma_{hD}N^{-}_{A0} - \Gamma_{IhA} - \Gamma_{IeA}$
$A_{33} = \Gamma_{IhA} + \Gamma_{IeA}$
$A_{34} = \Gamma_{IhD} - \Gamma_{IhA} + \Gamma_{IeD} - \Gamma_{IeA}$
$A_{41} = -\gamma_{eD}N^{+}_{D0}$
$A_{42} = \gamma_{hD}\left(N_D - N^{+}_{D0}\right)$
$A_{43} = 0$
$A_{44} = -\Gamma_{IeD} - \Gamma_{IhD}$

and the source terms are given in Table VI.

Table VI. Source Terms

$B_1 = s_{eD}I_1\left(N_D - N^{+}_{D0}\right) + s_{eD}I_1N^{-}_{A0} + I_1\alpha$
$B_2 = s_{hD}I_1N^{+}_{D0} + s_{hA}I_1\left(N_A - N^{-}_{A0}\right) + I_1\alpha$
$B_3 = -s_{hD}I_1N^{+}_{D0} - s_{hA}I_1\left(N_A - N^{-}_{A0}\right) + s_{eD}I_1\left(N_D - N^{+}_{D0}\right) + s_{eA}I_1N^{-}_{A0}$
$B_4 = s_{eD}I_1\left(N_D - N^{+}_{D0}\right) - s_{hD}I_1N^{+}_{D0}$

The analytic solution to this set of coupled equations is too cumbersome to present here. However, the numerical solution is straightforward.

4.4 Metastable Defects and Permanent Gratings in Semiconductors

One of the most striking behaviors associated with defects is optically-induced metastability, which occurs for a special class of defect. These are defects that can exist in two or more physical configurations with strikingly different physical properties. Two of the most studied metastable defects are the EL2 defect in GaAs, and the DX centers in n-type doped III-V and II-VI alloys. The EL2 defect is currently considered to be the arsenic antisite defect As_{Ga}, i.e. an arsenic atom on a gallium site in the GaAs lattice [67-69]. The DX centers are substitutional n-type dopant impurity atoms [70, 71]. Both of these defects can be transformed from a stable lattice configuration into a metastable configuration by the absorption of a photon. At low temperatures ($T < 100$ K) these metastable states can exist for hours or years, depending on the temperature.

The important feature of the metastable defects is that carriers are liberated from the defects during the process that ends in the metastable state. The liberated carriers can profoundly change the electrical properties in the semiconductor, especially the Fermi level. For instance, both the DX and EL2 defects are the source of persistent photoconductivity after illumination (as electrons are liberated in the first case and holes are liberated in the second case). At low temperatures, these carriers cannot recombine with the defects in their metastable condition, but remain in the conduction and valence bands. Because the carriers cannot trap at the defect sites, the material behaves as if the optical illumination had removed the defects from the material, i.e. the defects had been quenched. Optical illumination therefore changes the compensation in the material and can cause large changes in the Fermi level. The normal defect states can only be regenerated after the crystal temperature is raised above a characteristic annealing temperature. Once the defects return to their normal state they can again trap carriers.

The photoquenching of EL2 and DX defects proceeds through photon absorption cross-sections, and therefore leads to an exponential decay of the normal states as a function of exposure. The total defect density as a function of time is given by

$$N(I,t) = N_0 \exp\left(-\frac{I}{h\nu} s^* t\right) \tag{70}$$

where s^* is the photoquenching cross-section (which is a function of photon energy), N_0 is the initial defect concentration, I is the intensity, and the characteristic quenching time constant τ is given by

$$\frac{1}{\tau} = \frac{I s^*}{h\nu} \tag{71}$$

and the exposure is defined as the product of Intensity and time.

Metastable defects offer a unique avenue for generating permanent photorefractive gratings in semiconductors. The interference of two writing beams inside a semiconductor crystal containing metastable defects causes a spatial modulation of the compensation and therefore a spatial modulation of the Fermi level [72]. The defect density is given by

$$N(x, t) = N_0 \exp\left(-\frac{I_0}{h\nu}\left(1 + m\cos(Kx)\right) s^* t\right) \qquad (72)$$

When the compensation is changed by partial exposure, charge is permanently transferred to re-establish thermal equilibrium of charge carriers with the normal defects, setting up large space-charge fields that can exceed normal diffusion fields by more than an order of magnitude with correspondingly large electro-optic gratings [72, 73]. In extreme cases with high doping, the Fermi level becomes degenerate with the conduction band, producing permanent free-carrier gratings with large plasma-induced refractive index changes [74]. These gratings, once generated, can persist for years at low temperatures.

4.4.1 GaAs:EL2

The first demonstration of large metastable optical gratings in semiconductors was made using the GaAs:EL2 system [72, 73]. The Fermi level in this case can be modulated by nearly 750 meV, from midgap in GaAs to near the top of the valence band. The normal state of the arsenic antisite is a double donor with the first ionization level at E_V + 0.75 eV. The donor level compensates shallow acceptors and pins the Fermi Level near midgap. The arsenic antisite is one of the commercially most important defects, because it is the principle compensating deep level defect that makes semi-insulating GaAs substrates available for integrated circuits. At room temperature, and at higher temperatures needed for integrated circuit processing, the defect is stable. However, below 140 K the defect can be optically quenched and effectively removed from the material, rendering the material p-type as the Fermi Level drops to the acceptor levels.

The EL2 configuration that is optically quenchable is the neutral charge state. It is quenched by the absorption of a photon with a wavelength in a small range around 1.06 microns. This narrow absorption band corresponds to the intracenter photoexcitation of an electron into an excited state that undergoes large lattice relaxation into a metastable state. The metastable state is neutral and is electrically inactive. Because normal $EL2^0$ states do not participate in compensation, their loss by optical quenching would not alone change the electrical properties of the material.

However, holes from $EL2^+$ can be photoexcited into the valence band, leaving behind a neutral $EL2^0$ that can then be optically quenched. The net result of this sequence is the liberation of holes and the removal of EL2, forcing the Fermi level to drop towards the valence band.

The Fermi Level modulation experiment in GaAs:EL2 was performed using a 1.06 micron YAG:Nd laser to write the metastable gratings in a 1 cm long LEC GaAs crystal with a holographic cut. The electro-optic gratings were probed using a 1.3 micron YAG:Nd laser in a nondegenerate mixing geometry in a cryostat [40, 41]. The gratings are unaffected by photons with 1.3 micron wavelengths.

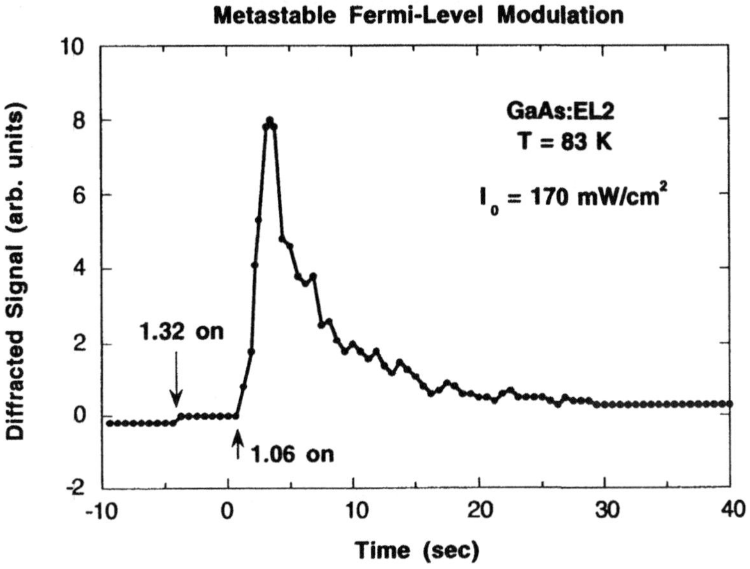

Fig. 15 Diffraction from metastable Fermi-Level modulation. A 1.32 micron laser probes gratings introduced by a 1.06 micron laser that quenches EL2 and spatially modulates the Fermi level [73].

The diffraction efficiency as a function of exposure time is shown in Fig. 15. The 1.32 μm wavelength probe was turned on first, followed by the 1.06 μm laser with an intensity of 170 mW/cm^2 that quenches EL2 with a 3 second time constant. The distinctive features in the figure are the fast rise time and slow decay of the photorefractive gratings. The peak diffraction efficiency was nearly an order of magnitude larger than from diffusion fields. At long times, the EL2 concentrations are quenched everywhere inside the crystal. The persistence of the gratings is shown in Fig. 16 when the sample was held at a temperature of 60 K. The sample was given an exposure with the 1.06 laser to write maximum amplitude gratings, and both the pump and probe lasers were turned off. The gratings persisted up to one hour, after which the 1.06 laser was turned back on and the entire crystal was quenched.

The role of space-charge fields and electro-optic gratings during the quenching of EL2 can be described as modulation doping using coherent illumination. The quenching of EL2 effectively prevents the deep defect from compensating shallow acceptor impurities. Therefore, the Fermi level drops to the top of the valence band in the bright fringes, but remains pinned at the EL2 energy in the dark fringes. The resulting charge transfer from the bright to the dark fringe produces band-banding that forms a p-i-p modulation doped structure, shown in Fig. 17.

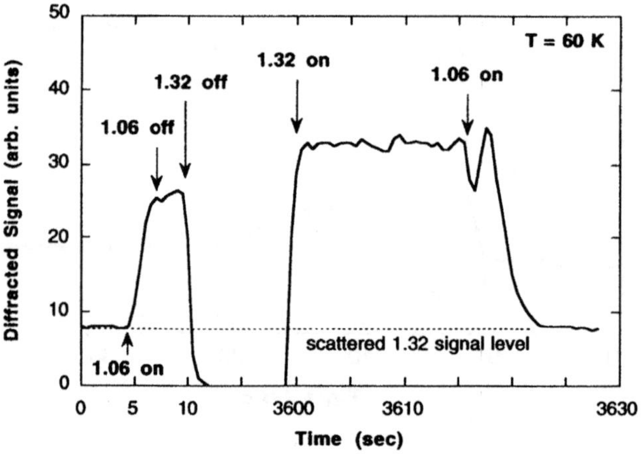

Fig. 16 Persistence of the gratings up to one hour when the quenching exposure is interrupted and the crystal temperature is held at 60 K [73].

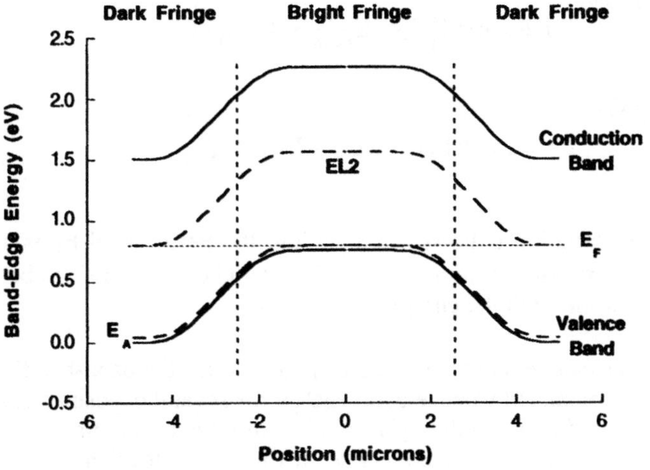

Fig. 17 P-i-p modulation-doped structure formed by the quenching of EL2 with an interference fringe spacing of 10 microns.

Fermi-Level modulation by coherent illumination can be calculated using a new set of steady-state "photorefractive" equations

$$\nabla^2 \phi(x) = -\frac{e}{\varepsilon \varepsilon_0}\left(N_{EL2}^{+}(x) + p(x) - N_A^{-}(x) - n(x)\right) \quad (73)$$

$$\begin{aligned} E_F &= \text{const.} \\ E_v(x) &= -e\phi(x) \\ E_c(x) &= E_g - e\phi(x) \\ E_{EL2}(x) &= E_{EL2} - e\phi(x) \end{aligned} \quad (74)$$

$$\begin{aligned} n(x) &= N_c \exp\left(\left(E_F - E_c(x)\right)/k_B T\right) \\ p(x) &= N_v \exp\left(\left(E_v(x) - E_F\right)/k_B T\right) \end{aligned} \quad (75)$$

$$N_{EL2}^{+}(x) = \frac{N_{EL2}(x)}{1 + \frac{1}{2}\exp\left(\left(E_F - E_{EL2}(x)\right)/k_B T\right)}$$
$$N_A^{-}(x) = \frac{N_A}{1 + \frac{1}{2}\exp\left(\left(E_v(x) + E_A - E_F\right)/k_B T\right)} \tag{76}$$

where the first equation is Poisson's equation, the Fermi level E_F is a constant, all other energies vary with position, and the defect and carrier densities become implicit functions of $\phi(x)$ through the energies.

These equations can be solved numerically to produce the band-bending for a given exposure. The calculated valence band energy for several exposures is shown in Fig. 18 assuming a modulation index of $m = 0.9$. Initial exposure has little effect on the band energy, but there is a sudden rise in the Fermi level modulation near half exposure ($t = 0.5\tau$). At full exposure, the Fermi level has dropped to the top of the valence band in the bright fringe, but remains pinned at EL2 in the dark fringe. For longer exposure times, EL2 is quenched throughout the sample.

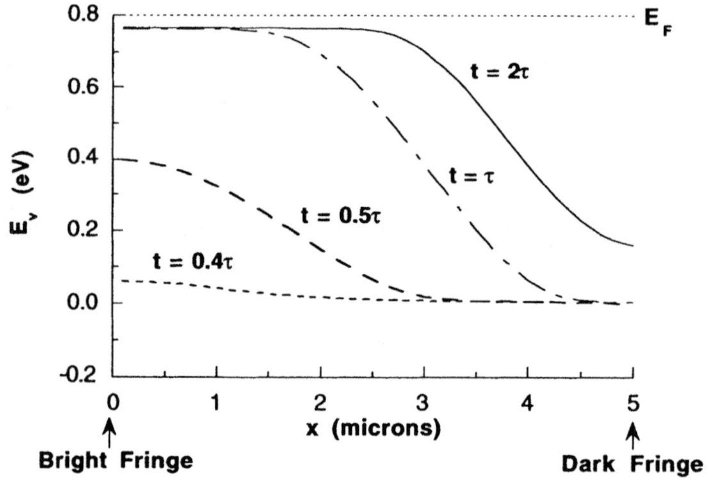

Fig. 18 Valence band energies as a function of position for several exposure times.

The space-charge electric fields associated with the Fermi-Level modulation and band bending produce electro-optic gratings. The calculated diffraction efficiency for full Fermi-level modulation ($\Delta E_v \approx 0.8$ eV) is shown in Fig. 19. The fast rise and slow decay observed experimentally in Fig. 15 is captured in the numerical simulation. The slow decay is connected to the weak irradiance in the center of the dark fringe. Experimentally, smaller diffracted signals were observed because the Fermi-Level modulation is sensitive to defect densities and fringe spacings.

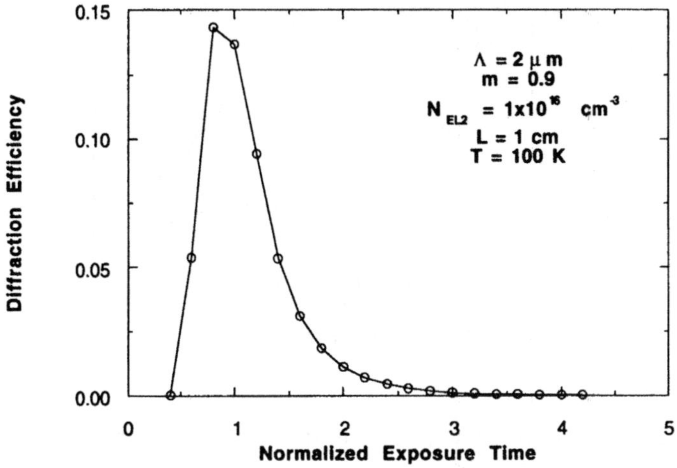

Fig. 19 Calculated diffraction efficiency as a function of exposure for Fermi-Level modulation of GaAs:EL2.

4.4.2 Permanent Plasma Gratings

The DX center has potentially greater applicability for Fermi level modulation because it is a common defect in many III-V and II-VI alloys and can be quenched by a continuous range of photon energies. In addition, the Fermi level after quenching can be degenerate with the conduction band, producing permanent free-carrier (or plasma) gratings [74]. Plasma gratings can produce significant refractive index changes that can exceed electro-optic gratings. Especially in the case of the DX centers, the space-charge fields are small because the material remains locally electrically neutral after quenching. The relative contributions from electro-optic gratings and plasma gratings are calculated and shown in Fig. 20 for EL2 and Fig. 21 for DX in AlGaAs. There are strong qualitative differences between the two cases, reflecting the pronounced differences in the properties of these two metastable defects. In GaAs:EL2, electro-optic gratings dominate quickly at short times, but weak plasma gratings may persist for considerably longer exposure times. In AlGaAs:DX, the plasma gratings dominate. However, in nearly all cases, these electro-optic and plasma gratings significantly exceed traditional photorefractive diffusion gratings.

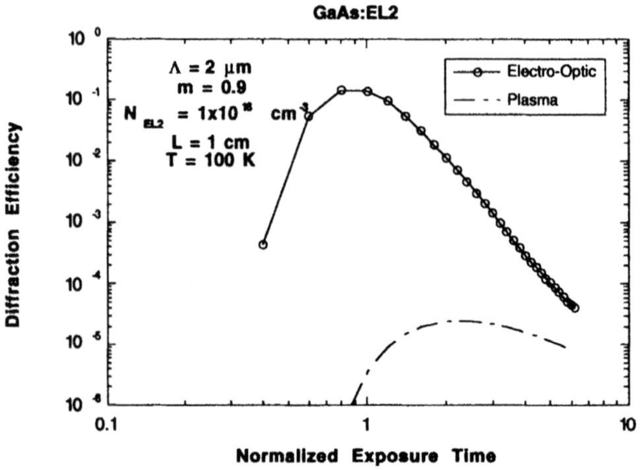

Fig. 20 Calculated contributions to the diffraction signal from electro-optic and plasma gratings.

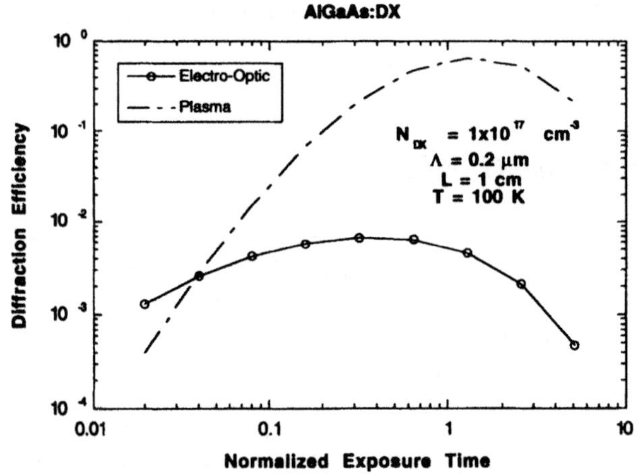

Fig. 21 Calculated contributions from electro-optic and plasma gratings in AlGaAs:DX. The electro-optic gratings are weak in this case, but the plasma gratings can produce significant diffraction efficiencies.

5. ELECTRO-OPTICS

When space-charge fields are established inside a photorefractive material, by the mechanisms discussed in the previous section, the electric fields modify the optical properties through electro-optic effects. Photorefractive phenomena in most photorefractive crystals rely on linear electro-optic effects, in which the refractive index depends linearly on the field. However, for resonant or near-resonant photoexcitation with interband transitions, quadratic effects can dominate the electro-optic response, in which the refractive index and absorption depend on the square of the field.

The index ellipsoid of an optical crystal is described by the expression

$$\frac{x^2}{n_{11}^2(E)} + \frac{y^2}{n_{22}^2(0)} + \frac{z^2}{n_{33}^2(0)} + \frac{2\,yz}{n_{23}^2(0)} + \frac{2\,xz}{n_{13}^2(0)} + \frac{2\,xy}{n_{12}^2(0)} = 1 \tag{77}$$

where x,y,z are arbitrary axes. The index is altered by an electric field, where the components depend on the field through

$$\begin{aligned}\Delta\left(\frac{1}{n_{ij}^2}\right) &= \frac{1}{n_{ij}^2(E)} - \frac{1}{n_{ij}^2(0)} \\ &= \sum_{k=1}^{3} r_{ijk}E_k + \sum_{k=1}^{3}\sum_{l=1}^{3} s_{ijkl}E_kE_l\end{aligned} \tag{78}$$

where the r_{ijk} are the linear electro-optic tensor components and s_{ijkl} are the quadratic electro-optic tensor components. The linear electro-optic tensor is only nonzero in crystals that lack a center of inversion symmetry.

The electro-optic tensors have permutation symmetries that make it possible to use a contracted notation. The contracted notation for the linear components is

$$r_{1k} = r_{11k} \qquad r_{2k} = r_{22k} \qquad r_{3k} = r_{33k}$$

$$r_{4k} = r_{23k} = r_{32k} \qquad r_{5k} = r_{13k} = r_{31k} \qquad r_{6k} = r_{12k} = r_{21k}$$

where k = 1,2,3. The tensors in the contracted notation are therefore expressed as asymmetric tensors. For instance the linear electro-optic tensor of $BaTiO_3$ with 4mm symmetry is

$$\bar{\bar{r}} = \begin{pmatrix} 0 & 0 & r_{13} \\ 0 & 0 & r_{13} \\ 0 & 0 & r_{33} \\ 0 & r_{51} & 0 \\ r_{51} & 0 & 0 \\ 0 & 0 & 0 \end{pmatrix} \tag{79}$$

As another example, the electro-optic semiconductors, such as GaAs and InP of the III-V semiconductor family and CdTe of the II-VI semiconductor family, have $\bar{4}3m$ symmetry with an electro-optic tensor given by

$$\bar{\bar{r}} = \begin{pmatrix} 0 & 0 & 0 \\ 0 & 0 & 0 \\ 0 & 0 & 0 \\ r_{41} & 0 & 0 \\ 0 & r_{41} & 0 \\ 0 & 0 & r_{41} \end{pmatrix} \tag{80}$$

For a complete description of the electro-optic tensors of all the crystal symmetry families, see Ref. [75].

For the study of mode-coupling and multi-wave mixing caused by a spatial modulation of the optical properties of a crystal, the relevant parameter in the wave equation is the dielectric tensor and the modulation of the dielectric tensor. The change in the dielectric tensor that is induced by the electro-optic effect is related to the tensor $\Delta(1/n^2)$ by the expression

$$\Delta\bar{\bar{\varepsilon}} = \frac{-1}{\varepsilon_0}\,\bar{\bar{\varepsilon}}\,\overline{\Delta\left(\frac{1}{n^2}\right)}\,\bar{\bar{\varepsilon}} \tag{81}$$

which can also be expressed as

$$\Delta\varepsilon_{ij} = -\varepsilon_0\, n_i^2\, n_j^2\, \Delta\left(\frac{1}{n_{ij}^2}\right) \tag{82}$$

in the principle coordinate system in which the dielectric tensor is diagonal. In isotropic crystals, such as GaAs, the change in the dielectric tensor therefore is

$$\Delta\varepsilon = -\varepsilon_0 n^2 \begin{bmatrix} 0 & r_{41} E_z & r_{41} E_y \\ r_{41} E_z & 0 & r_{41} E_x \\ r_{41} E_y & r_{41} E_x & 0 \end{bmatrix} \tag{83}$$

which is a symmetric tensor. It is important to point out that linear electro-optic gratings in a crystal can coexist with absorption gratings that are caused by separate physical processes of induced absorption or absorption bleaching. These absorption gratings contribute to the complex-valued dielectric gratings in the photorefractive crystal, and can also couple modes in multi-wave mixing.

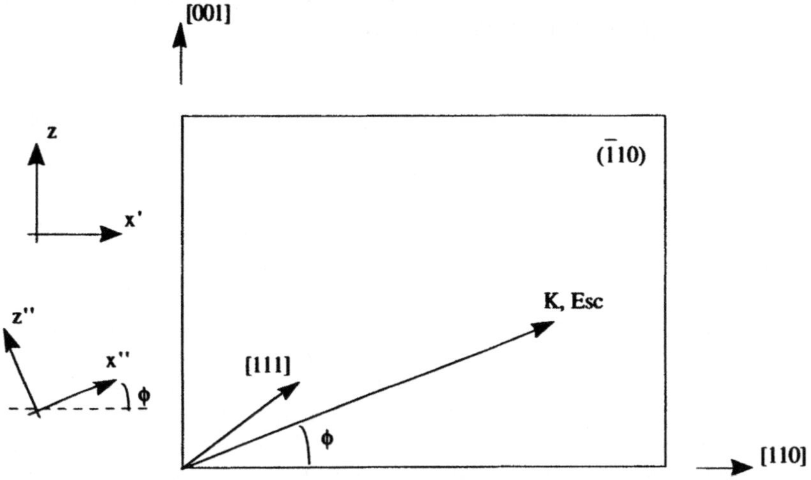

Fig. 22 Holographic cut for a photorefractive semiconductor with $\bar{4}3m$ symmetry showing the important crystallographic directions. Beam polarizations parallel and perpendicular to the grating vector define p-polarized and s-polarized beams during photorefractive mixing.

The traditional "holographic cut" for cubic crystals such as GaAs is a rectilinear crystal with [110],[001], and [$\bar{1}$10] crystallographic axes, shown in Fig. 22. In this configuration, the laser beams are incident on the ($\bar{1}$10) face, and the grating vector and the photorefractive space-charge electric field are perpendicular to the [$\bar{1}$10] direction. When the direction of the grating vector (and the space-charge field) are oriented relative to the [110] axis by an angle ϕ,the electric field is

$$\vec{E} = \vec{E}_0 \left[\frac{\cos\phi}{\sqrt{2}}, \frac{\cos\phi}{\sqrt{2}}, \sin\phi \right] \tag{84}$$

Using this equation in eq. (78) with the electro-optic tensor eq. (80) gives the new refractive index ellipsoid

$$\frac{x^2}{n_0^2} + \frac{y^2}{n_0^2} + \frac{z^2}{n_0^2} + \frac{2\,E_0}{\sqrt{2}} \cos\phi \; r_{41}\, yz + \frac{2\,E_0}{\sqrt{2}} \cos\phi \; r_{41}\, xz$$
$$+ 2\,E_0 \sin\phi \; r_{41}\, xy = 1 \qquad (85)$$

This ellipsoid has new principle axes that are not necessarily oriented with the electric field. A new coordinate system can be found that describes the normal modes of propagation. In our example with the holographic cut, we proceed by considering the (110) plane (whose normal is in the direction of propagation). The intersection of this plane with the index ellipsoid is given by

$$\left[\frac{1}{n_0^2} + E_0\, r_{41} \sin\phi\right] x'^2 + \frac{1}{n_0^2} z^2 + 2\,E_0\, r_{41} \cos\phi \; x'z = 1 \qquad (86)$$

where the x' axis is directed along the [110] axis, shown in Fig. 22. This equation describes the index ellipse for propagation parallel to the $[\bar{1}10]$ axis.

During two-wave and four-wave mixing, it is customary to describe the laser beam polarization relative to the plane of incidence, where polarization parallel to the plane is p-polarized and polarization perpendicular to the plane is s-polarized. These polarizations are not necessarily the normal modes of the index ellipsoid. In our example, the grating vector is oriented at an angle ϕ relative to the x' (or [110]) axis. New coordinates x" and z" that define the p-polarized and s-polarized directions are given by the transformation

$$x' = x'' \cos\phi - z'' \sin\phi$$
$$z = x'' \cos\phi + z'' \sin\phi \qquad (87)$$

which transform the index ellipse of eq. (86) into

$$\left[\frac{1}{n_0^2} + 3\,E_0\,r_{41}\cos^2\phi\,\sin\phi\right]x''^2 + \left[\frac{1}{n_0^2} + E_0\,r_{41}\sin\phi\left(3\sin^2\phi - 2\right)\right]z''^2 + \left[2\,E_0\,r_{41}\cos\phi\left(3\cos^2\phi - 2\right)\right]x''\,z'' = 1 \tag{88}$$

It is easy to see from this expression for the index ellipse that the s- and p-polarizations are only normal modes when $\cos\phi(3\cos^2\phi-2) = 0$. This occurs only for $\phi = 35.26^o, 90^o$ and 144.74^o, i.e. for **K** parallel to [111], [100] and $[\bar{1}\,\bar{1}1]$. Eq.(88) is solved for the change in the refractive index for s-polarized and p-polarized waves induced by the electric field E_0. The index changes are

$$\begin{aligned} &\text{p–polarized:}\ \Delta n = -\frac{1}{2}\,n_0^2\,3\,E_0\,r_{41}\cos^2\phi\,\sin\phi \\ &\text{s–polarized:}\ \Delta n = -\frac{1}{2}\,n_0^2\,E_0\,r_{41}\sin\phi\left(3\sin^2\phi - 2\right) \end{aligned} \tag{89}$$

Again, special orientations can be identified. When $\phi = 90^o$ (for K || [001]) then s-polarized waves experience an index grating, but p-polarized waves do not. This condition is often used when it is necessary to distinguish between mixing from absorption gratings and mixing from refractive index gratings. When $\phi = 35.26^o$ or 144.74^o (for K || [111] or $[\bar{1}\,\bar{1}1]$) then the maximum refractive index change is experienced by the p-polarized beam with a value 15% larger than for s-polarized waves at $\phi = 90^o$ [76]. Photorefractive mixing for general angles ϕ and general polarizations has been solved in Ref.[76]. The variety of behavior with crystal orientation and beam polarization produces many interesting effects, some of which have been demonstrated in magneto-photorefractive experiments [77].

This section has described how the space-charge electric field induced by spatially varying illumination produces a change in the refractive index. This completes the sequence of processes that transforms a spatially varying irradiance into a refractive index grating. Once the refractive index grating is formed, it can perform as a diffraction grating for optical fields propagating in a periodic dielectric medium.

6. WAVE PROPAGATION IN PERIODIC DIELECTRIC MEDIA

Consider a dielectric medium with a 1-dimensional periodic dielectric modulation in the z-direction given by

$$\varepsilon(\omega; z) = \varepsilon(\omega) + \frac{1}{2}\sum_{K\neq 0} \varepsilon_K \, e^{iKz} \tag{90}$$

where ε_K are Fourier components of the dielectric function, and all Fourier coefficients are allowed. The coefficient ε_K contains contributions from both electro-optic and absorption gratings.

Electromagnetic waves propagating in this medium scatter from the periodic gratings, generating additional diffracted waves. The theory of waves propagating in a periodic medium crosses many disciplines. This theory is especially relevant to the physics of crystalline materials with Bloch waves and band structures. The exact solutions of these problems can be obtained for cases of strong scattering, leading to nonlinear dispersion of the propagating waves. However, in nonlinear optical mixing, the scattering is generally weak and a perturbative approach is more appropriate when the periodic part of eq. (90) is small compared to the unperturbed dielectric function. This perturbative approach is the coupled-mode theory [75].

The wave equation for the electric field components of a plane electromagnetic wave propagating in the medium described by eq. (90) is

$$\left[\nabla^2 + \omega^2\mu\left(\varepsilon(\omega) + \frac{1}{2}\sum_{K\neq 0} \varepsilon_K \, e^{iKz}\right)\right]\sum_{k_z} \frac{\hat{e}_k}{\sqrt{|k_z|}} A_{k_z}(z)\, e^{i\left(k_z z - \omega t\right)} = 0 \tag{91}$$

where k_z is the z-component of the k-vector in the material, μ is the magnetic permeability, e_{kz} is the unit polarization vector and $A_{kz}(z)$ is the field envelope. The factor in the square root is required for normalization. The equation can be rewritten in terms of the field envelopes $A_{kz}(z)$ as

$$\sum_{k_z}\left[\frac{d^2}{dz^2} A_{k_z} + 2ik_z \frac{d}{dz} A_{k_z}\right] \frac{\hat{e}_k}{\sqrt{|k_z|}} e^{ik_z z} = -\frac{\omega^2\mu}{2} \sum_{k_z'} \sum_{K\neq 0} \varepsilon_K \, e^{iKz} A_{k_z'} \frac{\hat{e}_{k'}}{\sqrt{|k_z'|}} e^{ik_z' z} \tag{92}$$

where the Fourier coefficients ε_K of the dielectric function couple modes k_z and k'_z.

The second derivative of the field envelope in eq. (92) is usually small relative to the first derivative

$$\left|\frac{d^2}{dz^2} A_k\right| << \left|k \frac{d}{dz} A_k\right| \tag{93}$$

and can be neglected. This is called the slowly varying amplitude approximation. Using this approximation, eq. (92) becomes

$$2i\sum_{k_z} k_z \frac{\hat{e}_k}{\sqrt{|k_z|}} \left(\frac{d}{dz} A_{k_z}\right) e^{ik_z z} = -\frac{\omega^2 \mu}{2} \sum_{k'_z} \sum_{K \neq 0} \varepsilon_K e^{iKz} A_{k'_z} \frac{\hat{e}_{k'}}{\sqrt{|k'_z|}} e^{ik'_z z} \tag{94}$$

Taking the scalar product with the field of the k_z-mode leads to the differential equation

$$\frac{d}{dz} A_{k_z} = i \frac{\omega^2 \mu}{4} \sum_{k'_z} \sum_{K \neq 0} \varepsilon_K e^{iKz} \frac{\hat{e}^*_k \cdot \hat{e}_{k'}}{\sqrt{|k_z||k'_z|}} A_{k'_z} e^{i(k'_z - k_z)z} \tag{95}$$

that describes the spatial evolution of the field amplitude of the k_z-mode. Eq.(95) was obtained assuming plane waves. A more general expression that includes mode profiles is also possible [75].

In eq. (95), significant coupling between modes k_z and k'_z is only possible when the Bragg condition

$$K = k_z - k'_z \tag{96}$$

is satisfied. Otherwise, the rapid oscillations in the phase in the exponent on the right-hand side of eq. (95) cancel and the mode amplitude is unaltered. Slight deviations from phase matching are allowed if they are small, such that $\Delta kL < 1$, where

$$\Delta k = k_z - k'_z - K \tag{97}$$

and L is the thickness of the sample. The Bragg condition is equivalent to eq. (3) for the grating vector formed by two interfering waves. This is the condition of self-phase-matching: the same beams that write a nonlinear grating also satisfy the phase-matching condition for maximum coupling between the beams. Using the Bragg condition, the coupled mode eq. (95) is finally

$$\frac{d}{dz}A_{k_z} = i\frac{\omega^2\mu}{4}\sum_{k_z'}\sum_{K\neq 0} \varepsilon_K \frac{\hat{e}_k^* \cdot \hat{e}_{k'}}{\sqrt{|k_z||k_z'|}} A_{k_z'}\,\delta\left(K - k_z + k_z'\right) \qquad (98)$$

During photorefractive wave mixing, the dominant Fourier coefficient is associated with the grating vector ±K and only beam pairs that are connected through the grating vector are coupled. The coupled mode equations therefore further simplify to

$$\begin{aligned} \frac{d}{dz}A_1 &= i\frac{k_{z1}}{|k_{z1}|}C_{12}\,A_2 \\ \frac{d}{dz}A_2 &= i\frac{k_{z2}}{|k_{z2}|}C_{21}\,A_1 \end{aligned} \qquad (99)$$

where the coupling constants are

$$\begin{aligned} C_{12} &= \frac{\omega^2\mu}{4}\frac{\hat{e}_1 \cdot \overleftrightarrow{\varepsilon}_{-K}\hat{e}_2^*}{\sqrt{|k_{z1}k_{z2}|}} \\ C_{21} &= \frac{\omega^2\mu}{4}\frac{\hat{e}_2 \cdot \overleftrightarrow{\varepsilon}_{K}\hat{e}_1^*}{\sqrt{|k_{z1}k_{z2}|}} \end{aligned} \qquad (100)$$

which in general are not Hermetian, i.e. $\varepsilon_{-K} \neq \varepsilon^*_K$, because of absorption gratings. In the simple case of isotropic index and extinction gratings, the coupling constants are

$$C_{12} = \frac{\pi}{\lambda}\Delta\tilde{n}\,e^{-i\phi}$$

$$C_{21} = \frac{\pi}{\lambda}\Delta\tilde{n}\,e^{i\phi}$$

where $\Delta\tilde{n}$ is complex and includes absorption gratings. Note in this case $C_{12} \neq C^*_{21}$. The coupling constants C_{12} and C_{21} contain all of the details of the photorefractive grating formation processes, including the space-charge fields, the electro-optic effects, the photorefractive phase shift, and absorption gratings. For instance, with an electro-optic grating only, the coupling constant C_{12} can be expressed as

$$C_{12} = C_{21}^{*} = -\frac{1}{2}\frac{\omega^2}{c^2}\frac{n^4}{\sqrt{|k_{z1}k_{z2}|}}\left(\hat{e}_1 \cdot \overleftrightarrow{r} |E_1| \hat{e}_2^{*}\right)\left(\hat{e}_1 \cdot \hat{e}_2^{*}\right)\frac{A_1 A_2^{*}}{|m| I_0} e^{-i\phi} \tag{101}$$

in which the photorefractive phase shift ϕ has been expressed explicitly. To ensure coupling, the coupling constant in eq. (100) must be non-zero after the tensor properties of the dielectric function are included with the polarization of the electric fields.

It is important to make the distinction in eq.(99) between co-directional coupling and contra-directional coupling. When the z-components of the k-vectors have opposite sign then the modes experience contra-directional coupling. When the z-components of the k-vectors have the same sign, they experience co-directional coupling. The differences between these two cases are subtle, but important. An inspection of Fig. 2 shows that the coherent interference between two beams in two-wave mixing always produces contra-directional coupling. Even though the beams may enter from the same face or from opposite faces, the z-component of the k-vectors are always oppositely directed. On the other hand, co-directional coupling becomes important during degenerate four-wave mixing.

7. PHOTOREFRACTIVE TWO-WAVE MIXING

7.1 Counter-Propagating Two-Wave Mixing

In the counter-propagating geometry of two-wave mixing, the two beams enter from opposite faces and generate a grating vector parallel to the z-axis, shown in Fig 23. The two waves experience contradirectional coupling because the z-components of their k-vectors have opposite signs

$$\frac{k_{z1}}{|k_{z1}|} = -\frac{k_{z2}}{|k_{z2}|}$$

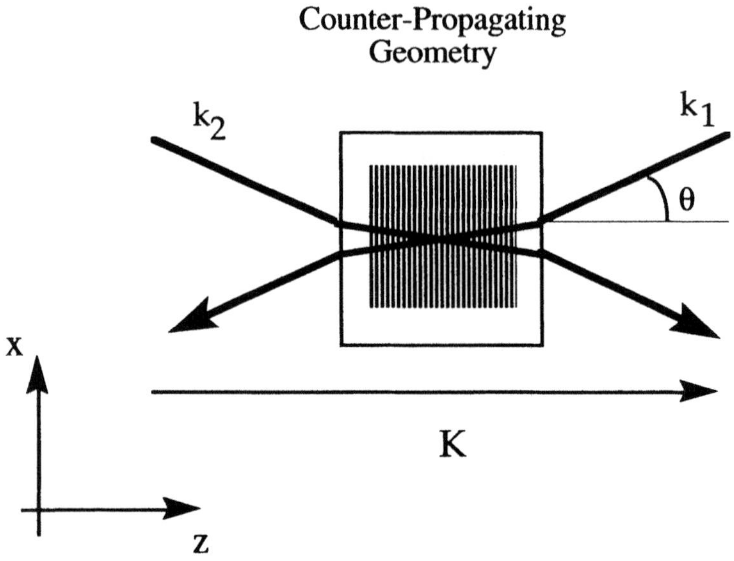

Fig. 23 The counter-propagating geometry for two-wave mixing is characterized by large grating vectors and small fringe spacing.

The coupled-mode equations, without absorption, become

$$\begin{aligned} \frac{d}{dz} A_1 &= -\frac{1}{2 I_0} \tilde{\Gamma} |A_2|^2 A_1 \\ \frac{d}{dz} A_2 &= \frac{1}{2 I_0} \tilde{\Gamma}^* |A_1|^2 A_2 \end{aligned} \tag{102}$$

where the complex gain coefficient (for symmetric incidence) is

$$\tilde{\Gamma} = i \frac{4\pi \, \Delta n_m}{\lambda \cos \theta'} e^{-i\phi} \tag{103}$$

and the normalized electro-optic modulation Δn_m is defined by

$$\Delta n_m = -\frac{n^3}{2} \left(\hat{e}_1 \cdot r \frac{\overline{|E_1|}}{|m|} \hat{e}_2^* \right) \left(\hat{e}_1 \cdot \hat{e}_2^* \right) \tag{104}$$

Because we are interested in the intensities I_1 and I_2, eq.(102) can be transformed using the expression [17]

$$A_{1,2} = \sqrt{I_{1,2}} \exp(-i\psi_{1,2}) \tag{105}$$

where $I_{1,2}$ and $\psi_{1,2}$ are functions of z. This transformation leads to

$$\begin{aligned} \sqrt{I_1} \exp(i\psi_1) \frac{d}{dz}\left[\sqrt{I_1} \exp(-i\psi_1)\right] &= -\frac{1}{2 I_0} \Gamma I_2 I_1 \\ \sqrt{I_2} \exp(i\psi_2) \frac{d}{dz}\left[\sqrt{I_2} \exp(-i\psi_2)\right] &= \frac{1}{2 I_0} \Gamma^* I_1 I_2 \end{aligned} \tag{106}$$

Taking the derivative, the equations become

$$\begin{aligned} -i\psi_1 I_1 \frac{d\psi_1}{dz} + \frac{1}{2}\frac{dI_1}{dz} &= -\frac{1}{2 I_0} \Gamma I_2 I_1 \\ -i\psi_2 I_2 \frac{d\psi_2}{dz} + \frac{1}{2}\frac{dI_2}{dz} &= \frac{1}{2 I_0} \Gamma^* I_1 I_2 \end{aligned} \tag{107}$$

Separating the real and imaginary parts gives the equations for the intensity

$$\begin{aligned} \frac{d}{dz} I_1 &= -\Gamma \frac{I_2}{I_0} I_1 \\ \frac{d}{dz} I_2 &= -\Gamma \frac{I_1}{I_0} I_2 \end{aligned} \tag{108}$$

and for the phase

$$\begin{aligned} \frac{d}{dz} \psi_1 &= \frac{\Gamma}{2 \tan\phi} \frac{I_1}{I_0} \\ \frac{d}{dz} \psi_2 &= -\frac{\Gamma}{2 \tan\phi} \frac{I_2}{I_0} \end{aligned} \tag{109}$$

When both beams are incident symmetrically on the same face with the same polarization and only a refractive index grating is present (no absorption grating) the photorefractive gain is given by

$$\Gamma = \frac{4\pi \Delta n_m}{\lambda \cos\theta'} \sin(\varphi) \tag{110}$$

The solutions of eq.(108) are

$$I_1(L) = I_1(0)\frac{1+\beta}{\beta+\exp(\Gamma L)}$$
$$I_2(0) = I_2(L)\frac{1+\beta}{1+\beta\exp(-\Gamma L)} \tag{111}$$

where $\beta = I_1(0)/I_2(L)$. The importance of the phase shift ϕ is clear. When $\phi = \pi/2$, there is maximum energy coupling between the two beams.

7.2 Co-Propagating Two-Wave Mixing

In co-propagating two-wave mixing the z-components of the k-vectors again have opposite signs, and the two waves experience contradirectional coupling with the grating vector directed along the z-axis. Because the two beams enter the same face, the intensity is now implicitly a function of x rather than z.

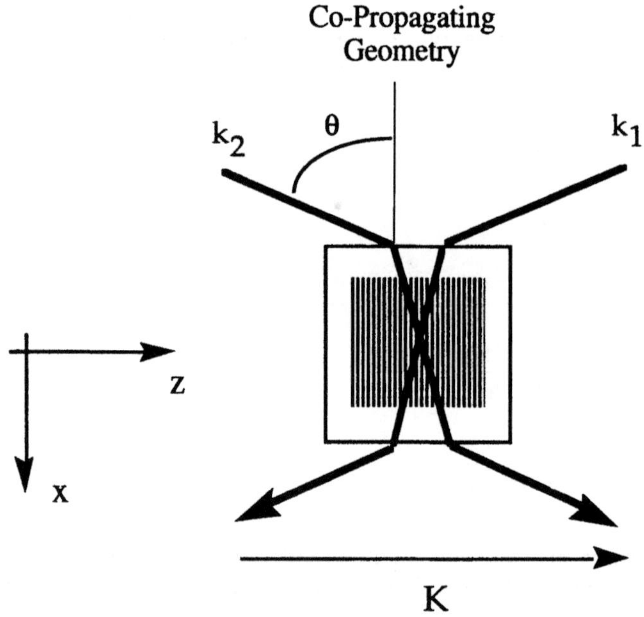

Fig. 24 In the co-propagating geometry, the grating vector is in the z-direction, but propagation is in the x-direction.

The derivatives in eq.(99) become

$$\frac{dA_1}{dz} = \frac{1}{\tan\theta'}\frac{dA_1}{dx}$$
$$\frac{dA_2}{dz} = \frac{-1}{\tan\theta'}\frac{dA_2}{dx}$$

and the coupled-mode equations are

$$\begin{aligned} \frac{d}{dx}A_1 &= -\frac{\tan\theta'}{2I_0}\Gamma|A_2|^2 A_1 - \frac{\alpha}{2}A_1 \\ \frac{d}{dx}A_2 &= -\frac{\tan\theta'}{2I_0}\Gamma^*|A_1|^2 A_2 - \frac{\alpha}{2}A_2 \end{aligned} \tag{112}$$

where linear absorption has been included, and the complex gain coefficient is defined as before in eq. (103). The coupled mode equations for the intensities I_1 and I_2 are

$$\begin{aligned} \frac{d}{dx}I_1 &= -\Gamma\frac{I_1 I_2}{I_0} - \alpha I_1 \\ \frac{d}{dx}I_2 &= \Gamma\frac{I_1 I_2}{I_0} - \alpha I_2 \end{aligned} \tag{113}$$

and Γ is defined as in eq. (110), when the beams are incident symmetrically on the sample.

The solutions to eq.(113) are

$$\begin{aligned} I_1(x) &= I_1(0)\frac{1+\beta}{\beta+\exp(\Gamma x)}\exp(-\alpha x) \\ I_2(x) &= I_2(0)\frac{1+\beta}{1+\beta\exp(-\Gamma x)}\exp(-\alpha x) \end{aligned} \tag{114}$$

where $\beta = I_1/I_2$, as defined in eq. (9). Experimentally, the photorefractive gain is often measured as the ratio of the intensity at the output of the crystal in the presence of coupling relative to the intensity at the output in the absence of coupling. This ratio is

$$\gamma = \frac{I_2(L)}{I_2(0)\exp(-\alpha L)} = \frac{1+\beta}{1+\beta\exp(-\Gamma L)} \tag{115}$$

When $\beta >> 1$, then

$$\Gamma = \frac{1}{L} \ln \gamma \tag{116}$$

7.3 Absorption Gratings

All photorefractive crystals have absorption, because absorption is the first step in the photorefractive process; photons are absorbed to generate carriers. Furthermore, the charge carriers trapped at defect sites that generate the spatially modulated electric fields (that induce the refractive index gratings) change the local absorption of those defects. Therefore, absorption gratings always accompany refractive index gratings in the photorefractive effect. Absorption gratings contribute to mode-coupling, but in a different way than index gratings.

Absorption gratings are easily included in the coupled-mode analysis by considering a complex refractive index. The complex gain coefficient from eq. (103) becomes

$$\tilde{\Gamma} = i\frac{4\pi}{\lambda}\left(\Delta n_m + i\frac{c}{2\omega}\Delta\alpha_m\right)\exp(-i\varphi) \tag{117}$$

where the index grating is now a complex grating. The normalized absorption grating magnitude is given by $\Delta\alpha_m$. The coupled intensity equations for co-propagating coupling become

$$\begin{aligned} \frac{d}{dx} I_1 &= -\Gamma_n \frac{I_1 I_2}{I_0} + \Gamma_\alpha \frac{I_1 I_2}{I_0} - \alpha I_1 \\ \frac{d}{dx} I_2 &= \Gamma_n \frac{I_1 I_2}{I_0} + \Gamma_\alpha \frac{I_1 I_2}{I_0} - \alpha I_2 \end{aligned} \tag{118}$$

where

$$\begin{aligned} \Gamma_n &= \frac{4\pi\,\Delta n_m}{\lambda \cos\theta'} \sin\varphi \\ \Gamma_\alpha &= \frac{\Delta\alpha_m}{\cos\theta'} \cos\varphi \end{aligned} \tag{119}$$

for equal polarizations. The coupling from the absorption grating appears symmetrically in the two equations in eq. (118). Therefore, absorption gratings cannot lead to nonreciprocal energy transfer between the two beams. Furthermore, the absorption grating is insensitive to the sign of the phase shift, unlike the coupling from the refractive index grating. On the other hand, Γ_α can be either positive or negative. If it is positive, then it causes an increase in the transmitted beam, which can be confused with photorefractive gain. The solutions to eq. (118) are

$$
\begin{aligned}
I_1(x) &= I_1(0)\frac{1+\beta}{\beta+\exp\left(\left(\Gamma_n-\Gamma_\alpha\right)x\right)}\exp(-\alpha x) \\
I_2(x) &= I_2(0)\frac{1+\beta}{1+\beta\exp\left(\left(-\Gamma_n-\Gamma_\alpha\right)x\right)}\exp(-\alpha x)
\end{aligned}
\qquad (120)
$$

for co-propagating two-wave mixing.

8. PHOTOREFRACTIVE FOUR-WAVE MIXING

More than two waves can interact inside a photorefractive crystal, leading to a rich variety of higher-order wave-mixing phenomena. Four-wave mixing has considerable importance for applications. Chief among these is the generation of phase conjugate light, also known as phase-reversed or time-reversed light [78-81]. In four-wave mixing, four distinct waves are identified that participate as pump or probe waves, or as waves generated by the others. There are two distinct classes of four-wave mixing that are classified according to the frequencies of the interacting waves.

Degenerate four-wave mixing uses laser beams that are coherent and have the same (or nearly the same) laser frequency. Degenerate four-wave mixing has several advantages, including self-Bragg matching and phase conjugation. In nearly degenerate four-wave mixing one or more or the coherent beams is frequency shifted by a small amount, leading to moving gratings inside the photorefractive material. Diffracted beams are Doppler-shifted when they scatter from the moving grating, but may still have many of the properties of degenerate mixing, depending on the magnitude of the frequency shift.

When the frequency shift is too large, the material cannot respond quickly enough and the four-wave mixing becomes nondegenerate four-wave mixing. During nondegenerate four-wave mixing two waves form photorefractive gratings that scatter a third probe wave. The probe wave must be incident at the Bragg angle to produce a diffracted signal. The advantage of nondegenerate four-wave mixing is that the probe signal does not have to be coherent with the pump waves, and in fact does not have to be from a coherent source. Also, the flexibility to work at several photon energies can be used to perform photorefractive spectroscopy on photorefractive materials.

8.1 Nondegenerate Four Wave Mixing

The geometry for nondegenerate four-wave mixing is shown in Fig. 25. Two pump lasers are incident on the sample with the grating vector directed along the z-axis. The probe laser is incident at the Bragg angle, given by

$$\sin\theta' = \sin\theta \frac{\lambda'}{\lambda}$$

forming a diffracted "fourth" wave. In general the probe wavelength can be larger or smaller than the pump laser wavelength. As in two-wave mixing, where all wave coupling was contra-directional, in nondegenerate four-wave mixing it is again contra-directional coupling because the k-vectors of the probe and scattered waves have opposite z-components.

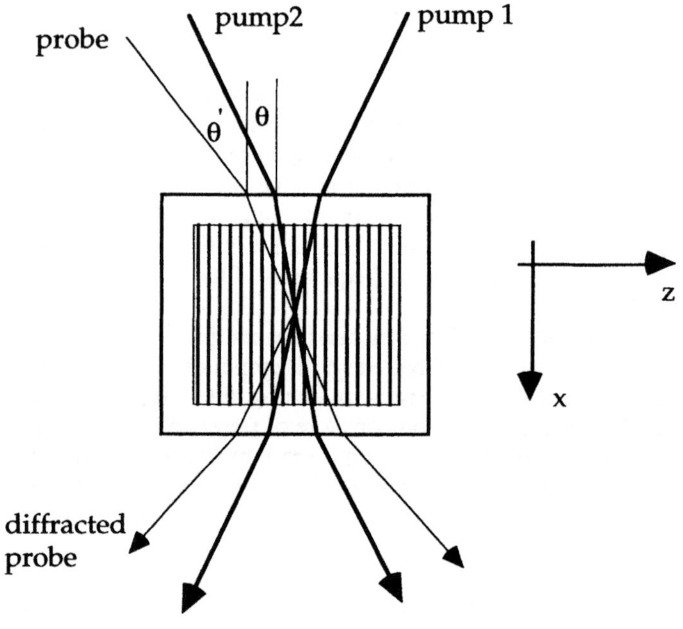

Fig. 25 In nondegenerate four-wave mixing the grating generated by two pump beams diffracts the probe beam incident at the Bragg angle.

The coupled-mode equations for the probe and diffracted waves are similar to the case of co-propagating two-wave mixing. The equations are [82]

$$\frac{d}{dx} A_3 = i\, C_{34} A_4$$
$$\frac{d}{dx} A_4 = i\, C_{43} A_3 \tag{121}$$

where the coupling coefficient C_{43} is given by

$$C_{43} = \frac{\pi\, \Delta\tilde{n}}{\lambda \cos\theta'} \tag{122}$$

The solution to eq. (121) is

$$A_3(x) = A_3(0)\cos\left(C_{34}\, x\right)$$
$$A_4(x) = A_4(0)\sin\left(C_{34}\, x\right) \tag{123}$$

for pure refractive index gratings and

$$A_3(x) = A_3(0)\cosh\left(\left|C_{34}\right| x\right) e^{-\alpha x / 2\cos\theta'}$$
$$A_4(x) = A_4(0)\sinh\left(\left|C_{34}\right| x\right) e^{-\alpha x / 2\cos\theta'} \tag{124}$$

for pure absorption gratings. The diffraction efficiency is defined as

$$\eta = \frac{I_4(L)}{I_3(0)} \tag{125}$$

and is given by

$$\eta = \frac{1}{2}\left[\cosh\left(2\,\mathrm{Im}\left(C_{43} L\right)\right) - \cos\left(2\,\mathrm{Re}\left(C_{43} L\right)\right)\right] \tag{126}$$

for complex gratings. When the diffraction efficiency is small, and both index gratings and absorption gratings contribute to the diffraction, the diffraction efficiency is

$$\eta \approx \left[\left(\frac{\pi\, \Delta n\, L}{\lambda \cos\theta'}\right)^2 + \left(\frac{\Delta\alpha\, L}{4\cos\theta'}\right)^2\right] \exp\left(-\alpha L / \cos\theta'\right) \tag{127}$$

The photorefractive phase shift does not participate in this expression, and the index and absorption gratings contribute in quadrature to the total diffraction efficiency.

8.2 Degenerate Four Wave Mixing

Degenerate four-wave mixing can lead to much more interesting behavior because all four waves can exchange energy among each other. A typical geometry is shown in Fig. 26. A forward or backward pump wave can form transmission or reflection gratings with the probe wave. The gratings diffract energy out of the pump wave into a beam that propagates back along the direction of the probe wave. This fourth wave is a phase conjugate wave [83]. The back pump can be directed onto the crystal externally, or can originate from internally scattered and reflected light in self-pumped phase conjugation geometries [84]. Although the transmission and reflection gratings can be present at the same time, it is possible, by using selected polarizations of the pump and probe waves, to generate a phase conjugate wave with either grating alone.

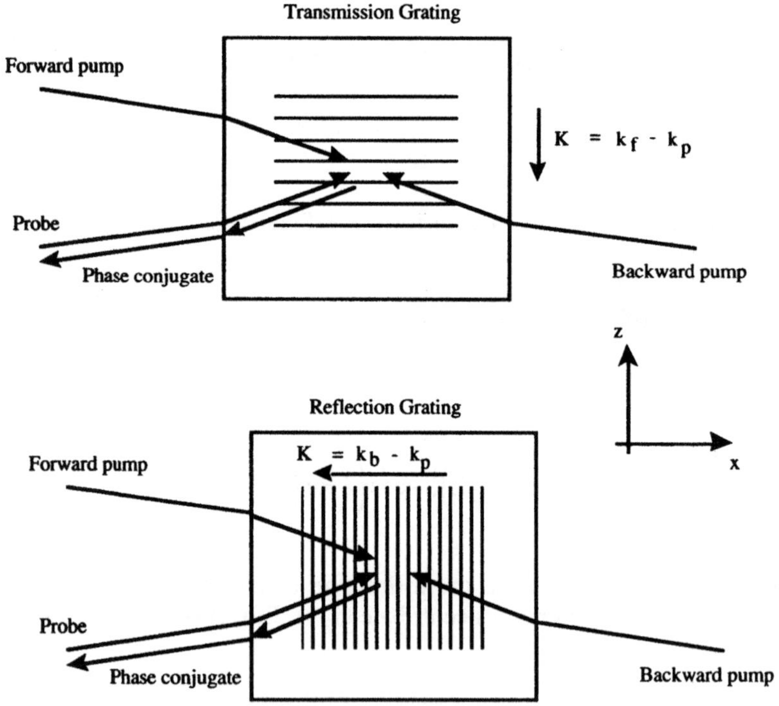

Fig. 26 Phase-conjugate grating geometries. The gratings can be selected by controlling the pump beam polarizations.

When only transmission gratings are present, the coupled mode equations for degenerate four-wave mixing are

$$\begin{aligned}
\frac{dA_p^*}{dx} &= \frac{\gamma}{I_0}\frac{nc\varepsilon_0}{2}\left(A_f A_p^* + A_b^* A_c\right)A_f^* \\
\frac{dA_c}{dx} &= \frac{\gamma}{I_0}\frac{nc\varepsilon_0}{2}\left(A_f A_p^* + A_b^* A_c\right)A_b \\
\frac{dA_f}{dx} &= -\frac{\gamma}{I_0}\frac{nc\varepsilon_0}{2}\left(A_f A_p^* + A_b^* A_c\right)A_p \\
\frac{dA_b^*}{dx} &= -\frac{\gamma}{I_0}\frac{nc\varepsilon_0}{2}\left(A_f A_p^* + A_b^* A_c\right)A_c^*
\end{aligned} \tag{128}$$

where the complex coupling constant is

$$\gamma = \frac{i\omega}{2c\cos\theta'} r_{eff}\, n^3 E_{sc} \exp(i\varphi) \tag{129}$$

If the coupling is not too strong, then the pump beams can be assumed not to vary significantly. In this undepleted pump approximation the first two coupled mode equations are solved to yield [85]

$$\begin{aligned}
A_p^*(L) &= A_p^*(0)\frac{1}{\exp(-\gamma L) + r}(1 + r) \\
A_c(0) &= A_p^*(0)\frac{r A_f / A_b^*}{\exp(-\gamma L) + r}\left(\exp(-\gamma L) - 1\right)
\end{aligned} \tag{130}$$

where $r = I_b/I_f$, and using the boundary condition $A_c(L) = 0$ [86]. The reflectance and transmittance are

$$R = \left|\frac{\sinh(\gamma L/2)}{\cosh\left[(\gamma L/2) + (\ln r)/2\right]}\right|^2 \tag{131}$$

and

$$T = \left| \frac{\cosh\left(\frac{\ln r}{2}\right)\exp(\gamma L/2)}{\cosh\left[(\gamma L/2) + (\ln r)/2\right]} \right|^2 \tag{132}$$

The maximum phase-conjugate reflectivity when $\phi = \pi/2$ is

$$R_{max} = \sinh^2(\gamma L/2) \tag{133}$$

and can exceed unity, with important consequences for phase conjugation applications.

References:

[1] A. Ashkin, G. D. Boyd, J. M. Dziedzic, R. G. Smith, A. A. Ballmann, H. J. Levinstein and K. Nassau, "Optically induced refractive index inhomogeneities in LiNbO3 and LiTaO3", Appl. Phys. Lett., vol. 9, p. 72 (1966)

[2] N. Bloembergen, *Nonlinear Optics* (New York, W. A. Bengamin, 1965)

[3] Y. R. Shen, *The Principles of Nonlinear Optics* (New York, Wiley, 1984)

[4] G. C. Valley, A. L. Smirl, M. B. Klein, K. Bohnert and F. Boggess, "Picosecond photorefractive beam coupling in GaAs", Opt. Lett., vol. 11, pp. 647-649 (1986)

[5] C. T. Chen, D. M. Kim and D. von der Linde, IEEE J. Quant. Electron., vol. QE-16, p. 126 (1980)

[6] A. Smirl, K. Bohnert, G. C. Valley, R. A. Mullen and T. F. Boggers, J. Opt. Soc. Am., vol. B6, p. 606 (1989)

[7] G. Pauliat and G. Roosen, J. Opt. Soc. Am., vol. B7, p. 2259 (1990)

[8] L. Disdier and G. Roosen, Opt. Commun., vol. 88, p. 559 (1992)

[9] F. S. Chen, J. T. LaMacchia and D. B. Frazer, Appl. Phys. Lett., vol. 13, p. 223 (1968)

[10] D. L. Staebler and J. J. Amodei, Ferroelectrics, vol. 3, p. 107 (1972)

[11] F. Micheron, C. Mayeux and J. C. Trotier, Appl. Opt., vol. 13, p. 784 (1974)

[12] P. Günter, "Holography, coherent light amplification and optical phase conjugation with photorefractive materials", Phys. Rep., vol. 93, pp. 199-299 (1982)

[13] T. J. Hall, R. Jaura, L. M. Conners and P. D. Foote, "The Photorefractive Effect-A Review", Prog. Quant. Electron., vol. 10, pp. 77-146 (1985)

[14] J. Feinberg, "Photorefractive Nonlinear Optics", Phys. Today, vol. 41, p. 46 (1988)

[15] P. Günter and J.-P. Huignard, Photorefractive Materials and Their Applications I and II, in Topics in Applied Physics **61**, (Springer-Verlag, Berlin, 1988)

[16] M. P. Petrov, S. I. Stepanov and A. V. Khomenko, *Photorefractive Crystals in Coherent Optical Systems* (Berlin, Springer-Verlag, 1991)

[17] P. Yeh, *Introduction to Photorefractive Nonlinear Optics* (New York, John Wiley & Sons, Inc., 1993)

[18] Gower and Proch, Optical Phase Conjugation, (Springer Verlag, Berlin, 1994)

[19] P. Hariharan, *Optical Holography* (Cambridge, Cambridge University Press, 1984)

[20] V. Markov, S. Odulov and M. Soskin, Opt. Laser Technol., vol. 95, p. (1979)

[21] R. S. Rana, D. D. Nolte, R. Steldt and E. M. Monberg, "Temperature dependence of the photorefractive effect in InP:Fe: role of multiple defects", J. Opt. Soc. Am., vol. B9, p. 1614 (1992)

[22] M. Lannoo and J. Bourgoin, *Point Defects in Semiconductors I* (Berlin, Springer-Verlag, 1981)

[23] J. Bourgoin and M. Lannoo, *Point Defects in Semiconductors II* (Berlin, Springer-Verlag, 1983)

[24] Lucovsky, Sol. St. Commun., vol. 3, p. 299 (1965)

[25] M. Lannoo, G. A. Barraff and M. Schlueter, "Self-consistent second-order perturbation treatment of multiplet structures using local density theory", Phys. Rev. B, vol. 24, p. 943 (1981)

[26] D. D. Nolte, N. M. Haegel and K. W. Goossen, Photo-Induced Space-Charge Effects in Semiconductors: Electro-Optics, Photoconductivity and the Photorefractive Effect, MRS Symposium Proceedings **261**, (MRS, Pittsburgh, 1992)

[27] V. L. Vinetskii and N. V. Kukhtarev, "Theory of the conductivity induced by recording holographic gratings in nonmetallic crystals", Sov. Phys.-Solid State, vol. 16, p. 2414 (1975)

[28] N. V. Kukhtarev, "Kinetics of hologram recording and erasure in electrooptic crystals", Sov. Tech. Phys. Lett., vol. 2, p. 438 (1976)

[29] N. Kukhtarev, V. Markov, S. Odulov, M. Soskin and V. Vinetskii, "Holographic Storage in Electrooptic Crystals. Beam Coupling-Light Amplification", Ferroelectrics, vol. 22, p. 949 (1979)

[30] G. C. Valley and M. B. Klein, "Optimal properties of photorefractive materials for optical data processing", Opt. Eng., vol. 22, p. 704 (1983)

[31] J. Feinberg, D. Heiman Jr., A. R. Tanguay and R. W. Hellwarth, "Photorefractive effects and light-induced charge migration in barium titanate", J. Appl. Phys., vol. 51, p. 1297 (1980)

[32] R. A. Mullen and R. W. Hellwarth, "Optical measurement of the photorefractive parameters of BSO", J. Appl. Phys., vol. 58, p. 40 (1985)

[33] J. M. C. Jonathan, R. W. Hellwarth and G. Roosen, "Effect of applied electric field on the bulidup and decay of photorefractive gratings", IEEE J. Quant. Electron., vol. QE-22, p. 1936 (1986)

[34] M. B. Klein and G. C. Valley, "Beam coupling in BaTiO3 at 422 nm", J. Appl. Phys., vol. 57, pp. 4901-4905 (1985)

[35] R. Orlowski and E. Krätzig, "Holographic method for determination of photoinduced electron and hole transport in electro-optic crystals", Solid State Commun., vol. 27, p. 1351 (1978)

[36] G. C. Valley, "Simultaneous electron/hole transport in photorefractive materials", J. Appl. Phys., vol. 59, pp. 3363-3366 (1986)

[37] F. P. Strohkendl, J. M. C. Jonathon and R. W. Hellwarth, "Hole-electron competition in photorefractive gratings", Opt. Lett., vol. 11, p. 312 (1986)

[38] S. Ducharme and J. Feinberg, "Altering the photorefractive properties of BaTiO3 by reduction and oxidation at 650 oC", J. Opt. Soc. Am., vol. B3, pp. 283-292 (1986)

[39] D. D. Nolte, D. H. Olson and A. M. Glass, "Nonequilibrium screening of the photorefractive effect", Phys. Rev. Lett., vol. 63, pp. 891-894 (1989)

[40] D. D. Nolte, D. H. Olson and A. M. Glass, "Deep level photodiffractive spectroscopy of semiconductors", Appl. Phys. Lett., vol. 56, pp. 163-165 (1990)

[41] D. D. Nolte and A. M. Glass, "Nonequilibrium charge transfer and low-temperature photorefractive effects", Opt. Quant. Electron., vol. 22, pp. S47-S60 (1990)

[42] D. D. Nolte, D. H. Olson and A. M. Glass, "Spontaneous current oscillations in optically pumped semi-insulating InP", J. Appl. Phys., vol. 68, p. 4111 (1990)

[43] V. F. Belinicher and B. I. Sturman, "The photogalvanic effect in media lacking a center of symmetry", Sov. Phys. - Usp., vol. 23, p. 199 (1980)

[44] S. G. Odulov, "Spatially oscillating photovoltaic current in iron-doped lithium niobate crystals", JETP Lett., vol. 35, p. 10 (1982)

[45] M. G. Moharam, T. K. Gaylord and R. Magnusson, "Holographic grating formation in photorefractive crystals with arbitrary electron transport lengths", J. Appl. Phys., vol. 50, p. 5642 (1979)

[46] V. M. Fridkin and R. M. Magomadov, "Anomalous photovoltaic effect in LiNbO3:Fe in polarized light", JETP Lett., vol. 30, p. 686 (1980)

[47] A. M. Glass, D. von der Linde and T. J. Negran, "High-voltage bulk photovoltaic effect and the photorefractive process in LiNbO3", Appl. Phys. Lett., vol. 25, p. 233 (1974)

[48] A. M. Glass, D. von der Linde, D. H. Auston and T. J. Negran, J. Electron. Mat., vol. 4, p. 915 (1975)

[49] D. von der Linde and A. M. Glass, "Photorefractive effects for reversible holographic storage of information", Appl. Phys., vol. 8, pp. 85-100 (1975)

[50] G. Chanussot and A. M. Glass, "A bulk photovoltaic effect due to electron-phonon coupling in polar crystals", Phys. Lett., vol. 59A, p. 405 (1976)

[51] G. Chanussot, Ferroelectrics, vol. 20, p. 37 (1978)

[52] H. G. Festl, P. Hertel, E. Krätzig and R. von Baltz, "Investigation of the photovoltaic tensor in doped LiNbO3", Phys. Stat. Solidi, vol. B113, p. 157 (1982)

[53] P. Günter, "Electric-field dependence of phase-conjugate wave-front reflectivity in reduced KNbO3 and BGO", Opt. Lett., vol. 7, p. 10 (1982)

[54] J. F. Lam, "Origin of phase conjugate waves in self-pumped photorefractive mirrors", Appl. Phys. Lett., vol. 46, p. 909 (1985)

[55] G. C. Valley and J. F. Lam, "Theory of photorefractive effects in electro-optic crystals", in Photorefractive Materials and Their Applications **61**, pp. 75-98 (Springer-Verlag, Berlin, 1988)

[56] J. P. Huignard and A. Marrakchi, "Coherent signal beam amplification in two-wave mixing experiments with photorefractive BSO crystals", Opt. Commun., vol. 38, pp. 249 - 254 (1981)

[57] S. I. Stepanov, V. V. Kulikov and M. P. Petrov, Sov. Tech. Phys. Lett., vol. 8, pp. 229-230 (1982)

[58] S. I. Stepanov, V. V. Kulikov and M. P. Petrov, "Running holograms in photorefractive BSO crystals", Opt. Commun., vol. 44, pp. 19 - 23 (1982)

[59] S. I. Stepanov and M. P. Petrov, Sov. Tech. Phys. Lett., vol. 10, pp. 572 - 573 (1984)

[60] S. I. Stepanov and M. P. Petrov, "Efficient unstationary holographic recording in photorefractive crystals under an external alternating electric field", Opt. Commun., vol. 53, pp. 292 - 295 (1985)

[61] N. Kukhtarev, V. Markov and S. Odulov, "Transient energy transfer during hologram formation in LiNbO3 in external electric field", Opt. Commun., vol. 23, p. 338 (1977)

[62] G. C. Valley, "Short-pulse grating formation in photorefractive materials", IEEE J. Quant. Electron., vol. QE-19, p. 1637 (1983)

[63] D. D. Nolte, D. H. Olson, G. E. Doran, W. H. Knox and A. M. Glass, "Resonant photodiffractive effect in semi-insulating multiple quantum wells", J. Opt. Soc. Am., vol. B7, p. 2217 (1990)

[64] M. C. Bashaw, T. P. Ma, R. C. Barker, S. Mroczkowski and R. R. Dube, "Introduction, revelation, and evolution of complementary gratings in photorefractive BSO", Phys. Rev. B, vol. 42, p. 5641 (1990)

[65] M. C. Bashaw, T. Ma, R. C. Barker, S. Mroczkowski and R. R. Dube, "Theory of Complementry Holograms arising from electron-hole Transport in Photorefractive Media,", J. Opt. Soc. Am. B, vol. 7, p. 2329 (1990)

[66] G. Picoli, P. Gravey, C. Ozkul and V. Vieux, "Theory of two-wave mixing gain enhancement in photorefractive InP:Fe: A new mechanism of resonance", J. Appl. Phys., vol. 66, p. 3798 (1989)

[67] D. J. Chadi and K. J. Chang, "Metastability of the isolated arsenic-antisite defect in GaAs", Phys. Rev. Lett., vol. 60, pp. 2187-2190 (1988)

[68] J. Dabrowski and M. Scheffler, "Theoretical evidence for an optically inducible structural transition of the isolated As antisite in GaAs: identification and explanation of EL2 ?", Phys. Rev. Lett., vol. 60, pp. 2183-2186 (1988)

[69] J. Dabrowski and M. Scheffler, "Isolated arsenic-antisite defect in GaAs and the properties of EL2", Phys. Rev. B, vol. 40, p. 10391 (1989)

[70] D. J. Chadi and K. J. Chang, Phys. Rev. B, vol. 39, p. 10063 (1989)

[71] P. M. Mooney, J. Appl. Phys., vol. 67, p. R1 (1990)

[72] D. D. Nolte, D. H. Olson and A. M. Glass, "Spatial modulation of the Fermi level by coherent illumination of undoped GaAs", Phys. Rev., vol. B40, pp. 10650-10652 (1989)

[73] D. D. Nolte and R. S. Rana, "Modulation Doping with Coherent Photons", in Proceedings of 20th Int. Conf. Phys. Semicond., p. 2247 (World Scientific, 1990)

[74] R. A. Linke, T. Thio, J. Chadi and G. E. Devlin, "Diffraction from optically written persistent plasma gratings in doped compound semiconductors", Appl. Phys. Lett., vol. 65, p. 16 (1994)

[75] A. Yariv and P. Yeh, *Optical Waves in Crystals* (New York, John Wiley & Sons, 1984)

[76] J. Strait, J. D. Reed and N. V. Kukhtarev, "Orientational dependence of photorefractive two-beam coupling in InP:Fe", Opt. Lett., vol. 15, p. 209 (1990)

[77] R. S. Rana, E. Oh, K. Chua, A. K. Ramdas and D. D. Nolte, "Magneto-photorefractive effects in a diluted magnetic semiconductor", Phys. Rev. B, vol. 49, p. 7941 (1994)

[78] A. Yariv, IEEE J. Quant. Electron., vol. QE-14, p. 650 (1978)

[79] R. W. Hellwarth, J. Opt. Soc. Am., vol. 67, p. 1 (1977)

[80] B. Y. Zeldovich, N. F. Pilipeetskii and V. V. Shkunov, Sov. Phys. - Usp., vol. 25, p. 713 (1982)

[81] H. J. Eichler, P. Gunter and D. W. Pohl, *Laser-Induced Dynamic Gratings* (Berlin, Springer-Verlag, 1986)

[82] H. Kogelnik, "Coupled wave theory for thick hologram gratings", Bell Sys. Tech. J., vol. 48, p. 2909 (1969)

[83] R. A. Fischer, *Optical Phase Conjugation* (New York, Academic Press, 1983)

[84] J. Feinberg, Opt. Lett., vol. 7, p. 486 (1982)

[85] J.-I. Sakai, *Phase Conjugate Optics* (New York, McGraw-Hill, 1992)

[86] B. Fischer, M. Cronin-Golomb, J. O. White and A. Yariv, "Amplified reflection, transmission and self-oscillation in real time holography", Opt. Lett., vol. 6, p. 519 (1981)

2.

THE PHOTOREFRACTIVE EFFECT IN FERROELECTRIC OXIDES

M.H. Garrett*, G.D. Fogarty§, G.D. Bacher†, R.N. Schwartz and B.A. Wechsler

*Deltronic Crystal Ind., Inc.
60 Harding Avenue, Dover NJ 07801
§Tufts University, EOTC
4 Colby St., Medford, MA 02155
†University of Southern California,
Dept. of Physics, University Park, Los Angelos, CA 90089
GM Hughes Research Laboratory
3011 Malibu Canyon Rd., Malibu, CA 90265

1. INTRODUCTION

Ferroelectric materials derive their name by analogy with ferromagnetic materials, of which have found great utility as memory storage media and magnetic field sensors. Crystals such as lithium niobate ($LiNbO_3$), barium titanate ($BaTiO_3$) potassium niobate ($KNbO_3$) are ferroelectric crystals that have the Perovskite structure, ABO_3. They undergo a structural phase transition in which the force of the local electric field caused by the ionic displacement is larger than the first order elastic restoring force [1]. This results in the formation of electric dipole moments, a net spontaneous dielectric polarization, and a noncentrosymmetric crystal. The change from a paraelectric to ferroelectric state is accompanied by either a continuous (second-order) or discontinuous (first-order, e.g. $BaTiO_3$) change in the spontaneous polarization at the structural phase transition temperature called, (by analogy with ferromagnetic materials), the Curie temperature, T_C. The polarity of these crystals results in strong anisotropic nonlinear properties, e.g., the electrooptic or Pockels effect. These large band-gap materials (E_g~3 eV) are insulators, but intrinsic or extrinsic defects with energy levels in the band gap make these materials photoconductive at visible and near-infrared wavelengths. As described in the introductory chapter of this book, these defects help produce the photorefractive modulation of the real and imaginary parts of the index of refraction. Index modulations or phase modulations can be used for optical information processing, e.g. correlators, filters or neural networks, and potentially (as has been speculated since the discovery of the photorefractive effect) may be a method by which a ferroelectric inorganic crystal can become a memory storage media, i.e. for volume holographic data storage. The capacity of this memory media could be on the order of a terabyte.

The many optical processing algorithms that use the photorefractive effect and related novel phenomena, e.g., self-pumped phase conjugation, the novelty filter, and photorefractive solitons, have evoked many envisioned uses for photorefractive materials that ultimately depend on photorefractive properties such as the response time, the amplitude of response, angular sensitivity, the storage time or dark decay time of the space-charge field and the wavelength of operation. The material scientist may control these properties by well known techniques such as growth-induced native defects, doping induced defects, and oxidation or reduction to modify the valence state of the defects. The influence of several dopants or post-processing techniques on the photorefractive behavior is either understood or predictable, as will be described within this chapter. For instance, doping of $BaTiO_3$ produces photorefractive crystals that are 'sensitive' from near its band edge around 400 nm out to at least 1.06 μm. In addition, with reduced-$BaTiO_3$, beam-coupling response times compatible with video frame rates have been demonstrated.

Ferroelectric photorefractive crystals other than $BaTiO_3$ have also demonstrated photorefractive behavior that will be beneficial to applications. Unlike silicon, there doesn't appear to be one optical material for most applications. It is likely that more than one ferroelectric photorefractive material will find its way into applications. Although the perovskite crystals belong to the same class, each has its unique photorefractive properties. For example, $KNbO_3$ a more closely packed structure than $BaTiO_3$ and grows at a lower temperature. As a result it is more difficult to dope $KNbO_3$ with high concentrations, and it naturally grows with less intrinsic defects. Nevertheless, $KNbO_3$ can have a large amplitude of response even with low doping densities due to its large electrooptic coefficients. In addition, when reduced, $KNbO_3$ is one of the fastest responding photorefractive materials in the visible spectrum. However, it is difficult to reduce in large sample sizes and is therefore a subject of current research efforts. Because of the idiosyncrasies of each material, the advantages and disadvantages of the various photorefractive crystals are only relevant when referenced to a specific application and is beyond the scope of this chapter to analyze. Therefore, we focus on one of the more popular, but not necessarily the best (depending on the application), photorefractive materials $BaTiO_3$

$BaTiO_3$ has photorefractive properties that in many cases are representative of the photorefractive behavior in all ferroelectrics. $BaTiO_3$ operates over a larger range of wavelengths than any other photorefractive ferroelectric crystal and has the largest known electrooptic coefficient of inorganic crystals. Its main detractors for use in applications have been its lower temperature phase transition at ~7°C and crystal growth yields. Nevertheless, both of these issues are purely technical. For example, currently, many computer chips need thermoelectric coolers (TEC) for proper operation and a TE heater could be easily used to stabilize the operation temperature of $BaTiO_3$. In addition, barium strontium titanate, $Ba_xSr_{1-x}TiO_3$, is a ferroelectric material used for pyroelectric detection of infrared radiation. Texas Instruments has developed an 80,000 element 1 in^2 $Ba_xSr_{1-x}TiO_3$ infrared detector that operates near the phase transition for commercial and military applications. Its phase transition temperature is compositionally adjusted to be near room temperature and the operating temperature is stabilized by a TE element. Furthermore, growth of $BaTiO_3$ can be facilitated by a continuous feed method that eliminates the need for melt cooling and makes the yield virtually unlimited. When there is a photorefractive application that demands $BaTiO_3$, these techniques will certainly be implemented.

Early in the characterization efforts of photorefractive materials many samples had "trap densities" determined by photorefractive measurements to be ~10^{16} cm^{-3}. At that time, it appeared that this number had little variation from crystal-to-crystal. This was surprising because it is well known that much higher defect densities exist. For example, native defect densities that are believed to be responsible for the visible absorptive properties of bismuth silicon oxide, ($Bi_{12}SiO_{20}$), have been measured by non-photorefractive means to be around 10^{21} cm^{-3}. The photorefractive "trap densities" were in most cases determined with theoretical fits that were for single deep level models. However, these models could not explain many experimental observations including asymmetry of the gain, the intensity dependence of the response time and light-induced absorption measurements.

Experimentally, we now determine the electrooptic and absorptive (or polarizability) grating contributions to the grating amplitude and an "effective" trap density. Experimentally and theoretically multiple trap levels contribute to the photorefractive effective trap density. (The influence of multiple traps was recognized in past studies on photoconductivity.) Using present multi-level trap models one determines the 'total effective trap density' which is the sum of the total effective trap density of both deep and shallow traps. Each effective trap density is dependent on the number density of compensating traps and the particular trap density. The effective trap density should not be confused with the total number of defects or traps in the material. Typically one dopes the melt from which crystals are grown with concentrations in the parts per million, 10^{16} atoms per cm^3, that are usually deep level dopants. The starting materials may also contain impurities on this order. Shallow traps associated with oxygen vacancies or other impurities are also usually present. Deep level compensators also exist that help modulate the deep level valence state. In addition, we know that native defects are inherent in all crystals. "Total effective trap densities" are now determined that usually vary between 10^{14}

cm^{-3} in 'nominally-undoped' crystals and 10^{18} cm^{-3} in 'heavily' doped samples. Shallow effective trap densities have been measured to be as high as 10^{18} cm^{-3}.

Researchers have also perceived the influence of the elasto-optic contribution to the index of refraction modulation that results in the photorefractive effect. The 'new' effective electrooptic coefficient is different from the old effective electrooptic coefficient which was usually computed using unclamped components. The new effective electrooptic coefficient takes into account unclamped and clamped components because of the compression and rarefaction of the lattice from the space-charge field. It can be very different in magnitude than the unclamped or clamped electrooptic coefficient.

To 'engineer' a ferroelectric photorefractive material it is first necessary to determine if and how a dopant and/or post-processing technique would affect the photorefractive properties using both experimental and theoretical tools. This task requires one to have good control of the growth process, the segregation coefficient of the dopant, post-processing into single crystal material, optical methods to characterize the photorefractive behavior of the resultant material, and theoretical modeling and analysis tools. Described in this chapter are measurements of the absorption, electron paramagnetic resonance (EPR), spark source mass spectroscopy (SSMS), light-induced absorption, electrooptic and absorptive gain dependence on intensity and grating wave vector, response time dependence on intensity, and dark decays of $BaTiO_3$:Co, $BaTiO_3$:Rh, and *n*-type $BaTiO_3$.

We also describe the solution growth process from which most ferroelectric crystals are grown. We outline two-level photorefractive models that more completely characterize photorefractive experimental observations and also find that light-induced absorption measurements indicate that in many cases at least three levels are necessary to explain the observed effects. We also describe a complete set of nonlinear tensor elements of $BaTiO_3$ including piezoelectric, electrooptic, elasto-optic coefficients which are used to compute the effective electrooptic coefficient that includes clamped and unclamped components. Elegant x-ray topographs are presented that show the uniformity of the lattice and its deformation due to the space-charge field.

2. $BaTiO_3$

2.1 Introduction

At room temperature $BaTiO_3$ belongs to the tetragonal space group and 4mm point group. Shown in Figure 1 is its unit cell. Octahedrally coordinated sites are typical of Perovskites. Impurities, usually from transition metal ions that are carried into the melt with the starting material titanium dioxide, TiO_2, can have energy levels in the band gap. These impurities usually lie on the titanium site. Generally, barium carbonate, $BaCO_3$, is used to provide barium atoms and this source material can have impurities that have isovalent states with it, e.g. strontium Sr, and thus may not have a great influence on the trap levels in the band gap but may affect the dielectric and electrooptic properties as is the case with the mixed crystal strontium barium niobate (SBN).

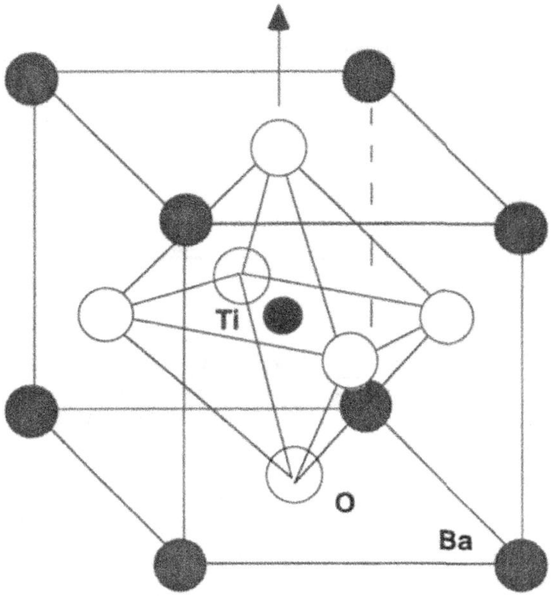

Figure 1. Unit cell of $BaTiO_3$. ● are titanium sites, ○ are oxygen sites and ● are barium sites. Covalent radii are: Ba=1.98 Å, Ti=1.32 Å, and O=0.73 Å. Lattice constants for $BaTiO_3$ are a=3.992 Å, and c=4.036 Å. The density is 6.02 g/cm^3.

Displacements, indicated by the arrow in Figure 1, of Ti^{4+} and O^{2-} produce a tetragonal space symmetry and ferroelectric spontaneous polarization in the direction of their displacement. Theoretical calculations at the center of the Brillouin zone using the self-consistent-field-multiple-scattering Xα (SCR-MS-Xα) technique of the band properties of $BaTiO_3$ show that the valence and conduction bands are associated with the TiO_6^{-8} clusters [2]. The top of the valence band is composed of 2p oxygen states and the conduction band is shown to consist of 3d titanium states for O_h and C_{4v} symmetry. Within the band gap, as shown in Figure 2, are defect states. Deep traps N_D, shallow traps M_T and 'inactive' compensating traps, e.g., Nd^{5+}, (not shown) help determine the number of filled traps, N_F. Deep traps are those that are not easily thermally ionized at room temperature, i.e., their energies are greater than the thermal energy k_BT away from a band. (Here, k_BT is Boltzmann's constant times the temperature T which is approximately ~0.025 eV at room temperature.) Usually, as-grown $BaTiO_3$ has p-type photoconductivity at visible wavelengths and thus the Fermi level lies below mid-band.

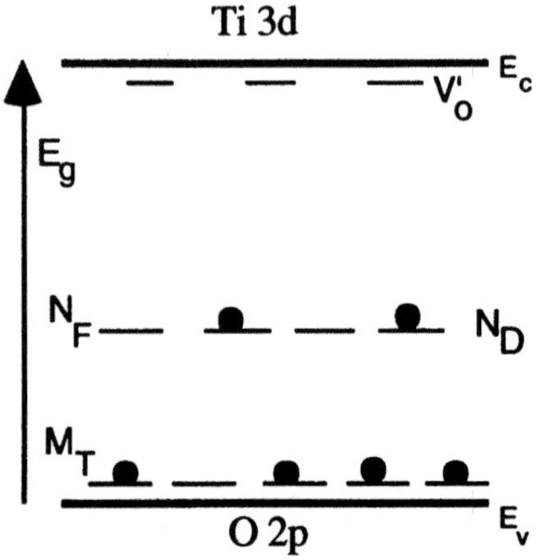

Figure 2. Energy level diagram with deep and shallow traps, N_D and M_T respectively. The uncompensated deep trap density is N_F. E_g is the energy of the band gap, E_v is the energy of the valence band and E_c is the energy level of the conduction band.

Deep traps, e.g., cobalt, are responsible for the memory effect of photorefractive gratings and have a long dark dielectric relaxation time. Oxygen vacancies and compensating deep traps are believed to modify the relative proportion of the deep trap valence states, e.g. Co^{2+}/Co^{3+}. Shallow traps near the bottom of the conduction band are believed to arise from oxygen vacancies [3]. This may be the case because at the growth temperature the atmosphere is slightly reducing P_{O2}~ 0.2 atm. Shallow traps lead to band tail absorption effects. Shallow traps near the top of the valence band may arise from native defects and impurities like sodium, Na [4]. Shallow traps can store charge, have space-charge modulations, can lead to an overall reduction of the deep trap modulation and, when thermalized, will erase deep trap gratings in the dark. Although iron is usually the most abundant impurity in the starting materials other transition metal ions may be incorporated into the crystal and contribute to the total active deep trap density of the crystal, e.g. chromium, magnesium, cobalt or nickel. Impurities are not necessarily incorporated as individual species but may create defect levels that are the result of complexes with oxygen. Rare earth ions such as cerium can also participate in the photorefractive effect in $BaTiO_3$.

Purification of the starting materials and the components used to grow the crystals is of utmost importance if one is to isolate the defects that are responsible for the photorefractive characteristics. Even so, Schunemann et al. [5] showed that $BaTiO_3$ grown from purified TiO_2 still retained relatively high beam-coupling gain. This could arise from native defects such as barium vacancies [6], because native defects (e.g. substitutions or vacancies) are inherent in the solution growth process.

2.2 Crystal Growth of $BaTiO_3$

The growth process profoundly influences the resulting photorefractive properties of $BaTiO_3$. The phase diagram of barium titanate was originally determined by Rase and Roy [7] and has been re-measured by Kirby and Wechsler [8] and is shown in Figure 3.

The solid lines of the phase diagram represent equilibrium conditions of phases. When $BaTiO_3$ is grown using its stoichiometric composition of 50 mole % BaO and 50 mole % TiO_2 it has the hexagonal space symmetry at a temperature of ~1620° C. Upon cooling this material to room temperature it undergoes several structural phase transitions at the temperatures indicated in Table 1. During cooling to below ~1432 °C the structural phase transition to cubic symmetry is made. Further cooling to below ~130° C will produce the cubic-to-tetragonal transition where the material becomes ferroelectric with C_{4v} symmetry and is therefore noncentrosymmetric. We note that the $BaTiO_3$ phase transitions exhibit a hysteresis of several degrees with heating or cooling through the transition.

	Hexagonal	Cubic	Tetragonal	Ortho-rhombic	Rhomo-hedral
Stability Temperature Range °C	1620 - 1432	1432 - 130	130 - 7	7- -90	< -90

Table 1. Temperatures ranges over which different structural phases are stable.

The hexagonal-to-cubic phase transition is destructive, resulting in severely cracked material, and is avoided by growing at a reduced temperature such that the initial

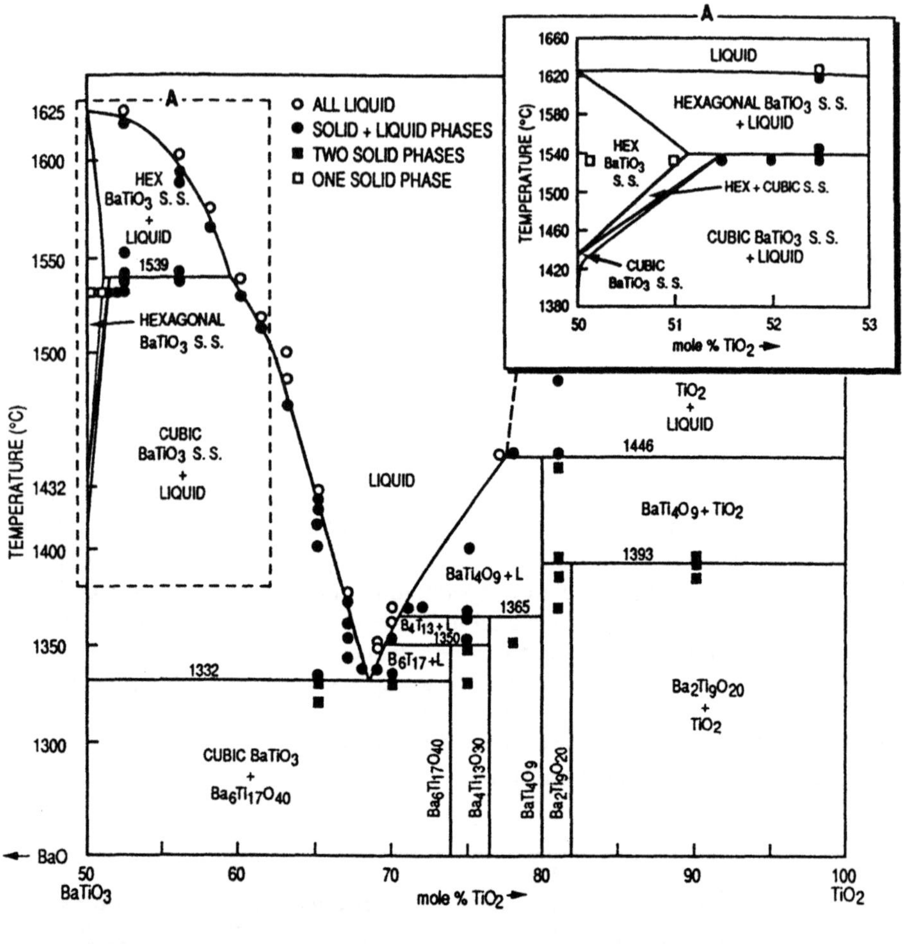

Figure 3. The phase diagram of the BaO-TiO_2 system [8]. Barium titanate, $BaTiO_3$, can be grown either from its stoichiometric composition or by adding a flux of excess TiO_2 which produces a solution with a lower growth temperature.

material is in the cubic phase. To accomplish this an excess of TiO_2 is added to the melt which suppresses the growth temperature to lower than 1432 °C. For instance, if the mole % of TiO_2 is 66.67 % then the starting growth temperature will be approximately 1380 °C. The melt is then cooled to precipitate $BaTiO_3$. The grown boule is extracted before the melt temperature reaches ~1332 °C where two solid phases coexist at the Eutectic point. Greatly oversimplifying the process, a seed of $BaTiO_3$ is brought into contact with the melt at the compositionally adjusted growth temperature and a boule of barium titanate is grown if the solution is cooled. This is called Top-Seeded Solution Growth [9], (TSSG). After growth, samples may be fabricated into single crystals as described in Reference [10].

The introduction of defects into the crystal, necessary for the photorefractive effect, is usually accomplished by adding dopants into the melt charge. However, native defects are inherent in the TSSG process. For example, because one grows $BaTiO_3$ from a melt of excess TiO_2 there will be an excess of titanium in the boule and a compensating loss of barium. For every titanium Ti^{4+} substitution onto a barium Ba^{2+} site charge neutrality requires that there will be a second loss of barium as a vacancy. These defects may influence the photorefractive properties [6]. The native defect concentration should increase as the boule is grown because the melt concentration of TiO_2 increases as the boule is grown. It is also difficult to control the dopant concentration uniformity because each dopant has a segregation coefficient usually smaller than one. (The segregation coefficient is a number that gives the fraction of dopant that will be in the crystal to that in the melt.) For example, if the segregation coefficient of a dopant is 0.4, this means that for every 10 parts per million atomic (ppma is the number of atoms per million per cm^3) in the melt, 40 % or 4 ppma will go into the crystal. As the boule grows the melt will become richer in dopant concentration relative to the crystal and, in addition, the actual concentration of available material to be extracted from the melt will decrease as the boule grows. The net effect is to increase the ppma of the impurity in the melt as the boule grows, and the concentration of the dopant in the boule will vary along the growth axis of the boule. The exact dopant concentration becomes difficult to control in TSSG growth.

TSSG is becoming a more widely used process to obtain crystalline materials because a great many materials can sometimes only be produced using this process and because it lowers the growth temperature to practical levels. The TSSG process is amenable to a modification that nearly completely eliminates the problems of correlating the cooling rate to the diameter, the sensitivity of the diameter to temperature changes, and the concentration evolution of intrinsic and extrinsic defects. It is possible to obtain solution growth at constant composition by the "continuous feed method" as has been demonstrated for $LiNbO_3$ [11]. That is, by measuring the weight of the material being extracted and by adding material to the melt in real-time to compensate for this loss (including the dopant extracted) then cooling the melt is unnecessary and the growth will occur at constant temperature and composition and allow for virtually unlimited extraction of material with minimized compositional variations. We also note that in the case of $BaTiO_3$ it may be possible to grow cubic material at a congruent composition (that transforms to tetragonal at T_c) thus avoiding the hexagonal-to-cubic destructive phase transition by the addition of strontium to the melt [12].

3. COBALT AND RHODIUM DOPING OF $BaTiO_3$

3.1 Cobalt Doping

Throughout this section we will analyze two sets of $BaTiO_3$ samples that were grown at the Massachusetts Institute of Technology (MIT). These include cobalt-doped and rhodium-doped $BaTiO_3$. Doping $BaTiO_3$ with cobalt makes the crystals' color a light brown to a dark orange-brown color depending on the dopant concentration with an absorption spectra shown in Figure 4. These samples were grown from Aran Isles Chemicals starting materials with 99.999% purity. After each boule was grown, cobalt was added to the melt in the concentration shown in the figure inset (neglecting the previous additions). Note that the peak in the absorption band is at about 500 nm or ~2.4 eV and that these samples are p-type.

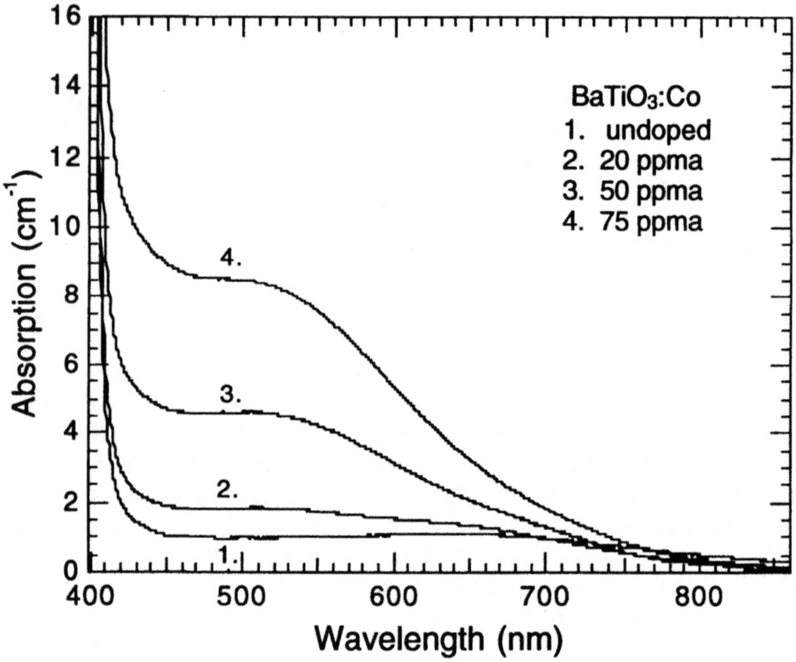

Figure 4. Absorption spectra of a series of cobalt-doped $BaTiO_3$.

3.2 Rhodium Incorporation

Table 2 contains the growth conditions of samples which have rhodium incorporated from a Rh/Pt wire dopant source. A 10% platinum/rhodium wire that was used to retain the seed was dipped into the melt before the growth of crystal MIT-3 resulting in a dark-blue colored crystal. Crystal MIT-2 was light bluish in color but was not intentionally doped with the Pt/Rh wire. Its coloration may have come from oxidation of the wire above the melt. Crystal MIT-1 was grown from a melt using purified TiO_2 and had no rhodium incorporation.

Sample MIT-1 was very clear and transparent. Its absorption spectra is shown in Figure 5. Iron levels measured by EPR were below 2 ppm in MIT-1. Shown in the next figures are the absorption spectra of samples listed in Table 2. In sample MIT-2 there is a broad absorption peak centered at 690 nm. This absorption in the red gives the crystal a bluish color, hence the name "Blue" $BaTiO_3$ was given to these materials. The 690 nm peak was first observed by Schunemann et al. [13] but its source was unknown. Other material scientists indicated to the first author of this chapter that they had noticed that the crystal became 'dark' in color when rhodium was accidentally incorporated into the melt. It was first recognized and measured by the first author of this chapter that the "Blue" crystals would have an enhanced infrared response. It was first reported by Warde et al. [14,15] from MIT that rhodium was the impurity responsible for Blue $BaTiO_3$. In sample MIT-3 the peak is shifted to 650 nm possibly from the convolution of several valence states or impurities.

Sample	TiO_2 Purity	$BaCO_3$ Purity	Source
MIT-1	Purified TiO_2	99.999%	Distilled/Aran Isles
MIT-2	99.999%	99.999%	Aran Isles
MIT-3	99.999%	99.999%	Aran Isles

Table 2. Growth conditions of three samples of $BaTiO_3$. MIT-1 was grown from a high purity melt. MIT-2, and MIT-3 both had accidental rhodium incorporation, among other elements.

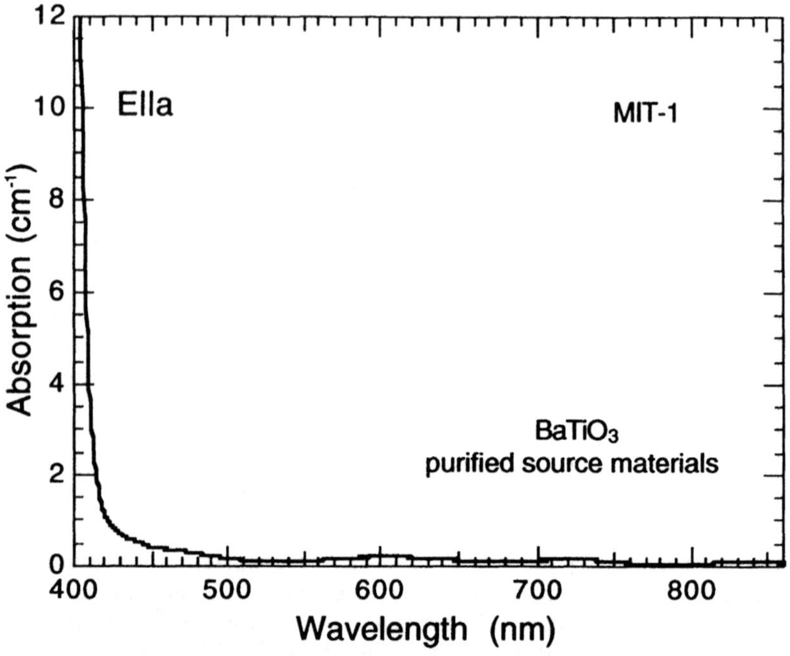

Figure 5. Absorption spectra of $BaTiO_3$ grown from a charge that used purified TiO_2, sample MIT-1.

Intentional doping experiments were performed to duplicate the accidental rhodium incorporation. The purity of the starting materials was 99.999% for all growth runs, except Growth IV, and starting materials were obtained from Aran Isles Chemicals. An initial boule was grown from each melt to confirm it was not Blue.

Platinum/rhodium wires with increasing rhodium concentrations were then introduced into each of five melts. Table 3 shows the wire's rhodium concentration and doping conditions for these growth experiments. The Pt/Rh wire was introduced into the melt prior to growth and soaked for a duration and at the temperatures shown in the table. The wire was then removed and a boule was grown. It was found, contrary to conventional wisdom, that inserting the platinum/rhodium wire into the melt did not always produce the dark coloration or Blue $BaTiO_3$. Nevertheless, as discussed in the next sections, spark source mass spectroscopy and electron paramagnetic analysis indicate that rhodium Rh^{4+} is the active element in the Blue $BaTiO_3$.

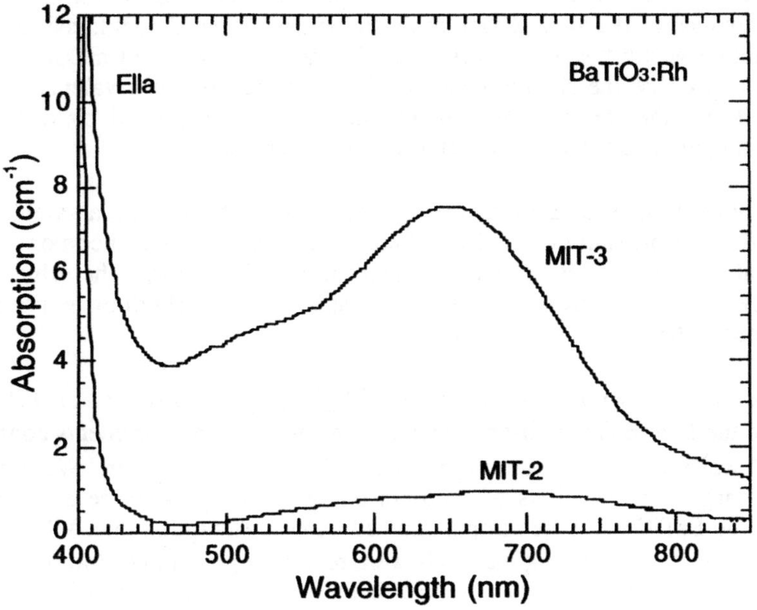

Figure 6. Absorption spectra of $BaTiO_3$:Rh.

Growth Run	% Rh in Wire	Soak Temp. °C	Soak Time (hr's)	Crystal Color
I	10	1460	48	straw yellow
II	20	1460	48	straw yellow
III	30	1380	24	straw yellow
IV	50	1450	48	Blue
V	pure	1460	48	straw yellow

Table 3. Growth runs and doping conditions that were used in an attempt to intentionally rhodium dope $BaTiO_3$

We note that Growth IV produced Blue $BaTiO_3$ and also had the most impure melt, using an order of magnitude less pure starting materials. We suspect that the purity of the starting materials may play a role in the incorporation and activation of Rh^{4+}. Rhodium metal melts at 1963°C and therefore must be oxidized to be incorporated into the melt. At the growth temperature of approximately 1385°C the atmosphere is slightly reducing ~0.2 atm. Rhodium oxidation may be aided by a transition-metal oxide reaction where the oxide is reduced, liberating oxygen and making oxygen available to oxidize the rhodium metal. Because the preferred valence state of rhodium is 3+, the rhodium 4+ valence state may be activated by deep level compensating impurities which can act as electron acceptors.

Inductively coupled plasma mass spectroscopy was performed on a sample from Growth III. This measurement indicated that indeed rhodium was incorporated into the crystal. However, as shown in Table 3, it did not have the Blue color. Potentially other impurities are masking its appearance or rhodium is not in the correct valence state.

Intentional doping with rhodium oxide, Rh_2O_3, by Wechsler et al. [16] also duplicated the Blue $BaTiO_3$ absorption spectrum and EPR measurements confirmed that Rh^{4+} was the active infrared species as will be discussed later. We note that relatively large concentrations of rhodium oxide (1600 ppma) were necessary to produce Blue $BaTiO_3$, but the sample still had a lower absorption than MIT-3. This may indicate that their starting materials were relatively pure and/or the segregation coefficient of rhodium is low for the growth and impurity conditions of the starting materials.

4. DEFECT CHEMISTRY

The defect chemistry of $BaTiO_3$ plays an important role in determining its electrical and optical properties. Although barium titanate is a line compound it is grown from a solution of excess titanium dioxide. This introduces the possibility to incorporate excess titanium and oxygen. Titanium on barium sites would produce barium vacancies to maintain charge neutrality. Titanium may also possibly be incorporated in a 3+ valence state at the high growth temperatures ~1400 C, (not the usual 4+), which will induce oxygen vacancies. In addition, if transition metal ions (which are usually acceptor ions with respect to the valence state of titanium) are present in the melt, they are usually incorporated on titanium sites and help make the material p-type in the as-grown state. It is also possible to alter the stoichiometry of the as-grown crystal by oxidation or reduction either during or after growth. The electrical and optical properties are modulated with oxidation/reduction and these directly affect the photorefractive properties of barium titanate, i.e., the dark conductivity, the photoconductivity, and the absorption. Next we will discuss defect modeling and see how acceptor-doping and oxygen-reduction control the dark conductivity and oxygen vacancy concentration in barium titanate. In the next section, we will describe the effect of reduction on the absorption spectra of barium titanate. We will describe how the photorefractive properties are optimized for specific reduction conditions in section 14.

The defect model described below is essentially that of Hagemann [17] and that discussed by Schunemann [18] and also Wechsler and Klein [6]. In the following discussion, equilibria are presented using Kröger-Vink notation [19]. Each species is identified by a symbol indicating the nature of the defect (e.g., V for vacancy A for acceptor); a subscript indicating the crystallographic site on which the defect is located; and a superscript denoting the charge of the site relative to that of the species normally occupying the site (^{x} = neutral, $'$ = -1, $^{\bullet}$ = +1).

Let us first consider the case of the undoped stoichiometric barium titanate with a Ba/Ti ratio of unity. A very simplified model can be based on the following assumptions: (1) intrinsic ionic disorder contributes negligibly to the defect structure; (2) all defects are fully ionized. At low oxygen partial pressures and elevated temperatures, oxygen vacancies would be created and balanced by the generation of mobile electrons according to the chemical reaction and law of mass action as,

$$O_o = V_o^{''} + \frac{1}{2}O_2(g) + 2e', \quad \left[V_o^{''}\right]n^2 = K_1 P_{O_2}^{-1/2}, \tag{1}$$

where K_1 is the equilibrium constant and is temperature dependent, P_{O_2} is the oxygen partial pressure, n is the density of free electrons, $V_o^{''}$ are doubly-ionized oxygen vacancies. Conservation of charge, mass and concentration has been observed. The first equation expresses the equality between oxygen on an oxygen site and the production of a doubly ionized oxygen liberating oxygen gas and leaving two electrons on the oxygen site for charge concentration. The second equation describes the relationship between the oxygen vacancy concentration, the free carrier density and the partial pressure of oxygen which are related by the rate constant K_1. The electroneutrality condition in this region is approximated by,

$$n \approx 2\left[V_o^{''}\right]. \tag{2}$$

such that the material would exhibit n-type electrical conductivity given by,

$$\sigma_e = ne\mu_e = K_1' P_{O_2}^{-1/6} e\mu_e, \tag{3}$$

where n is the density of free charge, e is the magnitude of the charge and μ_e is the electron mobility. This equation predicts that the conductivity of pure stoichiometric barium titanate is a function of the partial pressure of oxygen and would vary as the -1/6 power.

At even higher oxygen partial pressure, once all the oxygen vacancies were effectively filled, pure stoichiometric barium titanate is predicted to behave simply as an intrinsic semiconductor with the electron-hole pairs given by

$$\text{nil} = h^{\bullet} + e' \text{ and, } np = K_2 = K_2' \exp(-E_g / k_B T), \qquad (4)$$

where h is the hole density, E_g is the energy gap, k_B is Boltzmann's constant, K'_2 is the product of the densities of states of the conduction and valence bands, and nil is the thermodynamic standard state, defined here as the perfect crystal with all electrons in the lowest available energy state. This model therefore predicts two regions of behavior; a conductivity that is dependent on oxygen partial pressure and an intrinsic conductivity region.

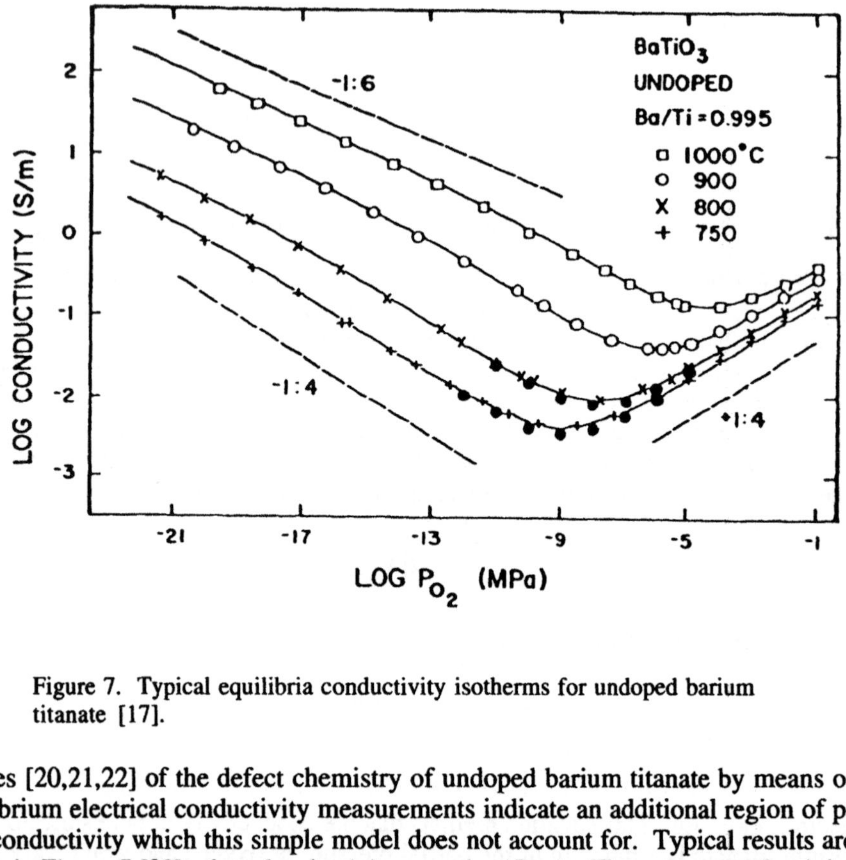

Figure 7. Typical equilibria conductivity isotherms for undoped barium titanate [17].

Studies [20,21,22] of the defect chemistry of undoped barium titanate by means of equilibrium electrical conductivity measurements indicate an additional region of p-type conductivity which this simple model does not account for. Typical results are shown in Figure 7 [22], plotted as log(σ) versus log(P_{O_2}). The p-type conductivity is represented at high oxygen partial pressures as an increase in conductivity with increased P_{O_2}, while at low P_{O_2} the conductivity changes to n-type. The -1/4 and +1/4 dependence of log(σ) versus log P_{O_2} observed for the n-type and p-type regions respectively are inconsistent with the above model that predicts a ±1/6 dependence.

These inconsistencies can be explained by assuming the presence of a species with a net negative charge, i.e. acceptors. In the region of partial oxygen pressure where the total concentration of oxygen vacancies is fixed by acceptor impurities, the electroneutrality relation is given by

$$2\left[V_o^{\cdot\cdot}\right] \approx \left[A'_{Ti}\right] >> \text{n and p}. \tag{5}$$

Substitution of equation (5) into equation (1) gives

$$n = K_3 P_{O_2}^{-1/4}, \tag{6}$$

which indicates a -1/4 power dependence with oxygen partial pressure.

At lower P_{O_2} and high temperatures, the concentration of oxygen vacancies will exceed the extrinsic levels induced by acceptors, and oxygen vacancies will be compensated by electrons, as indicated by the simple model in which the power coefficient is -1/6. At higher P_{O_2}, however, the extrinsically generated oxygen vacancies will begin to be partially filled according to the reaction,

$$V_O^{\cdot\cdot} + \frac{1}{2}O_2(g) = O_O^x + 2h^{\cdot}, \quad p^2/[V_O^{\cdot\cdot}] = K_4 P_{O_2}^{1/2}. \tag{7}$$

where p≡[h˙]. Solving Equation (7) yields,

$$p = K_5 P_{O_2}^{1/4}, \tag{8}$$

with a corresponding p-type conductivity,

$$\sigma_h = pe\mu_h = K_s P_{O_2}^{1/4} e\mu_h, \tag{9}$$

where p is the free hole density, μ_h is the hole mobility and the partial pressure of oxygen has the power coefficient of 1/4. At still higher oxygen partial pressures the oxygen vacancies could be sufficiently depressed to saturate the hole concentration at

$$p = [A']. \tag{10}$$

The defect chemistry of barium titanate becomes more complex when the acceptor impurities or dopants can exist in more than one stable valence state. Such is the case for transition metal ions, in particular Cr, Mn, Fe, Co and Ni substituted at titanium sites. To account for acceptors in variable valence states the defect model must be expanded to include the acceptor ionization reactions,

$$A_{Ti}'' = A_{Ti}' + e', \quad \frac{[A_{Ti}']n}{[A_{Ti}'']} = K_6, \tag{11}$$

for more reducing conditions and,

$$A_{Ti}' = A_{Ti}^{x} + e', \quad \frac{[A_{Ti}^{x}]n}{[A_{Ti}']} = K_7, \tag{12}$$

for more oxidizing conditions. The total electroneutrality condition is

$$n + [A_{Ti}'] + 2[A_{Ti}''] = p + [V_O^{\bullet}] + 2[V_O^{\bullet\bullet}]. \tag{13}$$

The consequence of acceptors existing in variable valence states is that there are two regimes in which the oxygen concentration is extrinsically controlled. Hagemann [23] measured the oxygen vacancy concentration as a function of oxygen partial pressure in undoped and acceptor-doped barium titanate, annealed in the 1000° C range as shown in Figure 8.

Two distinct regions can be recognized: at medium and high oxygen partial pressures the oxygen vacancy concentration is controlled by the acceptor dopant, in accordance with the oxygen compensation mechanism. For example, in cobalt-doped barium titanate, after reduction, for each two Co^{+3} ions changing to Co^{+2}, one oxygen vacancy is formed until nearly all of the valence states are +2. In addition, there is a constant oxygen vacancy concentration contribution because nearly all acceptors are in the same valence state. This means that the formation of additional vacancies, by lowering the partial pressure, is hampered because energetically favorable acceptor states are no longer available for trapping the electrons, which are always generated together with the vacancies. In this regime, the charge balance condition requires

$$[A_{Ti}'] \approx 2[V_O^{\bullet\bullet}], \tag{14}$$

for the transition region from the tetravalent to the trivalent state or

$$[A_{Ti}'] + 2[A_{Ti}''] \approx 2[V_O^{\bullet\bullet}], \tag{15}$$

if trivalent and divalent acceptors are present. However, if the oxygen partial pressure is lowered beyond the limits set by acceptor amount, the intrinsic defects, i.e. oxygen vacancies and mobile electrons, become increasing important. In this intrinsic regime (similar to highly reduced pure stoichiometric barium titanate), the vacancy concentration is much larger than the acceptor concentration and the condition of charge balance is

$$n \approx 2[V_O^{\bullet\bullet}], \tag{16}$$

which produces n-type conductivity. In Figure 8 the intrinsic region is to the left of the knee of the change in slope, and for the sample characterized by Hagemann it is at a partial pressure less than 10^{-10} atmospheres at 1000° C. Partial pressures above 10^{-10} were used to produce high-sensitivity acceptor-doped $BaTiO_3$:Co, as will be described later in this chapter.

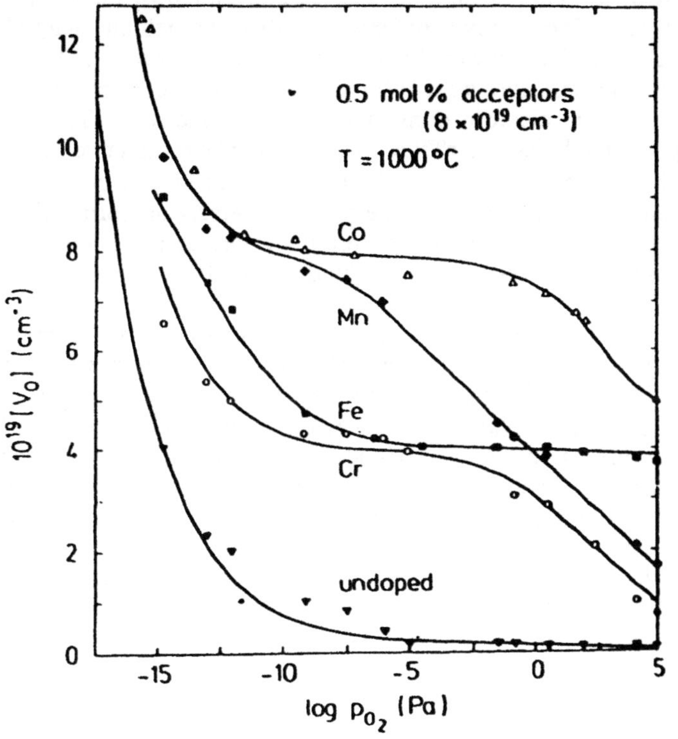

Figure 8. Oxygen vacancy concentration dependence with oxygen partial pressures of undoped and acceptor doped barium titanate [17].

5. REDUCTION OF $BaTiO_3$

Reduction or oxidation are methods by which one may alter the valence states of the defects. With these techniques one may optimize the relative concentrations of the valence states of defects, introduce new defect states in the band gap, and move the position of the Fermi level. The changes to the trap properties from reduction or oxidation directly affect the photorefractive properties of the crystal.

One method to reduce a sample is to heat the crystal in a reducing atmosphere like Argon, or Nitrogen. One may also use CO/CO_2, and the oxygen partial pressure P_{O_2} may be adjusted by changing the temperature. At elevated temperatures the carbon dioxide bonds break and liberate oxygen. The amount of oxygen in the gas may be lower than in the crystal and, after a sufficient amount of time the oxygen will diffuse out of the crystal until it is in equilibrium with the oxygen partial pressure in the gas. The removal of oxygen will leave behind charged oxygen vacancies with charges that are weakly bound when doubly-ionized. These charges are trapped by the dopants and, with sufficient reduction, the Fermi level will rise, eventually producing an electron dominated or *n*-type material. The mobility of electrons is higher than holes in $BaTiO_3$ and thus the photorefractive response time should diminish. As the Fermi level rises the dark conductivity also increases. Because the amplitude of the space-charge field is inversely proportional to the sum of the dark and photoconductivity, its amplitude decreases as the Fermi level increases. This is the response-time-to- amplitude trade-off.

Wechsler and Klein [6], have described how the density of ions and carriers and the beam-coupling gain and rise time are affected by the oxygen partial pressure of $BaTiO_3$:Fe, see Figure 9. They assume that only two possible valence states of the Fe-ion are possible and that holes and electrons could be photogenerated. Their analysis predicts three sign changes of the dominant carrier with changes in the oxygen partial pressure. The photo-induced transition at 514.5 nm or 2.4 eV would most likely induce one type of carrier because the band gap is ~3.0 eV. Thus, Wechsler and Klein argue that only a single change in sign of the dominant photocarrier would be observed when the crystal is reduced. The model presented by Wechsler and Klein [6] describes the donor and acceptor concentration in the dark and does not include the alteration in species valence states with illumination.

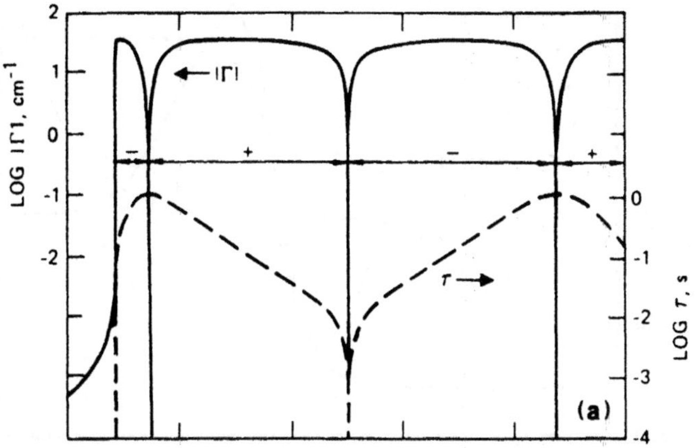

Figure 9. Plot of the gain and response time as a function of the oxygen partial pressure for $BaTiO_3$:Fe[6]. Note that several sign changes in the dominant carrier are predicted.

Recently, a similar analysis by Lahlafi et al. [24], but including the change in concentration of multivalent ions Fe^{2+}-Fe^{3+}, Fe^{3+}-Fe^{4+}, $V_O^{\circ\circ}$-V_O°, and V_O°-V_O^{x}, as a function of the light intensity was made. The equilibrium conditions of the density of iron ions and charge carriers as a function of oxygen partial pressure is predicted with the gain dependence on P_{O_2}. This is shown in Figure 10. When the illumination is 100 mW/cm^2 they found that the dominant photocarrier would change sign one time as a function of oxygen partial pressure. One sign change is consistent with that observed experimentally over eighteen orders in magnitude of P_{O_2} [25].

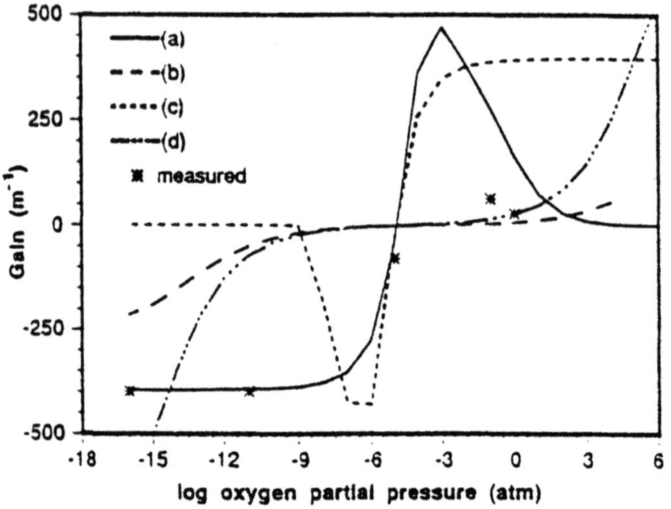

Figure 10. Plot of the gain as a function of P_{O_2} for iron-doped $BaTiO_3$ [24]. Note that only one sign change in the dominant carrier is predicted.

Schunemann et al. [13] first showed that reduction of $BaTiO_3$ to P_{O_2} ~10^{-15}-10^{-17} atm could produce a more sensitive crystal in the visible. An absorption feature at ~470 nm appeared after reduction. The sample reduced by Schunemann et al. [13] originally had an absorption feature at 690 nm and thus is suspected to have incorporated rhodium. Note also that the starting materials of their growth contained other transition metal ions and therefore it is likely that there is a concomitant reduction of all impurities with levels in the band gap. (*We note that if one uses a higher partial pressure, ~10^{-12} atm, the crystal is blue in color due to high absorption of free carriers and is so conductive it is not possible to pole the spontaneous polarization.*)

At high partial pressures of oxygen the singly-ionized oxygen electron is preferentially trapped at an oxygen site [26]. However, at lower partial pressures, oxygen is doubly-ionized and electron tapping occurs on the transition metal ions. These and most samples of $BaTiO_3$ contain trace impurities of Fe, Cr, V, and Ni. These ions will trap free electrons when the crystal is highly reduced. At these high reduction conditions they are converted to their divalent state 2+, (and possibly in the case of cobalt to a +1 valence state), in the process and their states will lie deeper in the band gap, closer to the valence band. The heavily reduced MIT-2 rhodium-doped $BaTiO_3$ will be predominantly in the Rh^{2+} state. The heavily reduced cobalt-doped sample will have valence states Co^{2+} and Co^{1+}. The increase in the Co^{1+} concentration is reminiscent of the optimization of iron-doped lithium niobate where the Fe^{2+}/Fe^{3+} concentration is optimized and a similar absorption peak is induced through reduction [27].

Sample MIT-2, which contains iron and rhodium, was reduced to $P_{O_2} \sim 10^{-16}$ atm. The absorption spectra of sample MIT-2 is shown in Figure 11 as-grown and reduced. The peak at 470 nm is nearly identical to that observed by Schunemann et al. [13].

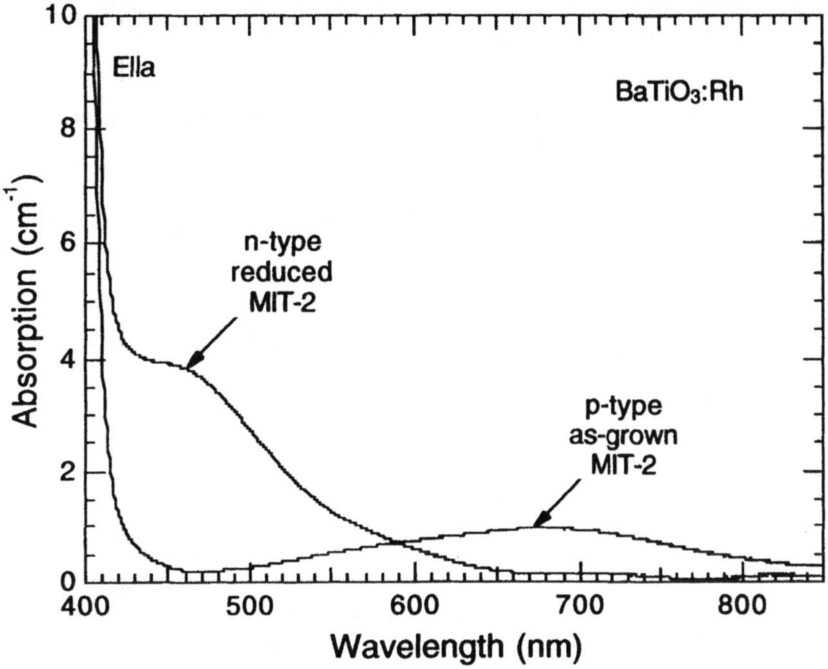

Figure 11. Absorption spectra of $BaTiO_3$:Rh as grown and reduced to $P_{O_2} \sim 10^{-16}$ atm using an atmosphere of 10% CO/CO_2.

As will be described later, this sample has a ~30 millisecond response time and demonstrated video frame-rate beam-coupling operation, i.e., 30 Hz at an intensity of 1 W/cm^2 and a wavelength of 514.5 nm using a 45°-cut crystallographic orientation.

The 20 ppm cobalt-doped $BaTiO_3$ was also reduced and produced similar absorption spectra with a peak at 470 nm when reduced to the proper level as shown in Figure 12. Trace number 3 is the reduced cobalt-doped sample which has an absorption peak at 470 nm. All other samples were also cobalt-doped, [Co]= 20 ppm in the melt, and from the same boule (except sample #1 which was nominally undoped.) Note that when the temperature is increased and the type of atmosphere is the same (10%

CO/CO_2) the partial pressure of oxygen increases. In other words, the higher the temperature the more oxidizing. The graph therefore justifies our comments that at an oxygen partial pressure around 10^{-12} atm the sample is dark blue and has an extremely high conductivity. This condition corresponds to spectrum number 6. *The blue color is presumably caused by absorption from the high free carrier density and is not to be confused with rhodium-doping.* It is not until the sample is reduced even further, $P_{O_2} < 10^{-14}$ atm, that the absorption peak at 470 nm begins to appear.

The photorefractive behavior of these two reduced samples, MIT-2 and the 20 ppma $BaTiO_3$:Co, will be compared in a following section on n-type $BaTiO_3$.

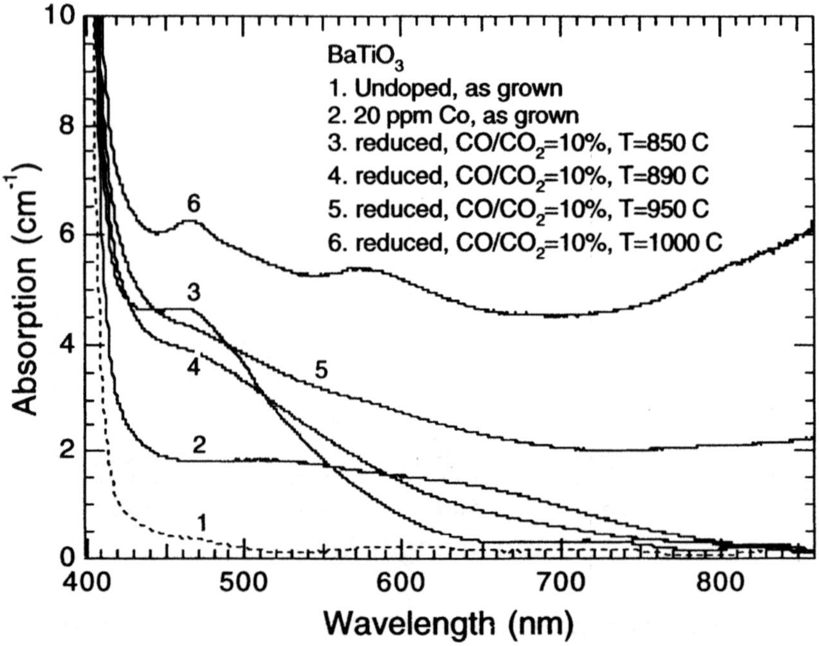

Figure 12. Absorption spectra of reduced-$BaTiO_3$:Co using an atmosphere of 10% CO/CO_2. The first sample spectrum (1) is from nominally undoped $BaTiO_3$. The second spectrum (2) is $BaTiO_3$:Co with the concentration of cobalt equal to 20 ppma in the melt with respect to the $BaTiO_3$ in the melt. The next spectra (3-6) are of this cobalt-doped sample but with different reduction conditions as indicated.

6. ELECTRON PARAMAGNETIC RESONANCE IN TRANSITION-METAL-DOPED $BATIO_3$

6.1 Deep Centers: Background

The defect centers (i.e., intrinsic defects such as oxygen vacancies or extrinsic defects such as transition metal dopants) believed to be responsible for the photorefractive effect are characterized as *deep centers*. These centers have ionization energies in the range of a few tenths of an electron volt to approximately half the band gap, in contrast to shallow centers which have ionization energies in the 50 to 150 meV range relative to the band edges. A reliable theoretical description of deep centers in semiconducting materials has been one of the major challenges faced by solid-state theory over the last three decades. The interested reader should consult the monographs and articles by Jaros [28], Pantelides [29], Zunger [30], and Lannoo and Bourgoin [31,32] for a more complete discussion. We present here a few basic concepts to make this section self-contained.

Introducing an impurity or defect center into a perfect crystalline lattice destroys the translational symmetry of the solid and introduces localized states in the band gap, i.e., states with spatially decaying wavefunctions. Effective-mass theory, which has been very successful for describing the electronic structure of shallow centers, is unsuitable for deep centers, which are characterized by strong short-range potentials. Embedded-cluster calculations [28,29] and Green's function methods [28-32] have been extensively developed for treating the electronic structure of deep centers. Both of these one-electron techniques have been applied to the study of deep centers in covalent semiconductors, whereas cluster calculations appear to be the method of choice for calculating the electronic properties of transition-metal ion impurities in ferroelectric oxides [33,34].

In the cluster approach the infinite crystal is simulated by a small number of atoms, i.e., the crystal is treated like a large molecule [35]. In this technique all the eigenstates of the perturbed crystal, that is, both the bound states and propagating states, are obtained from a single calculation in which the infinite crystal is truncated. The primary advantage of this method is that it is fairly straightforward to implement. Furthermore, for ionic host crystals, this approach is more easily modified to include contributions to the energy from lattice distortion and polarization. The drawbacks of this method are that the results are very sensitive to the cluster boundary conditions and that the precise energy positions of the bound states in the band gap are not described well by a small cluster because these states (ionization levels) are a bulk property. In the Green's function technique [28-32] the infinite perfect crystal is first treated, utilizing the translational symmetry of the periodic lattice. The changes caused by the deep level are then calculated by taking advantage of the fact that they are localized in real space. The primary advantage of the Green's function method is that it treats the deep level with the same accuracy as that chosen for the perfect crystal. The disadvantage of the method is that it is much more difficult to implement than cluster calculations.

Of fundamental importance is the relationship of the calculated one-electron deep center states to the observed optical processes.[30] There are two types of electronic transitions involved; those between different multiplets of the same oxidation (charge) state, and those that alter the oxidation state of the deep center. The former, which conserve the oxidation state, are called *internal transitions*, and the latter, which change the formal charge state of the center, are called *photoionization transitions*. Internal transitions, also referred to as crystal field transitions [36,37], resemble transitions in free ions, in that the initial and final electronic states belong to the same charge state, but the system changes its electronic configuration from one many-electron state to another. For transition-metal ion centers in centrosymmetric environments, the crystal field transitions have weak oscillator strengths as a consequence of their origin in $d \rightarrow d^*$ transitions, which for electric dipole transitions are strongly allowed only between states of different parity. These transitions occur through environmental perturbations which mix in states with opposite parity. Such mixing may be induced by either the absence of a center of symmetry or the destruction of the center of symmetry by local asymmetrical vibrational modes. In ionic solids, internal transitions do not lead to a significant redistribution of charge around the transition-metal ion center, i.e., the initial and final one-electron wave functions are almost equally localized on the ion and, to a lesser extent, its first-nearest neighbors of opposite charge (ligands).

Photoionization transitions, on the other hand, involve a change in the charge state and therefore lead to a significant redistribution of charge about the transition metal ion center. To be specific, consider a single-acceptor ionization process. The notation that is generally accepted [30] is as follows: A transition metal ion that enters the lattice substitutionally on a cation site is designated by A^q, where q refers to the *formal charge* and represents the charge relative to the ideal lattice. In this notation, for example, Co^{2+} substituting for Ti^{4+} in $BaTiO_3$ would be designated as A^{2-} and corresponds to an acceptor state because its formal charge relative to the lattice is negative. An example of a donor state in $BaTiO_3$ is Fe^{5+}, which would be designated as A^+, because its formal charge relative to the lattice is positive. It should be noted that the *formal oxidation state* represents the actual local charge on the impurity ion, for example, 2+ for Co^{2+}.

For the single-acceptor photoionization process, an electron from the fully occupied valence band is transferred to an impurity level ζ' and is represented by the following reaction [30,33,38]

$$[A^o, (VB)^m, \zeta^n] \rightarrow [A^-, (VB)^{m-1}, \zeta'^{(n+1)}]. \tag{17}$$

Here m represents the number of electrons in the valence band (VB) and ζ, ζ' denote one-electron multiplet configurations populated by n and n+1 electrons, respectively. The ionization energy associated with this process, i.e., "hole emission to the valence band", is the change in the total many-electron energy (free energy) for the system (ion + host crystal) and is denoted by $E(A^-/A^0)$ or simply E(-/0) [30,33,38]

$$E(-/0) = E_T\{A^-, (VB)^{m-1}, \zeta'^{(n+1)}\} - E_T\{A^0, (VB)^m, \zeta^n\}. \tag{18}$$

In the above equation, $E_T\{\cdots\}$ represents the fully *relaxed* total many-electron energy of the ground-state multiplet of the corresponding charge states. The relaxation referred to here is associated with the fact that the removal or addition of an electron to a deep center creates a new electronic distribution and requires the surrounding atoms to displace to a new equilibrium nuclear configuration with lower total energy.

A more physical way of describing the ionization energy E(-/0) is to note that the photoionization transition under consideration is described by the process:

$$A^0 + h\nu \rightarrow A^- + h^+_{VB}, \tag{19}$$

where h^+_{VB} represents a hole in the valence band and hν is the energy of the photon producing the transition. E(-/0) therefore represents the difference in energy between the relaxed state in which the embedded ion is in charge state A^- and a hole resides in the valence band, and the relaxed state for which the ion is in charge state A^0 and the valence band has its full complement of electrons.

It should be clear that the ionization energy discussed above, i.e., E(-/0) is not to be confused with the localized single-particle electronic energy levels obtained, for example, from an embedded-cluster calculation.[33] The impurity/defect one-electron energy levels (relative to the host crystal valence and conduction bands) represent the energy required to remove an electron from a level without change of the local configurational coordinate of the defect center. Ionization energies, as discussed above, represent differences in *total* energies for the system in which, for example, an electron is promoted from a localized impurity/defect center state to an extended electronic state associated with the conduction band of the host crystal. Because in general the localized excited states of the impurity/defect center are "resonant" with the extended states of the host crystal, photoionization transitions are expected to be rather broad and featureless.[39]

6.2 Cobalt-Doped BaTiO3

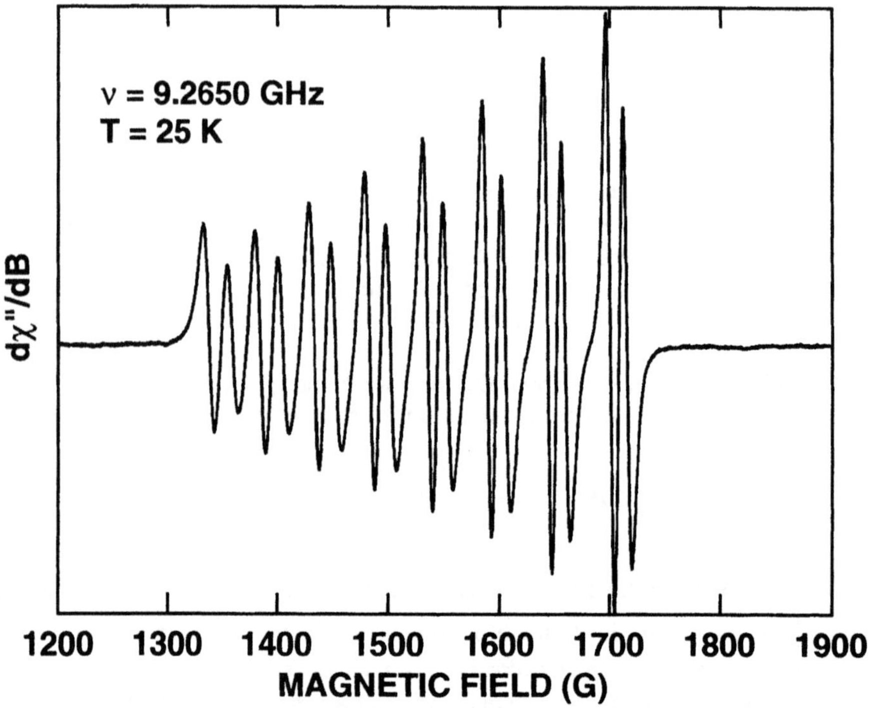

Figure 13. EPR spectrum of Co^{2+} in as-grown single crystal $BaTiO_3$:Co. The crystal was grown from a melt containing 200 ppma Co relative to Ba. The magnetic field **B** lies in the plane perpendicular to the **c**-axis.

Among the 3*d* transition-metal dopants studied in $BaTiO_3$, cobalt has proven to be quite interesting, having been shown to lead to reproducible high gain, relatively fast response time, and improved infrared sensitivity.[40-42] The paramagnetic resonance of Co^{2+} ($3d^7$, S = 3/2) in $BaTiO_3$ single crystals was first reported by Zdánsky' *et al.*[43] In a more recent paper by Aguilar [44], the EPR of Co^{2+} was reexamined and in addition, in a highly reduced Co-doped $BaTiO_3$ single crystal, the high-spin Co^{1+} ($3d^8$, S = 1) ion was identified. In addition to the EPR studies of cobalt ions in $BaTiO_3$ crystals, Blazey and Müller[45] have also observed low-spin Co^{4+} ($3d^5$) in $SrTiO_3$. More recently, Possenriede, *et al.*[46] have observed the low spin Co^{4+}-V_o center in $BaTiO_3$:Co. From the analysis of the EPR spectra it is clear that cobalt ions incorporated in $BaTiO_3$ are substitutional for the titanium ions located in the oxygen octahedra.

Shown in Figure 13 is a representative X-band EPR spectrum of the Co^{2+} ion in single crystal $BaTiO_3$:Co grown at Hughes Research Laboratory (HRL) measured at 25 K. A characteristic feature of cobalt ion spectra is the extended magnetic hyperfine structure arising from the 100% abundant isotope ^{59}Co which has a nuclear spin I = 7/2.

The ground state orbital of the free Co^{2+} ($3d^7$) ion is ^4F. In an octahedral crystalline field the ^4F state splits into three levels with symmetry designation $\Gamma_4(^4T_{1g})$, $\Gamma_5(^4T_{2g})$, and Γ_2 ($^4A_{2g}$), with the Γ_4 level lowest in energy.[47,48] This lowest Γ_4 triplet is further split by crystalline field terms of lower symmetry (e.g., tetragonal or trigonal) and by spin-orbit coupling into levels belonging to the irreducible representations Γ_6, Γ_8 and $\Gamma_7 + \Gamma_8$ with fictitious angular momenta J' = 1/2, 3/2, and 5/2, respectively. Because the J' = 3/2 and 5/2 levels are only a few hundred reciprocal centimeters above the lowest Kramer's doublet J' = 1/2 ground level, the spin-lattice relaxation time T_1 is quite short at room temperature, and therefore paramagnetic resonance is observed only at low temperatures [49]. Although the true spin of the system is S = 3/2, even in crystal field environments with symmetry lower than cubic, the strong resonance lines of the Co^{2+} ion can usually be interpreted by means of the following spin Hamiltonian with fictitious spin S' = 1/2 and nuclear spin I = 7/2:[48]

$$H = \beta \mathbf{S}\cdot\mathbf{g}\cdot\mathbf{B} + h\mathbf{S}\cdot\mathbf{A}\cdot\mathbf{I}. \tag{20}$$

Here **g** and **A** are the g- and hyperfine tensors, h is Planck's constant, **B** is the magnetic field, and **S** and **I** are the electron and nuclear spin angular momentum operators, respectively. The spin-Hamiltonian parameters for the Co^{2+} spectrum shown in Figure 13 are $g_{\parallel} = 4.263$, $g_{\perp} = 4.347$, $|A_{\parallel}| = 0.0100$ cm^{-1} and $|A_{\perp}| = 0.0105$ cm^{-1}, where the parallel and perpendicular directions are relative to the <111> axis [33,38,46]. As discussed by Possenriede *et al.*[46], the two hyperfine octets arise from resonance's in rhombohedral [111] and those in $[\bar{1}11]$, $[1\bar{1}1]$ and $[11\bar{1}]$ domains, not all of them being present with equal weights.

An additional pair of hyperfine octets with $g_{eff} \approx 2.310$ is observed in $BaTiO_3$:Co powder samples prepared by the solid state synthesis (SSS) method and processed at 1473 K in 99% CO [50]. The intensity of this feature in the highly reduced powder sample (see Figure 14) is low compared to the Co^{2+} spectral lines (in the $g \approx 4$ region), indicating that the concentration of this species is small. Based on the work of Aguilar [44] in reduced single crystalline $BaTiO_3$:Co, this spectral feature is assigned to high spin Co^{1+}. The average hyperfine splitting observed for the powder sample is $|A_{av}| \approx 0.0105$ cm^{-1}, whereas Aguilar[44] reported values of $|A| \approx 0.0055$ cm^{-1} and $g_{eff} \approx 2.117$, for a magnetic field orientation parallel to [100].

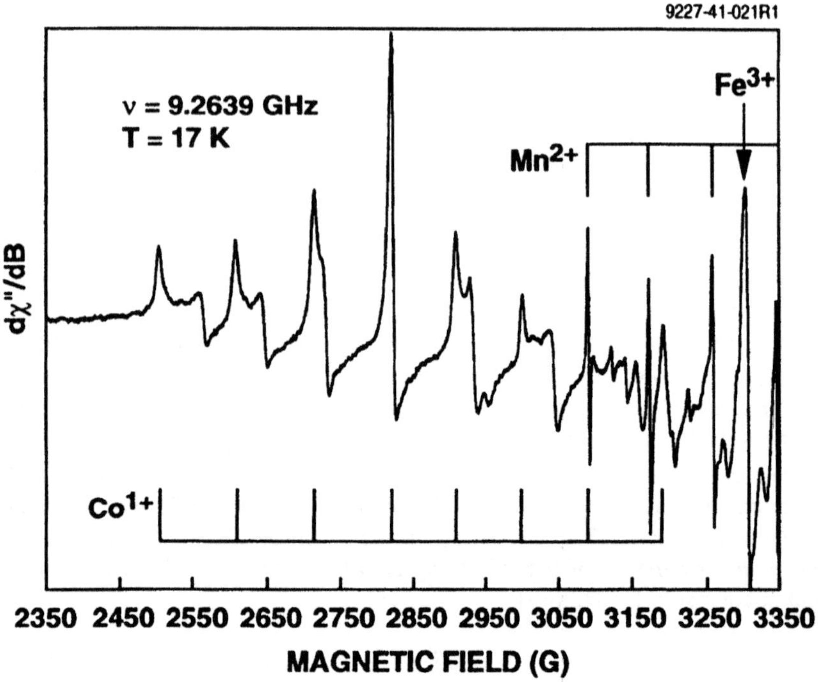

Figure 14. Polycrystalline EPR spectrum of SSS sample of $BaTiO_3$:Co (400 ppma Co) processed in 99% CO at 1473 K.

Detailed photo-EPR studies of $BaTiO_3$:Co crystals have provided insight into possible excitation and decay channels, as well as recombination and trapping mechanisms [38]. As-grown crystals illuminated at 25 K with light in the range 488-1010 nm show an increase in concentration of the Co^{2+} center relative to that observed at the same temperature in unilluminated crystals. In addition, a photoinduced paramagnetic center, stable below 50 K and assigned to the O^- ion (hole-like intrinsic center), is observed in $BaTiO_3$:Co under illumination with light in the wavelength range 488 to ~690 nm. The photo-EPR response of $BaTiO_3$:Co was also measured at low temperatures with periodically chopped laser light. A model consistent with the observed transient photoresponse explains that at low temperatures, holes liberated during the photoionization of Co^{3+} to Co^{2+} are trapped at O^{2-} sites which have in their local vicinity acceptor-like defects. These measurements provide direct evidence that the deep levels associated with the Co-dopant and intrinsic oxygen-related defects are involved in photoionization and trapping and recombination processes in $BaTiO_3$.

6.3 Rhodium-doped $BaTiO_3$

In recent reports [15,16,51] it has been observed that $BaTiO_3$ doped with rhodium displays enhanced photorefractive performance in the near infrared. EPR measurements confirm the presence of Rh^{4+} in as-grown or oxidized $BaTiO_3$:Rh single crystals.

Rhodium, a 4*d* transition-metal, may exist in several charge states ($4d^{n+}$) and spin configurations in $BaTiO_3$. Because the 4*d* electrons are less tightly bound than 3*d* electrons, the strong-crystal field [52] approximation is applicable for interpreting the experimental spectra of Rh ions. $Rh^{3+}(4d^6)$ is the most stable charge state and hence the most extensively studied species. Its ground state in perfect O_h symmetry is $^1A_{1g}[(t_{2g})^6]$, with singlet excited states $^1T_{1g}[(t_{2g})^5(e_g)]$ and $^1T_{2g}[(t_{2g})^5(e_g)]$ [53,54]. For these states the spins are all paired; therefore, diamagnetic behavior is expected. Rh^{2+} and Rh^{4+} have been less extensively studied. Both have paramagnetic ground states and both have been observed by EPR spectroscopy [55,56]. In the strong crystal field approximation [52] the ground state of $Rh^{2+}(4d^7)$ is $^2E[(t_{2g})^6(e_g)]$ [55,56], and for $Rh^{4+}(4d^5)$ the ground state is $^2T_{2g}[(t_{2g})^6]$ [55].

Shown in Figure 15 is an EPR spectrum of $BaTiO_3$:Rh (grown at HRL) recorded at 25 K with the magnetic field **B** applied along the [111] direction. We note that this EPR spectrum is also found in sample MIT-3 as reported by Garrett *et al.* [57]. This spectrum is described by a spin Hamiltonian with effective electron spin S = 1/2 and nuclear spin I = 1/2 and is assigned to Rh^{4+} centers on the Ti^{4+} site [16]. The two observed sets of hyperfine doublets (^{103}Rh, I = 1/2, 100% natural abundance) are consistent with axial symmetry along the [111] rhombohedral domain. The measured spin Hamiltonian parameters, $g_{||}[111] = 1.4565$, $A_{||}[111] = 0.0026_5$ cm^{-1} and $g_{\perp}[111] = 1.3865$, $A_{\perp}[111] = 0.0025_8$ cm^{-1}, are consistent with those reported by Possenriede *et al.*[46] for Rh^{4+}. No evidence was found for Rh^{2+}; Rh^{3+} is diamagnetic and cannot be observed by EPR. Therefore, its presence can neither be confirmed nor ruled out on the basis of these measurements.

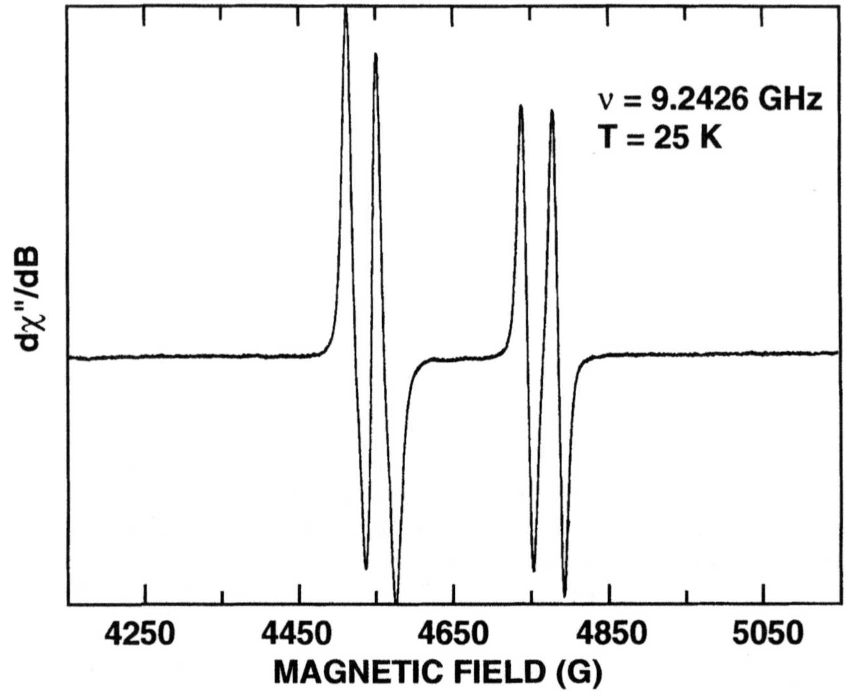

Figure 15. EPR spectrum of Rh^{4+} in as-grown single crystal $BaTiO_3$:Rh. The crystal was grown from a melt containing 800 ppma Rh relative to Ba. The magnetic field **B** is parallel to [111].

Photo-EPR measurements have been made at a temperature of 25 K using near infrared light provided by a Ti:Al_2O_3 laser. For example, under illumination with 915 nm light the intensity of the Rh^{4+} and the Fe^{3+} signals decreased significantly. The signal returned to its original strength when the light was turned off. This suggests that the photoconductivity may involve, at least in part, ionization of a hole according to the charge transfer process:

$$Rh^{4+} + h\nu \rightarrow Rh^{3+} + h^{+}_{VB}, \tag{21}$$

and,

$$Fe^{3+} + h^{+}_{VB} \longrightarrow Fe^{4+},$$

or

$$Fe^{3+} + h\upsilon \rightarrow Fe^{2+} + h^{+}_{VB}.$$

In addition, for as-grown/oxidized $BaTiO_3$:Rh, it is found that hole photoconductivity is dominant throughout the visible-near infrared spectral region [16]. These observations strongly suggest that the charge transport mechanism involves the (Rh^{3+}/Rh^{4+}) ionization level, which presumably lies within ~1.1 eV of the valence band maximum of $BaTiO_3$. Given the low concentration of Rh in these crystals and the inherently weak oscillator strength of $4d$ - $4d$ transitions, the strong absorption band located around 640 nm is most likely caused by charge transfer transitions involving the oxygen 2p states, which make up the valence band edge of $BaTiO_3$, and the Rh^{3+} and/or Rh^{4+} states which lie in the band gap.

7. SPARK SOURCE MASS SPECTROSCOPY

To evaluate the impurities and dopants incorporated into the crystal it is necessary to use several analysis techniques including spark source mass spectroscopy (SSMS) and electron paramagnetic resonance (EPR). The detection limits of SSMS are around 1 ppma and depend on the element. In some cases elements cannot be identified. To utilize SSMS it is necessary to completely dissolve the crystal into solution that can be ionized by a spark source. Obviously, incomplete dissolution of the crystal can cause detection amplitude errors. The elements are accelerated in a magnetic field and detected by a time-of-flight mass spectrometer. Although this analysis technique can identify many elements, in most instances the concentrations of the elements that are multivalent are near the detection limit of SSMS. In addition, the reliability of this technique is subject to question because double-blind tests have given inconsistent results. Nevertheless it can be helpful for impurity identification.

Shown in the next figure are the SSMS data for crystals, MIT-1, MIT-2, MIT-3. MIT-1 was grown from distilled or purified TiO_2 and 99.999% pure $BaCO_3$ starting materials. Crystals MIT-2 and MIT-3 had accidental impurity incorporation. In searching for the impurity that made crystals MIT-2,3 more sensitive in the infrared, where an absorption feature at 650-690 nm sequentially increases, we have plotted only the concentration of those element that increase from MIT-1 to MIT-3 in Figure 16. This plot reveals that only three elements sequentially increased in these samples and that the concentration of rhodium is rather low given the high absorption. This implies that when rhodium is in $BaTiO_3$, in the correct valence state, it can have a high excitation cross-section. The iron impurity concentration detected here is similar to that determined by EPR.

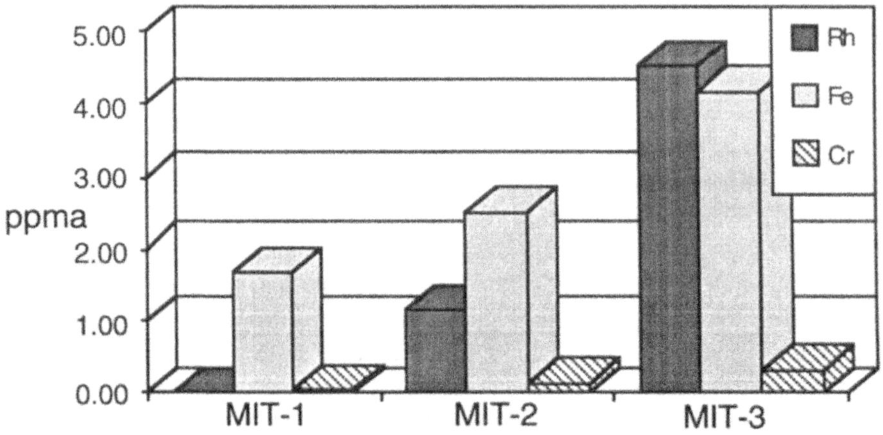

Figure 16. Chart of the elements that were detected by SSMS which sequentially increased from MIT-1 to MIT-3.

8. X-RAY TOPOGRAPHY

8.1 Technique

The technique of high-resolution x-ray diffraction imaging, or high-resolution topography, is similar in concept to, but far more capable than, classical x-ray topography [58]. A large area, monochromatic, collimated beam of synchrotron x-radiation is used to illuminate a crystalline sample at the correct Bragg angle for a chosen energy and set of crystal planes. The diffracted beam is recorded with an x-ray sensitive video camera or film. Deviations from perfection in the crystal lattice which cause small changes (≅1 arc second) in the local Bragg angle will appear as contrast changes in the image. From observation of contrast change as a function of diffraction angle near the Bragg angle, and from a set of diffraction images from carefully chosen sets of different crystal planes, it is possible to measure both the magnitude and nature of lattice defects. (Burgers vectors for dislocations can be determined, and small changes in lattice constant and orientation measured.) Such quantitative information is difficult or impossible to obtain from either white-beam topographs (where frequently the diffracting planes are unknown) due to the polychromatic nature of the incident beam, or classical topography with bremstrahlung sources, because of poor collimation. Among other experimental techniques, electron microscopy offers superior spatial resolution, but at the cost of a very small field of view, and poor resolution of changes in atomic displacements. In addition, the sample thinning required for electron microscopy introduces damage, which causes confusion in the interpretation of the resulting images because it is often difficult to determine if a particular feature has been introduced during thinning [58].

Synchrotron sources with appropriate optics, such as the National Institute of Standards and Technology beam line X23A3 at the National Synchrotron Light Source (NSLS), are the sources of choice for high-resolution diffraction imaging. The double crystal monochromator at X23A3 provides a tunable, highly monochromatic beam of x-rays (1 eV bandwidth at 10 keV), with a cross-sectional area of several square centimeters, and divergence below 1 arc second. The monochromaticity and collimation allow detection of changes in the Bragg angle below 1 arc second, corresponding to changes in lattice constant below 1 ppm (part per million), and the large area of the beam allows examination of an entire typical sample without scanning. Images may be recorded with a variety of media, such as nuclear emulsion plates (1 μm spatial resolution), or one of a variety of video cameras (up to 30 micron resolution in real-time), allowing both detailed high spatial resolution analysis, and quick survey work. Interesting high-resolution imaging work has been done at this beam line, on a variety of materials, photorefractive and otherwise [59,60,61]. There are a number of synchrotron facilities worldwide at which various types of topography are conducted [58].

8.2 Surface Quality and Lattice Regularity

More than a dozen samples of barium titanate from a variety of crystal growers, representing a good cross-section of the material available, have been examined by x-ray topography. All of the samples have shown excellent long-range lattice regularity (Figure 17). Images in Bragg geometry (reflection) show no more than a few arc seconds of lattice warping. One undesired, but often prominent, feature is residual strain from surface scratches (Figure 20). Many samples have dense networks of surface scratches which make imaging of interior detail difficult or impossible. In general, samples polished with diamond grit, even well below micron size, show substantial surface damage; while samples polished with sub-micron colloidal silica have much higher quality surfaces, with only a few isolated scratches. The residual strain from the scratches may persist even after the scratch itself has been polished away. Additionally, a-face surfaces appear to polish to a higher and more consistent quality than c-faces.

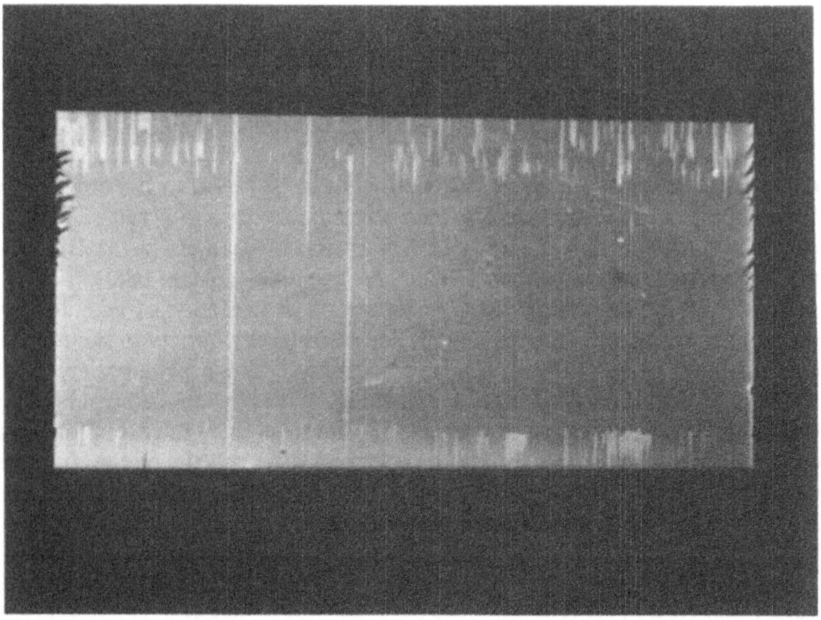

Figure 17. (402) Bragg geometry (reflection) image at 10 keV of barium titanate crystal showing excellent lattice regularity and surface polish. The bright areas are 180 degree domains.

Despite the general excellence of the lattice, there are a number of defects observed. Small angle grain boundaries are very common, and are misoriented from the bulk on the order of 10 arc seconds. Pure edge dislocations (Figure 21), and pure screw dislocations (Figure 18) are also observed, but at low densities (no more than 20 of each in a 5x3x1 mm^3 crystal). These densities are too low to be optically significant. In fact, apart from the surface scratches, no significant optical scatterers have been observed. This suggests that size of the interior defects responsible for seeding beam fanning lies below the current 1 micron spatial resolution limit of x-ray topography.

Two types of striations also have been observed. The first of these, as seen in Figure 18, are typically of low contrast, indicating that the associated lattice distortion is small (on the order of arc seconds) and only in one direction (in this case [001]), have a period of several microns, and appear to be at least partially correlated with screw dislocations. They do not appear to be of optical significance. The other striations are of greater interest, as they are related to faceted growth of the boule.

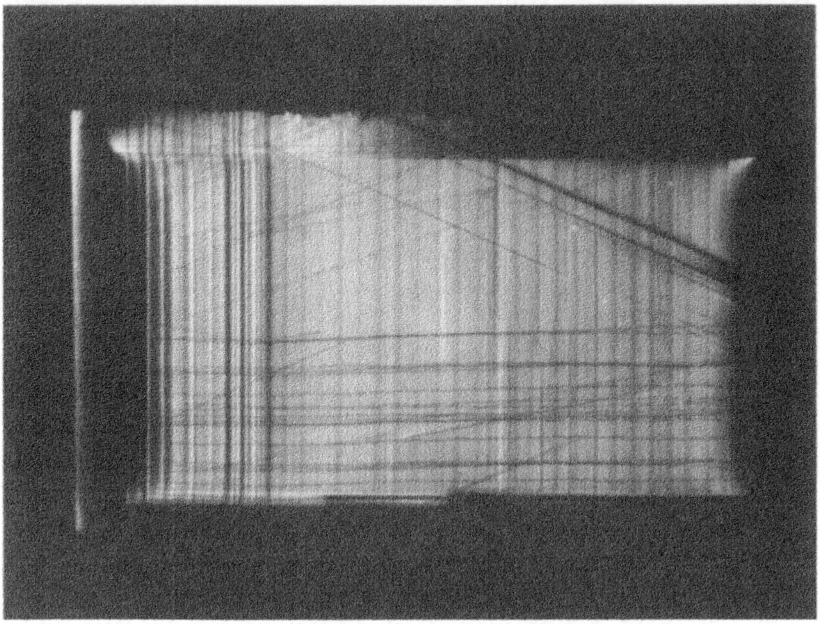

Figure 18. (002) Laue geometry (reflection) image at 12 keV of barium titanate crystal showing striations and screw dislocations (slightly wavy horizontal lines near bottom of image). The period of the striations is approximately 8-10 μm.

Figure 19. (400) Bragg geometry (reflection) image at 10 keV of barium titanate crystal showing growth facet intersections. Note the large number of surface scratches caused by incomplete polishing.

8.3 Facet Intersections

Among the most interesting images obtained of barium titanate have been those showing evidence of faceted growth and facet competition. The faceting is clearly evident in Figure 19. There is a central [001] facet aligned perpendicular to the growth direction, surrounded by four facets of unknown orientation. Close inspection of the boundaries between these surrounding facets reveals periodically wandering boundaries, indicating that the growth rates of the various facets fluctuated with time. Changes in the direction of the boundaries are clearly correlated with the striations, indicating that small periodic changes in the growth rate are responsible for the associated strain. Similar features also have been observed in bismuth silicon oxide (BSO) [59].

It also now is believed that the small angle grain boundaries observed in most samples are facet intersections, with the boule cut so that only one or two intersections are visible in any given sample. The faceting behavior shown in Figure 19 thus appears to be quite common.

8.4 Domains

Ferroelectric domains have been observed in every barium titanate crystal examined by x-ray topography. Ninety degree domains are readily visible due to the usually substantial difference in Bragg angle between the domain and the bulk for most observed diffractions. Because they rarely occur in well-poled crystals, and are readily visible by other means, they are not of primary interest. The morphology of the anti-parallel, or 180 degree, ferroelectric domains, is more illuminating.

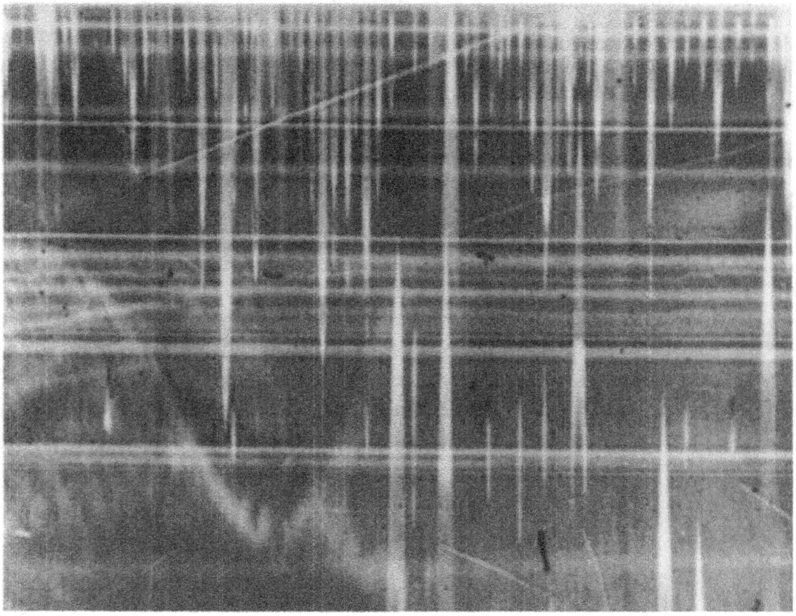

Figure 20. Enlarged portion of a (002) Laue geometry (transmission) image at 12 keV of a barium titanate crystal showing details of 180 degree domain shape. Domain diameters are in the range 20-60 μm. The horizontal lines are edge dislocations.

The 180 degree domains typically appear as "spiked" or "needle-shaped" in both reflection and transmission images (see Figures 18,21, and 22). They occur both at the surfaces and in the interior of all the samples, but the concentration is greatest near the surfaces, and in particular near the c-faces. The appearance of the domains in transmission images (Figure 20) indicates that those originating from the c-faces of the crystal are usually conically shaped, with the base of the cone at the crystal surface, and the point directed along either the positive or negative c-axis toward the interior of the crystal. Some of the domains may extend completely through the sample to the other c-face, but in well-poled samples the domains usually are confined to the region within 1 millimeter of the c-face. Domains in the interior of the crystal, far from either c-face, are typically double-cone shaped, with the points directed toward the c-faces. The fringes visible in transmission images of domains are caused by the *Pendellösung* effect, an x-ray diffraction effect in which the diffracted x-ray amplitude oscillates as a function of depth within the domain [62], analogous to the manner in which the coupling efficiency of an optical waveguide coupler oscillates with propagation distance [63].

The visibility of the 180 degree domains is surprising, because the bulk and domain regions are crystallographically identical, and would be expected to have identical Bragg angles. They are observed only in diffraction images in which the normal to the diffracting lattice planes has a component along the c-axis. Because barium titanate is non-centro-symmetric, the structure factors determining the reflections from the (hkl) and ($\bar{h}\bar{k}\bar{l}$) reflections are not equal [64]. This inequality causes a shift in the coupling between the dynamical x-ray wave fields in the crystal, and results in contrast in the image. The visibility condition for the domains also indicates that the strain at a domain wall is either small, or relaxes over a distance much smaller than a micron away from the wall.

It should be noted that the domains in unpoled crystals are not required to be crystallographically oriented. In topographs of a sample which had been accidentally de-poled by cooling through the tetragonal-orthorhombic phase transition, the domains were only approximately c-oriented, and wandered considerably through the interior of the sample. In that sample there also appeared to be a roughly equal distribution of +c and -c domains. Interestingly, the domains in all poled crystals appear to be rigidly c-oriented.

8.5 Imaging gratings

The sensitivity of x-ray diffraction imaging to small changes in lattice constant has been used to obtain the first *in-situ* measurements of photorefractive gratings. The photorefractive space-charge field produces, through the piezoelectric effect, a periodic lattice strain, and hence a periodic change in the x-ray Bragg angle, which causes contrast in a diffraction image. Gratings have been observed in a variety of orientations relative to the c-axis, at periods from 7 to 70 microns, and in both x-ray reflection and transmission [65,66].

Figure 21 shows a reflection image of a 50 micron period grating in barium titanate. There is an obvious phase change in this image as the grating crosses the 180 degree domain boundaries at the top of the figure. This phase shift is due simply to the sign change of the strain field caused by reversal of the c-axis with respect to the space-charge field. This phase change also simultaneously verifies that the lattice distortion is caused by the space-charge field, and confirms the identification of the 180 degree domains.

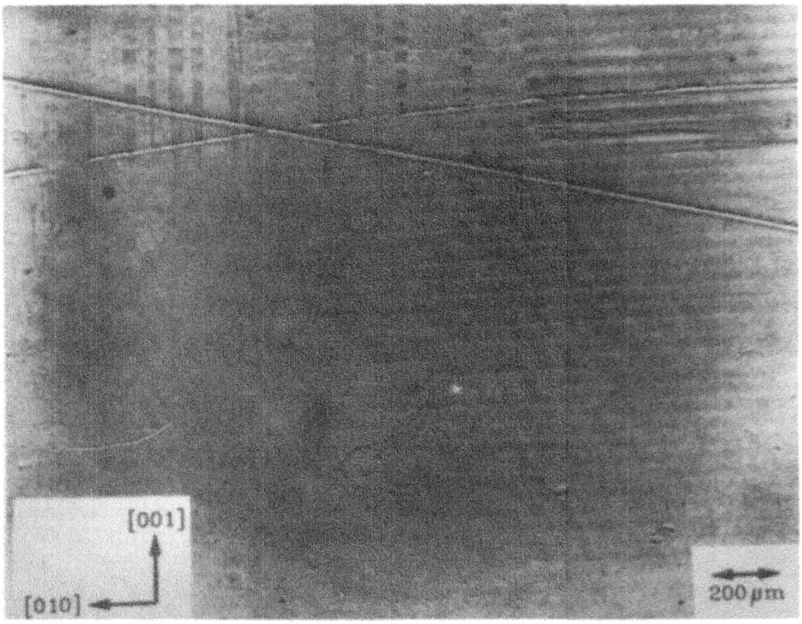

Figure 21. Enlarged portion of a (402) Bragg geometry (reflection) image at 12 keV of a barium titanate crystal with a 40 μm period photorefractive grating, formed by the interference of two beams split from a 532 nm YAG laser. Note the phase shift at the domain boundaries at the top of the image.

Careful calibration of multiple images of an individual grating taken at known exposures, and at known locations on the x-ray rocking curve (Bragg peak), allows recovery of the amplitude of the strain field, and hence the space-charge field profile and amplitude. Figure 22 shows a typical profile of the Bragg angle changes caused by a long period (40 μm) grating. Although the grating is, as expected, non-sinusoidal in shape, preliminary results from such analysis give space-charge field amplitudes which are significantly higher than expected for the long period gratings observed where the diffusion field should dominate. However, such results depend critically on how the elastic and piezoelectric behavior of the crystal are modeled [67].

It has also recently become known that the lattice distortions caused by the piezoelectric effect are significantly different near the surface than in the bulk [68]. A final judgment on the true amplitude of the space-charge field may have to wait until the most recent results on the surface lattice distortions are applied to reducing the x-ray data.

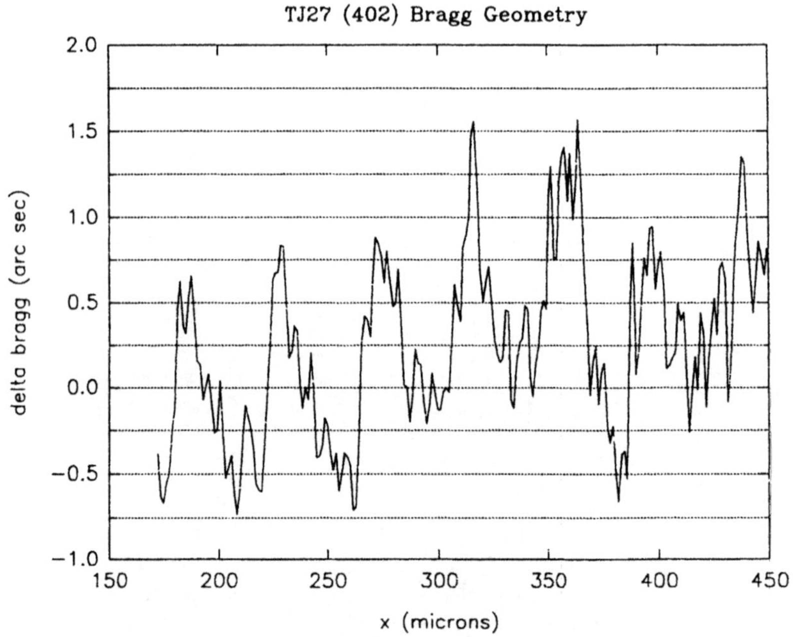

Figure 22. Plot of the changes in Bragg angle caused by a 40 μm period grating in a (402) reflection image at 12 keV.

9. EFFECTIVE ELECTROOPTIC COEFFICIENTS

The photorefractive space-charge field causes a refractive index change

$$\Delta\left(\frac{1}{n^2}\right)_{ij} = r_{ijk}E_k^{sc}, \tag{22}$$

where the r_{ijk} are the elements of the electrooptic tensor and we use Einstein summation convention where the repeated indices imply summation. This is more simply written as

$$\Delta n = -\frac{1}{2} n_o^3 r_{eff} E^{sc}, \tag{23}$$

where r_{eff} is a combination of the electrooptic tensor elements, with the particular combination depending on the orientation of the grating vector with respect to the crystal axes. For any particular geometry, the simple calculation of r_{eff}, neglecting piezoelectric crystal deformations, gives

$$r_{eff} = \frac{\varepsilon_{ii}\left(r_{ijk} k_{g_k}\right)\varepsilon_{jj}}{k_g n_o^3 n_\lambda}, \tag{24}$$

where the ε are the elements of the dielectric tensor, k_g is the magnitude of the grating wavevector, and n_λ is either n_o or n_e for interacting beams of ordinary or extraordinary polarization, respectively. For two-wave mixing in a material with symmetry 4mm (e.g., barium titanate), the above reduces to

$$r_{eff} = r_{13} \sin\left(\frac{\alpha+\beta}{2}\right), \tag{25}$$

for ordinary polarization, and

$$r_{eff} = \left(n_e^4 r_{33} \sin\alpha \sin\beta + 2n_e^2 n_o^2 r_{42} \cos^2\left(\frac{\alpha+\beta}{2}\right) + n_o^4 r_{13} \cos\alpha \cos\beta\right) \frac{\sin\left(\frac{\alpha+\beta}{2}\right)}{n_e n_o^3}, \tag{26}$$

for extraordinary polarization, where α and β are angles the wavevectors of the interacting beams make with the c-axis.

Although the simplicity of the above treatment makes it attractive, it leaves unanswered the question of whether to use the unclamped (stress-free), or clamped (strain-free) electrooptic coefficients. As the clamped and unclamped values differ by more than a factor of 2 in barium titanate, the distinction is critical. In fact, neither set of values can be used as is. The space-charge field causes compression and rarefaction of the crystal lattice (see Figure 23), and these lattice changes influence the refractive index via the elasto-optic effect. Several researchers have perceived the importance of considering the elasto-optic contribution to the refractive index change from photorefractive space-charge fields in crystals [67,69,70]. To properly find the effective electrooptic coefficient that determines the optical indicatrix changes under the influence of a space-charge field, one must know the refractive indices n_{ij}, the clamped electrooptic coefficients $r^S{}_{ijk}$, the piezoelectric tensor e_{ijk}, the elasto-optic (Pockels) tensor at constant electric field $p^E{}_{ijkl}$, and the tensor of the elastic

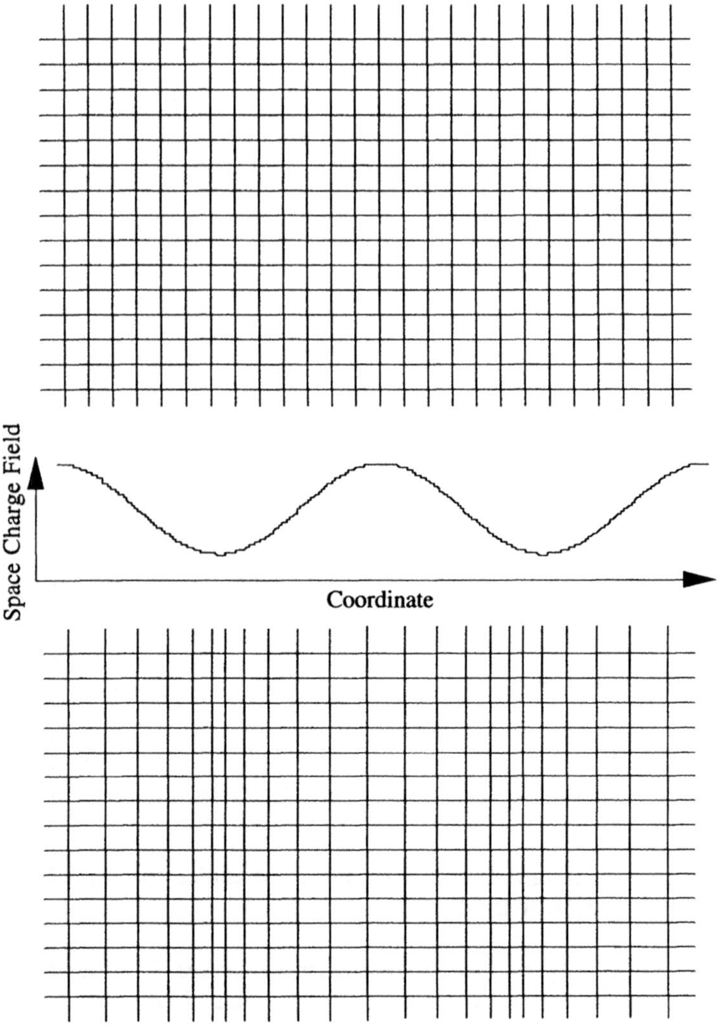

Figure 23: Schematic of the lattice distortions produced by a photorefractive grating. The top grid illustrates the undistorted lattice, and the bottom grid illustrates the lattice distorted by the space-charge field.

constants at constant electric field $C^E{}_{ijkl}$. Furthermore, the effective dielectric constants must be determined. To do so it is necessary to know the piezoelectric tensor e_{ijk}, the elastic stiffness tensor $C^E{}_{ijkl}$ and the clamped dielectric constant ε_{ij}.

The complete set of coefficients has been determined by Zgonik et al. [71] for $BaTiO_3$ and are listed in Tables 4 and 5.

Table 4. "Basic set of the material parameters: dielectric ($\varepsilon^S{}_{ij}$), elastic stiffness ($c^E{}_{ijkl}$), piezoelectric stress (e_{ijk}), elasto-optic ($p^E{}_{ijkl}$), clamped electro-optic ($r^S{}_{ijk}$) tensor elements of $BaTiO_3$ as determined by fitting the experimental data. All materials constants are given at room temperature (23°C) a the wavelength of λ_0=633 nm [71]."

Parameter	Value	Inaccuracy	Unit
$\varepsilon^S{}_{33}$	56	±3	
$c^E{}_{1111}$	222	±10	10^9 N m^{-2}
$c^E{}_{1122}$	108	±18	10^9 N m^{-2}
$c^E{}_{1133}$	111	±8	10^9 N m^{-2}
$c^E{}_{3333}$	151	±7	10^9 N m^{-2}
$c^E{}_{1313}$	61	±3	10^9 N m^{-2}
$c^E{}_{1212}$	134	±6	10^9 N m^{-2}
e_{311}	-0.7	±0.6	C m^{-2}
e_{333}	6.7	±0.3	C m^{-2}

Table 4 (continued)

e_{131}	34.2	±1.8	C m^{-2}
p^E_{1111}	0.50	±0.04	
p^E_{1122}	0.106	±0.01	
p^E_{1133}	0.20	±0.014	
p^E_{3311}	0.07	±0.007	
p^E_{3333}	0.77	±0.04	
p^E_{1313}	1.0	±0.2	
p^E_{1212}[a]	-	-	
r^S_{113}	10.2	±0.6	10^{-12} m V^{-1}
r^S_{333}	40.6	±2.5	10^{-12} m V^{-1}
r^S_{131}	730	±100	10^{-12} m V^{-1}

"[a] The only missing value of the complete set: the elasto-optic tensor element p^E_{1212}, is less important for applications and could not be determined from the existing measurements. Other experimental methods or sample-cuts will be necessary to determine it [71]."

Table 5. "Additional material parameters of $BaTiO_3$ crystal: stress-free dielectric tensor ε^T_{ij}, elastic stiffness tensor at constant dielectric polarization c^D_{ijkl}, elastic compliance tensor at constant electric field s^E_{ijkl}, elastic compliance tensor at constant dielectric polarization s^D_{ijkl}, piezoelectric d_{ijk} tensor, and stress-free electro-optic tensor r^T_{ijk}. These materials constants are calculated from the basic set given in Table 4, therefore also describing the properties at room temperature (23°C) at the wavelength λ_0=633 nm [71]."

Parameter	Value	Inaccuracy	Units
ε^T_{11}	4400	±400	
ε^T_{33}	129	±5	
c^D_{1111}	223	±10	10^9 N m^{-2}
c^D_{1122}	109	±5	10^9 N m^{-2}
c^D_{1133}	102	±5	10^9 N m^{-2}
c^D_{3333}	240	±10	10^9 N m^{-2}
c^D_{1313}	121	±5	10^9 N m^{-2}
s^E_{1111}	7.4	±0.3	10^{-12} m^2 N^{-1}
s^E_{3333}	13.1	±1.5	10^{-12} m^2 N^{-1}
s^E_{1122}	-1.4	±0.3	10^{-12} m^2 N^{-1}
s^E_{1133}	-4.4	±0.5	10^{-12} m^2 N^{-1}
s^E_{1212}[a]	1.9	±0.2	10^{-12} m^2 N^{-1}

Table 5 (continued)

s^E_{1313}[a]	4.1	±0.4	10^{-12} m^2 N^{-1}
s^D_{1111}	6.4	±0.3	10^{-12} m^2 N^{-1}
s^D_{3333}	5.6	±1	10^{-12} m^2 N^{-1}
s^D_{1122}	-2.3	±0.4	10^{-12} m^2 N^{-1}
s^D_{1133}	-1.7	±0.3	10^{-12} m^2 N^{-1}
s^D_{1313}[a]	2.1	±0.4	10^{-12} m^2 N^{-1}
d_{311}	-33.4	±2	10^{-12} C V^{-1}
d_{333}	90	±5	10^{-12} C V^{-1}
d_{131}[a]	282	±20	10^{-12} C V^{-1}
r^T_{113}	8	±2	10^{-12} m V^{-1}
r^T_{333}	105	±10	10^{-12} m V^{-1}
r^T_{131}	1300	±100	10^{-12} m V^{-1}

[a] "Note that we strictly use the tensor notation which is much more practical than the matrix contracted notation and helps avoiding errors in computer calculations. A factor of four $4s_{ijij}=s_{kk}$ and two $2d_{iji}=d_{ik}$ relates the tensor elements to the matrix [71]."

We follow here the effect of lattice distortions given by Günter and Zgonik [67]. The total refractive index change can be written as

$$\Delta\left(\frac{1}{n^2}\right)_{ij} = r^S_{ijk} E^{sc}_k + p'_{ijkl} u_{kl}, \tag{27}$$

where $r^S{}_{ijkl}$ are the clamped electrooptic tensor elements, p'_{ijkl} are the elements of the modified elasto-optic tensor (which is anti-symmetric under interchange of the last two indices), and the u_{kl} are defined as

$$u_{kl} = \frac{\partial u_k}{\partial x_l}, \tag{28}$$

where u_k are the components of the vector strain field relating each point in the distorted crystal to its original, undistorted location. In barium titanate the anti-symmetric part of p'_{ijkl} is small compared to the symmetric part, so it is possible to use the conventional elasto-optic tensor p_{ijkl} instead [67].

The u_k can be found by setting the divergence of the stress tensor T_{ij} equal to zero[67]:

$$\frac{\partial T_{ij}}{\partial x_j} = C^E_{ijkl} \frac{\partial S_{kl}}{\partial x_j} - e_{kij} \frac{\partial E_k}{\partial x_j} = 0, \tag{29}$$

where $C^E{}_{ijkl}$ are the elastic tensor elements at constant electric field, e_{kij} are the piezoelectric stress tensor elements

$$e_{kij} = d_{klm} C^E_{lmij}, \tag{30}$$

and S_{kl} are the strain tensor elements

$$S_{kl} = \frac{1}{2}\left(u_{kl} + u_{lk}\right). \tag{31}$$

The result in the general case, for a space-charge field of the form

$$E_k(r) = E^{sc}_k \cos\left(k_{g_j} r_j\right), \tag{32}$$

gives lattice distortions

$$u_k = \left(C^E_{ijkl} n_j n_l\right)^{-1} \frac{E^{sc}}{k_g} e_{kij} n_k n_j, \tag{33}$$

where **n** is the unit vector in the direction of the grating vector, and $()^{-1}$ denotes matrix inversion. The general form of the modulation of the optical indicatrix is given as,

$$\Delta\left(\frac{1}{n^2}\right)_{ij} = \left(r^S_{ijk} n_k E^{SC} + p'_{ijkl}\, n_l u_k\right)\cos(\mathbf{K}\cdot\mathbf{r}). \tag{34}$$

A simple example illustrates the effect of the lattice distortions. For the case of ordinary polarization, with the grating vector aligned along the c-axis, (*typically used for photorefractive characterization measurements because it minimizes beam fanning*), taking the lattice distortions into account in the above manner gives the refractive index changes [67],

$$\Delta\left(\frac{1}{n^2}\right)_{11} = \left(r^S_{113} + \frac{p_{1133} e_{333}}{C^E_{3333}}\right) E^{SC}. \tag{35}$$

Taking the appropriate values from Tables 4 and 5 we find an effective electrooptic coefficient of 19 pm/V, compared with either 8 pm/V for the unclamped case or 10.2 pm/V for the clamped case. (We note that the experimental value for r_{13}=19.5 pm/V.) It is clear that the lattice deformations have a significant effect on the nonlinear refractive index changes, although the exact corrections must be calculated individually for each particular material and interaction geometry

We point out that the above analysis is valid only at large (compared to a grating period) distances from any crystal surface. In thin samples, surface boundary conditions must also be considered [68], as certain crystal deformations forbidden far from the surface become important.

10. DEEP-AND-SHALLOW TRAP PHOTOREFRACTIVE MODEL

The photorefractive effect has been extensively analyzed using the single-level model of Kukhtarev et al. [72] and the hole-electron model of Strohkendl et al. [73] and Valley [74]. Two-active trap photorefractive models have been employed to explain observations not predicted by the single-level model including: the sublinear dependence of the response time on intensity [75,76], light-induced dark decays [76], and intensity dependence of the effective Debye screening wavevector [77]. Light-induced absorption effects have also been observed that cannot be explained with a single-active-level model [78]. A single-level model also predicts symmetric beam amplification or depletion which we do not observe in our measurements of beam coupling. This will be explained further in the section on photorefractive measurements.

Brost et al. [78] solved numerically for the photorefractive effects of two-active traps and Tayebati [79] analytically solved modified Kukhtarev equations including deep and shallow traps. Also, Tayebati and Mahgerefteh [80] have analytically formulated the deep and shallow trap model of the photorefractive effect that explains the anomalous intensity dependent properties mentioned above and also the existence and functional behavior of absorption gratings. A general single carrier multi-level model has been given by Cudney et al [81] and applied to $BaTiO_3$ with two traps. The general model is described in section 13.2. of this chapter. In addition, a hole-electron and shallow trap model has been developed by Tayebati [82].

10.1 Beam-Coupling Theory

When two coherent optical beams of wavevectors $\mathbf{k_1}$ and $\mathbf{k_2}$ interfere within a photorefractive crystal they create an intensity interference pattern I(x) with grating wavevector $k_g=|\mathbf{k_1}-\mathbf{k_2}|$ where $I(x)=I_0 Re[1+m\cdot exp(ik_g x)]$ and I_0 is the total average intensity.

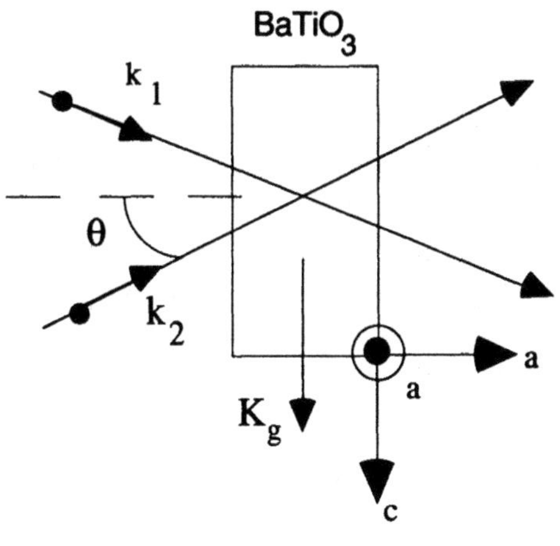

Figure 24. Diagram of the configuration of the crystal and optical beams used to measure photorefractive properties of $BaTiO_3$.

For small intensity modulation, m<<1, the solution to the modified Kukhtarev equations, including the effects of hole shallow traps, gives the magnitude of the deep and shallow trap charge-gratings N_1 and M_1 respectively as [80],

$$-eN_1 = -emN_E \frac{k_g^2}{\left(k_g^2 + k_0^2\right)} + emM_E \frac{1}{(1 + s_T I_0 / \beta)} \frac{k_{0D}^2}{\left(k_g^2 + k_0^2\right)} \quad \text{and,}$$

$$eM_1 = emM_E \frac{1}{(1 + \beta / s_T I_0)} \frac{k_g^2}{\left(k_g^2 + k_0^2\right)} - emM_E \frac{1}{(1 + s_T I_0 / \beta)} \frac{k_{0D}^2}{\left(k_g^2 + k_0^2\right)}. \tag{36}$$

The thermal excitation of holes is assumed to arise only from the shallow traps with rate β. The square of the Debye screening wavevector is,

$$k_0^2 = k_{0D}^2 + k_{0T}^2, \tag{37}$$

with

$$k_{0D}^2 = \frac{e^2 N_E}{\varepsilon k_B T} \quad \text{and} \quad k_{0T}^2 = \frac{e^2 M_E}{\varepsilon k_B T}, \tag{38}$$

where k_{0D} and k_{0T} are the deep and shallow trap screening wavevectors respectively, ε is the relevant dielectric constant, k_B is Boltzmann's constant, T the absolute temperature and e is the magnitude of the electric charge. The intensity dependent effective deep and shallow trap densities N_E and M_E are respectively defined as,

$$N_E = \frac{(N_F - M_0)(N_{DF} + M_0)}{N_D} \quad \text{and} \quad M_E = \frac{M_0(M_T - M_0)}{M_T}, \tag{39}$$

where N_D is the total deep trap density, N_F is the uncompensated deep trap density, N_{DF} is the difference between the total deep trap density and the uncompensated deep trap density, M_0 is the mean concentration of empty hole shallow traps induced by light, and M_T is the total density of shallow traps. The total charge grating, $e(N_1 - M_1)$, determines the space-charge field. Because the second terms in the expressions of N_1 and M_1 (eqn. (36)) are equal, they add up to zero and do not contribute to the net charge grating. For this reason they are called screening gratings [80]. This term is called the "offsett" by Cudney et al. [81].

In Eqns. (36) the first two terms, called in-phase gratings [80], are both in phase with the intensity pattern. They give rise to the space-charge field that produces electrooptic gain from the coupling due to an index modulation that has a $\pi/2$ spatial phase shift with respect to the intensity pattern. From Poisson's equation, the space-charge field is

$$E_1 = -im\left(\frac{k_B T}{e}\right)\left[\frac{k}{1+k^2/k_0^2}\right]\eta(I), \tag{40}$$

where the intensity dependent term has a value where $0<\eta(I)\leq 1$ and is equal to,

$$\eta(I) = \frac{1}{k_0^2}\left[k_{0D}^2 + \frac{k_{0T}^2}{1+\beta/s_T I_0}\right]. \tag{41}$$

The electrooptic gain is given as

$$\gamma_{eo} = i\left(\frac{2\pi r_{eff}}{n_o \lambda m}\right)E_1, \tag{42}$$

where $i=\sqrt{-1}$ and r_{eff} is the effective electrooptic coefficient. For ordinary polarization and a crystallographic orientation as shown in Figure 24, r_{eff} is equal to $n_o^4 r_{13}$ where r_{13}=19.5 pm/V [82] and n_o=2.488, is the ordinary index of refraction [84].

The screening gratings can be observed in absorptive coupling measurements. In these measurements the absorptive gain is given by,

$$\gamma_{abs} = \frac{\hbar\omega}{m}\left(s_D N_1 - s_T M_1\right), \tag{43}$$

where s_D is the deep trap excitation cross-section, and $\hbar\omega$ is the photon energy. In this expression, (Eqn (43), incorporating Eqns. (36)) the contribution of the screening gratings is nonzero. In particular, at large grating spacings, where the in-phase gratings vanish (electrooptic coupling) the screening gratings remain finite and give rise to nonzero absorptive coupling. In this limit, (i.e., small k_g), the absorptive gain γ_{abs} is proportional to $(s_D - s_T)N_1$. Therefore, the sign of γ_{abs} is determined only by the relative magnitudes of s_D and s_T. We note that the "absorptive gain" does not contribute to amplification or depletion of the signal in the same way as the electrooptic gain. The $\pi/2$ spatial phase shift of an index or phase grating allows for the coherent diffraction of the pump into the signal. An absorption grating is in phase with the intensity pattern. When the sign of the absorptive coupling is positive there will be less absorption where the light is most intense (light-induced transparency). On the other hand if the absorptive coupling is negative there will be more absorption where the light is most intense (light-induced absorption). These effects will produce an increase or decrease respectively in the transmission of the signal (and pump) beam and thus will change the apparent coupling.

The intensity-induced absorption/transparency is also determined by the relative magnitudes of s_D and s_T. Optical absorption for two trap levels can be described as [80]:

$$\alpha = \hbar\omega\left[s_D N_F + (s_T - s_D)M_0\right] \tag{44}$$

For $s_T > s_D$, the absorption increases with increasing intensity and the sign of the absorptive coupling gain γ_{abs} at large grating spacings is negative as reported by Cudney et al. [81]. In contrast, when $s_D > s_T$ light-induced transparency is observed with increasing intensity and the sign of the absorptive coupling gain γ_{abs} at large grating spacings is positive.

11. BEAM-COUPLING MEASUREMENT TECHNIQUE

Several optical characterization techniques are used to help understand photorefractive crystals. Transmission and absorption spectra give us information on the spectral effect of doping. Light-induced absorption/transparency helps us understand the existence and behavior of multiple traps. Dark decay measurements of the space-charge field help one determine the dark storage time or dielectric relaxation time and also the affect of shallow traps on the erasure of the grating. Measuring the response time dependence on intensity, (or the I^X problem), also helps with the understanding of proportions of deep and shallow traps. The response time dependence on grating period helps one understand the type of conductivity or mobility of the carriers. Beam-coupling measurements are especially fruitful where one can determine if there are deep and shallow traps, holes and/or electrons, electrooptic and absorptive coupling, the relative excitation cross-sections of the traps, the magnitude of the effective electrooptic coefficient, and the effective deep and shallow trap densities.

A beam-coupling experimental set up is shown in Figure 25. Pump beam I_P, and signal beam I_S, are incident on the crystal. To make accurate beam-coupling measurements relative to theory one must have the beam polarization perpendicular to the spontaneous polarization (or c-axis), or ordinarily polarized, and the grating wave vector k_g parallel to the c-axis. In this case the r_{13} electrooptic coefficient is used and beam-fanning is minimized. Furthermore, the sample thickness should be from 1-3 mm to avoid grating curvature and to insure complete beam overlap. The signal beam is usually much smaller in diameter than the pump beam to have overlap throughout the volume and to insure that the intensity variation which the signal beam experiences is small, (this is an uncertainty in intensity-dependent measurements.)

The transmitted intensity of the signal beam including gain γ, and loss or absorption α, are given in Eqn. (115) of Chapter One and in the undepleted pump regime, $I_s(0)\exp(\gamma L) << I_p(0)$, reduces to

$$I_s(L) = I_s(0)\exp[(\gamma - \alpha)L], \tag{45}$$

where, $I_S(0)$, is the signal intensity at the entrance face of the crystal, $I_S(L)$, is the transmitted signal intensity through the interaction length $L=l/\cos\theta_{int}$ where l is the crystal thickness and θ_{int} is the internal half-angle. In the past, one would simply measure the signal intensity transmission before and after the pump beam illuminated the crystal. The ratio of these measurements eliminated the bulk absorption and were solved for the gain coefficient. However, if the crystal has light-induced absorption or transparency then bulk absorption changes $\Delta\alpha$ may contribute to a significant fraction of the change in the signal intensity, see Figure 27, when the pump illuminates the crystal.

A beam-coupling experimental method to separate the electrooptic and absorptive contributions to the photorefractive effect and eliminate bulk absorption changes in the measurement relies on measuring the *asymmetry* of the gain. The intensity two beam-coupling gain of the signal may be written,

$$I_s = I_0 \exp\{[\pm\gamma_{eo} \pm \gamma_{abs} \pm \Delta\alpha] * L\}, \tag{46}$$

where $\Delta\alpha = \alpha_I - \alpha_{I\approx 0}$ is the difference of the absorption at intensity I and at low intensity where I~0. The electrooptic gain coefficient γ_{eo} will be positive or negative depending on the direction of the spontaneous polarization relative to the pump and signal beam and the sign of the charge carriers. The absorptive gain γ_{abs} will be positive if the sample has light-induced transparency and negative if the sample has light-induced absorption, and it is spatially symmetric for 180° rotations. The change in the absorption, $\Delta\alpha$, is spatially symmetric for 180° rotations. Although the absorption changes and absorptive gain both depend on excitation cross-sections and thermal excitation rates, the absorptive gain is also dependent on the grating wave vector, and the Debye screening wave vector (see equations (43) and (36)). Thus, the absorptive coupling cannot be predicted by the light-induced absorption. However, at large grating periods there is a direct relationship between the sign of the light-induced absorption change and the sign of the absorptive coupling as will be discussed in sections 12 and 13 of this chapter.

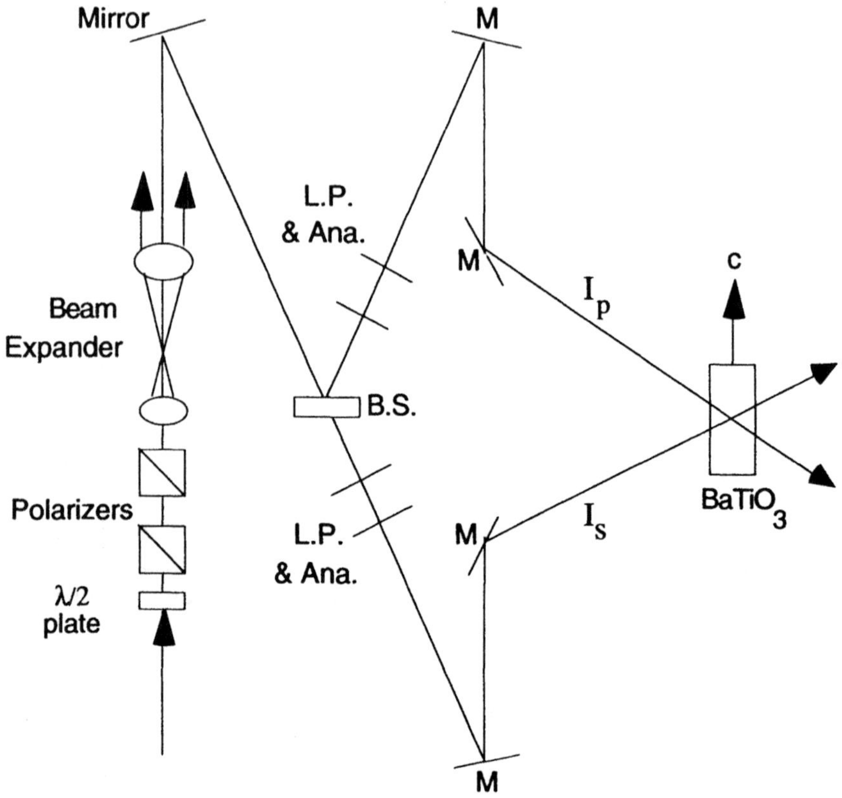

Figure 25. Beam-coupling experimental set-up. The polarization and power of the laser light is controlled by the half-wave ($\lambda/2$) plate and two cube polarizers. It is then expanded and divided into the pump I_p and the signal I_s. The signal beam is much smaller in diameter than the pump beam. The power ratio of the pump-to-signal is set by the linear polarizer (L.P.) and analyzer (Ana.) combination in each beam path. The beams are directed to the crystal by mirrors (M).

To make gain measurements that eliminate changes in the signal caused by $\Delta\alpha$ one continually illuminates the crystal with the signal and the pump [85]. Initially, the pump and signal are incoherent with each other and no beam-coupling is observed. (Incoherence may be established by a path length difference or by rapidly vibrating one of the beams much faster than the grating writing time and over a distance greater than the grating period such that a photorefractive grating cannot be written.) Under these conditions the signal intensity is,

$$I_s = I_0 \exp[\pm\Delta\alpha L]. \tag{47}$$

When the vibration is turned off the beams become coherent and the coupling either amplifies or diminishes the signal and it's transmitted intensity is,

$$I_s^{wcp} = I_0 \exp[(\pm\gamma_{eo} \pm \gamma_{abs} \pm \Delta\alpha)*L]. \tag{48}$$

The ratio of Equation (48) to (47) gives,

$$\frac{I_s^{wcp}}{I_s} = \exp[(\pm\gamma_{eo} \pm \gamma_{abs})*L]. \tag{49}$$

Taking the natural log of both sides gives,

$$(\pm\gamma_{eo} \pm \gamma_{abs}) = \frac{1}{L}\ln\left(\frac{I_s^{wcp}}{I_s}\right). \tag{50}$$

If the crystal is rotated by 180° or the relative positions (intensity) of the pump and signal are changed the exponential intensity gain becomes,

$$I_s^{wcp'} = I_0 \exp[(m\gamma_{eo} \pm \gamma_{abs} \pm \Delta\alpha)*L], \tag{51}$$

where the sign of the electrooptic gain γ_{eo} coefficient is reversed. Taking the ratio of Equation (7) to Equation (3) one finds,

$$\frac{I_s^{wcp'}}{I_s} = \exp[(m\gamma_{eo} \pm \gamma_{abs})*L], \tag{52}$$

and again taking the natural log gives,

$$(m\gamma_{eo} \pm \gamma_{abs}) = \frac{1}{L}\ln\left(\frac{I_s^{wcp'}}{I_s}\right). \tag{53}$$

If we add Equations (6) and (9) and divide by two we obtain the absorptive gain coefficient. If we subtract Equations (6) and (9) and divide by two we obtain the electrooptic gain coefficient. In practice one makes three measurements equivalent to determining Equations (3), (4) and (7). The crystal is not rotated but the relative intensity of the signal and pump are changed by the polarizer and analyzer in each beam path (see Figure 26).

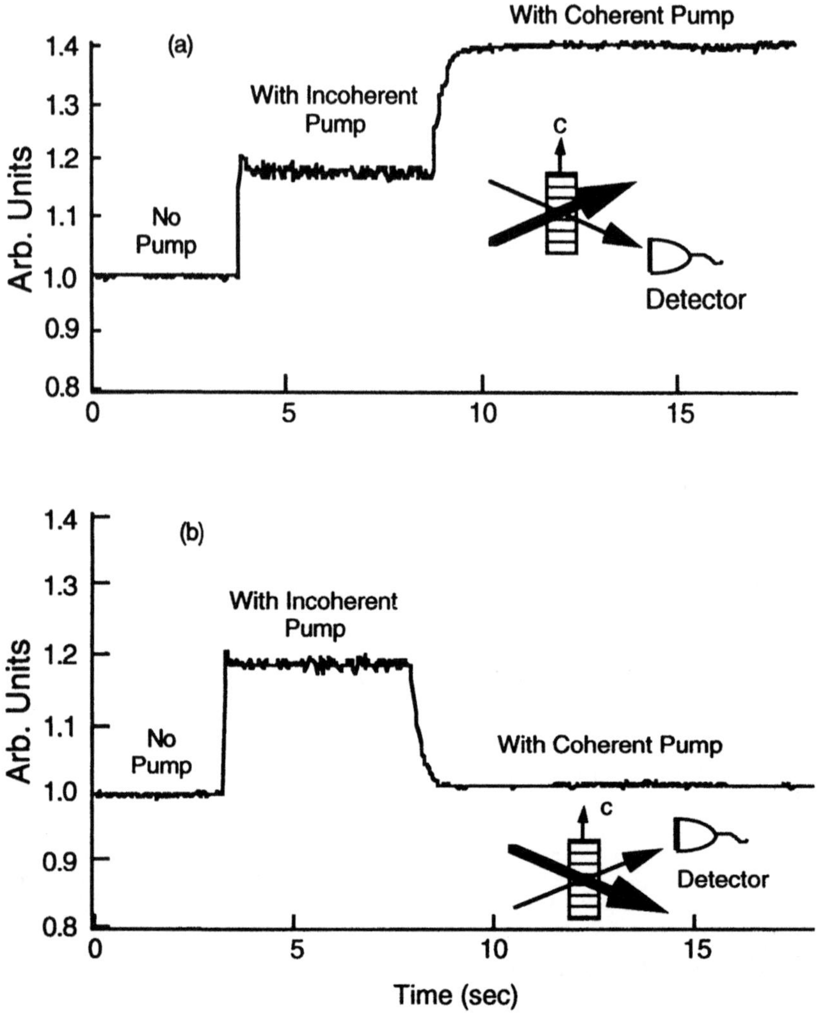

Figure 26. Two-beam measurements. Initially the signal is transmitted through the sample. With the incoherent pump and signal present simultaneously, reduced n-type crystal MIT-1 displays light-induced transparency. When the vibration is removed and coherence obtained the signal has amplification in a). When the relative intensity of the pump and signal is reversed the signal has depletion in b).

We first point out that if the old technique of just measuring the transmitted signal with and without the pump present were used to determine the gain coefficient, a large error would have been made because this crystal, at the intensity used, has high light-induced transparency. Notice that the light-induced transparency is symmetric in the two measurements but there is a very slight asymmetry of the gain. The asymmetry can be much larger, as revealed in the following graphs of the absorptive gain coefficient in the next sections of this chapter. To make these measurements accurately the laser and optical set-up must be very stable, free of air turbulence, and several measurements must be averaged.

We note that one may also use a crystal orientation, as described by Pierce et. al. [85], in which there is only absorptive coupling, or polarizability gratings. This method is also very effective in determining photorefractive parameters. It is more fully described in Section 13.2 of this chapter.

12. CHARACTERIZATION OF $BaTiO_3$:Co

With the formulation of the response time dependence on intensity, $\tau \propto I^{-x}$, described more completely by Mahgerefteh and Feinberg [75], one can classify crystals into two categories, Type A and Type B [75]. Type A is a sample that is dominated by a deep level and its intensity power coefficient is near 1, i.e., $\tau \propto I^{-1}$. Type B crystals have intensity power coefficients close to 0.5. Actually, there can be a gradual variation between the two extremes. In addition, shallow levels can thermalize in milliseconds or less after turning off the illumination and erase deep-trap gratings. The magnitude of the intensity power coefficient depends on the relative proportions of deep and shallow traps and their properties. From the deep-and-shallow trap model we can understand the sublinearity as follows. The response time τ of the space-charge field is proportional to the dielectric constant divided by the conductivity σ,

$$\tau_{pr} \propto \frac{\varepsilon}{\sigma} \propto \frac{\varepsilon}{e\rho_0(I)\mu}, \tag{54}$$

where the conductivity is proportional to the free-carrier density ρ_0 and the carrier mobility μ. The free-carrier density is equal to the product of the charge carrier generation rate $s_D I_0(N_{DF}-M_0)$, and the recombination time τ_r of the deep level, and inversely proportional to the empty trap densities (N_F+M_0). If there are no empty shallow traps, then no photoexcited carriers can be stored in the shallow level, i.e. $M_0=0$. The free-carrier density is linearly dependent with intensity, $n_0 \propto I$, where $x=1$. Because empty shallow traps can store the photoexcited carriers from the deep level, the stored carrier density M_0 in the shallow traps can become comparable to or larger than the original deep level trap density N_F as the intensity is increased. In this case, the recombination time of the free carrier to the traps (N_F+M_0) decreases appreciably. When there are empty shallow traps the free carrier density becomes a sublinear function of intensity ($x<1$) and thus there is a sublinear intensity dependence of the response time. (A more detailed formulation is given by in References [75,80].) Shown in Figure 27 is a log-log plot of the response rate versus intensity with linear fits to determine the intensity coefficient. We observe that as the cobalt concentration increases so does the power coefficient, which indicates that the deep traps are dominating the photorefractive behavior.

The behavior of deep and shallow traps is more clearly observed in the following plot of the dark decay of the space-charge field (see Figure 29). A grating is written in the crystal with 514.5 nm light. A weak He-Ne probe beam, optically chopped, diffracts off the grating and is detected by a photomultiplier and lock-in. When the grating writing beams are turned off the He-Ne beam probes the decay of the diffraction. As shallow traps thermally depopulate they erase the deep trap gratings, which have a much longer dark decay time. The data is normalized, but the absolute magnitude of the diffraction increases with increasing cobalt concentration (see beam-coupling gain Figure 32). Thus, the relative ratio of deep-to-shallow traps increases. This implies that the shallow trap concentration is constant from crystal-to-crystal within this series. This means that even though we observe a larger value of the intensity coefficient x, with higher deep-level dopant concentrations, the shallow traps still exist and their effects are simply overwhelmed by the increasing number of deep traps. Concurrently, we observe that even for the highest dopant level of 75 ppm the decay time is really a combination of decay times, i.e. double exponential decay rates. The "dark decay time", which is proportional to the dielectric constant divided by the resistivity, must therefore be clearly referenced to the correct initial amplitude. This decay behavior is the origin of the term "relaxation semiconductor."

The light-induced absorptive changes series of cobalt-doped samples are shown in Figure 30. These were determined with an incoherent pump and probe of 514.5 nm. If there were only two traps we would expect a monotonic increase in the absorption with intensity which is observed for the 75 ppm sample. However, some samples show light-induced transparency at lower intensities and, as the intensity is increased, changes to light-induced absorption. This indicates that more than two traps participate in the charge transfer processes. Light-induced absorption/transparency depends on the relative excitation cross-sections of the full and empty states of the defects. If we are dealing with two defects, deep and shallow, there are four possible cross-sections that determine light-induced absorption/transparency effects. This makes interpretation of the data difficult. Dark decay measurements that reveal multiple decay time constants can help to clarify the situation.

The light-induced absorption of the 75 ppm doped sample shows a near-linear or monotonic dependence over this intensity range, indicating that as the cobalt concentration is increased single-trap properties dominate. At low intensities the less heavily doped samples have light-induced transparency caused by the transfer of charge from the deep levels to the shallow levels. Because the deep traps are losing charge to the shallow levels their relative absorption diminishes. We note that at > 500 mW/cm^2 all the samples have light-induced absorption because the shallow traps that are being populated are also absorbing the radiation and thus increasing the total

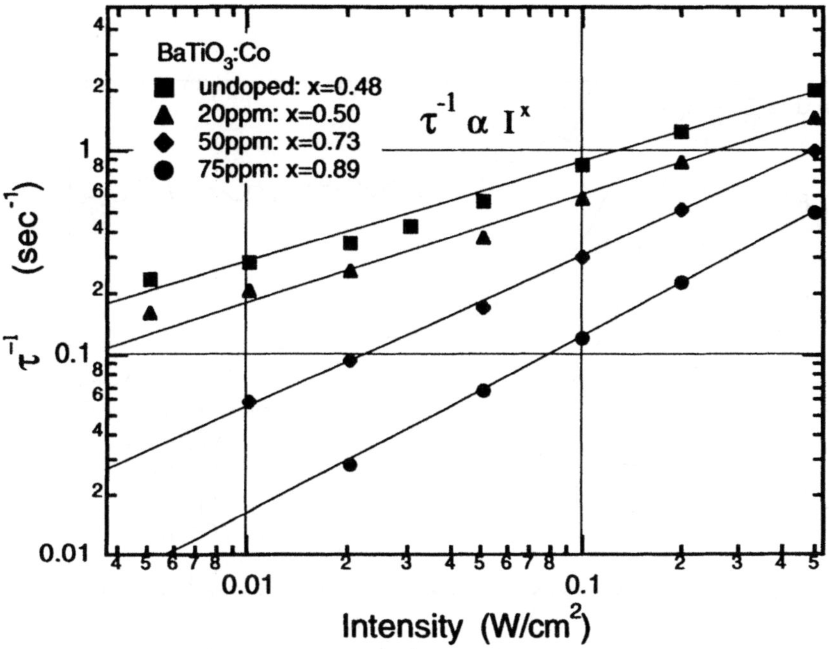

Figure 27. Log-log plot of the response rate, $1/\tau$, dependence on the intensity for the cobalt-doped series of $BaTiO_3$:Co crystals. As the cobalt concentration increase the crystal "Type" changes from Type B to Type A.

absorption. From the equation for the optical absorption coefficient we can then argue that $s_T > s_D$, but we see that over the whole range of intensities this interpretation does not hold true and implies that more than two traps are involved.

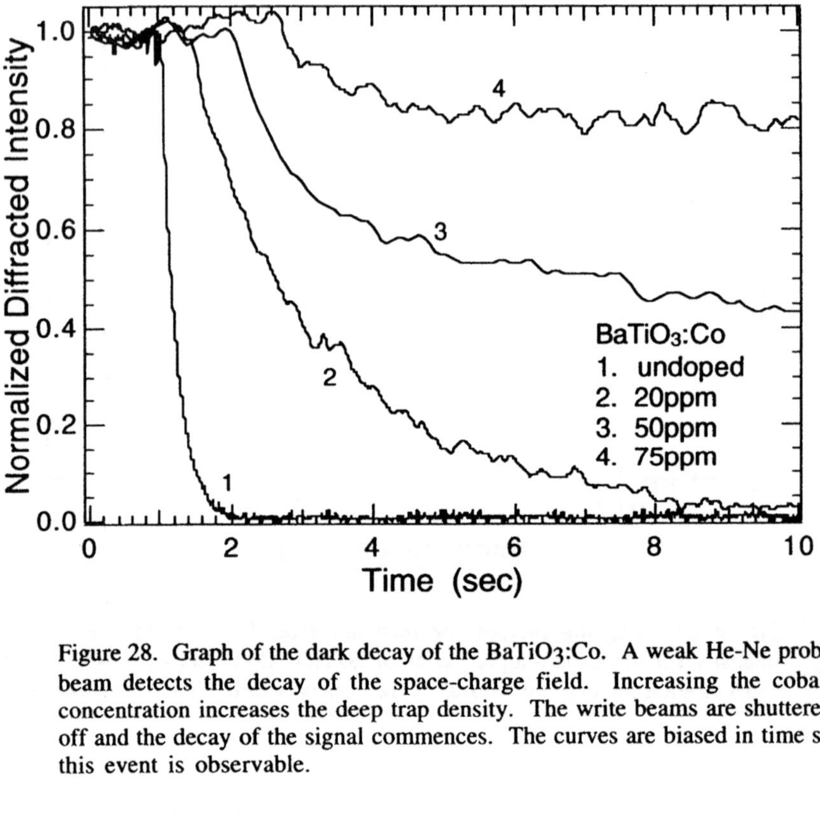

Figure 28. Graph of the dark decay of the $BaTiO_3$:Co. A weak He-Ne probe beam detects the decay of the space-charge field. Increasing the cobalt concentration increases the deep trap density. The write beams are shuttered off and the decay of the signal commences. The curves are biased in time so this event is observable.

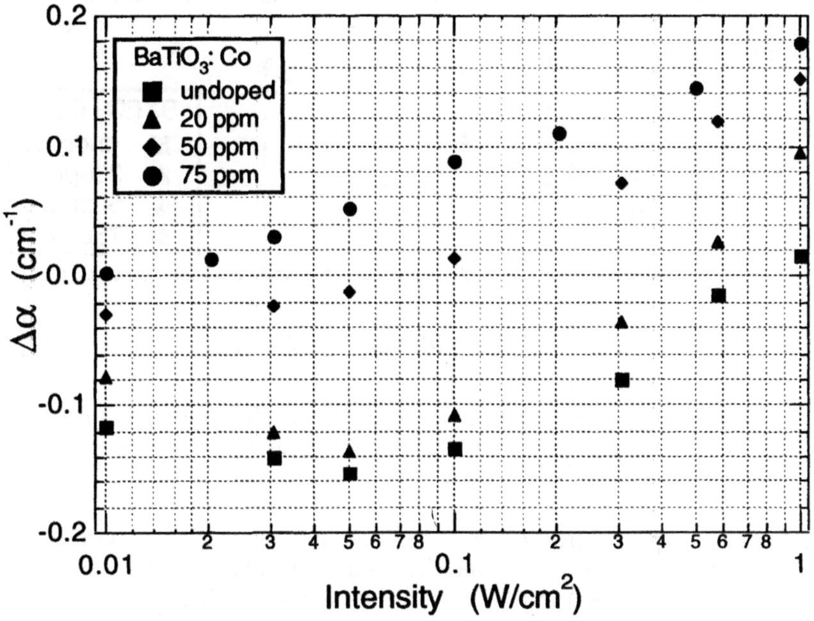

Figure 29. A incoherent pump and signal at λ=514.5 nm were used to determine the light-induced absorption/transparency of the cobalt-doped series of $BaTiO_3$.

The intensity dependence of the electrooptic gain coefficient for the cobalt series is shown in Figure 30. We observe that as the doping concentration increases the gain becomes less sensitive to the intensity. The heavily doped crystals require much lower intensities to overcome thermal erasure of the gratings from the charges trapped in shallow levels. This agrees with time response data showing shallow traps effects to be more important in lightly cobalt-doped crystals.

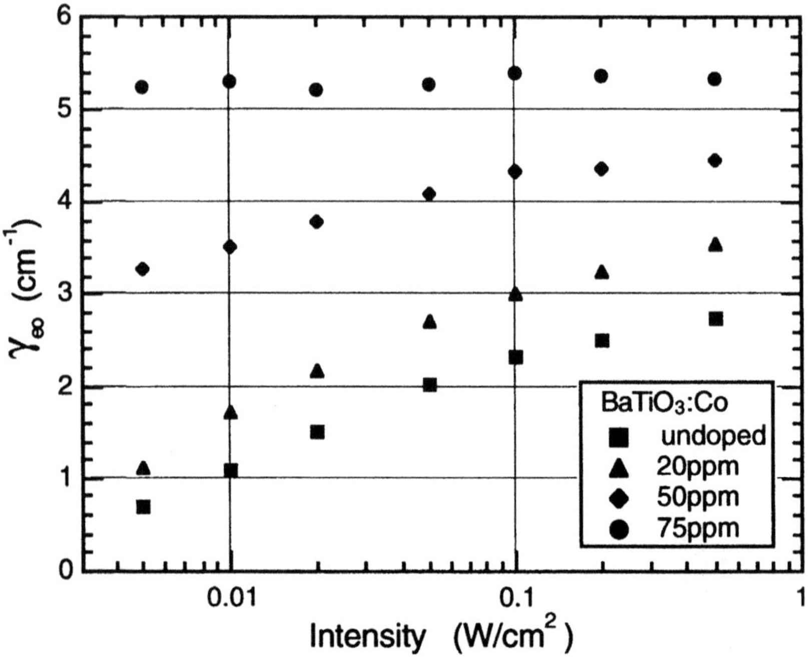

Figure 30. Plot of the electrooptic gain coefficient as a function of the log of the total intensity with 514.5 nm Argon-ion laser illumination at a grating wave vector of 10.47 μm^{-1}.

The electrooptic and absorptive gains as functions of grating wavevector were determined with the previously described beam-coupling experimental method and are plotted in Figures (32) and (33). The solid lines are theoretical fits that were made using the deep-and-shallow trap model, assuming a value of 19.5 pm/V for the electrooptic coefficient. The electrooptic gain coefficient decreases to zero at small k_g as expected from the deep-and-shallow trap theoretical formulation. Note that the absorptive coupling is actually a loss and the sign of γ_{abs} is actually negative. We note that the absorptive component is a maximum of 15% of the total gain in our measurements shown here.

At large grating wave vectors we observe that the absorptive coupling increases with increased cobalt-doping. For large k_g we have $s_D N_1 > s_T M_1$ and since we previously concluded that $s_T > s_D$, (for the intensity for these measurements), the increase in the absorptive coupling with higher cobalt-doping must arise from higher values of the deep trap screening charge-grating N_1, which is a reasonable conclusion since we added cobalt to the melt (a deep level dopant). The absorptive 'gain' coefficient for most values of k_g of the cobalt-doped crystals is actually an absorptive loss. This is because the crystal has light-induced absorption at the intensities examined. When the beams are *incoherent* the intensity is $2I_0$. Whereas when the beams become *coherent* the intensity in the bright regions becomes $4I_0$, diminishes in the dark regions, and thus the total intensity has increased, but light induced-absorption causes the transmission in the regions where the light is most intense to decrease. Therefore the beams experience more absorption when they are coherent and $\gamma_{abs}<0$. We observe that the absorptive gain coefficients of Figure 33 are negative for most grating wave vectors in agreement with the light-induced absorption values at 500 mW/cm^2.

We may reformulate the expression for the electrooptic gain coefficient in the following way,

$$K_g/\gamma_{eo} = A + (A \times B)\, K_g^2 \ , \tag{55}$$

where,

$$A = \frac{e\lambda}{2\pi n^3 r_{eff} k_B T} \frac{1}{\eta(I)} \text{ and } A \times B = \frac{e\lambda}{2\pi n^3 r_{eff} k_B T} \frac{1}{\eta(I)} \frac{1}{k_0^2} \ . \tag{56}$$

If one plots beam-coupling data as K_g / γ_{eo} as a function of K_g^2 [74] then the deep-and-shallow trap model predicts a linear fit of Eqn. (55), the intercept value being equal to A in Eqn. (55). On the other hand, the hole-electron model [73,74] may show a deviation from a straight line depending on the grating wave vector, absorption strengths and diffusion lengths of the carriers. Shown in Figure 34 is such a plot where no deviation from a straight line is observed, which implies no hole-electron competition. Nevertheless, some of the offset or shift at the intercepts of this graph could be attributed to changes in the relative conductivity of electrons and holes [74]. Therefore, it is not possible to eliminate the possibility of hole-electron competition in the crystals examined here.

In general, we have observed that the addition of cobalt to $BaTiO_3$ increases the deep trap density relative to the shallow trap density. If the dopant density is high enough the deep traps overwhelm the shallow traps and tend to mask their influence on the photorefractive properties. Nevertheless, the shallow traps appear to be always present. The deep-and-shallow trap model more correctly fits the experimental photorefractive data, but evidence exists in the light-induced absorption/transparency data for at least three active traps. Indeed, the photorefractive models have idealized discrete energy levels for the traps, but there is a distribution of trap energies as revealed in their absorption spectra. A summary of photorefractive properties, derived from the data and fitting of the deep-and-shallow trap model of the cobalt-doped series of $BaTiO_3$, are shown in the next tables. We note that even though the gain increases with increasing cobalt-dopant the photorefractive sensitivity diminishes. This is because the absorption and response time also increase. Perhaps a better way to characterize the response time is to determine the "time to equivalent gain" [86]. From this latter perspective, the 75 ppm crystal has the highest sensitivity. Nevertheless, we note that it does not have net gain when using r_{13} because the absorption is high. But, if the sample is cut in a 45°-cut [87] so as to access the higher electrooptic coefficient r_{42}, the gain is calculated to be ~65 cm^{-1} and in this case the absorption is much less than the gain. 45°-cut crystals will be discussed in a later section of this chapter.

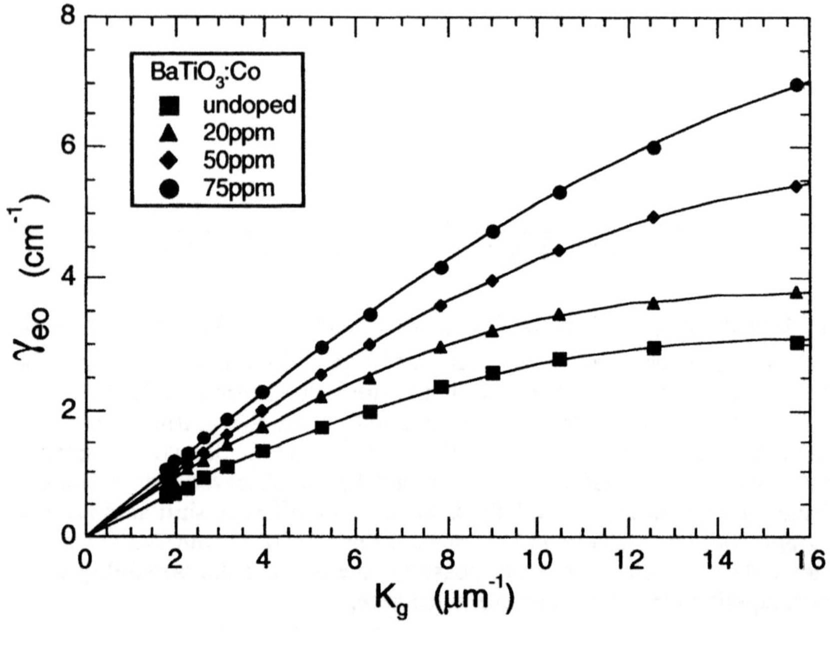

Figure 31. Electrooptic beam-coupling gain coefficient dependence on grating wave vector k_g for a series of cobalt-doped $BaTiO_3$ crystals. The total intensity for this measurement was 500 mW/cm^2 and the wavelength of illumination was 514.5 nm.

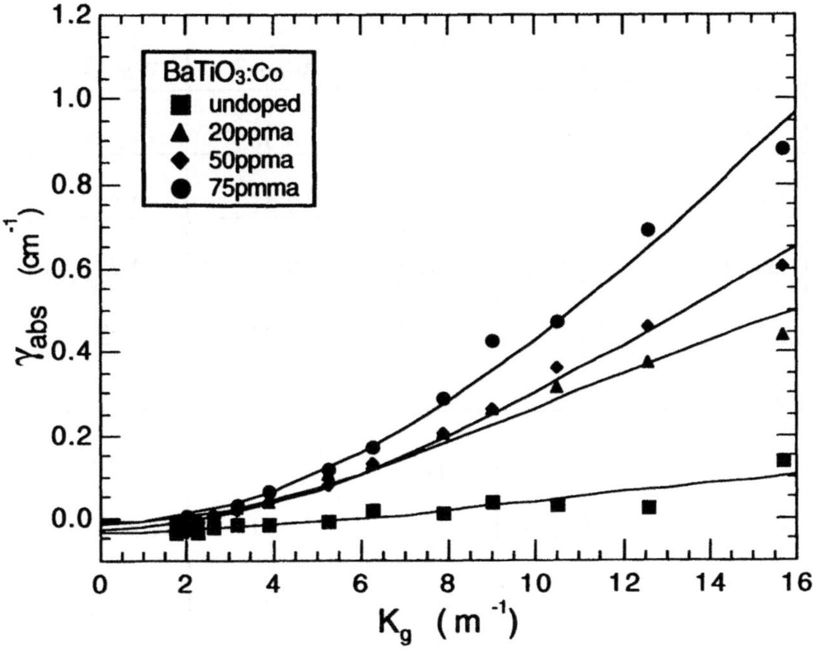

Figure 32. Graph of the absorptive coupling coefficient as a function of the grating wave vector for a series of cobalt-doped $BaTiO_3$. (500 mW/cm^2, 514.5 nm)

(a)

514.5 nm	α (cm^{-1}) @ $\mu W/cm^2$	$\Delta\alpha$ (cm^{-1}) 0.5 W/cm^2	η(I) 0.5 W/cm^2	$\tau_{pr} \propto I^{-X}$ 0.5 W/cm^2
undoped	0.46	-0.016	0.38	x=0.48
20 ppm	1.84	0.010	0.51	0.50
50 ppm	4.59	0.120	0.56	0.73
75 ppm	8.40	0.160	0.63	0.89

(b)

514.5 nm	N_T (10^{16} cm^{-3})	γ_{abs} (cm^{-1})	γ_{eo} (cm^{-1})	Sensitivity (cm^3/kJ)
undoped	6.4	0.97	2.6	0.098
20 ppm	6.7	0.65	3.6	0.023
50 ppm	11.5	0.50	4.5	0.008
75 ppm	20.0	0.10	5.3	0.002

Tables 6(a) and (b). Photorefractive parameters of the series of cobalt-doped $BaTiO_3$ crystals determined at an intensity of 500 mW/cm^2 and a wavelength of 514.5 nm. The background absorption coefficient is α, $\Delta\alpha$ is the light-induced change in the absorption, $\eta(I)$ is the beam-coupling intensity dependent factor, τ_{pr} is the 1/e erasure time or response time, N_T is the total effective trap density, γ_{eo} is the absorptive gain coefficient, γ_{eo} is the electrooptic gain coefficient, and S is the sensitivity equal to $\gamma_{eo}\lambda/4\pi\alpha I\tau_{pr}$.

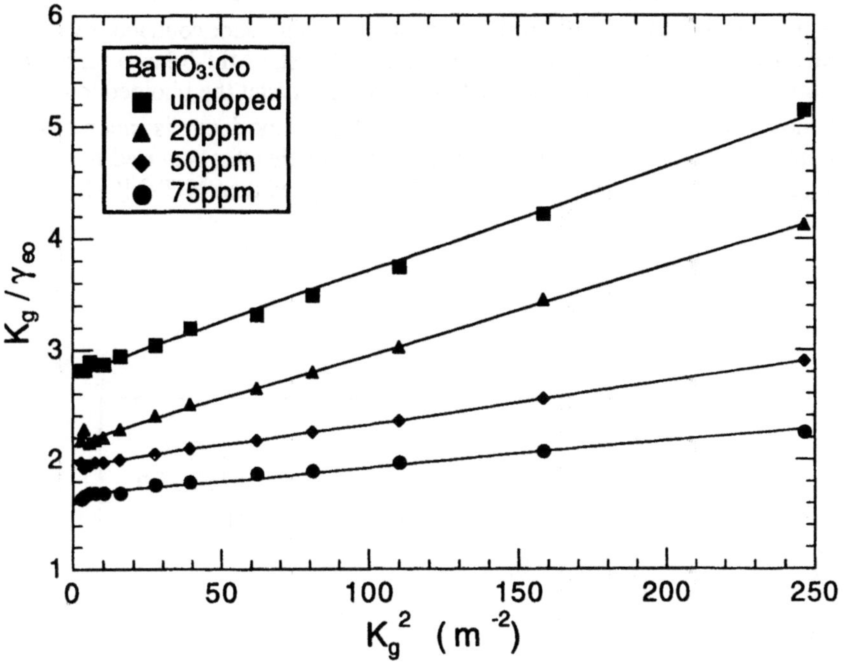

Figure 33. Plot to determine if hole-electron effects are present in the cobalt-doped series of $BaTiO_3$:Co crystals. No linear deviation is observed at a wavelength of 514.5 nm and intensity of 500 mW/cm^2.

13. CHARACTERIZATION OF $BaTiO_3$:Rh

Crystals MIT-2 and MIT-3 have increased absorption in the red part of the visible spectrum, (a peak at approximately 650-690 nm), due to rhodium incorporation. As a result, they are more sensitive at near-infrared wavelengths. Shown in Figure 34 is the SPPC (self-pumped phase conjugate) reflectivity at near-infrared wavelengths of a $BaTiO_3$:Rh sample from the boule labeled MIT-3 as measured by Ross et al. [57]. High reflectivity s are found until, at a wavelength of 1.06 m, the SPPC reflectivity decreased to 38% [88]. In addition, the SPPC reflectivity of a crystal from the series of rhodium-oxide doped $BaTiO_3$:Rh was reported by Wechsler et al. [76] with a similar wavelength dependence and reflectivity of 80%. In Figure 36 is the beam coupling gain of MIT-3 measured by Ross et al. [88] with extraordinary polarized light at a wavelength of 800 nm.

Crystals MIT-2 and MIT-3 have different relative proportions of deep and shallow traps. Crystal MIT-2 is dominated by shallow traps and a donor level that is nearly compensated. It shows a highly sublinear time response versus intensity with re response time scaling as $I^{-0.5}$ at λ=514.5 nm, and has high dark conductivity due to the thermal excitation of charges in shallow traps. Crystal MIT-3 has a large deep trap density that dominates the shallow traps. In this crystal the photoconductivity varies almost linearly with intensity, ($I^{-0.9}$), and the dark decay is much slower. Both samples had p-type photoconductivity at the wavelengths reported here. We first examine the multiple trap properties of MIT-2 [89] and then MIT-3 [57].

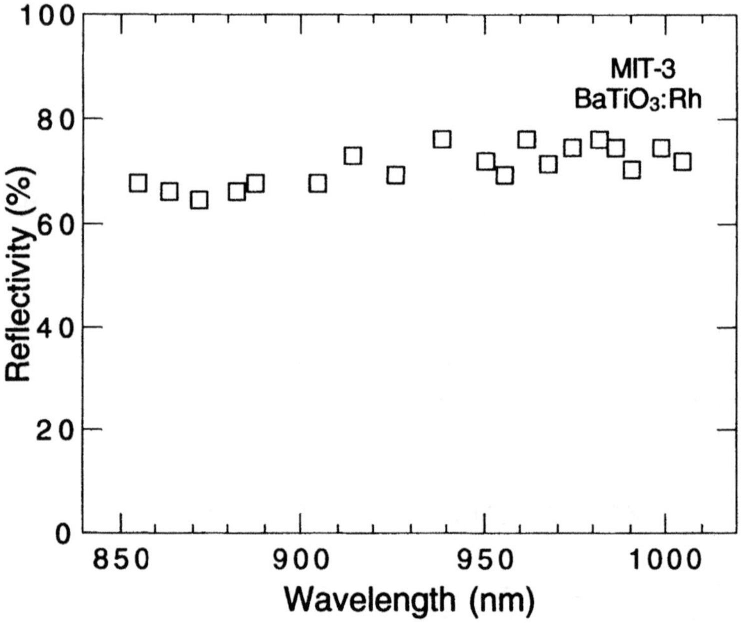

Figure 34. Self-pumped phase conjugate reflectivity as a function of wavelength for $BaTiO_3$:Rh from boule MIT-3 [57].

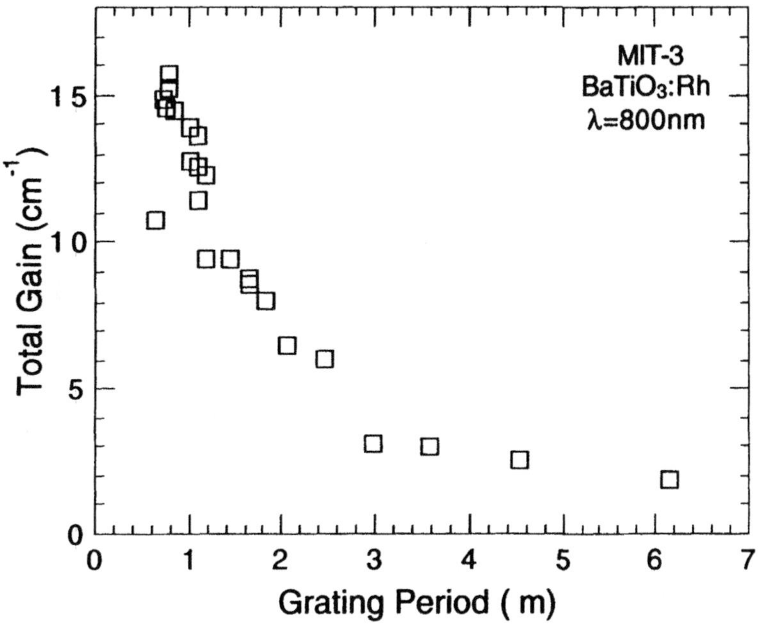

Figure 35. Beam-coupling gain at λ=800 nm, using light of extraordinary polarization, of a sample from boule MIT-3 as reported by Ross et al. [88]

13.1 Characterization of MIT-2

From SSMS and EPR we observe that rhodium and iron are present in MIT-2 and thus we expect to find light-induced absorption changes for wavelengths of light that induce and detect the interaction of the two levels. The change in the absorption for λ=514.5 nm of an optically chopped, phase detected, weak probe beam (5 W/cm^2) as a function of the incoherent pump intensity at λ=514.5 nm for MIT-2 is shown in Figure 36 [89]. We find that this sample has light-induced transparency. Considering the formula for the absorption with two traps [80],

$$\alpha = \hbar\omega\left[s_D N_{DF} + (s_T - s_D)M_0\right],$$

the light-induced transparency occurs because $s_D > s_T$. From the data we observe that it is a smoothly increasing function of intensity that begins to saturate somewhere above 1 W/cm^2. Reflecting this participation of shallow traps, the photorefractive response time is a sublinear function of intensity with x=0.48.

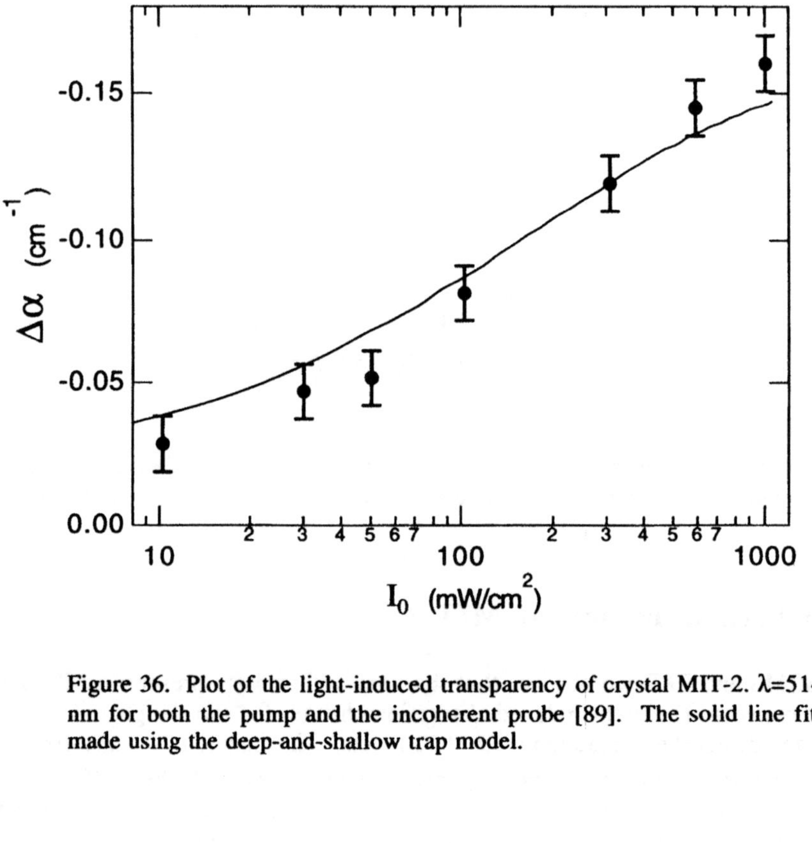

Figure 36. Plot of the light-induced transparency of crystal MIT-2. λ=514.5 nm for both the pump and the incoherent probe [89]. The solid line fit is made using the deep-and-shallow trap model.

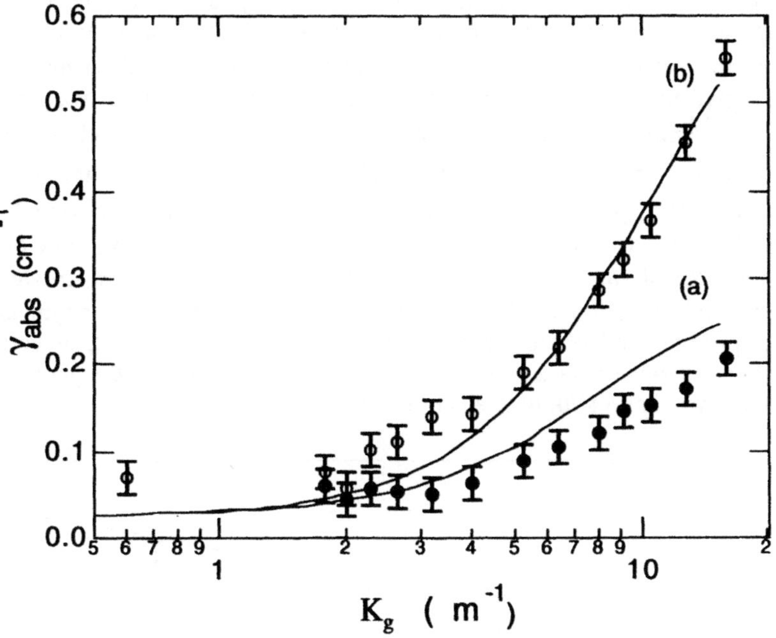

Figure 37. Positive absorptive gain coefficient of MIT-2 at λ=514.5 nm for intensities of (a) 50 mW/cm^2 and (b) 500 mW/cm^2 as a function of the grating wave vector k_g [57]. The solid line fit is made using the deep-and-shallow trap model.

The absorptive gain coefficient of MIT-2 at two intensities is shown in Figure 37. In contrast to the cobalt-doped crystals, MIT-2 has positive absorptive coupling. The absorptive gain coefficient is positive because the full traps are concentrated in the darker regions of the interference pattern. Thus, the absorption is higher in the darker regions and lower in the bright regions, so both the pump and probe gain energy when a grating forms. The total optical energy loss will be less than when the beams are incoherent due to its light-induced transparency. The absorptive coupling does not vanish as k_g approaches zero as described in the Section 10 on the deep-and-shallow trap model. The largest grating period measured was 10 m and the absorptive coupling remained finite and positive. In the limit of small k_g, where $N_1=M_1$, and where the absorptive gain is proportional to the difference between the absorption coefficients, we again find that $s_D>s_T$. The saturating intensity dependence of the positive absorptive coupling is shown in Figure 38.

Shown in Figure 40 are the electrooptic gain coefficients as a function of grating wave vector for two pump intensities, 50 and 500 mW/cm^2. Its dependence on intensity is shown in Figure 41 measured at a grating wave vector of 10.47 m^{-1}.

Simultaneous theoretical fits were made to the data of Figures 37-41 using a standard fitting routine that minimizes the least square error. The low-intensity absorption coefficient was set to 2 cm^{-1}. Some general observations could be made. For example, regardless of the initial values, the density of full deep traps was found to be N_F<7 x 10^{14} cm^{-3}. The density of shallow traps was always found to be larger than M_T>7x10^{17} cm^{-3} and the thermal excitation cross-sections of the deep and shallow traps were β_D~7 4 s^{-1}, β_T ~0.4 0.3 s^{-1}, and we consistently found that s_D>s_T in all cases. (In other words, deep sites absorb more than shallow ones at λ=514.5 nm) Small values of N_F mean that the material is well compensated, i.e. there are few full deep traps [57]. The intensity-dependent behavior is typical of a shallow-trap-dominated photorefractive crystal. We also note that, at best, the simultaneous fits to the data are representative, which may indicate the participation of more than two traps and/or hole-electron competition [57].

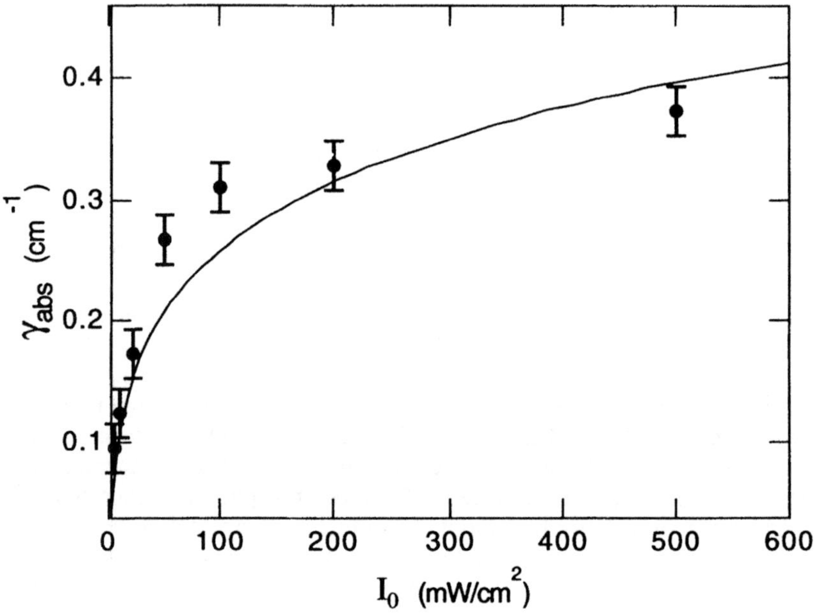

Figure 38. Plot of the positive absorptive gain dependence on intensity of crystal MIT-2 for λ=514.5 nm and a grating wave vector k_g=10.47/ m [57].

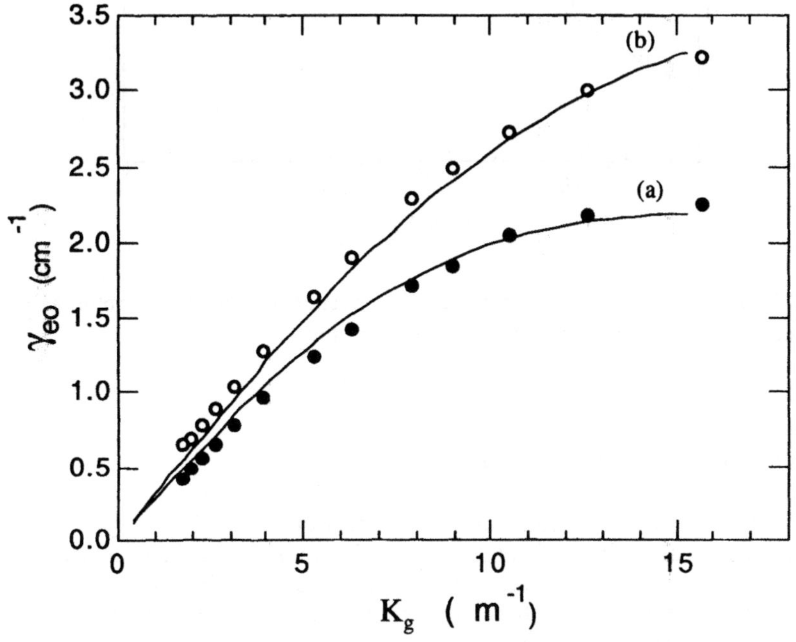

Figure 39. Plot of the electrooptic gain coefficient of MIT-2 at pump intensities of (a) 50 mW/cm^2 and (b) 500 mW/cm^2 for λ=514.5 nm. [57].

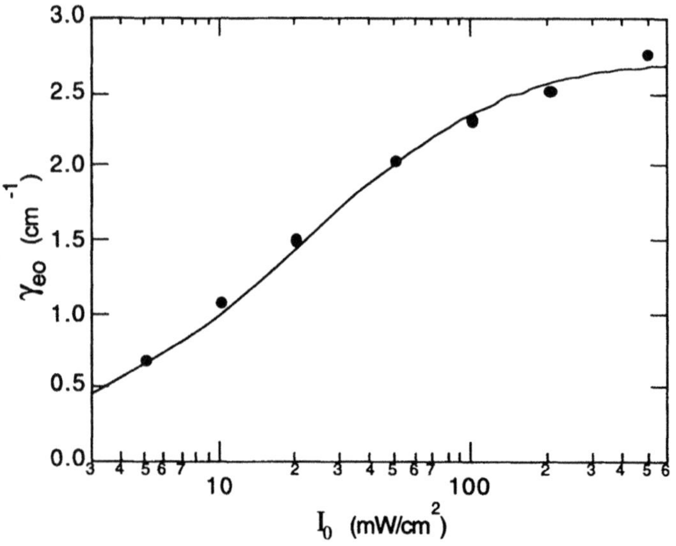

Figure 40. Plot of the intensity dependence of the electrooptic gain coefficient of crystal MIT-2 at λ=514.5 nm [57].

13.2 Characterization of MIT-3

Crystal MIT-3 has a deep blue color and excellent optical quality. It shows exceptionally low scattering of visible light, which makes even intense laser beams difficult to see inside this crystal. As we show below, this crystal exhibits a pronounced decrease in its absorption when illuminated. If there are two photo-active levels, increasing the light intensity monotonically changes the absorption [90]. However, if there are more than two photo-active levels, then local extrema can appear in the absorption as the light intensity is increased [91]. The appearance of these extrema unambiguously indicates the presence of more than two trap levels.

The crystal's absorption caused by a single probe beam was measured first. Then, the crystal was left in the dark for about 48 hours. We then monitored the transmission of a very weak probe beam ($I < 0.5$ mW/cm^2) through the crystal. Immediately after the probe was first turned on, the transmission of the crystal slowly increased with a time constant of about 1 minute. Experiments were performed with two different color probe beams. With a red probe beam at λ = 633 nm, the change in the absorption after 5 minutes was $\Delta\alpha = -0.08$ cm^{-1}. With a blue probe beam at λ = 488 nm, the change in the crystal's absorption was $\Delta\alpha = -0.05$ cm^{-1} (no pump beam present during either of these experiments). That such weak probe beams can cause a measurable change in the absorption of the crystal implies that the optically-populated level is not efficiently emptied by thermal excitation.

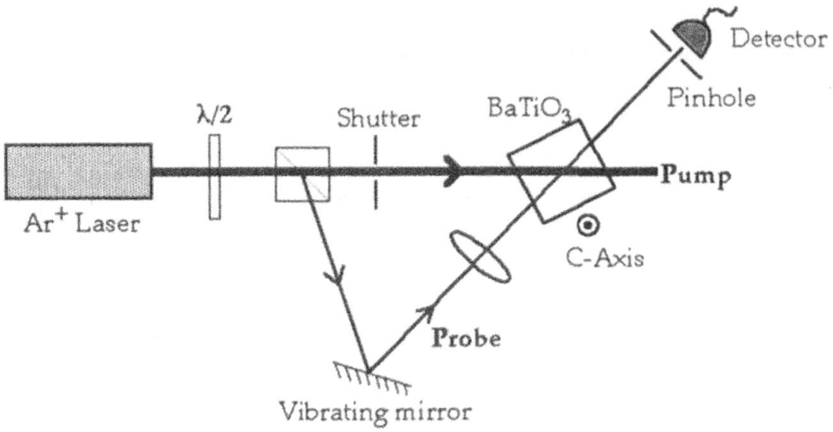

Figure 41. Experimental set up for measuring light-induced absorption in $BaTiO_3$. The vibrating mirror jitters the phase of the weak probe beam to prevent the formation of any stationary interference pattern between the probe and pump beams. Both the pump and probe beams are polarized in the plane of the figure. [57]

We then measured the light-induced change in the absorption of a weak probe beam caused by the presence of a strong pump beam in the MIT-3 crystal using the set up shown in Figure 41. We first measured the steady-state transmission through the crystal of a weak ($I = 0.5$ mW/cm^2) o-polarized probe beam at $\lambda = 488$ nm. The crystal was then illuminated with a strong ($I \sim 1$ W/cm^2) o-polarized pump beam also at $\lambda = 488$ nm, and again measured the steady-state transmission of the probe beam (we made these two laser beams completely incoherent with each other to prevent the formation of any type of gratings in the crystal). We also repeated this experiment using the same blue pump beam but now with a red probe beam at $\lambda =$ 633 nm.. These measurements were computer controlled to increase their repeatability. We made the crossing angle of the two beams large enough (full angle > 10) to eliminate any scattered pump light from reaching the probe beam detector. Because this experiment was sensitive to thermal excitation of charges, we tried to maintain the crystal at a constant temperature (T = 19.2 C) using a Peltier cooler. However at the highest pump intensities the heat dumped into the crystal overwhelmed our feedback system, causing the temperature of the crystal to vary slightly (~ 0.5 C).

Figure 42 shows the light-induced change in the absorption versus the intensity of the blue pump beam (the change in the crystal's absorption caused by the probe beam alone has been subtracted from this data). With the red probe beam the transmission increased monotonically with the pump beam intensity. In contrast, with the blue probe beam the transmission showed a distinct peak at a pump intensity near I = 300 mW/cm^2. We explain below that such behavior rules out a single-carrier, two-level model for this crystal. We emphasize that the underlying charge distribution caused by the blue pump beam is the same here for either the red or the blue probe beam. Extrapolating the blue probe data to higher pump beam intensities predicts a change from light-induced transmission to light induced absorption at I_{pump}> 20 W/cm^2.

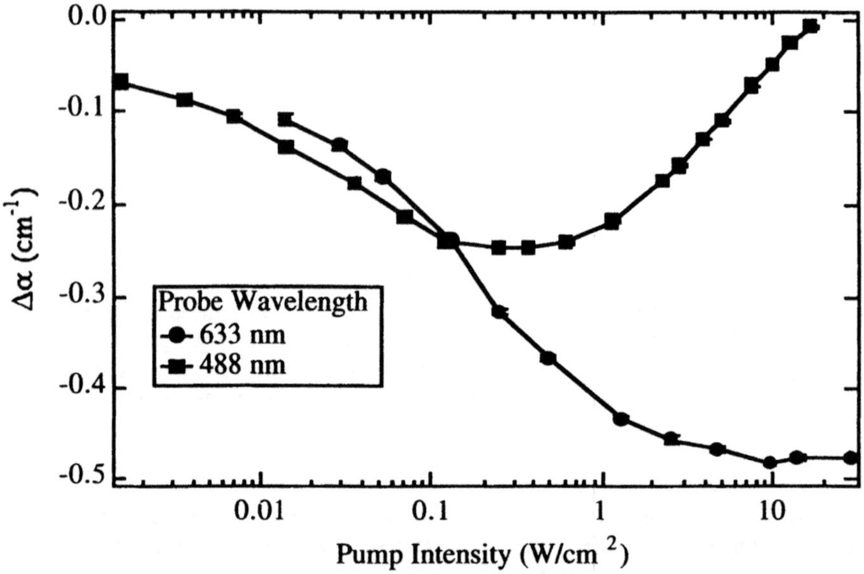

Figure 42. Change in the absorption of the crystal MIT-3 caused by a strong pump beam at λ = 488 nm on weak probe beams at either λ = 488 nm (squares) or λ = 633 nm (circles). The pump beam was not coherent with the probe beams. The local minimum seen in the data using the blue probe beam implies that there are more than two active trap levels in this crystal. [57]

We can explain the light-induced change in absorption seen in Figure 42 by postulating the presence of (at least) three trap levels, which we call the deep, middle and shallow trap levels. Consider a weak light beam whose photons have sufficient energy to excite charge from any of these levels to the nearest band. When such low-intensity light is first turned on, charges are excited from all of the levels. Any charges that recombine into the shallow level are quickly re-excited by optical or thermal processes, so at low intensities we can ignore this shallow level because it is almost empty. The result, then, of low-intensity illumination is a net shift of charges from the deep level to the middle level. Ignoring the shallow level ,the crystal's absorption is

$$\alpha_0 = \hbar\omega\left(N_D^F s_D^F + N_D^E s_D^E + N_M^F s_M^F + N_M^E s_M^E\right) + \alpha_{host} ,$$

where N denotes the density of a particular species of trap site and s is its excitation cross-section. The subscript differentiates the deep and middle levels, while the superscript differentiates the empty and full sites. Here we include the possibility that the two trap states, full or empty, of a particular defect can have different absorption.

If an absorbed photon now moves one additional charge from the deep level to the middle level, the absorption becomes

$$\alpha_1 = (N_D^F - 1)s_D^F + (N_D^E + 1)s_D^E + (N_M^F + 1)s_M^F + (N_M^E - 1)s_M^E + \alpha_{host} . \quad (57)$$

For induced transparency, $\alpha_1 < \alpha_0$, yielding

$$s_D^F - s_D^E > s_M^F - s_M^E . \quad (58)$$

Equation 58 states that the absorption difference between a full trap and an empty trap in the middle level is less than the corresponding difference in the deep level at that wavelength.

We now apply the above equations to the crystal MIT-3. Holes were identified as the dominant photorefractive charge carriers at blue/green wavelengths in this crystal. The good photorefractive response of this crystal and its appreciable low-intensity absorption in the near infrared require that there be a hole-occupied trap level located no more than $hc/\lambda = 1.8$ eV from the valence band edge. We call this the middle trap level. In thermal equilibrium the deep level will be almost full of holes. Light at $\lambda = 488$ nm can excite holes from any of the three levels. Any holes that fall into the shallow level are quickly re-excited either by light or heat, eventually ending up in either the deep or middle levels. Because the deep level is almost full of holes, most of the excited holes will fall into the middle level where more trapping sites are available. Low-level illumination therefore transfers holes from the deep level to the middle level, while the shallow level remains almost empty of holes. If the absorption difference between a full and an empty middle trap is less than the absorption difference between a full and an empty deep trap at a particular wavelength, then the crystal's absorption at that wavelength will decrease [81,92], as seen in Eq. 4. As the light intensity is increased further, optical excitation from the deep and middle levels eventually overwhelms all processes emptying the shallow level, and the shallow level begins to fill with holes. At high intensity the hole population in the shallow level becomes comparable to the hole population of the other levels. This, coupled with the highest absorption difference occurring between full and empty shallow traps, causes the crystal's absorption to increase again at high light intensity.

The blue probe beam data of Figure 42 shows clear evidence of at least these three trapping levels. The light-induced transparency is seen to first increase with intensity, as the middle trap level is populated, and then decrease at higher intensity, as the shallow trap level becomes populated. We explain the monotonic increase in $\Delta\alpha$ with pump intensity seen by the *red* probe beam in Figure 42 by postulating that for red light there is little difference between the absorption of a shallow or a middle-level trap. In that case, because the absorption of the probe beam is the same for charges in either the middle or shallow levels, they act as a single level, and the three-level system acts instead like a two-level system.

We confirm this three-level theory by measuring the speed with which light changes the absorption of the MIT-3 crystal. We measure the transmission of a weak probe beam (I_{probe} = 0.5 mW/cm^2, λ = 488 nm) through the crystal. At t = 0 we turn *off* an intense pump beam (I_{pump} = 1 W/cm^2). Figure 43 shows that when the pump beam is switched off the probe beam transmission momentarily increases (within a few milliseconds) but then slowly decreases with an exponential time constant of τ = 10 1 seconds.

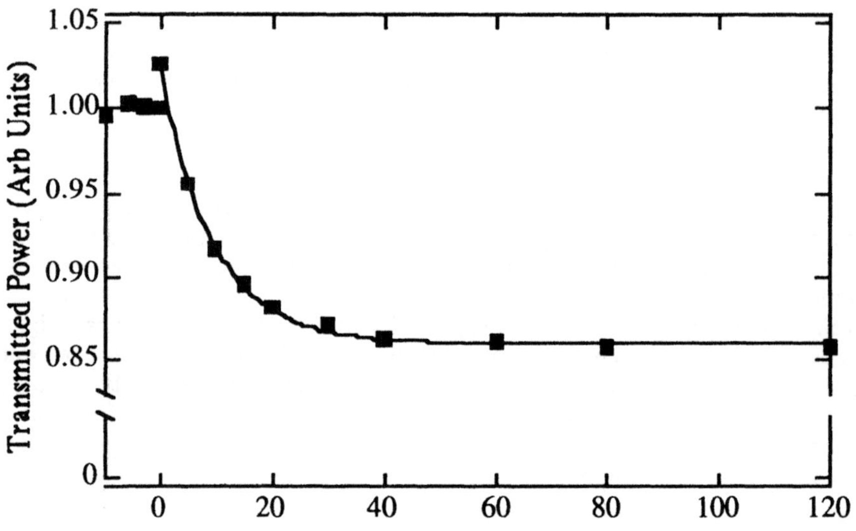

Figure 43. Probe beam transmission in $BaTiO_3$ crystal MIT-3 versus time. The strong pump beam is blocked at t = 0. The initial rapid rise (τ < 50 ms) is caused by the rapid thermal decay of the population of the highly absorbing shallow level. The slower decay (τ = 10 sec) is caused by the optical depopulation of a deeper, less absorbing level. The rate of this slower decay increases with probe beam intensity. The solid line is an exponential fit. [57]

We explain this behavior as follows. When the probe beam is first turned on, the deep level is full. The weak probe beam pumps charges from the deep level that eventually settle into the middle level. This redistribution of charge between the deep and middle levels causes the absorption to slowly decrease until equilibrium is eventually reached. Turning on an intense pump beam, with its much higher intensity, causes charges to quickly redistribute, bringing the shallow levels into play. When the pump beam is then blocked the shallow traps rapidly empty by thermal excitation, causing the fast decrease in the absorption seen just after $t = 0$ on Figure 43. However, with the pump beam now blocked there is an excess of charge trapped in the middle level that must be returned to the deep level, primarily via optical excitation of the weak probe beam. As the deep level slowly refills from the middle level, the absorption again increases and the transmitted light decreases, as shown at longer times in Figure 44.

The fact that a blue pump beam *decreases* the absorption of a red probe beam initially surprised us. Because we attribute the $\lambda = 690$ nm absorption feature of our MIT-3 crystal to the excitation of charges from a middle trap level, we thought that pumping charges from the deep level into the middle level with blue light should increase the absorption of red light. However, our light-induced absorption measurements do not reveal the absorption of a given trap level; instead they measure the *difference* between the absorption of a full and empty trap site. The light-induced transparency data seen in Figure 43 implies that the absorption difference in the red between a full and an empty site is greater for the deep level than for the middle level.

From these light-induced absorption experiments we conclude that the enhanced infrared sensitivity of crystal MIT-3 arises from holes in the crystal's middle trap level that can efficiently absorb infrared light. Pre-illumination of this crystal with blue light should increase the population of holes in the middle level, and so make the crystal even more sensitive in the infrared. Ross *et al.* [57] observed just such behavior.

An important parameter of a photorefractive crystal is its effective photorefractive trap number density, N_{eff}. This density of charge carriers determines the maximum spatial frequency and the maximum diffraction efficiency of photorefractive gratings that can be written in the crystal. N_{eff} is customarily measured by crossing two light beams to create a photorefractive grating with its wavevector along the crystal's c-axis, and then noting the variation in the grating's strength as the beam-crossing angle is varied [93]. However, for $N_{eff} > 10^{17}$ the beam crossing angles required for this technique become inconveniently large.

By taking advantage of barium titanate's large anisotropy one can measure even a large N_{eff} without having to resort to inconvenient beam-crossing angles. One writes gratings with wavevectors aligned not parallel to the c-axis but perpendicular to it. The applicable dc dielectric constant of the crystal then increases from 85 to 4400, which reduces the inverse Debye screening length, $k_o = (N_{eff}e^2 / \varepsilon\varepsilon_o k_b T)^{1/2}$, by the factor $(\varepsilon_{\parallel} / \varepsilon_{\perp})^{1/2} = 0.14$ [71], and so decreases the range of required beam-crossing angles (actually, the Sine of the half angle) by the same factor. Although for grating wavevectors aligned perpendicular to the c-axis there is no longer any electrooptic coupling between the two light beams, the spatially periodic rearrangement of full and empty trap sites makes the crystal's polarizability spatially periodic, and this polarizability grating couples the two light beams [93,94,95]. Because the optical polarizability is in general a complex number, both an absorption grating (from the imaginary part of the polarizability) and an index grating (from the real part of the polarizability) can exist. The absorption grating changes the amplitude of the light beams; the phase grating changes the phase of the light beams.

In the following we briefly describe the multilevel model of Cudney et al. [81] which is more general than the two level model presented in section 10. With proper assumptions, the model of Cudney et al. [81] reduces to the deep-and-shallow trap model of Tayebati and Mahgerefteh [80].. Using the undepleted pump approximation and taking into account the bulk absorption of the crystal, the complex amplitude gain for a weak probe beam caused by the presence of a strong pump beam is

$$E_p(l) = E_p(0)\exp[(g_r + ig_i)l] \tag{59}$$

where E_p is the amplitude of the probe beam electric field, and g_i and g_r are the imaginary and the real amplitude gain coefficients. Note that with an absorption grating the pump and the probe beams both either gain or lose energy. (Energy is still conserved; the absorption of the crystal is being rearranged by light, and this absorption change affects both beams the same way.) This is in contrast to electrooptic two-beam coupling, where one beam grows at the expense of the other.

Both the real and imaginary parts of the polarizability grating must have the same spatial phase, because both gratings are tied to the same distribution of full and empty trap sites. Additionally there can be no spatial shift between the light pattern and the charge pattern in this geometry, because there is no preferred direction for such a shift along an a-direction in a $BaTiO_3$ crystal.

No matter how many trap levels are active, this complex coupling gain g depends on the grating wavevector k_g according to [81],

$$g = \frac{G\left(\frac{k_g}{k_0}\right)^2 + \Phi}{1 + \left(\frac{k_g}{k_0}\right)^2} \left|\hat{e}_p \cdot \hat{e}_r\right|. \tag{60}$$

In Eq. 6 the term with G arises from charge transport along the direction of the grating wavevector, while Φ (also called the эoffset term in Ref. [81] or the amplitude of the screening gratings, see section 10) arises from charge redistribution between the trap levels. In general both *G* and Φ are complex. Also, *G*, Φ, and k_o have an intensity dependence that varies according to the total number of trap levels [81].

We measured the complex trap gain coefficient *g* in the $BaTiO_3$ crystal MIT-3 using a phase modulation technique [95]. We cross two ordinary polarized beams at λ = 488 nm in the crystal, with both beams polarized in the plane of the grating wavevector. Neither the grating wavevector nor the polarization of either beam has a projection along the c-axis of the crystal, so there is no electrooptic coupling between the beams. The pump beam intensity was 4.7 W/cm^2; the probe beam intensity was 50 times weaker. We eliminated any effects from light-induced absorption by continuously illuminating the crystal with both beams. We measured the intensity of the transmitted probe beam first with coupling and then without coupling by vibrating one of the beam-steering mirrors quickly enough to wash out any grating in the crystal [93]. Figure 44 shows both the real and imaginary parts of g as the crossing angle between the two light beams is varied, along with a fit of the data to Eq. 6. The curve fit yielded an effective number density of N_{eff} = (1.8 0.1) $x10^{17}$ cm^{-3}, which is twice as large as the number density in any of our nominally undoped $BaTiO_3$ crystals at the same light intensity. The curve fit also gave G = (0.44 0.02)+(0.79i 0.02i) cm^{-1} and Φ=-(0.04 0.01)+(0.03i 0.01i) cm^{-1}.

Although our beam-coupling measurements reveal a large effective number density of photorefractive trap sites in the $BaTiO_3$ crystal MIT-3, the blue-light photoconductivity of this rhodium-doped crystal is 2-3 times smaller than in our nominally undoped $BaTiO_3$ crystals. This reduced photoconductivity may, in fact, be *caused* by the large density of mid-level trapping sites. A charge excited by blue light from deep traps can recombine into the middle trap sties, which are thermally inaccessible. If the charge had instead recombined into a shallow level, it could have been thermally re-excited to continue its wandering. The net effect is that, in a crystal having a large density of mid-level traps, more photons are required on the average to move a charge a given distance, so the photoconductivity is decreased.

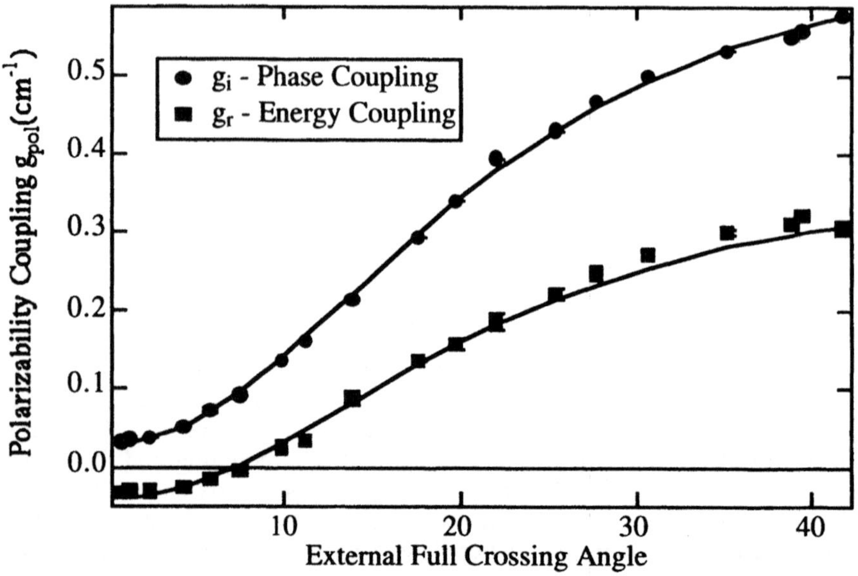

Figure 44. Two-beam coupling gain caused by polarizability gratings vs. the grating wavevector in the $BaTiO_3$ crystal MIT-3 at λ = 488 nm. Both the phase and the amplitude of the probe beam are altered by the pump beam. The solid lines are fits to Eq. 6, and they reveal an effective number density of traps of $N_{eff} = (1.8 \quad 0.1) \times 10^{17}$ cm^{-3}. [57]

The offset term Φ in Eq. 6 arises from the redistribution of charges between different levels in the same spatial location, and is related to the slope of the light-induced absorption coefficient $d\Delta\alpha/dI$ according to [96]:

$$2\,\mathrm{Re}[\Phi(I)] = -I\frac{d\Delta\alpha(I)}{dI}. \tag{61}$$

Figure 45 shows that the energy coupling part of our measured offset Φ agrees qualitatively with the offset computed from Eq. 4 and our induced absorption data of Figure 42.

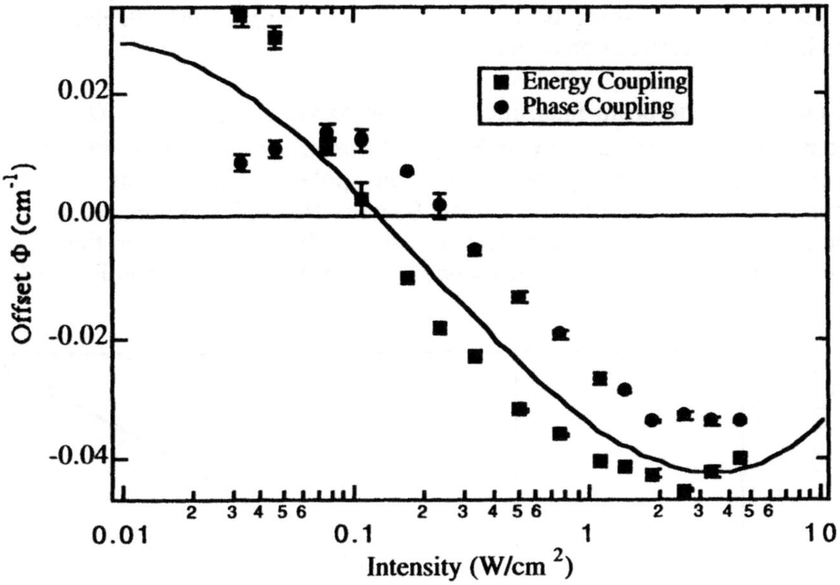

Figure 45. ϶Measurement of the beam coupling offset of polarizability gratings vs. total light intensity at a fixed, small beam-crossing angle ($\theta_{full\ external} = 2$) in the $BaTiO_3$ crystal MIT-3 at $\lambda = 488$ nm. The real part of this coupling is related to the light induced absorption; the offset calculated from the blue-probe data of Figure 42 is here shown by a solid line. [57]

The electrooptic coupling strength γ_{eo} in crystal MIT-3 was measured using two beam coupling. For a grating of wavevector k_g, the electrooptic coupling gain per length is

$$\gamma_{eo} = G_{eo}(I)\frac{k_g}{1+\frac{k_g^2}{k_0^2(I)}}. \tag{62}$$

Here, G contains the effective Pockels coefficient, a fractional poling parameter, and a hole-electron competition factor [97]. Note that both the G and the inverse screening length k_0 can be intensity dependent caused by the presence of multiple active trap levels.

The electrooptic coupling of the blue MIT-3 crystal was measured using the same experimental setup used to measure the polarizability coupling, except that we first rotated the crystal by 90 to orient its c-axis along the grating wavevector. We also rotated the polarization of the pump and probe beams so that they remained ordinary rays. Although extraordinary rays produce larger coupling they also produce larger problems, including strong beam fanning [73] and strong pump depletion, which complicate the interpretation of the data.

We used the same beam intensities as in the previous experiments (I_{pump} = 4.7 W/cm^2, pump to probe ratio of 50:1). We restricted our experiment to small crossing angles where $k_g << k_0$ so that any polarizability grating present was dominated by the Φ term in Eq. (6). Because this offset grating is tied to the position of the light intensity pattern we can simply subtract it from the measured coupling to obtain the purely electrooptic coupling gain γ_{eo} shown in Figure 46.

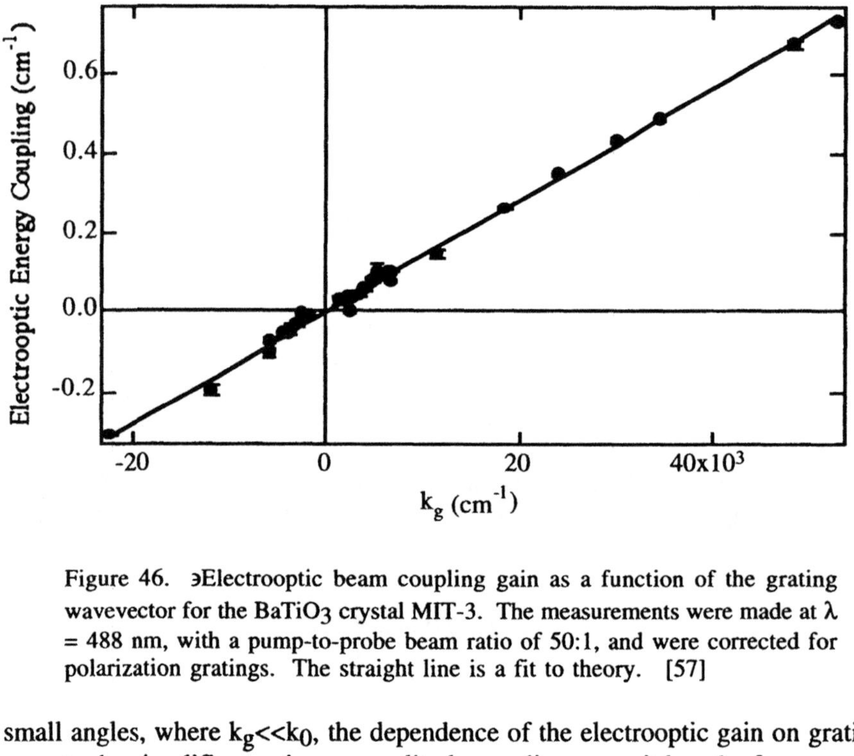

Figure 46. ϶Electrooptic beam coupling gain as a function of the grating wavevector for the $BaTiO_3$ crystal MIT-3. The measurements were made at λ = 488 nm, with a pump-to-probe beam ratio of 50:1, and were corrected for polarization gratings. The straight line is a fit to theory. [57]

At small angles, where $k_g << k_0$, the dependence of the electrooptic gain on grating wavevector k_g simplifies to give an amplitude coupling per unit length of

$$\gamma_r^{eo} = \eta(I)\frac{\omega}{2c}\frac{k_B T}{e} n_{ord}^3 r_{13} k_g, \qquad (63)$$

where we have explicitly spelled out the form of the proportionality constant and grouped the intensity dependence, fractional poling, and hole-electron competition into one coefficient $\eta(I)$. Fitting Eq. 9 to the data of Figure 46 gives $\eta(I)r_{13}$ = 5.6

pm/V. This is less that 30% of the value r_{13} = 19 pm/V measured using dc fields applied to a $BaTiO_3$ crystal [71].

Other $BaTiO_3$ crystals have also shown small electrooptic coupling efficiencies, though to a lesser extent, even at intensities high enough rule out any reduction in coupling from shallow trap levels [95]. To check for the possible effects of shallow traps here, we reduced both the pump and probe intensities by a factor of ~50 and repeated our electrooptic coupling experiment. The slope of the resulting photorefractive gain curve was only 7% lower than that of Figure 46, implying that shallow traps are not causing a significant reduction in the coupling here. At this lower intensity of 100 mW/cm^2, the time response of the crystal became quite slow, with a diffracted signal requiring ~12 sec to decay to $1/e$ of its initial value. Varying the intensity over two orders of magnitude revealed an only slightly sublinear dependence of the erasure rate on intensity, further indicating that thermal excitation of charges from shallow trap levels is not important here [75]. Although competition between electrons and holes can also reduce the electrooptic coupling [82], there was no sign of any angular dependence of the coupling to confirm this hypothesis [98]. However, we investigated beam-coupling at shorter wavelengths and did observed significant hole-electron competition in this crystal.

14. OPTIMIZING THE PHOTOREFRACTIVE SENSITIVITY

14.1 n-type $BaTiO_3$

To optimize the photorefractive sensitivity, or index change per absorbed energy, one needs to maximize the index change and minimize the response time. As previously discussed in the section on defect chemistry, the valence state of the active species can be adjusted through oxidation/reduction. With this technique the Fermi level is adjusted by the level of oxidation/reduction. Oxygen vacancies provide free carriers that are trapped at defect centers. $BaTiO_3$ crystal MIT-2 was reduced to a level, (P_{O2} 10^{-16} atm), that produced an absorption peak at ~ 470 nm. The crystal has a light orange color and has electron-dominated photoconductivity. The as-grown and reduced absorption spectra are shown in Figure 47. This is nearly the identical spectra as that produced by Schunemann et al. [13] (We shall show later that nearly the same spectra can be produced with cobalt-doped barium titanate). As described in the section on Defect Chemistry, heavily reduced acceptor-doped samples will have doubly-ionized oxygen vacancies lying just below the bottom of the conduction band. In general, the ionized charge is captured by acceptors with states that lie just above the top of the valence band. If the density of these states is high enough and their location close enough to a band, there will be band tail effects, i.e., the band edge extends into the forbidden gap.

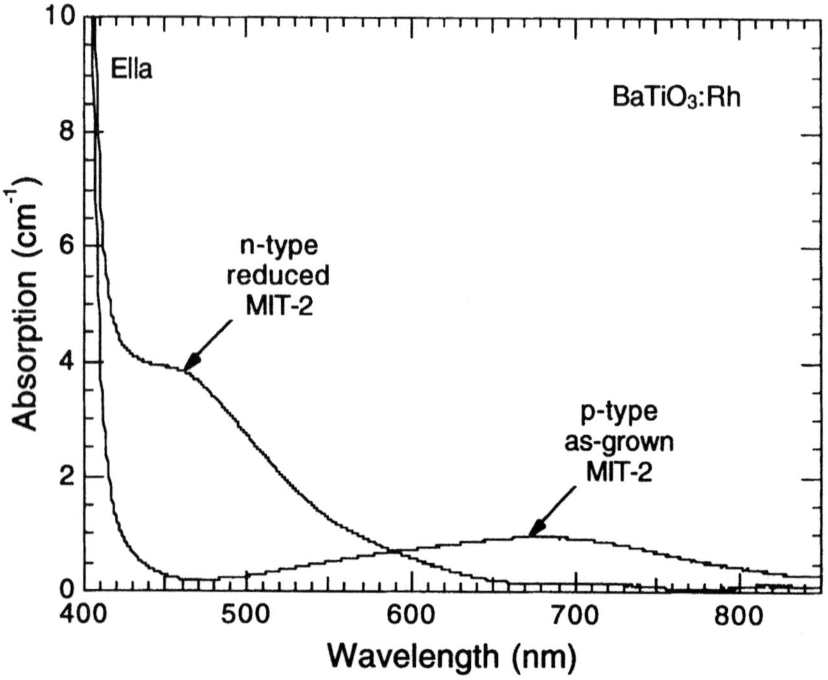

Figure 47. Absorption spectra of a crystal from boule MIT-2 in the as-grown and reduced state.

Light-induced changes in absorption as a function of wavelength were used to observe the trap levels present in the reduced MIT-2 sample. Shown in Figure 48 is the spectral dependence of the absorption/transparency of crystal MIT-2 after reduction in an oxygen partial pressure of $P_{O_2} \sim 10^{-16}$ atm. These spectra were measured with a weak probe beam that originated from a monochromator, was optically chopped, and phase detected with a lock-in. The sample was illuminated by a strong incoherent pump beam at $\lambda = 514.5$ nm at three different intensities.

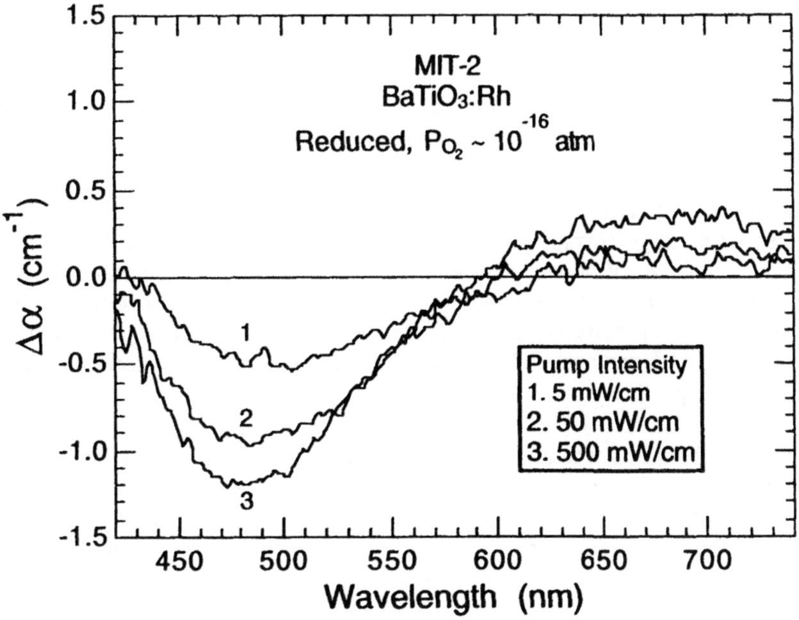

Figure 48. Spectral dependence of the light-induced absorption/transparency of reduced sample MIT-2.

The probe beam detected two Δα absorption peaks, at ~470 nm and ~690 nm, each with a different sign as the intensity-dependent absorption changes from light-induced transparency to light-induced absorption. We observed that the two peaks of the light-induced absorption have their peaks at the same wavelengths as the low intensity absorption peaks of the as-grown and reduced samples (see Figure 47). Initially, at around $\lambda = 500$ nm, the signal experiences light-induced transparency when the pump beam illuminates the sample. That is, the signal is less absorbed when the pump (at $\lambda = 514.5$ nm) is turned on. Recall that charge must be transferred between at least two levels for an absorptive change to be detected. This means that for blue wavelengths of the pump and probe the absorption difference between full and empty states of the deep level are greater than that of the shallow levels,

$$s_D^F - s_D^E > s_M^F - s_M^E. \tag{64}$$

At longer wavelengths one observes that the signal experiences light-induced absorption when the incoherent λ=514.5 nm pump illuminates the sample. Therefore, charge is being transferred from a deep level (~2.4 eV to the conduction band, e.g., from Fe) to shallow levels and a mid-level (the rhodium level). In this case, the near-infrared beam detects that the absorption difference of the full and empty traps of the shallow levels are greater than the absorption difference of the full and empty traps of the deep levels,

$$s_M^F - s_M^E > s_D^F - s_D^E. \qquad (65)$$

Photoexcitation takes place from a band of traps just above the top of the valence band, the latter of which are associated with the 470 nm peak, to the conduction band. Then, the charge is trapped by shallow oxygen traps and at the rhodium mid-level which is at ~1.8 eV above the top of the valence band (for optical excitation). The transfer of charge to these levels makes the crystal more absorbing to the red probe beam. These results were also confirmed using a coherent laser source at wavelengths of 514.5 nm and 633 nm. They experience light-induced transparency and absorption respectively.

The absorption changes are temporary and the crystal returns to its low intensity state within seconds after the pump is turned off. This indicates that a relatively shallow level is involved in the charge redistribution (oxygen vacancies and the thermal ionization of the rhodium traps). We also note, as should be expected, that the higher the pump intensity the greater the amount of intensity-dependent change. We observe that the two peaks of the light-induced absorption have the same wavelengths as the low intensity absorption peaks of the as-grown and reduced samples.

When reduced, MIT-2 behaves similarly to the as-grown state, with shallow traps that cause the grating to be rapidly erased in the dark. Both the as-grown and reduced sample have a response time that has a sublinear dependence on intensity. Reduced samples are interesting because their photorefractive response time is shorter than in the as-grown state as seen from Figure 51. This may be due to enhanced photoconductivity that could arise from increased absorption and/or a higher electron carrier mobility [99]. The "free carrier lifetime" may also increase with a higher shallow trap concentration. This allows the carrier to recombine to shallow levels where they are easily thermalized and can diffuse further before recombining to a deep level. For the reduction level and the impurity concentrations in MIT-2 there was a decrease of the response time by a factor of 2.8 relative to the as-grown sample. The reduced sample also had a slight increase in its electrooptic gain coefficient as shown in Figure 52. The net result is an increase in the sensitivity.

The absorption spectra of an as-grown and reduced cobalt-doped $BaTiO_3$ are shown in Figure 53. We note that it is nearly identical to the rhodium-doped reduced MIT-2 except that the absorption peak at ~470 nm is higher. This gives the crystal a dark orange color. The absorption feature at 470 nm could be an intrinsic phenomena. However, we cannot rule out the possibility that the rise in the Fermi level from reduction makes most deep level dopants in $BaTiO_3$ produce a similar peak at 470 nm. That is, there is a broad band of occupied traps below the Fermi level available for photoexcitation. We tried to reduce a $BaTiO_3$ sample that was grown from purified starting materials (MIT-1) which has a relatively low deep-trap concentration. We could not produce the absorption peak at 470 nm!

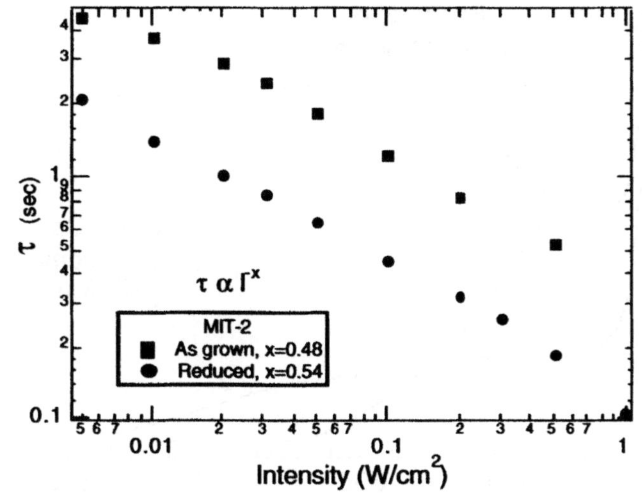

Figure 49. Plot of the grating response time as a function of the illumination intensity for sample MIT-2 as-grown and reduced to P_{O_2} ~ 10^{-16} atm.

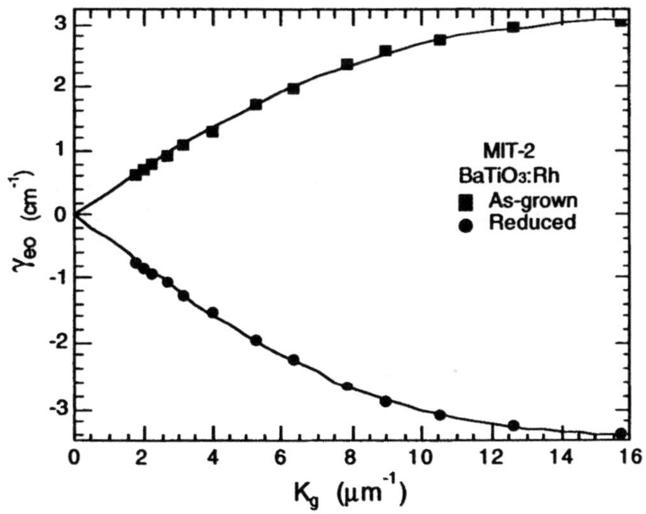

Figure 50. Plot of the electrooptic gain coefficient of crystal MIT-2, as-grown and reduced. The intensity was 500 mW/cm^2 and the wavelength was 514.5 nm. The polarization was ordinary. Solid lines are theoretical fits to the deep-and-shallow trap model.

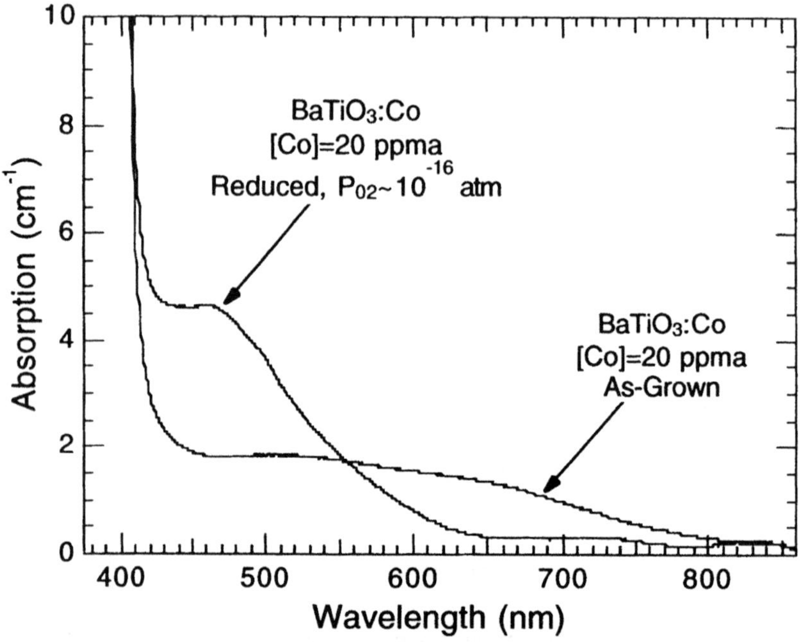

Figure 51. Absorption spectra of as-grown and reduced cobalt-doped $BaTiO_3$:Co. Dopant concentrations refer to that in the melt with respect to the $BaTiO_3$ concentration in the melt.

The beam-coupling gain of the reduced cobalt-doped $BaTiO_3$:Co was about 15% higher than MIT-2 (Probably due the higher dopant concentration which gives it a slightly higher absorption at 470 nm). The reduced cobalt-doped $BaTiO_3$ sample had a relatively fast response time of about 21 milliseconds at 1 W/cm^2 at $\lambda = 514.5$ nm in a 45 -cut, (approximately the same response time as the reduced 0 -cut). One may compute the theoretical limit of the response time as described by Yeh [100]. His formulation considers the number of photons that must be absorbed to produce a grating of 1% diffraction efficiency. He does not take into account the transport time of carriers to build the grating and it is thus a lower limit to the response time. With this formulation the response time of barium titanate could be as fast as 0.1 to 1.0 msec at 1 W/cm2 (depending on the quantum efficiency), in the 45 -cut orientation where high gain is experienced.

14.2 45 -cut $BaTiO_3$

It is possible to optimize the photorefractive sensitivity for $BaTiO_3$ by changing the crystallographic cut. It was found by Garrett et al. [42] that when $BaTiO_3$ was fabricated into a 45 -cut [87] that the gain increased dramatically but the response time did not increase as one might expect because the dielectric constant also increases. A 45 -cut $BaTiO_3$:Co can be fabricated out of a 0 -cut by extracting a sample that has the normal to its orthogonal faces at 45 with respect to the c-axis. That is, one rotates around the a-axis by 45 and cuts perpendicular to the a-c plane. A diagram of the crystals is shown in Figure 52. The advantage of this cut is that the grating wave vector can more easily be at an angle with respect to the c-axis and therefore accesses r_{42}, the largest electrooptic coefficient. Shown in Figure 54 is a plot of the beam-coupling gain coefficient as a function of the angle β between the grating wave vector k_g and the polar or c-axis for various beam-crossing half-angles θ. This was first plotted by Ford et al. [87].

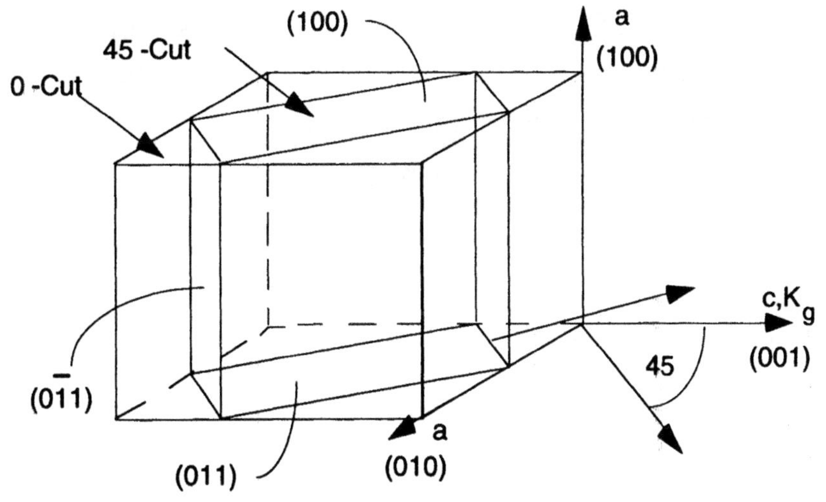

Figure 52. Diagram of the 0 and 45 crystallographic cuts used to access the highest electrooptic coefficient in beam-coupling for $BaTiO_3$.

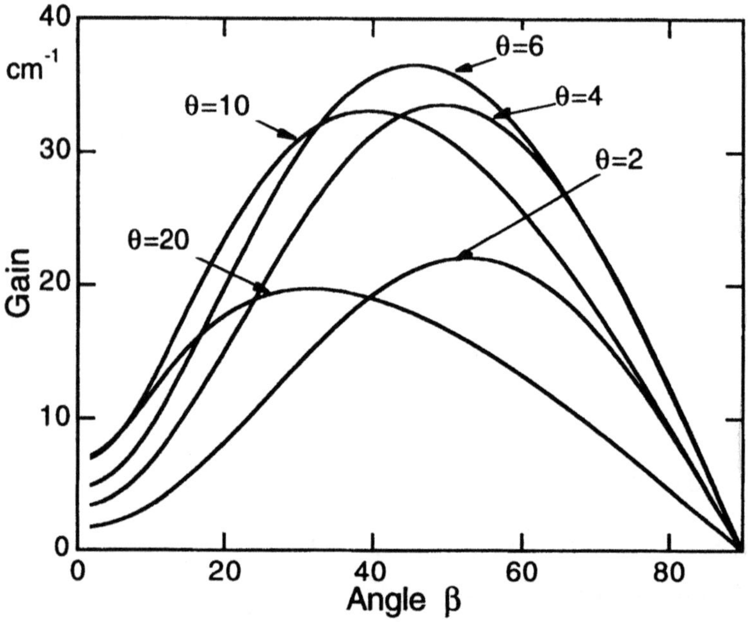

Figure 53. Plot of the gain coefficient as a function of β , the angle between the grating wave vector k_g and the polar or c-axis, for various half-angles between the signal and pump beam (Unclamped electrooptic coefficients were used to create this plot, $\eta(I)=1$, and the effective trap density set to 2 x 10^{16} cm^{-3}).

We note that the figure of merit for electrooptic materials, $n^3 r_{eff}/\varepsilon$, is larger than many photorefractive inorganic crystals when $BaTiO_3$ is in the 45° crystallographic cut. To obtain this value correctly one must use the proper value of the effective electrooptic coefficient and the dielectric constant. (Many authors have simply divided the r_{42} by ε_{33} and multiplied by the cube of the index of refraction giving a value of ≤ 6.33.) The effective electrooptic coefficient can be determined using the formulas given in Section 9 of this chapter. The magnitude of the dc dielectric constant in the direction of the grating wavevector is determined by the transformation

$$\varepsilon = \hat{k}_g \cdot \overline{\varepsilon} \cdot \hat{k}_g , \tag{66}$$

where ε is the relative dc dielectric tensor with unclamped components $\varepsilon_{11}=\varepsilon_{22}=3600$ and $e_{33}=150$ [84], and $\hat{k}_g$ is the normalized grating wavevector. The clamped electrooptic coefficient is 1640 pm/V. These numbers give $n^3 r_{eff}/\varepsilon$ equal to 9.45 pm/V. (A improved computation of the figure of merit would take into account the clamped and unclamped components but we conserve tradition in the latter computation simply correcting the relative amplitudes of the coefficients.)

A simple formulation for the response time, τ [101] shows that τ is directly proportional to the dc dielectric constant, inversely proportional to the mobility-lifetime product $\mu\tau_r$ and excitation cross-section s, i.e., $\tau \propto (\varepsilon\varepsilon_0/\mu\tau_r s)$. This implies that the response time should increase in the 45°-cut relative to the 0°-cut. However, it was found that there was no appreciable difference in the response time between the two orientations [42]. (This may imply that the mobility-lifetime product is changing in proportion to the dielectric constant.) The net result is that the gain increases substantially in the 45°-cut and the response time remains the same producing a much higher photorefractive sensitivity [42].

In Table 7 are the experimental values of the photorefractive sensitivity for several different oxides. We note that the magnitude of the gain determined for the reduced 45°-cut $BaTiO_3$:Co and SBN:Ce may be limited by fanning and these values depend on the thickness of the sample.

The gain of the $BaTiO_3$:Co reduced sample was measured in a 0°-cut before the sample was fabricated into a 45°-cut. Table 8 gives a comparison of the beam-coupling parameters with crystallographic cut, beam polarization, and sample thickness. The 45°-cut sample was reduced in thickness (0.87 mm) to minimize the effects of beam-fanning which diminishes the amount of light that can be transferred into the signal beam. Nevertheless, we observed a significant amount of fanning independent of the beam ratio.

It should be noted that there are several methods and parameters used to compare the performance of photorefractive crystals [102]. One "apples-to-apples" comparison is to measure the rise time to equivalent gain proposed by Vazquez et al. [86]. This method takes into account that the rise time needed to create higher space-charge fields is also longer.

Crystal (λ=514.5 nm)	α (cm^{-1})	γ (cm^{-1})	τ (ms)	I (W/cm^2)	S (cm^3/kJ)
$BaTiO_3$:Mn, 0°-cut	-	2.5	1000	2.5	0.002
$BaTiO_3$:Mn 0°-cut, 120°C	-	2.2	15	2.5	0.05
$BaTiO_3$:Co 0°-cut,	2.2	3.6	34	1.0	0.198
$BaTiO_3$:Co 45°-cut, reduced	2.2	38.7	21	1.0	3.44
SBN:Ce	1.0	36	333	1.0	0.44
$KNbO_3$:Fe, reduced	1.0	2.5	3.0	1.0	3.41
$Bi_{12}SiO_{20}$	0.8	0.2	0.1	1.0	10.0

Table 7. Experimentally determined values of the photorefractive sensitivity S for several photorefractive oxides. α is the absorption coefficient, γ is the gain coefficient, τ is the response time, I the total intensity and S the sensitivity.

Orientation, Polarization	r_{eff} (pm/V)	Gain (cm^{-1}) Theory	Gain (cm^{-1}) Exp.	Thickness (mm)	Beam Ratio
45°, ordinary	13.7	0.68	1.1	0.87	800
0°, ordinary	19.5	3.65	3.62	4.52	800
0°, extraordina ry	89.6	14.7	10.0	3.40	10^4
45°, extraordinary	1208. 5	32.5	26.7	0.87	10^4

Table 8. Comparison of the beam-coupling gain and response time for 0° and 45°-cut $BaTiO_3$:Co - reduced using extraordinary and ordinary polarized light at λ=514.5 nm and 1 W/cm^2. The grating period was 1 µm. The sample thickness and beam ratio was changed to maximize the coupling. Fanning limited the maximum gain obtainable. The r_{eff} of the 45°-cut crystals was calculated using the old formulation of the effective electrooptic coefficient incorporating clamped terms only.

A sample from boule MIT-2 was first reduced in an atmosphere where P_{O_2} ~10^{-16} atm, poled and then cut and fabricated into a 45°-cut orientation. The fast response time of the sample made it possible to shutter the signal beam at 30 Hz. When the shutter was closed the pump beam erased the gratings. Figure 54 shows the rise time of the signal for a single shutter opening and the repeated opening of the shutter that gives a 30 Hz frame rate for beam-coupling. Using a similar sample, a motion picture signal was input into a liquid crystal display that was between a polarizer and an analyzer. The image was Fourier transformed onto the $BaTiO_3$ sample and interfered with a pump beam that amplified the signal. It was then detected by a TV camera and replayed on a television monitor. The total power into the system was 45 mW and the replay faithfully reproduced the input video signal [103].

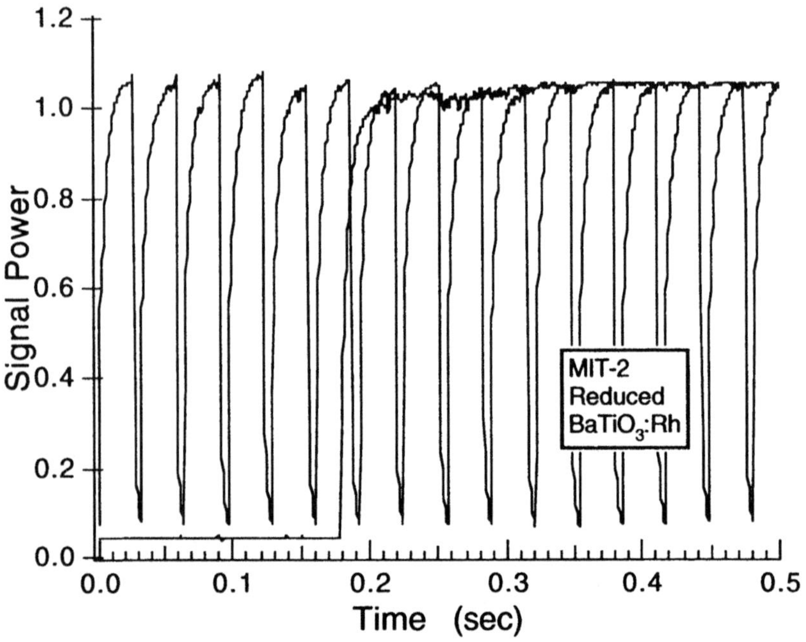

Figure 54. Plot of the signal-beam rise time for two bean-coupling from reduced sample MIT-2. The signal beam is shuttered on and off at a 30 Hz frame rate. The single trace shows the write time of the beam-coupling. The erasure of the grating by the pump is not identically to zero because there is some scattered light providing a bias to the detector. The reduced sample MIT-2 is in a 45°-cut, illuminated with a total intensity of 1 W/cm^2 at a wavelength of 514.5 nm.

15. CONCLUSIONS

Extensive studies by many researchers, both experimental and theoretical, have helped illuminate the mechanisms of the photorefractive effect. Multi-trap models more closely fit experimental observations. The origins of the defects that produce these traps are more clearly discerned, controlled and manipulated. The photorefractive sensitivity can be manipulated with doping and reduction techniques. Doping has increased the bandwidth of sensitivity for the photorefractive effect into the near-infrared. The ferroelectric characteristics of Perovskites has been examined on a microscopic level. The modulation of the lattice by space-charge fields has been observed with x-ray topography and theoretically analyzed. The photorefractive effect itself has become an investigative tool to determine trap properties. There are still many areas of future investigation that include: extending the sensitivity to optical communication channels at λ=1.3 and 1.5 μm; minimizing the influence of shallow traps that cause erasure of gratings in the dark; control of the doping to obtain absorption uniformity; control of the valence states of the traps through co-doping and optimization of the crystal growth through techniques such as continuous feed and the Stepanov technique which can produce high optical quality material with spatially uniform properties. The future of photorefractive ferroelectric oxides clearly depends on the use of these materials in applications. It is from the needs of these applications that the aforementioned areas of research will be initiated.

Acknowledgment

The first author of this chapter gratefully acknowledges the support of Professor C. Warde and Dr. H. P. Jenssen of MIT and collaborations with Dr. J. Y. Chang and Dr. P. Tayebati on several topics described herein.

References:

[1] C. Kittel, *Introduction to Solid State Physics*, J. Wiley & Sons, Inc., pg. 417, (1976).

[2] F.M. Michel-Calendini, H. Chermette, and J. Weber, J. Phys. C**13**, 1427 (1980).

[3] H.-J. Hagemann and D. Hennings, J. Am. Ceram. Soc. **64**, 590 (1981).

[4] H. Kröse, E. Possenriede, R. Scharfschwerdt, T. Varnhorst, O. F. Schirmer, H. Hesse, C. Kuper, presented at European Materials Research Society Meeting, Strasbourg, France, May 1994. To be published in Optical Materials journal.

[5] P.G. Schunemann, D.A. Temple, R.S. Hathcock, H.L. Tuller, H.P. Jenssen, D.R. Gabbe, and C. Warde, J. Opt. Soc. Am. B **5**, 1685 (1988).

[6] B.A. Wechsler and M. B. Klein, J. Opt. Soc. Am. B **5**, 1711 (1988).

[7] D.E. Rase and R. Roy, J. Am. Ceram. Soc. **38**, 110 (1955).

[8] K.W. Kirby and B.A. Wechsler, J. Am. Ceram. Soc. **74**, 1841 (1991).

[9] V. Belruss, J. Kalnajs and A. Linz, Mat. Res. Bull. **6**, 899 (1971).

[10] M.H. Garrett, J.Y. Chang, H.P. Jenssen, and C. Warde, Ferroelectrics **120**, 167 (1991).

[11] Y. Furukawa, M. Sato, K. Kitamura, and F. Nitanda, J. Cry. Growth **128**, 909 (1993).

[12] K. Buse, F. Baller, R. Pankrath, H. Hesse, and E. Krätzig, Sol. State. Comm. **88**, 587 (1993).

[13] P.G. Schunemann, T.M. Pollak, Y.-Y. Teng and C. Wong, J. Opt. Soc. Am. B **5**, 1702 (1988).

[14] C. Warde, T.W. McNamara, M. H. Garrett, and P. Tayebati, SPIE Conference CR48-07, San Diego, CA. July (1993).

[15] T. W. McNamara, S.G. Conahan, I. Mnushkina, M. H. Garrett, H.P. Jenssen, and C. Warde, *SPIE Critical Review Proceedings Vol. CR-48*, editors P. Yeh and C. Gu, 1994.

[16] B.A. Wechsler, M. B. Klein, C.C. Nelson, and R.N. Schwartz, Opt. Lett. **19**, 536 (1994).

[17] H.-J. Hagemann, "Akzeptorionen in $BaTiO_3$ und $SrTiO_3$ und ihre Auswirkung auf die Eigenschaften von Titanatkeramiken", Ph. D Thesis, Rheinisch-Westfalischen Technischen Hochschule, Aachen (1980).

[18] P.G. Schunemann, Master Thesis, Massachusetts Institute of Technology, Cambridge, MA (1987).

[19] F.A. Kroger and H.J. Vink, "Relations between the concentrations of imperfections in crystalline solids", in Solid State Physics **3**, F. Seitz and D. Turnbull, eds., Academic Press, 307 (1956).

[20] S.A. Long and R.N. Blumenthal, J. Amer. Ceram.. Soc. **54**, 515 (1971).

[21] N.G. Eror and D.M. Smyth, J. Solid State Chem.. **24**, 235 (1978).

[22] N.-H. Chan, R.K. Sharma and D.M. Smyth, J. Amer. Ceram.. Soc. **64**, 556 (1981).

[23] H.-J. Hagemann and D.Hennings, J. Amer. Ceram.. Soc. **64**, 590 (1981).

[24] A. Lahlafi, G. Godefroy, G. Ormancey, and P. Jullien, J. Opt. Soc. Am. B **10**, 1276 (1993).

[25] Observations made by the first author of this chapter.

[26] P. Coufová, Czech. J. Phys. B **18**, 1038 (1968).

[27] D.L. Staebler and W. Phillips, App. Opt. **13**, 788 (1974).

[28] M. Jaros, *Deep Levels in Semiconductors* (Adam Hilger Ltd., Bristol, 1982).

[29] S. T. Pantelides, ed., *Deep Centers in Semiconductors* (Gordon and Breach Science Publishers, New York, 1986); Rev. Mod. Phys. **50**, 797 (1978).

[30] A. Zunger, Solid State Phys. **39**, 276 (1986).

[31] M. Lannoo and J. Bourgoin, *Point Defects in Semiconductors I*, Vol. 22 of Springer Series in Solid-State Sciences (Springer-Verlag, Berlin, 1981).

[32] J. Bourgoin and M. Lannoo, *Point Defects in Semiconductors II*, Vol. 35 of Springer Series of Solid-State Sciences (Springer-Verlag, Berlin, 1983).

[33] R. N. Schwartz, B. A. Wechsler, and D. Rytz, J. Am. Ceram. Soc. **73**, 200 (1990).

[34] M. B. Klein and R. N. Schwartz, J. Opt. Soc. Am. B **3**, 293 (1986).

[35] K. A. Johnson, Adv. Quantum Chem. **7**, 143 (1973).

[36] D. S. McClure, Solid State Phys. **9**, 399 (1959).

[37] J.S. Griffith, *The Theory of Transition-Metal Ions* (Cambridge University Press, 1961).

[38] R.N. Schwartz, B.A. Wechsler, and R.A. McFarlane, Phys. Rev. B **46**, 3263 (1992).

[39] A.M. Stoneham, *Theory of Defects in Solids* (Oxford University Press, Oxford, 1975).

[40] D. Rytz, B.A. Wechsler, M.H. Garrett, C.C. Nelson, and R. N. Schwartz, J. Opt. Soc. Am. B **7**, 2245 (1990).

[41] D. Rytz, R.R. Stephens, B.A. Wechsler, M.S. Keirstead, and T. M. Baer, Opt. Lett. **15**, 1279 (1990).

[42] M.H. Garrett, J.Y. Chang, H.P. Jenssen, and C. Warde, Opt. Lett. **17**, 103 (1992).

[43] K. Zdánsky', H. Arend, and F. Kubec, Phys. Stat. Solidi **20**, 653 (1967).

[44] M. Aguilar, Solid State Commun. **50**, 837 (1984).

[45] K.W. Blazey and K.A. Müller, J. Phys. C: Solid State Phys. **16**, 5491 (1983).

[46] E. Possenriede, P. Jacobs, and O.F. Schirmer, J. Phys.: Condens. Matter **4**, 4719 (1992).

[47] A. Abragam and M.H.L. Pryce, Proc. Roy. Soc. (London) **A206**, 173 (1951).

[48] W. Low, Phys. Rev. **109**, 256 (1958).

[49] G.M. Zverev and A.M. Prokhorov, Soviet Phys. JETP **16**, 303 (1963).

[50] R.N. Schwartz and B.A. Wechsler, Phys. Rev. B **48**, 7057 (1992).

[51] G.W. Ross, P. Hribek, R.W. Eason, M.H. Garrett, and D. Rytz, Opt. Commun. **101**, 60 (1993).

[52] J.S. Griffith, *The Theory of Transition-Metal Ions* (Cambridge U. Press, London, 1961), Chap. 9.

[53] C.K. Jørgensen, Acta Chem. Scand. **10**, 500 (1956).

[54] C.K. Jørgensen, Acta Chem. Scand. **11**, 151 (1957).

[55] A. Raizman, J.T. Suss, and S. Szapiro, Phys. Lett., **32A** 30 (1970).

[56] M.G. Townsend, J. Chem. Phys. **41**, 3149 (1964).

[57] M.H. Garrett, C. Warde, H.P. Jenssen, S. MacCormack, D. Bacher, J. Feinberg and R.N. Schwartz, submitted for publication JOSA B, June 1994.

[58] B. Steiner, M. Kuriyama, R.C. Dobbyn, Prog. Crystal Growth and Charact. **20**, 189 (1990).

[59] B. Steiner, U. Laor, M. Kuriyama, G.G. Long, R.C. Dobbyn, J. Crystal Growth **87**, 79 (1988).

[60] B. Steiner, M. Kuriyama, R.C. Dobbyn, U. Laor, D. Larson, M. Brown, J. Appl. Phys. **66**, 559 (1989).

[61] M. Kuriyama, B. Steiner, R.C. Dobbyn, U. Laor, D. Larson, M. Brown, Phys. Rev. B **38**, 12421 (1988).

[62] B.W. Batterman, H. Cole, *Rev.*. Mod. Phys. **36**, 681 (1964).

[63] A.B. Buckman, Guided-Wave Photonics, Saunders College, New York, (1988).

[64] J. Gronkowski, Phys. Rpts., 206 (1991).

[65] G. Fogarty, M. Cronin-Golomb, B. Steiner, U. Laor, OSA 1993 Technical Digest Series **11**, 610 (1993).

[66] G. Fogarty, M. Cronin-Golomb, B. Steiner, Synchrotron Radiation News, **6** 11 (1993).

[67] P. Günter, M. Zgonik, Optics Letters **16**, 1826 (1991).

[68] G. Fogarty and M. Cronin-Golomb, unpublished.

[69] A.A. Izvanov, A.E. Mandel, N.D. Khatov, and S.M. Shandorov, Optoelectronics, data processing and instrumentation **2**, 80 (1986), trans. of Avtometriya **2**, 79 (1986).

[70] S.I. Stepanov, S.M. Shandorov, and N.D. Khatkov, Sov. Phys. Solid State **29**, 1754 (1987).

[71] M. Zgonik, P. Bernasconi, M. Duelli, R. Schlesser, P. Günter, M.H. Garrett, D. Rytz, Y. Zhu and X. Wu, accepted for publication Phys. Rev. B **50**, 1 (1994).

[72] N.V. Kukhtarev, V.B. Markov, S.G. Odoulov, M.S. Soskin and V.L. Vinetskii, Ferroelectrics **22**, 949 (1979).

[73] F.P. Strohkendl, M.C. Jonathan, and R.W. Hellwarth, Opt. Lett. **11**, 312 (1986).

[74] G.C. Valley, J. Appl. Phys. **59**, 3363 (1986).

[75] D. Mahgerefteh and J. Feinberg, Phys. Rev. Lett. **64**, 2195 (1990).

[76] G.A. Brost and R.A. Motes, Opt. Lett. **15**, 1194 (1990).

[77] F.P. Strohkendl, J. Appl. Phys. **65**, 3773, (1989).

[78] G.A. Brost, R.A. Motes, and J.R. Rotge, J. Opt. Soc. Am. **5**, 1879 (1988).

[79] P. Tayebati, Ph. D. Dissertation, University of Southern California, Los Angeles, CA (1989).

[80] P. Tayebati and D. Mahgerefteh, J. Opt. Soc. Am. B **8**, 1053 (1990).

[81] R.S. Cudney, R.M. Pierce, G.D. Bacher, and J. Feinberg, J. Opt. Soc. B **8**(6), 1326 (1991).

[82] P. Tayebati, J. Opt. Soc. Am. B **9**, 415 (1992).

[83] S. Ducharme, J. Feinberg and R.R. Neurgaonkar, IEEE J. Quant. Elect. QE-23 **12**, 2166 (1987).

[84] S.H. Wemple, M. Didomenico, Jr., and I Camlibel, J. Phys. and Chem. Solids **29**, 1797 (1968).

[85] R.M. Pierce, R.S. Cudney, G.D. Bacher, and J. Feinberg, Opt. Lett. **15**, 414 (1990).

[86] R.A. Vazquez, R.R. Neurgaonkar, and M.D. Ewbank, J. Opt. Soc. Am. B **9**, 1416 (1992).

[87] J.E. Ford, Y. Fainman, and S. H. Lee, Appl. Opts. **28**, 4808 (1989).

[88] G.W. Ross, P. Hribek, R.W. Eason, M.H. Garrett, and D. Rytz, Technical Digest of the *Photorefractive Materials, Effects, and Devices PRM 93*, THAO4, Kiev, Ukraine, Aug. 11-15, 51 (1993).

[89] M.H. Garrett, P. Tayebati, J.Y. Chang, H.P. Jenssen, and C. Warde, J. Appl. Phys. **72**, 1965 (1992).

[90] A. Motes and J.J. Kim, J. Opt. Soc. B **4**, 1379 (1987).

[91] G.D. Bacher , Ph.D. dissertation , University of Southern California, Los Angeles (1994).

[92] G.A. Brost, R.A. Motes, and J.R. Rotge, J. Opt. Soc. Am. **5**, 1879 (1988).

[93] R.M. Pierce, R.S. Cudney, G.D. Bacher and J. Feinberg, Opt. Lett. **15**, 414 (1990).

[94] A.V. Alekseev-Popov, A.V. Kynaz'kov and A.S. Satkin, Sov. Tech. Phys. Lett. **9**, 475 (1983).

[95] R.S. Cudney, G.D. Bacher, R.M. Pierce, and J. Feinberg, Opt. Lett. **17**, 67 (1992).

[96] R.S. Cudney, Ph. D. dissertation, University of Southern California, Los Angeles (1992).

[97] M.B. Klein in *Photorefractive Materials and Applications I,* edited P. Günter and J.-P. Huignard, Topics in Applied Physics Vol. 61, (Springer-Verlag Berlin, 1988).

[98] M.B. Klein and G.C. Valley, J. Appl. Phys. **57**, 4901 (1985).

[99] G.C. Valley, J. Opt. Soc. Am. B **14**, 14 (1987).

[100] P. Yeh, Appl. Opt. **26**, 602 1987

[101] D. Rytz, M.B. Klein, R.A. Mullen, R.N. Schwartz, G.C. Valley, and B. A. Wechsler, Appl. Phys. Lett. **52**, 1759 (1988). See also, P. Günter, Phys. Rev. **93**, 199 (1982).

[102] G.C. Valley and M.B. Klein, Opt. Eng. **22**, 704, (1983).

[103] Private communication between the first author of the chapter and Dr. T. Chang. of Rockwell International Science Center, Thousand Oaks, CA.

3.

PERMANENTLY FIXED VOLUME PHASE GRATINGS IN FERROELECTRICS

Anthony S. Kewitsch and Amnon Yariv
Department of Applied Physics
California Institute of Technology
Pasadena, California 91125

Mordechai Segev
Department of Electrical Engineering
Princeton University
Princeton, NJ 08544

1. INTRODUCTION

Volume phase holograms have been stored in a wide range of materials: silver halide photographic emulsions, dichromated gelatin films, glass, photorefractive polymers, organic materials and ferroelectric crystals. Each of these materials poses a unique set of technological problems when recording optical interference patterns as permanent index gratings. In this chapter we focus on hologram recording in ferroelectric oxide crystals, such as $LiNbO_3$ and $Sr_xBa_{1-x}Nb_2O_6$ (SBN:x). When these materials are doped to optimize their photorefractive properties, optical interference patterns redistribute charge among traps and generate local electronic space-charge fields which are replicas of the optical interference patterns (but generally shifted in phase and spatially distorted). An index change arises from the spatially modulated space-charge field. The complex microscopic details of this process are included within a measured linear electrooptic tensor r_{ijk} which relates the index change to the dc electric field in the non-centrosymmetric phase. Because these index gratings are electronic in origin, they are bleached by the same optical field that reconstructs them. This dynamic property is extremely useful for real time holography applications; however, for applications such as holographic data storage, permanent index gratings are desirable.

The goal of hologram fixing techniques is to transform this grating initially composed of a spatially modulated density of optically active species (electrons, holes) into a modulated density of optically inactive species (ion cores, vacancies, defects). This has been achieved with varying degrees of success in $LiNbO_3$, $LiTaO_3$, $Ba_2NaNb_5O_{15}$, SBN, and $BaTiO_3$. The common fixing stage in these oxides is the conversion of a dynamic grating composed of electrons and/or holes into a replica composed of stable, displaced ions. Upon reconstruction of the hologram, the electronic grating is partially erased and the ionic grating persists. These ions are typically optically inactive and have a relatively low mobility. The obstacle to fixing is the simultaneous requirement that ions must be mobile enough to redistribute under the influence of local space-charge fields, yet be resistant to optical erasure. One technique to achieve this is to uniformly heat the crystal during or after the hologram recording process so that ions can drift under the influence of electronic space-charge fields. A second technique uses the hysteresis associated with the reversal of ferroelectric domains to provide memory even after the electronic space-charge fields have diminished. This hysteresis may be increased to a significant level by applying external electric fields and by recording the grating at a temperature near the ferroelectric Curie temperature. As in a ferromagnetic memory in which the magnetization can be reversed locally on a surface by a magnetic recording head, periodic electric fields can spatially modulate the polarization throughout the entire volume of the ferroelectric.

In this chapter we focus on fixing techniques based on the modulation of optically inactive ions, either in the form of a small displacement (e.g. polarization switching) or transport of the ion to other atomic sites (e.g. ionic drift). We focus on the fundamental physical principles rather than the technology of fixing, which can be found in the original literature referenced herein. Two applications of polarization gratings are discussed in detail: selective, page-addressable fixing of volume holograms [1] and tunable quasi-phase matching [2]. We do not discuss related fixing methods based on multiple photorefractive species [3], enhanced non-destructive readout [4], two color methods [5], two-photon recording [6], or techniques in materials other than ferroelectric oxides. Furthermore, we do not describe here the extensive work on analytical modeling of the fixing process [7, 8, 9].

2. MATERIAL PROPERTIES OF COMMON FERROELECTRICS

Fixing has been most thoroughly studied in three common photorefractive ferroelectrics: $LiNbO_3$, SBN:75 and $BaTiO_3$. Their material properties span a broad range, which is responsible for the rich diversity of fixing phenomena reported. We highlight some of the defining material properties of this group.

$LiNbO_3$ is the primary material displaying the ionic drift thermal fixing mechanism. Photorefractive charge transport in lithium niobate is dominated by the very strong photovoltaic effect [10]. This effect may produce space-charge fields in excess of 100 kV cm^{-1}. Lithium niobate possesses an extremely large coercive field (3 10^5 V cm^{-1}) [11] and spontaneous polarization (70 μC cm^{-2}) [12]. The low frequency dielectric constant ε_{33} is 32 at room temperature. It exhibits a second-order ferroelectric-paraelectric phase transition at the Curie temperature of 1200 °C, so the polarization is relatively stable even at elevated temperatures. Lithium niobate belongs to the 3m symmetry class, with two orientations of the spontaneous polarization (±180°).

$BaTiO_3$ and SBN display a domain fixing mechanism at room temperature. The perovskite $BaTiO_3$ belongs to the 4mm symmetry class and undergoes three successive first-order ferroelectric phase transitions at 120, 5 and -90 °C. It possesses 4 possible orientations of polarization (±90°, ±180°) and a strong electrooptic effect ($r_{42} \approx 1600$ pm V^{-1}) at room temperature. In contrast to $LiNbO_3$, it has a large low frequency dielectric constant, $\varepsilon_{33} = 3600$.

SBN:75 also belongs to the 4mm symmetry class but has a disordered tungsten bronze structure [13]. It displays strong relaxor or glassy [14, 15] ferroelectric characteristics, as evidenced by a diffuse phase transition about the dielectric maximum temperature ($T_c \approx 55$ °C) and low coercive field (of the order of 1 kV/cm). SBN:75 is highly electrooptic ($r_{33} \approx 1400$ pm V^{-1}) and has only two allowed spontaneous polarization directions (±180°) with a spontaneous polarization of 15 to 25 μC cm^{-2}. The large low frequency dielectric constant, $\varepsilon_{33} = 3400$, reduces the space-charge field in this material to typically less than 1 kV cm^{-1}. The optical quality of SBN generally suffers from growth striations, while $LiNbO_3$ and $BaTiO_3$ possess relatively good optical quality.

3. PHOTOFERROELECTRIC PHENOMENA

Fixing techniques encompass several photoferroelectric phenomena [16] in addition to the well known photorefractive effect. Some relevant examples are photoassisted domain switching and the photodomain effect. These phenomena occur in a special class of ferroelectrics which are also photoconductive, hence the term photoferroelectric. These materials share several properties in common with semiconductors (e.g. optically excitable impurity centers), yet are distinct from semiconductors because of their relatively low values of electronic mobility (~ 1 cm^2 s^{-1} V^{-1}) and relatively large band gaps (> 3 eV).

Photoferroelectric phenomena are treated theoretically by including the electronic contribution to the thermodynamic free energy in addition to the lattice free energy. By minimizing this free energy, the equilibrium configuration of the system can be estimated. The effect of optical illumination on the ferroelectric properties, such as

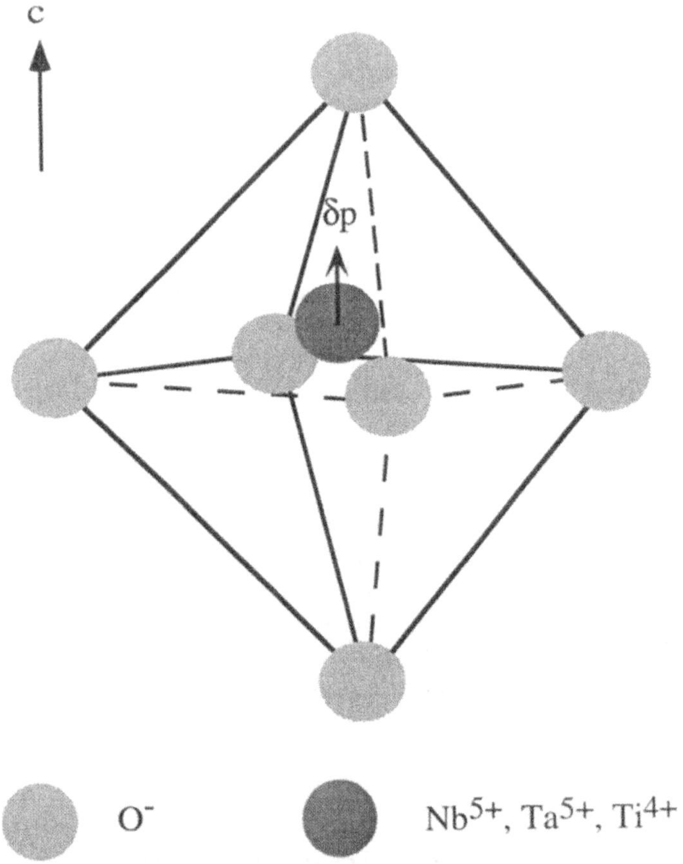

Fig. 1 Octahedral cage surrounding the transition metal ion, a characteristic unit of ferroelectric oxides. The spatially varying space-charge field repolarizes the lattice, introducing a perturbation δp in the local polarization.

the spontaneous polarization, Curie temperature and domain structure, can then be predicted analytically [16]. Qualitatively, optically excited electrons partially screen the Coulomb interaction between individual dipoles, from which the long range ferroelectric order originates. This reduces the Curie temperature by an factor dependent on the light intensity. Similarly, the spontaneous deformation, dielectric constant and piezoelectric constant are modified. In principle, these effects may produce an intensity dependent modulation of the index of refraction in phase with the optical interference pattern, in addition to the standard phase shifted photorefractive grating.

Numerous observations of photoferroelectric phenomena have been reported in the literature. For instance, the external electric fields required to pole and depole SBN decrease under illumination [17]. Accordingly, light increases the number of Barkhausen pulses [18, 19] observed in the ferroelectric SbSI [20] under applied fields. These pulses are current transients measured across the crystal as domains invert. The kinetics of domain nucleation and growth can be inferred [21] from their shape, duration, amplitude and number. This phenomenon, in which light reduces the coercive field, is called photoassisted domain switching.

In addition, uniform illumination alters the ferroelectric structure in the absence of external fields. For instance, light reduces the tetragonal to cubic phase transition temperature by 2.6 °C in $BaTiO_3$ [22]. Illumination of SbSI at a wavelength of strong optical absorption accelerates the transition from a monodomain to a polydomain state [23]. The maximum decrease of the remnant polarization occurs for illumination in the wavelength region of peak photosensitivity. This observation is called the photodomain effect and is attributed to screening of the ferroelectric polarization by nonequilibrium, optically excited carriers and to the formation of screening fields near the electrodes.

Photorefraction is perhaps the most familiar example of a photoferroelectric effect, in which nonequilibrium charge carriers are spatially redistributed and trapped to generate local space-charge fields. In the case of a ferroelectric oxide, as represented in Fig. 1, a space-charge field directed along the c axis perturbs the position of the transition metal ion within the oxygen octahedral cage, introducing a polarization perturbation δp. The position of this ion along the c axis contributes to the remnant polarization, so there is a close relationship between the space-charge fields and the magnitude and sign of the unit cell polarization. This polarization perturbation alters the index of refraction by typically 10^{-5} to 10^{-3} through the electrooptic effect and generates volume phase gratings.

4. ELECTROOPTIC EFFECT

Photorefractive gratings are produced by a modulated space-charge field that alters the index of refraction through the electrooptic effect. The index change created by the total polarization field in the noncentrosymmetric phase may be expressed as a linear electrooptic effect or equivalently as a quadratic electrooptic effect "biased" by the remnant polarization [24]. The Fourier component of the index change at the grating vector $\mathbf{k_g}$ is:

$$\Delta\left(\frac{1}{n^2}\right)_{ij}(\mathbf{k_g}) = 2\, g_{ijkl}\,(P_o)_k\, \Delta P(\mathbf{k_g})_l \quad , \tag{1}$$

where g_{iljkl} is the quadratic electrooptic tensor, P_o is the average remnant polarization and $\Delta P(\mathbf{k_g})$ is the Fourier component of the static ($\omega = 0$) polarization modulation at $\mathbf{k_g}$. This polarization modulation may be generated by a static space-charge field $\Delta \mathbf{E}(\mathbf{k_g})$, for instance. The modulation of the polarization $\Delta P(\mathbf{k_g})$ is expressed as the Fourier transform of the spatial variation of polarization $\Delta P(\mathbf{r})$:

$$\Delta P(\mathbf{k_g}) = \frac{1}{(2\pi)^{3/2}} \int_{-\infty}^{\infty} \Delta P(\mathbf{r})\, e^{-i\,\mathbf{k_g}\cdot\mathbf{r}}\, d\mathbf{r} \tag{2}$$

where the polarization perturbation includes three contributions:

$$\Delta P(\mathbf{r}) = \Delta P^{\text{optical excitation}}(\mathbf{r}) + \Delta P^{\text{charge transport}}(\mathbf{r}) + \Delta P^{\text{ferroelectric distortion}}(\mathbf{r})\,. \tag{3}$$

These three terms may be estimated analytically from:

$$\Delta P(\mathbf{r}) = \sum_{i}^{N} \int (\mathbf{R}_i + \mathbf{r}' - \mathbf{r})\, \Delta\rho_i\, (\mathbf{R}_i + \mathbf{r}' - \mathbf{r})\; d\mathbf{r}' \tag{4}$$

where the sum is over the N lattice sites i, $\mathbf{r}'$ is the position relative to the address of the i-th lattice site $\mathbf{R}_i$, and ρ_i is the total charge density of the atoms associated with the lattice site i.

The explicit decomposition of Eq. 3 clarifies the origin of the index gratings. The first term on the right represents the change in dipole moment of donors and acceptors upon optical excitation. The second term represents the change in polarization arising from both electron transport (due to drift, diffusion, pyroelectric or photovoltaic mechanisms producing photorefractive space-charge fields) and ionic transport (typically due to drift induced by electronic space-charge fields). The charges undergo transport between distant lattice sites (a distance on the order of the grating spacing, typically ~ μm). The third term represents the change in lattice polarization arising from the displacement of ions. In this case the center of charge is displaced only slightly within the unit cell from the unperturbed position (typically << 1 Å). The electric dipole generated by the first two terms polarizes the lattice through this third term. Note that the polarization is understood to be spatially averaged on a scale much larger than a unit cell, yet smaller than the optical wavelength of interest.

The standard electrooptic effect described by Eq. 1 treats only that polarization perturbation corresponding to the third term of Eq. 3. Measurements of the electrooptic effect are performed by applying an external electric field to the crystal and measuring the resulting net index change. The external electric field will modify the crystalline polarization by slightly displacing the positions of ions. As long as negligible current from the external source flows through the crystal, the drift of electrons across the crystal will also be negligible. Therefore, the electrooptic effect by definition will not include the direct index changes resulting from electron and ion optical excitation or drift; namely, the first two terms of Eq. 3.

The electrooptic effect describes the interaction of a low frequency electric field E^{Ω} and an optical field E^{ω}, generating an optical polarization $\Delta P^{\omega,\Omega}$ that is shifted in phase relative to the incident optical wave:

$$\Delta P^{\omega,\Omega} = \chi^{\omega,\Omega} E^{\omega} E^{\Omega}. \tag{5}$$

The proportionality constant is a nonlinear susceptibility $\chi^{\omega,\Omega}$, defined as

$$\chi^{\omega,\Omega} = \left[\frac{\partial^2 P^{\omega,\Omega}}{\partial E^{\omega} \partial E^{\Omega}} \right]_{E_0^{\omega}, E_0^{\Omega}} . \tag{6}$$

To determine the coefficient $\chi^{\omega,\Omega}$ responsible for the linear electrooptic effect, the changes ΔP caused by optical frequency electric fields E^{ω} must be considered. For instance, electron clouds in the vicinity of ion cores will feel very strong Coulombic forces from the nucleus. These electron orbitals will be only weakly perturbed by the optical field, which is much weaker than the ion core attraction. On the other hand, electrons which spend a greater fraction of their time far from the core and experience a screened core potential will exhibit stronger polarizability at optical frequencies.

Bonds can be separated into two archetypes: purely covalent or ionic [25]. The character of bonds between elements with different electronegativities falls between these two extremes. The stronger the covalent character of the bond, the greater fraction of the time the electron will be found between the ion cores where the core attraction is significantly weaker. Therefore, covalent bonding orbitals are more highly polarizable than ionic bonding orbitals under moderate optical fields. As a consequence, purely covalent crystals (e. g. diamond) have larger indices of refraction than ionic crystals (e. g. NaCl).

The nonlinear optical properties of the ferroelectric niobates are dominated by the distorted NbO_6 octahedron [24]. The primary contributions to the linear and nonlinear optical polarizabilities are the highly covalent Nb-O bonds. The highly ionic Li-O bond susceptibility [26] makes a negligible contribution, because the valence electrons are tightly bound. Therefore, in the case of $LiNbO_3$ or SBN, it is a good approximation to consider only the Nb-O bond contribution to the nonlinear polarizability (for $BaTiO_3$ and $LiTaO_3$, the dominant contribution to the susceptibility arises from the covalent Ti-O and Ta-O bonds [26]). This contribution to the electrooptic effect can be estimated by considering the dependence of the linear optical susceptibility of the Nb-O bond on rotation and stretch [27, 28]. From this

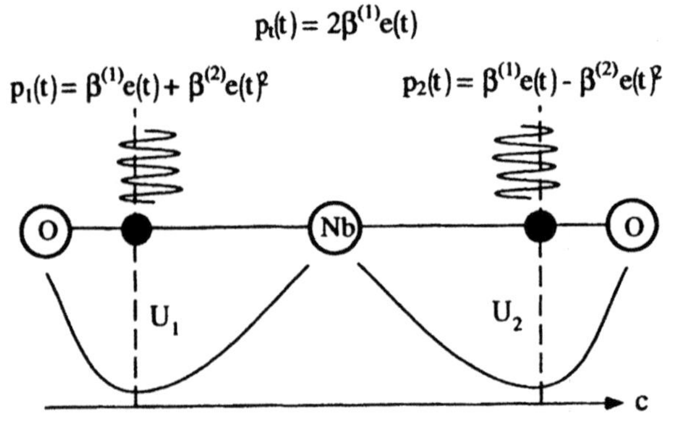

(a) inversion symmetry

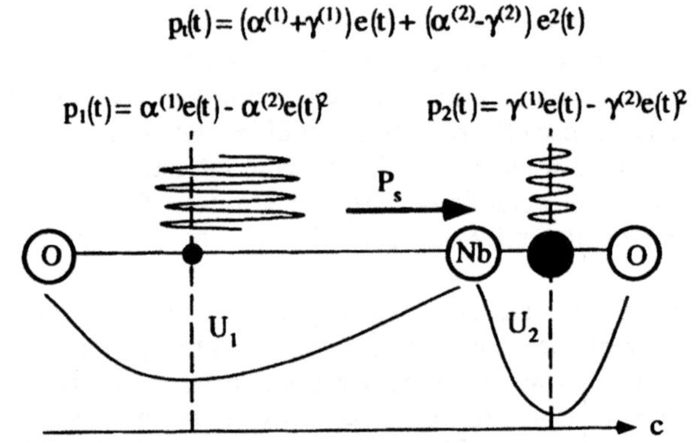

(b) no inversion symmetry

Fig. 2 Nb-O bond along the c axis. The black circles indicate bond charge, U_1 and U_2 are the coulomb potentials experienced by the bond charge. The coulomb potentials are distorted as the bonds are partially ionic. (a) In the paraelectric phase the Nb ion occupies the central position, so the total induced polarization $p_t(t)$ of the two bonds includes only even orders of the optical field e(t). (b) In the ferroelectric phase, the displacement of the Nb ion breaks the inversion symmetry, so $p_t(t)$ includes both even and odd orders.

microscopic viewpoint, the displacement of the Nb ion along the c axis changes the polarization $\Delta P(\mathbf{k_g})$ and induces an index change through the electrooptic effect (Fig. 2). The material possesses a linear electrooptic effect only when inversion symmetry is broken (.i. e., $u_1 \neq u_2$), so $\alpha^{(2)}$ and $\gamma^{(2)}$, the second order polarizabilities of adjacent Nb-O bonds, are not equal (Fig. 2b). Note that the distortion of this Nb-O bond length also produces the ferroelectric polarization in materials such as $LiNbO_3$ and SBN. Thus, the ferroelectric and nonlinear optical properties are intimately related. The decomposition of Eq. 3 on the atomic bonding level links the third term, describing ferroelectric distortion, to the electrooptic effect in these materials. This association is fundamental to understanding the relationship between nonlinear optical and ferroelectric properties, and between dynamic and remnant ferroelectric domain gratings, as will be discussed in section 6.

The origins of index gratings resulting from internal space-charge fields are microscopically more complicated than the phenomenon described by the electrooptic effect, in which external fields rather than space charge densities modulate the index of refraction. Indeed, from Eq. 3 there are two other contributions. The first term represents the change of dipole of the ions upon optical excitation. This makes a negligible contribution to the steady state index change because the density of electrons in the conduction band at any instant in time is very small relative to the cumulative charge redistribution that occurs upon the generation of steady state photorefractive space-charge fields. This is true at moderate optical intensities (< kW cm^{-2}) in the highly resistive ferroelectric oxides described here.

The second term in Eq. 3 describes the change in polarization caused by the optical redistribution of electrons among deep traps. By definition, because these photorefractive traps are deep, the electrons are strongly bound to their host atom, and as a consequence these trapping sites typically make a small contribution to the nonlinear susceptibility relative to the highly covalent Nb-O bonds. Note that in a strained, centrosymmetric crystal this contribution can play an important role through a Jahn-Teller distortion [29]. However, the materials under consideration possess a strong linear electrooptic effect, so the second term is neglected. This term also has a contribution from the motion of ions (such as protons in the case of thermal fixing in $LiNbO_3$) to new lattice sites by drift induced by electronic space-charge fields. These ions compensate the electronic grating to reduce the net internal field.

The first two contributions of Eq. 3 polarize their surroundings and perturb the lattice polarization through the third term to produce an index change. In effect, the photogenerated space-charge fields and ionic space-charge fields can be treated as *external* fields, which modify the index of refraction through the standard electrooptic effect.

5. FERROELECTRIC DOMAIN STRUCTURE

The ferroelectric domain structure has important implications on the nonlinear optical properties of the material. From the microscopic description of the electrooptic effect presented in section 4, it is apparent that an electrooptic grating corresponds to a modulation of the ferroelectric polarization. As described latter, this modulation corresponds not only to small perturbations of the Nb-O bond length, but also to bipolar switching between the two stable polarization orientations.

From electrostatic considerations, the potential and field of a polarized object is equivalent to that field produced by a bound volume charge density ρ_b

$$\rho_b = - \nabla \cdot \mathbf{P_S} \tag{7}$$

and a bound surface charge σ_b

$$\sigma_b = \mathbf{P_S} \cdot \hat{\mathbf{n}} \,. \tag{8}$$

where $\mathbf{P_S}(r)$ is the local spontaneous polarization averaged over a volume much larger than a unit cell, and $\hat{\mathbf{n}}$ is the unit normal vector to the surface of the crystal. The bound charge cannot be removed from a domain interface (where $\mathbf{P_S}$ changes); it is permanently attached to the ions responsible for the spontaneous polarization. In contrast to this bound charge, free charge can be optically excited and redistributed, producing photorefractive space-charge fields, for instance. Note that this free charge spends the great majority of its time trapped, because the recombination time (~ 10 to 100 ps) in oxide ferroelectrics such as SBN is extremely short (in this sense, the term *free* charge is misleading). If sufficient mobile charge is available, the bound charge is compensated by an equal and opposite free charge density ρ_f and surface charge density σ_f. Compensating charge is electrostatically trapped at domain walls for which the divergence of the polarization $\mathbf{P}$ is non-zero (i.e. head-to-head rather than antiparallel).

Compensating electric charge plays an important role in minimizing depolarization fields at domain walls in ferroelectrics, analogous to the closure requirements of ferromagnetic domains. To treat charge compensation at domain walls, consider the simplest case of a ferroelectric phase with only two allowed polarization directions (0 and 180 degrees to the c axis). Figure 3a illustrates a typical volume grating configuration in which the grating vector is parallel to the c axis. The free carrier density required to compensate the non-zero divergence of the polarization for a domain grating composed of head-to-head domain walls is given by

$$N_{comp} = \frac{2\pi \, \Delta P}{q \, \Lambda_g} \tag{9}$$

where ΔP is the polarization modulation. The carrier density required to produce a field equal to the coercive field is

$$N_{coercive} = \frac{2\,\pi\,\varepsilon\,E_{coercive}}{q\,\Lambda_g}, \tag{10}$$

where Λ_g is the grating period, q is the charge of an electron, and ε is the dielectric permittivity. For grating periods of 1 micron, P_S equal to 25 $\mu C\ cm^{-2}$ and $E_{coercive}$ equal to 1 kV cm^{-1} (typical for SBN:75), the compensating charge density required on the domain walls for bipolar modulation ($\Delta P = 2\ P_s$) is $N_{comp} \sim 2\times10^{19}\ cm^{-3}$, and the coercive charge density is $10^{17}\ cm^{-3}$. Therefore, significant charge densities (Fig. 3a) are required to completely screen depolarization fields at head-to-head domain walls and canted domain walls making small angles with the c axis (Fig. 3b). Thermally ionized shallow acceptors and illumination in the intrinsic or extrinsic bands contribute to these screening fields. In general, the average modulation ΔP may be much smaller than the spontaneous polarization, which reduces the compensating charge requirements described by Eq. 9.

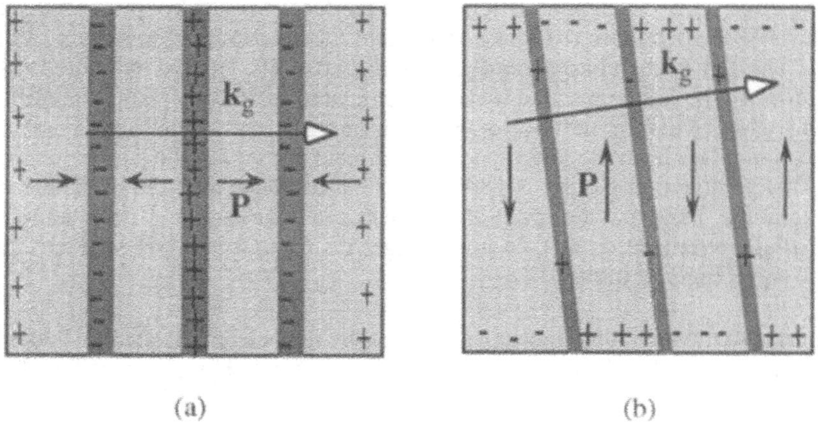

Fig. 3 (a) Head to head domain structure with a free compensating surface charge density $2\mathbf{P_S}$. (b) Canted domain structure with compensating surface charge density reduced by factor of sin θ, where θ is the angle between domain wall and the polarization $\mathbf{P_s}$. The shaded regions indicate domain walls, and the + and - indicate free compensating charge.

The divergence of the polarization at a domain wall is also reduced by a factor of sinθ for canted domain walls, where θ is the angle between the direction of polarization and the domain interface, as illustrated in Fig. 3b. Therefore, for $\theta = 0$, no charge is required at the wall for perfect bipolar modulation of the spontaneous polarization. The primary charge compensation occurs at the crystal c faces. These antiparallel walls are expected to be relatively stable.

6. PHENOMENOLOGY OF FIXING

The primary fixing techniques in ferroelectrics are based on the following mechanisms:

(i) Ionic drift to compensate electronic space-charge fields.
(ii) Periodic domain reversal induced by photorefractive space-charge fields.
(iii) Photoassisted domain switching.

The ionic fixing method is illustrated schematically in Fig. 4. In this case the buildup of the space-charge grating is described by the conventional band transport model of chapter 1. In $LiNbO_3$, for example, the photovoltaic effect is the dominant charge transport mechanism. This electronic space-charge grating (a) produces an index grating consistent with equations 1 and 3. δp is the change in polarization caused by a change in position of the transition metal ion under the influence of the space-charge field. This polarization change produces the electrooptic index grating. At elevated temperatures (b) ions compensate the electronic space-charge field by drift. The net space-charge field is diminished. The crystal is cooled (c) and subsequently the hologram is uniformly illuminated, partially erasing the electronic grating and revealing the out-of-phase ionic grating. The permanent index modulation arises from the ionic space-charge field perturbing the lattice polarization (d). This polarization grating is 180 degrees out of phase with the initial electronic index grating. Note that the space-charge field and polarization as depicted in Fig. 4 are spatially averaged on a scale much larger than a unit cell, yet smaller than the optical wavelength of interest.

In contrast to the ionic drift mechanism, fixing by ferroelectric hysteresis is based on polarization memory. The electrooptic grating exhibits hysteresis as the ions not only distort slightly, but also switch to produce inverted domains. Figure 5 illustrates an idealized ferroelectric hysteresis curve in polarization P for a voltage $V = dE$ applied across the c faces of the crystal, where d is the distance between electrodes and E is the electric field. P_s is the spontaneous or saturated polarization, P_r is the remnant polarization in the absence of an applied field, and E_c is the coercive field required to invert the polarization within half of the volume. This model serves to motivate the fixing methodology only on the most basic level. For instance, the electronic space-charge field ΔE creates a replica modulation of the local spontaneous polarization, $\Delta P_{dynamic}$. Upon optical erasure of the electronic space-charge field, a remnant polarization modulation $\Delta P_{remnant}$ persists. This is the fixed hologram. The goal of polarization fixing methods is to maximize the hysteresis. Note that the

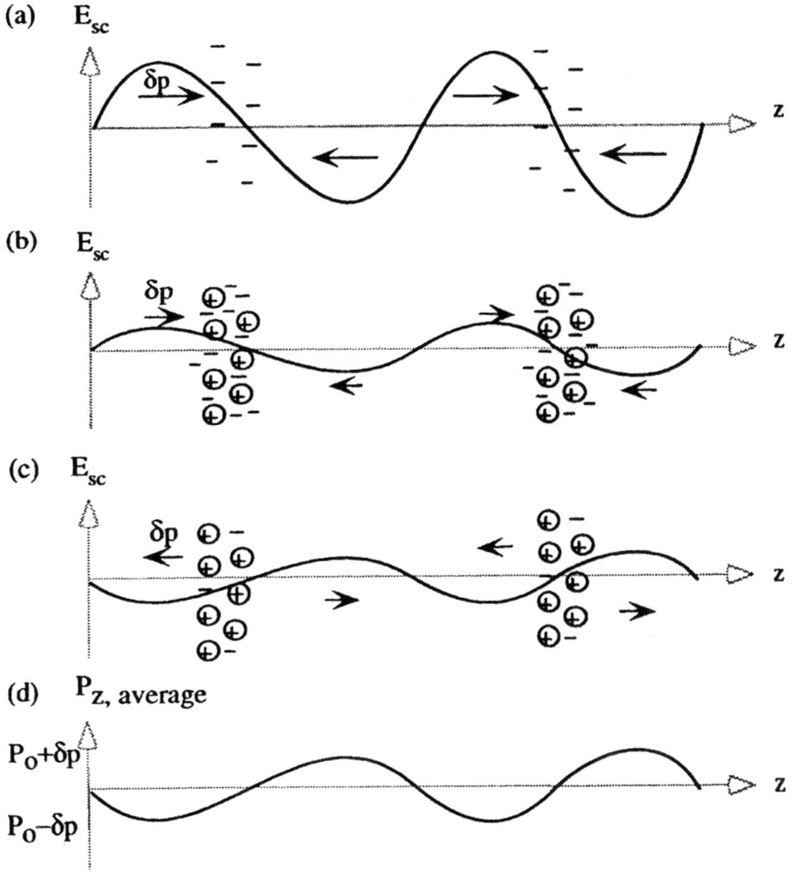

Fig. 4 Ionic drift fixing method: (a) Generation of spatially modulated electronic space-charge field, which induces index grating through the lattice distortion δp. (b) Upon heating, ions (denoted by circled +) drift to compensate electronic space-charge field. (c) After cooling, the crystal is uniformly illuminated to erase the electronic grating and to reveal 180 degree out of phase ionic grating. (d) The permanent modulation of the polarization arising from the net space-charge field (ionic - electronic) produces the index grating.

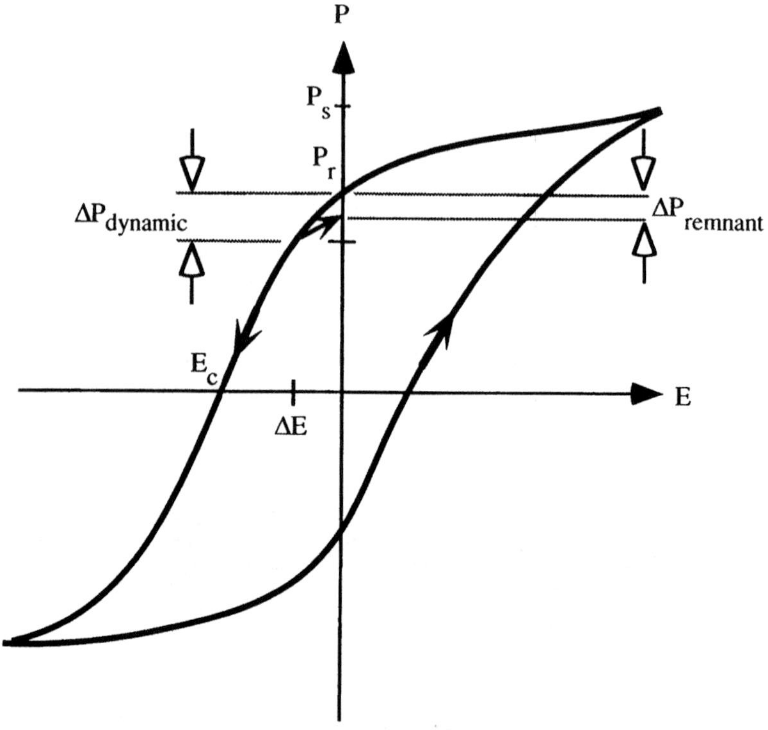

Fig. 5 Typical polarization hysteresis curve.

polarization hysteresis diagram represents a *macroscopic* polarization. However, because the space-charge field modulation has a period of microns, which is much greater than the unit cell, we apply the macroscopic picture to a small region in the crystal corresponding to a single grating period. There are two issues to keep in mind when applying this diagram. First, the scale of the hysteresis curve will depend on the size of the domain region under consideration. Second, Fig. 5 is experimentally measured with a voltage source providing effectively an infinite supply of charge. However, the photoinduced space-charge fields are a source of limited charge, and charge compensation requirements play an important role in the polarization modulation depth, as discussed in section 5 of this chapter.

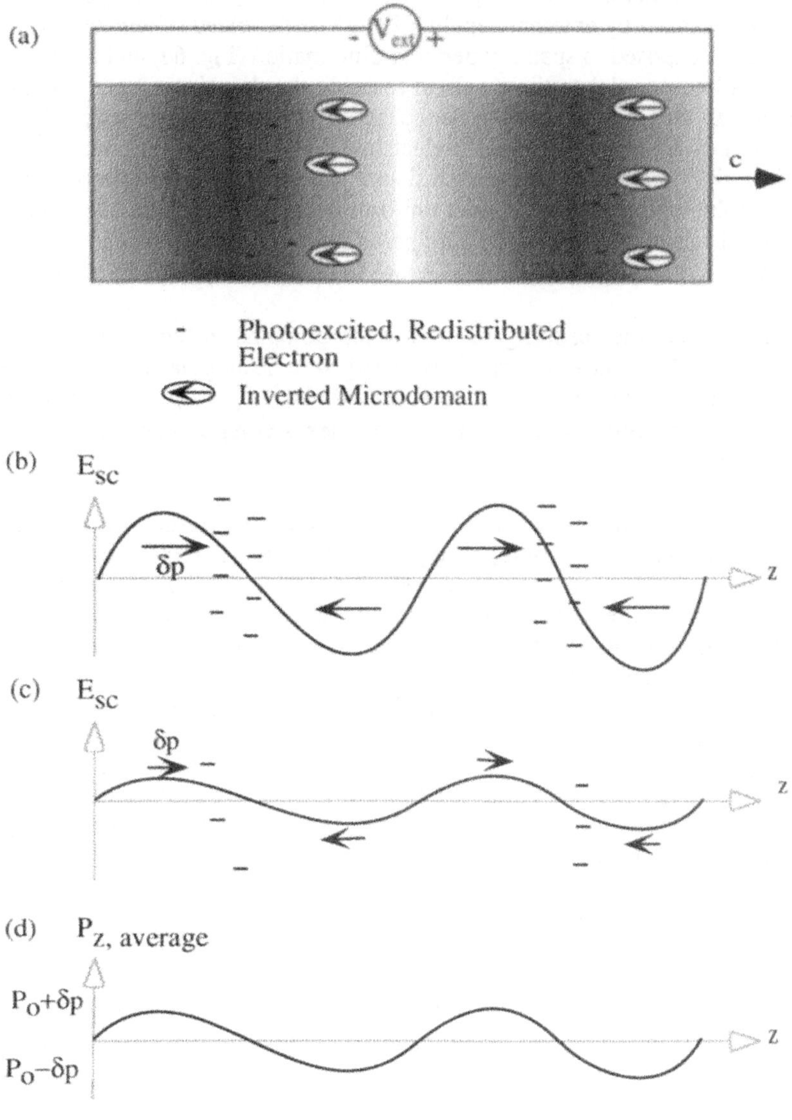

Fig. 6 Polarization fixing by photoinduced space-charge fields: (a) Experimental setup and overview (b) Generation of spatially modulated electronic space-charge field and resulting index grating caused by distortion δp. (c) Uniform illumination partially erases electronic space-charge field; however, a remnant polarization modulation δp persists due to ferroelectric hysteresis. (d) Remnant polarization modulation responsible for fixed domain grating.

The method to generate ferroelectric domain gratings by space-charge fields (and in some cases assisted by external depoling fields) is illustrated schematically in Fig. 6. The crystal is exposed to spatially periodic illumination (Fig. 6a) and a depoling field may be applied parallel to the c axis to assist in the domain reversal process. The total field can then be estimated to be the sum of the external and space-charge field (this is strictly true only in linear dielectrics). The initial electronic space-charge field (b) induces an index change consistent with a change in polarization δp, the dynamic polarization grating. Under uniform illumination (c), the electronic space-charge field is partially depleted. The polarization perturbation relaxes, but due to ferroelectric hysteresis not all the ions return to their pre-exposure locations.

Figure 6a depicts the resulting inverted microdomains in the regions of negative space-charge field, producing a spatially averaged remnant polarization modulation in Fig. 6d. The polarization modulation is biased about P_o. If perfect bipolar domain modulation were achieved, the fundamental index grating component would vanish, according to Eq. 1.

The phase relationship between the dynamic electronic grating and the remnant grating (either domain or ionic) is apparent upon reconstruction of the hologram (Fig. 7) while the dynamic grating is optically bleached. If the two gratings are in phase ($\Delta\phi = 0$), the diffraction efficiency does not pass through zero upon reconstruction, but decays to the remnant value. However, if the dynamic and remnant gratings are out of phase ($\Delta\phi = \pi$) and the dynamic grating is stronger than the remnant grating, then the diffraction efficiency will go through zero when the two gratings are equal in strength but opposite in sign, and will subsequently grow as the dynamic grating is completely extinguished. Note that this latter case corresponds to the revealing characteristic exhibited following ionic fixing, where the electronic grating is compensated by an ionic grating of opposite sign and smaller magnitude during the heating stage.

A third method to fix gratings is based on photoassisted domain switching. This effect plays an important role in the formation of ferroelectric domain gratings, most notably in configurations where the photoinduced space-charge field is perpendicular to the c axis (the direction of easy poling). In this geometry external electric fields are essential to reverse domains, because the photorefractive space-charge fields have a negligible projection along the c axis.

In regions of illumination, the field required to invert domains is reduced, due to the presence of photoexcited charge, which screens the ferroelectric dipole interaction and provides charge to screen the depolarization fields generated upon domain inversion. This lowers the local coercive field in the illuminated regions. Domain reversal is then expected in the illuminated regions, so the polarization modulation is exactly out of phase with the intensity pattern. This is illustrated schematically in Fig. 8.

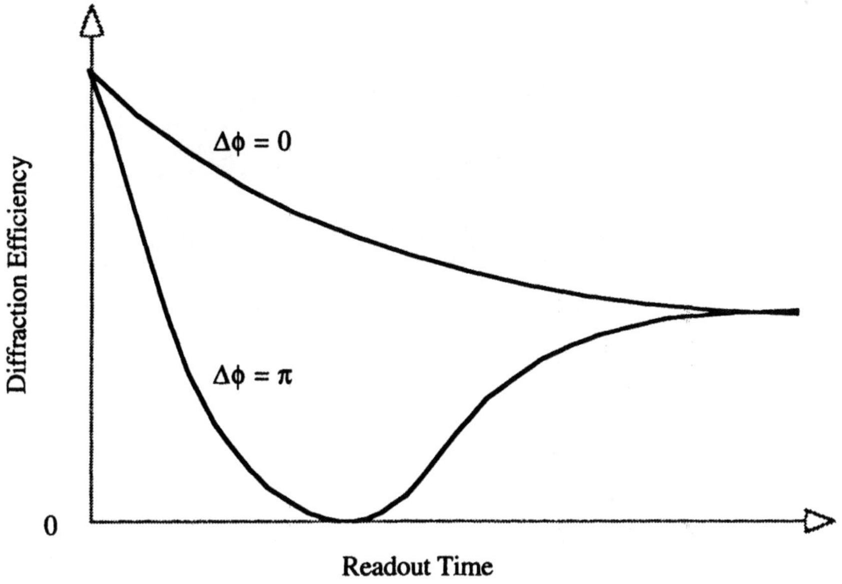

Fig. 7 Revealing characteristics, depending on relative phase, $\Delta\phi$, of electronic and fixed index gratings.

Several experimental observations of these three fixing mechanisms have been reported. The experimental work on thermal fixing by ionic drift will be discussed first, followed by a description of ferroelectric hysteresis fixing.

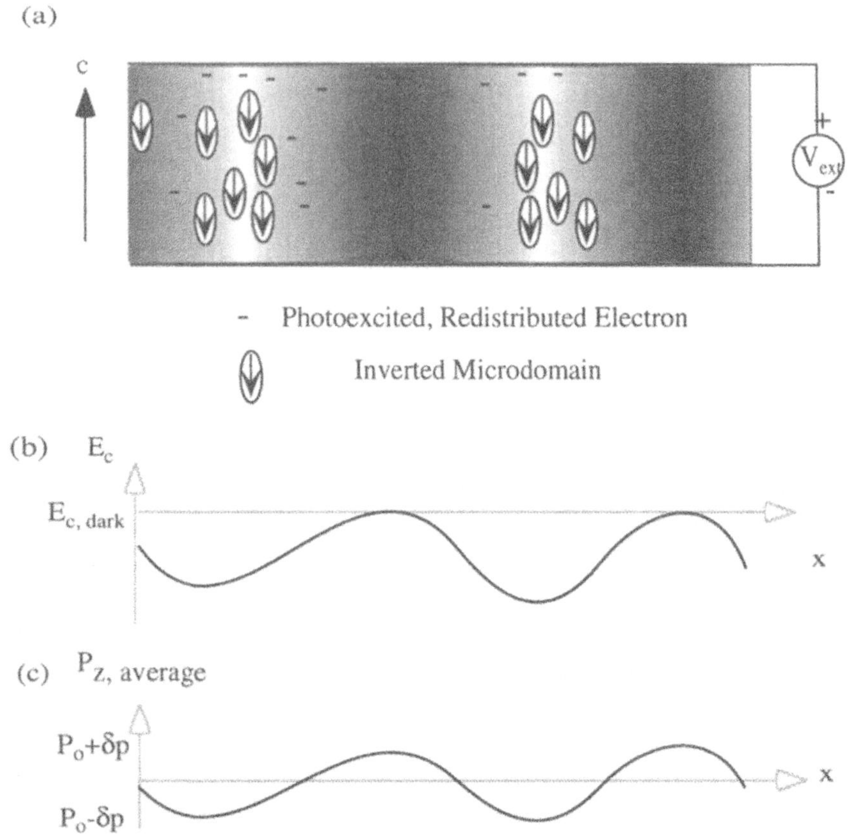

Fig. 8 Polarization fixing by photoassisted domain reversal. (a) Experimental setup and overview (b) Light intensity pattern modulates the crystal coercive field. (c) An external depoling electric field reduces the spontaneous polarization more significantly along grating planes with low coercive field. As in Fig. 6d, this remnant polarization modulation produces a fixed domain grating.

7. THERMALLY ASSISTED IONIC DRIFT

Fixing by ionic drift was first demonstrated in $LiNbO_3$ in 1971 by Amodei and Staebler [30]. They discovered that by heating the crystal to 100-200 °C either during or after the hologram exposure and by subsequently cooling, a permanent grating 180 degrees out of phase with the electronic grating was generated. They proposed that at these elevated temperatures, the electronic space-charge field generated by optically excitable ions causes optically inactive ions or vacancies to drift under the force of the local space-charge grating. This compensates the local fields, as in Fig. 3. The electronic grating remains trapped in the absence of light and at the elevated temperatures because the recombination centers are sufficiently deep (>eV). When the crystal is cooled to room temperature in the dark, the ionic grating is rendered immobile. The electronic grating is subsequently bleached because of photovoltaic charge transport, causing the diffraction efficiency to vanish when the electronic and ionic grating are equal in magnitude and opposite in sign (Fig. 6, $\Delta\phi = \pi$). Further erasure of the electronic grating reveals the fixed ionic grating.

Thermal fixing has since been demonstrated in $LiNbO_3$ of various doping combinations, such as Cu, Fe, Mn, Si (see, for example, [9] and references therein) as well as $KNbO_3$ [31], $LiTaO_3$:Fe [32] and $Ba_2NaNb_5O_{15}$:(Fe,Mo) [33]. Similar thermal fixing results under an external ac electric field have been reported in the non-ferroelectric $Bi_{12}TiO_{20}$ [34]. Several studies have followed Amodei and Staebler's original work to determine the ionic species responsible for the fixed grating. It is believed that the drift of protons are primarily responsible for fixing in $LiNbO_3$ [35]. The 1.1 eV activation energy of OH^- ions [36] agrees with those values (1.1 eV [37] to 0.93 eV [9]) determined from fixed hologram lifetimes, based on an Arrhenius type thermal activation (described by Eq. 14). The role of protons is further supported by the change in the OH^- infrared absorption band in thermally fixed regions [38]. This ion is believed to enter the crystal lattice during crystal growth [35]. Subsequent proton migration induced by the space-charge field takes place at about 140 °C, well below the temperature at which electron modulation in deep traps is thermally degraded ($\geq$ 180 °C) [35]. At room temperature, the probability of ion redistribution by an Arrhenius type activation process is exceeding low (10^{-17}). The dark storage time is estimated to be in excess of 100 years [38]. A comprehensive discussion of experimental results related to thermal fixing in $LiNbO_3$ may be found in Reference [9].

Fixed diffraction efficiencies as large as 32% have been obtained by recording the electronic grating at a temperature at which ions are mobile [40]. This allows mobile protons to continuously neutralize the electronic space-charge field. Because the recording beams remain on, the electronic grating continues to grow until it reaches the limiting space charge value. This technique is effective at grating periods larger than the Debye screening length (e. g. in the diffusion-limited regime of the photorefractive band transport model). A large enhancement in fixing efficiency is produced by short circuiting the crystal while recording, then developing the grating under open circuit conditions. The largest fixed diffraction efficiency reported in $LiNbO_3$ is 98% for a reflection spectral filter at 1.54782 μm [41].

8. FIXING BY FERROELECTRIC POLARIZATION REVERSAL

Evidence for polarization fixing can be traced back to the early work on optical damage, a phenomenon now synonymous with the photorefractive effect yet actually encompassing a variety of photoferroelectric phenomena. For instance, antiparallel domains at the periphery of a focused laser beam were reported in $BaTiO_3$ [42] following illumination of an initially poled crystal. Since then, Barkhausen current transients, which indicated domain switching [19] have been observed in $LiNbO_3$ under external fields of 10^4 V cm^{-1} at room temperature and during hologram recording with high-intensity beams [43]. Because of the strong photovoltaic charge transport mechanism in $LiNbO_3$, the photorefractive fields may be as large as 10^4 to 10^5 V cm^{-1}[10]. For this reason it has been proposed that local ferroelectric switching [43] may also contribute to fixing. To support this claim, the optical generation of needle-shaped 180 degree domains in $LiNbO_3$:Fe has been experimentally reported following illumination with a single beam at 440 nm, at room temperature [44, 45]. Distortion of the crystal structure in the region of illumination has also been observed in $LiNbO_3$:Fe by x ray diffraction [46]. Following optical erasure of the space-charge field, the strain of 0.03% disappears. Apparently, this strain reflects the coupling of the space-charge field to the lattice constant through the piezoelectric effect.

These experimental observations suggest that the interaction of light and domains is indeed a common phenomenon shared by many photoferroelectrics. To date, however, efficient ferroelectric domain gratings have been reported only in La doped lead zirconate-titanate (PLZT), SBN and $BaTiO_3$. The results may be classified into the following representative classes: (1) cooling through a ferroelectric phase transition, (2) electrical fixing, (3) high-intensity fixing and (4) short-exposure fixing. We describe these techniques in this order.

8.1 Fixing by Cooling Through a Phase Transition

Micheron and Trotier [46] reported the formation of domain replicas of electronic volume holograms in SBN:75 single crystals while cooling through the paraelectric-ferroelectric phase transition (T_c = 45 °C in their sample). They proposed that in the polar phase, local ionic displacements cancel the photoinduced electronic space-charge field which had been recorded in the high-temperature phase. Upon cooling, the orientation of the nucleating domains is influenced by the light-induced space-charge field. They obtained similar fixing results upon illumination with very high optical energy densities (100 W/cm^2), apparently because of optical heating and photoassisted domain switching.

8.2 Electrical Fixing

The observations of electrical fixing fall into two distinct categories, depending on whether the grating vector is nearly parallel or perpendicular to the c axis, as depicted in Figs. 3a and 3b, respectively. When the grating vector has a large projection along the c axis, the photoinduced space-charge field and external field work in tandem to invert domains (Fig. 6). However, if the grating vector is nearly normal to the c axis, the space-charge field plays only an indirect role in domain reversal; the presence of an external depoling field and photoassisted domain reversal (Fig. 8) is essential. This typically produces an antiparallel domain grating if the only allowed polarization directions are ±180 degrees. We will denote the two fixing mechanisms as space charge induced or photoassisted and point out the different origins of these effects. Note that from aspect ratio considerations of typical ellipsoidal domains, an antiparallel domain grating will also have a large number of head-to-head interfaces, assuming the length-to-width ratios of inverted domains (along the c axis) is less than about 100. Therefore, Fig. 3b represents an unrealistic idealization of the domain grating microstructure, for the individual domains do not extend across the entire thickness of the crystal (typically mm).

8.2.1 Space-Charge Induced Domain Switching

Electrical fixing of volume holograms in both SBN:75 and $BaTiO_3$ in the head-to-head domain configuration (Fig. 3a) has been demonstrated during the early work on volume holographic memories [48, 49, 50] and has been confirmed by more recent work [51, 52, 53]. In principle, this technique has the desirable characteristic that the crystal does not need to be heated as in thermal fixing techniques and the fixing can be achieved in a relatively short time (1 second) with low intensity optical beams. The electrical field fixing threshold is nearly equal to the coercive field E_c = - 970 V cm^{-1} (negative values represent depoling fields) and the minimum fixing time is the current-limited polarization switching time [48],

$$t_{min} = 2\, Q_r/i_{max} \tag{11}$$

where Q_r = A P_r, A is the area of the electrode, P_r is the remnant polarization, and i_{min} is the maximum current supplied by the high-voltage source. For their experiments t_{max} was 10 msec. A fixed diffraction efficiency of 52% is achieved with a fixing pulse of 1.25 kV cm^{-1} applied for 0.5 seconds. This domain grating is stable for more than 10 hours during reconstruction with a 1 W cm^{-2} reference beam. The domain pattern is unstable when the fixing voltage is too large compared to the coercive voltage or when the fixing pulse is applied for a time much longer than t_{min}, because significant depolarization of the entire crystal bleaches out the grating modulation. The fixed hologram can be erased by applying a field much larger than the coercive field.

In these experiments the grating vector is oriented nearly parallel to the c axis, so the space-charge field is also parallel to the c axis. The applied depoling field can then bias the space-charge field closer to the coercive field, increasing the ferroelectric hysteresis. As apparent from the hysteresis diagram (Fig. 5), the slope dP/dE is a maximum near the coercive field, so significant polarization modulation is expected in this region. In addition to field-induced domain switching, some fraction of domain reversal may also result from photoassisted domain switching.

8.2.2 Photoassisted Domain Switching

Electrical fixing by photoassisted domain switching was first demonstrated in photoconductive PLZT ceramics by Smith and Land [54]. They found that uniform illumination assisted both the poling and depoling of the sample (the poling field was applied to transparent conducting electrodes). The coercive field and hence the polarization was reduced in the illuminated regions. A photograph was stored in the poled plate by imaging it onto the ceramic while a depoling electric field was simultaneously applied. The regions of maximum illumination were completely depoled following the exposure. The experimental geometry is similar to that depicted in Fig. 8. The depoled regions produce transmittance minima because the formation of 71° and 109° domain walls enhances the light scattering, creating a negative of the original image.

The photoassisted domain switching effect has also been observed in SBN:60 by Kahmann et. al. [17]. In agreement with Smith and Land [54], they observed depolarization in the illuminated regions following the application of a negative fixing pulse. Because antiparallel domain walls are relatively narrow in SBN, they do not scatter light as strongly as PLZT. The grating is instead revealed by beam coupling topography [55]. Similar observations of polarization gratings in SBN:75 appear to be related to this effect [56, 57]. However, in this latter work, domain reversal is believed to occur preferentially at the intensity minima of the optical interference pattern.

8.3 High Intensity Fixing

The standard treatment of the steady state photorefractive effect is based on the often unrealistic single band approximation. A central conclusion of this model is that the steady state space-charge field is intensity independent. However in practice, when recording holograms with high intensities in the region of strong optical absorption, significant local heating occurs (2-5 °C for 10 W cm^{-2}) and multiple trap levels may be populated. Heating generates bias pyroelectric fields which can enhance the drift transport mechanism and increase the photorefractive space-charge fields, analogous to the enhancement produced by external electric fields [58]. In materials such as SBN with a ferroelectric phase transition near room temperature, illumination decreases the polarization viscosity so it is easier to modulate the polarization. In addition, intensity dependent photoferrelectric phenomena may play an increasingly important role (i. e. the photodomain effect). For instance, Thaxter and Kestigian [50] observed that hologram lifetimes increased with exposure to a typical value of several minutes. Kewitsch et. al. [1] reported similar results displaying an exposure threshold for fixing of 100 J cm^{-2}. The fixed hologram displayed an effectively infinite lifetime. This exposure threshold is primarily attributed to local heating in the region of intense illumination.

The occurence of domain switching is apparent from the noise in diffraction efficiency and displacement current across the crystal as a hologram is recorded. Figure 9 is a standard experimental setup for writing transmission gratings. The time dynamics of the hologram diffraction efficiency, as monitored by a Bragg matched HeNe laser, for different read and write intensities are depicted in Figs. 10a and b. For gratings written with intensities of 1 W for more than 100 sec, a significant fixed grating persists upon readout (Fig. 10b). The chaotic fluctuations in the diffraction efficiency during exposure is a signature of the domain reversal process and is called optical Barkhausen noise [59]. This noise arises primarily from the current transients generated by domain switching. During the first several minutes of high-intensity exposure, significant local heating occurs, producing non-uniform pyroelectric fields that depolarize the crystal. As domains reverse, they produce a depolarization current proportional to dP/dt. This time varying current noise introduces noise in the dynamic energy transfer in the two wave mixing process. Indeed, the noise in diffraction efficiency is strongly correlated in time to the noise in current across the crystal (Fig. 11). The cross-correlation function is given in Ref. [59]. The noise subsides as a steady state temperature is reached and as fringe stability is maintained. However, a weaker Barkhausen noise in both current and diffraction efficiency continues for the duration of the writing process, producing a significant depolarization of the crystal. The amplitude of the current fluctuations is proportional to the photorefractive space-charge field.

The domain reversal caused by optical heating is revealed by etching the "a" face of a domain-fixed SBN:75 crystal with HCl for 15 minutes. Upon examination under a polarizing optical microscope, a distinct etch pattern corresponding to the 1 mm beam diameter is revealed (Fig. 12) [59]. Circular rings are apparent on the periphery of the illuminated region. These rings are similar in form to the space-charge field patterns observed by Clarke et. al. [58]. The etch patterns confirm that domains are reversed in the vicinity of intense illumination.

In addition to the random depolarization that occurs in the illuminated region of the crystal, the photoinduced space-charge field inverts domains periodically to generate a ferroelectric domain grating. The role of the space-charge field in domain reversal can be isolated experimentally by first illuminating the crystal with an extraordinary and an ordinary beam. The presence of this intense illumination heats the crystal and inverts domains. Once the crystal reaches thermal steady state, the thermal Barkhausen noise subsides. Because the writing beams are orthogonally polarized, the do not produce an optical standing wave or a photorefractive space-charge field. Next, the electric field vector of one of the beams is rotated so that both beams are ordinarily polarized. Photorefractive space charge is then generated, and the Barkhausen current noise is simultaneously monitored.

Figure 13 illustrates the growth of diffraction efficiency (upper trace) upon the hologram exposure and the subsequent (0.6 sec latter) Barkhausen current events (lower trace). The magnitude of the space-charge field is ~ 50 V/cm, as estimated from the peak measured diffraction efficiency. The domain grating follows the photoinduced space-charge field, as indicated by the rapid occurrence of noise on the time scale of the photorefractive buildup time. This domain grating is *dynamic*, in contrast to the more typical *remnant* domain grating which is used to permanently fix domain holograms. The dynamic domain grating relaxes immediately upon the removal of the space-charge field. Ferroelectric hysteresis provides memory so that a fraction of the dynamic domain grating is retained as a remnant grating (Fig. 5). The resulting periodic modulation of the polarization is revealed microphotometrically [9] in Fig. 14. Figure 14a illustrates the initially monodomain state following electrical poling. A polarization grating is recorded with a period of 10 microns and revealed in Fig. 14b. This grating may be subsequently erased as in Fig. 14c by applying large electric fields, leaving a random multidomain crystal. A domain grating written with an exposure greater than 100 J cm^{-2} is difficult to erase electrically. In this case, the grating can be erased by heating the crystal above 350 ^{o}C to thermalize the domain structure, and unlock the energetically stable antiparallel domains.

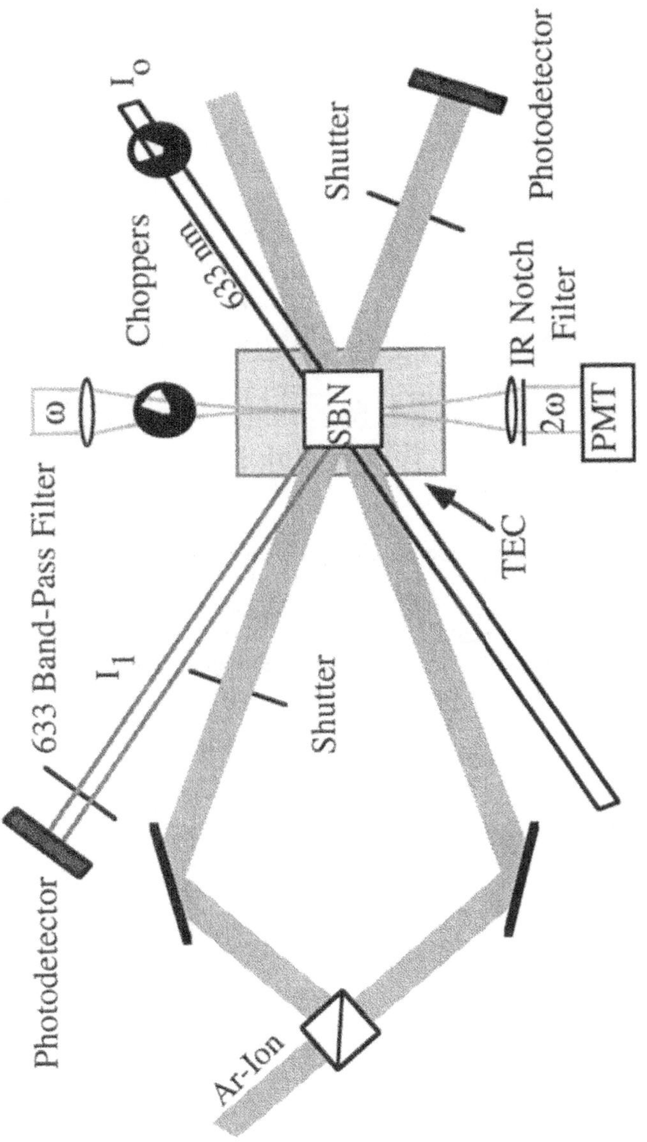

Fig. 9 Experimental setup for recording domain gratings and monitoring grating diffraction efficiency and quasi-phase matched second harmonic generation. Source: Reference [2, 61].

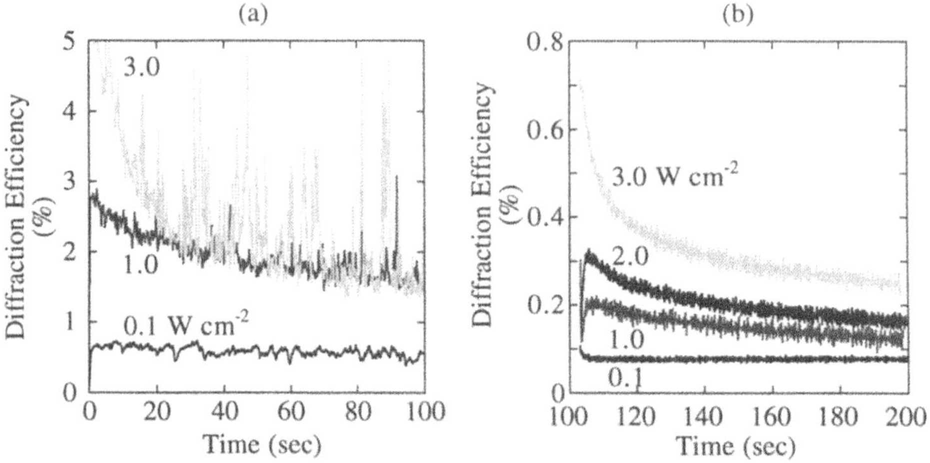

Fig. 10 (a) Diffraction efficiency during writing process for various total optical intensities of hologram writing beams. The noise at high intensities is caused by optical Barkhausen pulses. (b) Diffraction efficiency during readout process, with curves labeled according to writing beam intensity as in (a). Note that the magnitude of remnant polarization grating increases with exposure energy.

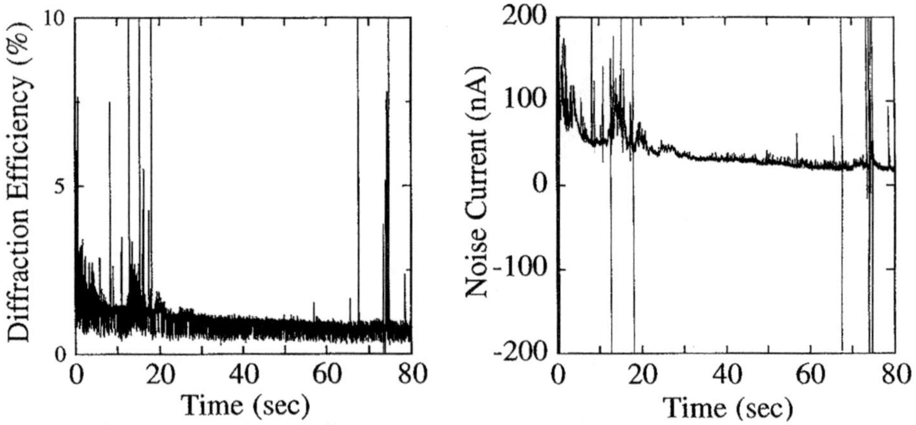

Fig. 11 Temporal correlation of Barkhausen noise in current and diffraction efficiency upon recording domain grating. Source: Reference [59].

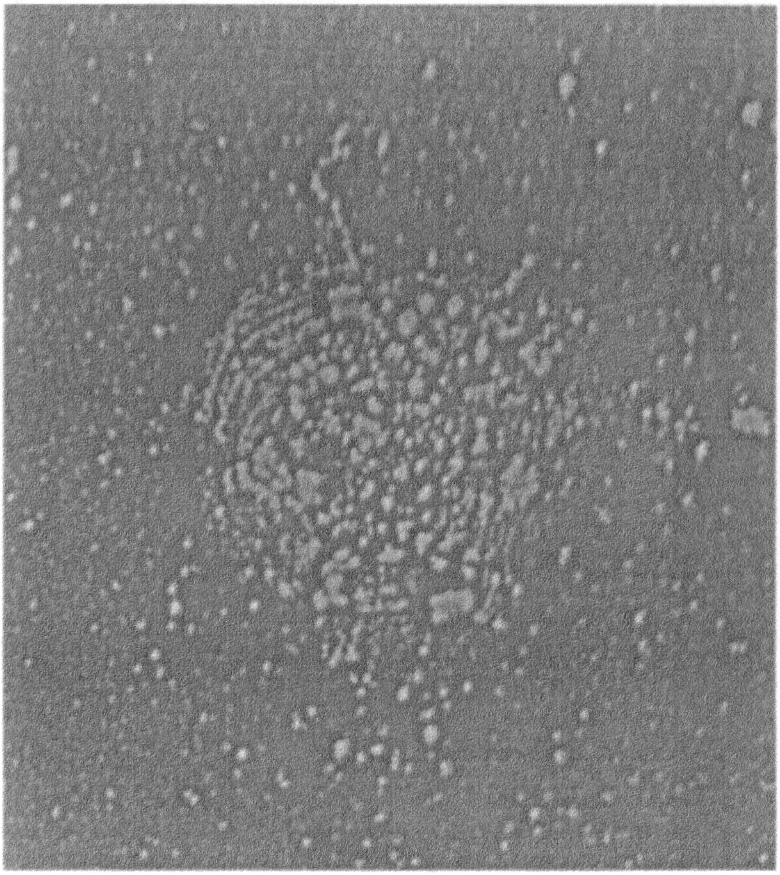

Fig. 12 Etched "a" face of SBN:75 crystal following optical illumination, as revealed by optical microscope. Source: Reference [59].

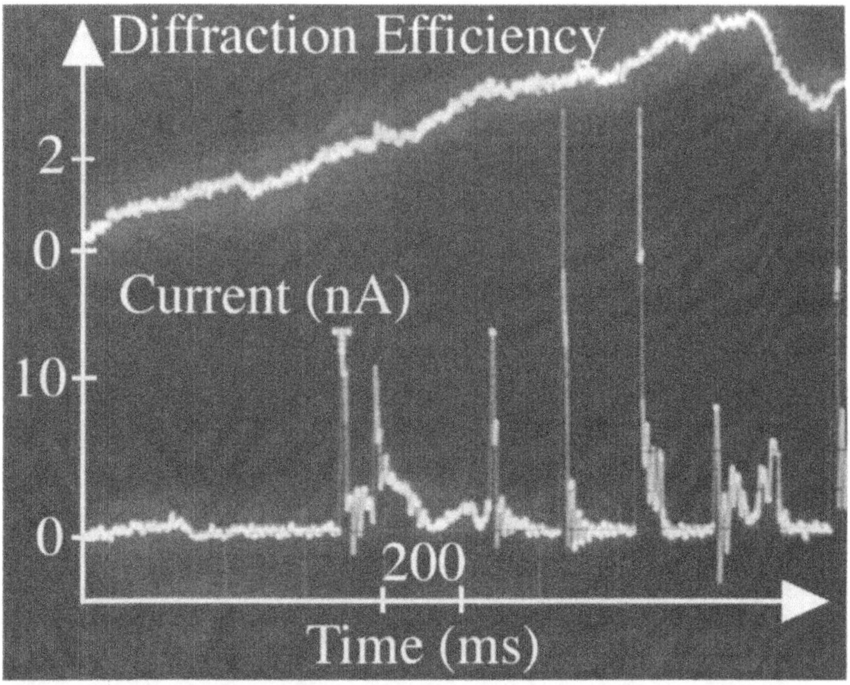

Fig. 13 Oscilloscope trace of diffraction efficiency in % (top) and Barkhausen current noise (bottom) during the first 2 seconds upon recording a ferroelectric domain grating in SBN:75. The rapid occurrence of domain switching indicates the presence of a dynamic domain grating. Source: Reference [59].

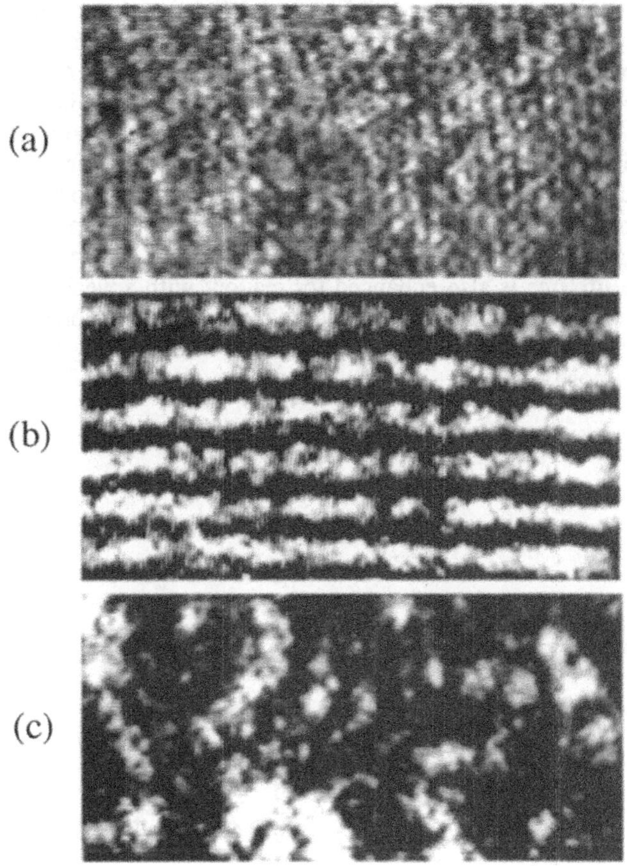

Fig. 14 Microphotometrically revealed domain grating with period of 10 microns in SBN:75. (a) poled crystal (b) domain grating (c) electrically erased grating. Source: Reference [61].

The experimental study of reference [1] indicates that the fixed diffraction efficiency during recording with high-intensity beams (> 1 W cm^{-2}) increases for more than 8 hours upon continuous writing. Figure 15a depicts the first 2 hours of recording. The monotonic increase in fixed diffraction efficiency occurs because domains continuously orient to produce a field in opposition to the electronic space-charge field, allowing the electronic grating to grow further until it reaches the limiting space charge value. A typical fixed diffraction efficiency is 3%. Figure 15b illustrates the long term decay of a fixed hologram written at room temperature with a total intensity of 4 W cm^{-2} for 1 hour. Extrapolating this curve, the hologram is expected to survive at least several hundred days during continuous intense readout at 25 °C. By cooling the crystal to below 20 °C, the ionic distortion associated with the domain grating is permanently frozen-in, further enhancing the hologram lifetime. The lifetime is also enhanced for readout with a weak plane wave. This minimizes undesirable domain reversal caused by the accumulation of space charge on the periphery of a focused reference beam and local heating which reduces the polarization viscosity.

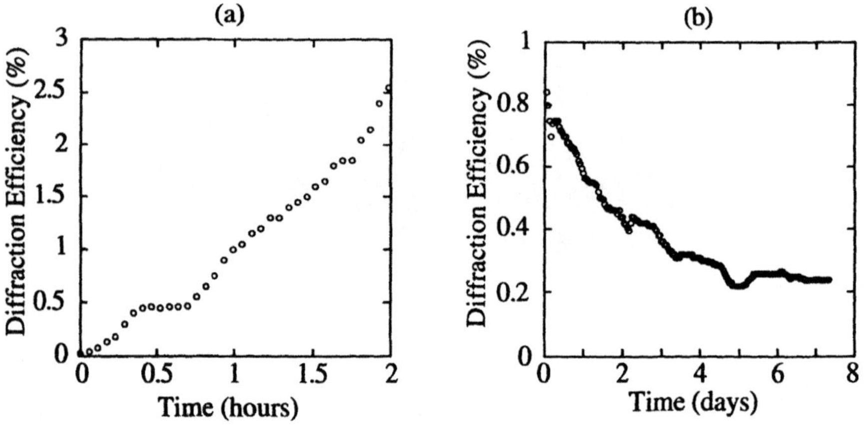

Fig. 15 (a) Long term hologram recording, illustrating a monotonic increase of remnant diffraction efficiency. (b) Decay of remnant hologram diffraction efficiency at 25 °C during continuous reconstruction with intense illumination (1 W cm^{-2}, 488 nm). Source: Reference [1].

These high-intensity recordings are performed in initially poled crystals. Following a continuous recording process of two days, the macroscopic spontaneous polarization as measured from the electrooptic coefficient decreased by a factor of 10. This severe depolarization requires optical exposures on the order of 10^5 J cm^{-2}.

Domain gratings may also be optically recorded in the canted domain wall configuration of Fig. 3b using an experimental setup similar to Fig. 9. This is demonstrated by recording a domain grating at an angle of $\theta = 7$ degrees to the c axis. The charge compensation requirements at the domain walls are then reduced by a factor of $\sin\theta$, from Eq. 9. The total diffraction efficiency while writing the grating is illustrated in Fig. 16. Curve (a) illustrates the diffraction efficiency while the first grating was written. Following this recording, the half wave voltage, which is inversely proportional to the effective linear electrooptic coefficient r_c, increased by a factor of 3 in the region where the hologram was recorded. Thus, the macroscopic polarization in the region of the hologram has decreased by 3. A second hologram was subsequently recorded in the same volume and the diffraction efficiency (b) dramatically decreased because of the depolarization of the crystal. In this canted domain grating configuration, with $\mathbf{k}_g$ nearly perpendicular to c, the crystal rapidly depolarizes in the region of illumination. This rapid depolarization is indeed expected because of the reduced compensating charge requirements for canted domain gratings. By re-poling the crystal with 3 kV cm^{-1} for approximately 1 minute the crystal is refreshed, and a third hologram, recorded in the same volume, displays the same diffraction efficiency as curve (a). In this geometry the space-charge field is nearly perpendicular to the c axis, so the photorefractive space-charge fields do not play a central role in domain reversal. Thus, these gratings are produced by a combination of optical heating and photoassisted domain reversal.

Therefore, several phenomena can play a role in high-intensity fixing. Optical heating, photoassisted domain switching and the photodomain effect all contribute to domain reversal. A significant fraction of the domain reversal is spatially random, in addition to the periodic modulation at grating periodicity K_g. After long exposures, this inevitably leads to a reduction in the spatially averaged linear electro-optic coefficient and a degradation in diffraction efficiency.

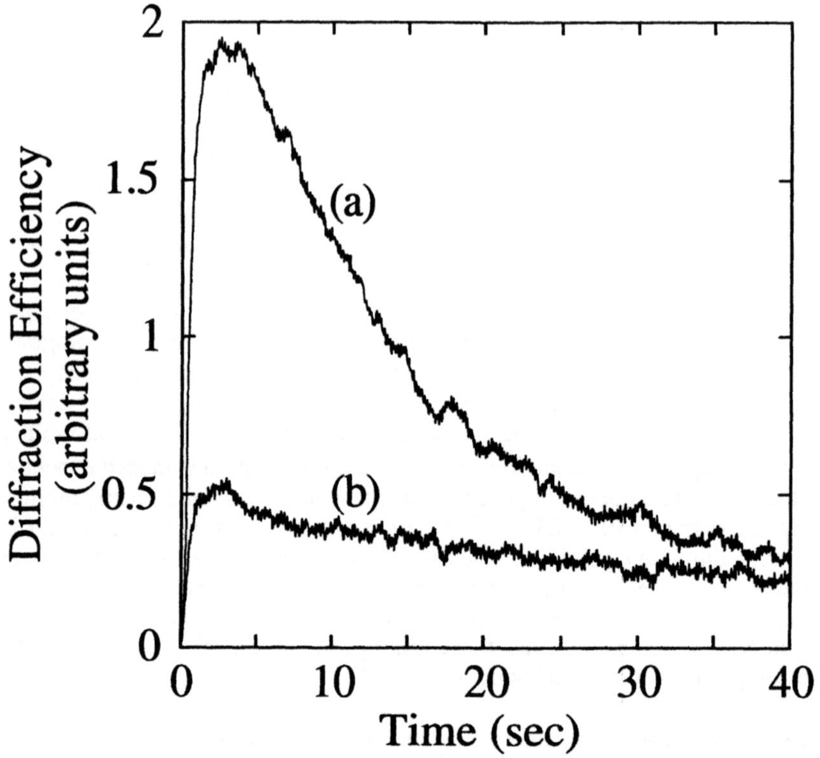

Fig. 16 Diffraction efficiency of canted domain grating. (a) is the first exposure, and (b) is the second exposure with a phase-shifted interference pattern in the same volume as (a). The reduction in diffraction efficiency between (a) and (b) arises from the reduction of r_c, the effective electrooptic coefficient, by a factor of 3.

8.4 Short Exposure Fixing

While high intensities enhance the fixing efficiency, low intensities and short optical exposures can also produce fixed holograms if the recording stage is followed by a dark development stage which increases ferroelectric hysteresis. Note that the polarization modulation increases as the product of magnitude and duration of the electric field, until polarization saturation is reached. This concept was applied to increase the polarization modulation caused by photoinduced space-charge fields [59]. The amplitude of the spatially periodic electric field can be estimated from the measured diffraction efficiency of the grating during Bragg-matched reconstruction. In the symmetric transmission grating configuration with the grating vector $\mathbf{k_g}$ parallel to the c axis, the amplitude of the net space-charge field, for small diffraction efficiencies and small half angle of the writing beams, is [58]:

$$E_{SC} = \frac{2\lambda e^{\alpha L/2}\sqrt{\eta}}{\pi L n_o^3 r_{33}} , \tag{12}$$

where α is the absorption at the reconstruction wavelength λ in vacuum, L is the interaction length between the readout beam and the volume grating, n_o is the unperturbed index of refraction of the medium, r_{33} is the dominant electrooptic coefficient for SBN:75 and η is the diffraction efficiency at the Bragg peak, defined as I_1/I_o, where I_1 and I_o are the intensities of the diffracted and incident He-Ne beams, respectively.

The remnant polarization modulation depends on the magnitude of the space-charge field. To isolate the role of this field, a short exposure technique is performed as follows: an arbitrary periodic space-charge field is generated in the crystal by low intensity optical beams for typically a second. The exposure is followed by a dark development stage of typically minutes, during which time the domains stabilize in their reversed orientation. The dark development stage is effective because the space-charge fields persist in the dark due to the low dark conductivity. Following a dark development stage of 10 minutes, the remnant polarization modulation exhibits a linear dependence on the initial field modulation (Fig. 17a). The polarization modulation is expressed in terms of an equivalent bound charge field to permit a quantitative comparison to the photorefractive space-charge field. The fixing efficiency, defined as the ratio of the remnant bound charge field to the initial space-charge field, is typically 1%. These light induced domain patterns are produced in the absence of external electric fields or thermal effects.

The hysteresis is further enhanced by increasing the time duration of the dark development stage, or by applying an external depoling field during the development. For instance, a 1 second depoling field pulse of 3 kV cm^{-1} applied after the optical exposure increases the fixed diffraction efficiency of the domain grating by a factor of two. Because the optical exposure is very short and the electric field is applied in the dark, photoassisted domain reversal (Fig. 8) does not play a major role. Also, longer dark development times increase the diffraction efficiency of fixed holograms by more than a factor of 100, as illustrated in Fig. 17b. For this set of experimental conditions (grating period = 4 μm and T = 25 °C) the fixed grating buildup time is approximately 50 minutes. The polarization modulation, estimated from Fig. 17b by the relation $\Delta P = \varepsilon\, \Delta E$, is approximately $10^{-4}\, P_s$. The induced polarization change increases in proportion to the product of the depoling space-charge field and time duration of the field (until the onset of saturation). These findings are consistent with the fixing mechanism described in Fig. 6.

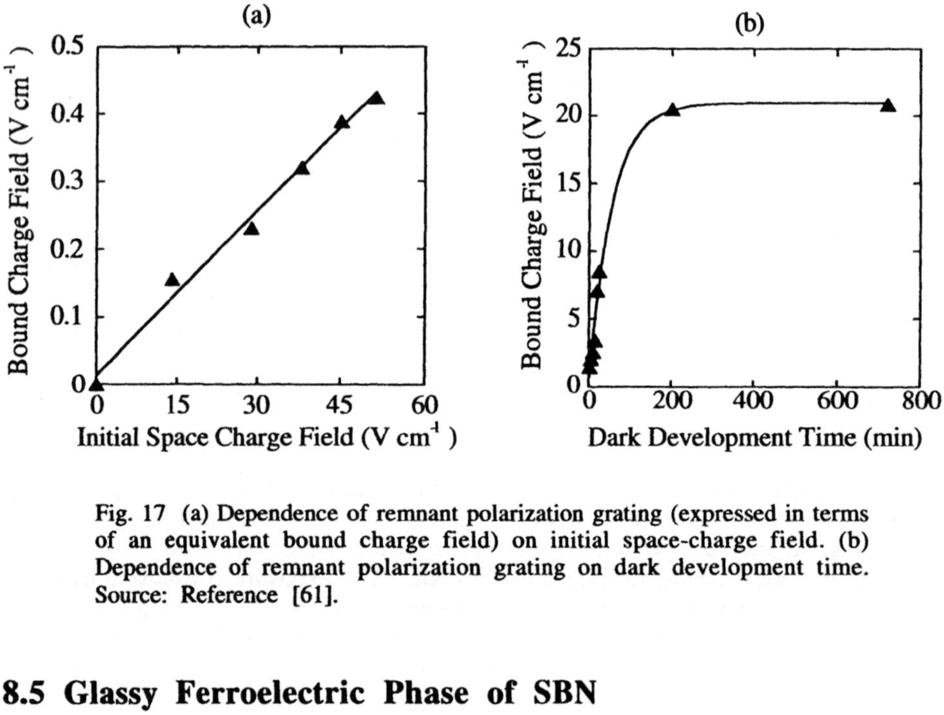

Fig. 17 (a) Dependence of remnant polarization grating (expressed in terms of an equivalent bound charge field) on initial space-charge field. (b) Dependence of remnant polarization grating on dark development time. Source: Reference [61].

8.5 Glassy Ferroelectric Phase of SBN

SBN possesses glassy ferroelectric properties that play a central role in the generation of optically induced domain gratings. SBN:75 is a relaxor [14, 15] ferroelectric characterized by a diffuse phase transition and a low coercive field . In the ferroelectric phase, the Nb^{5+}, Sr^{2+} and Ba^{2+} ions occupy one of two stable positions along the c axis (Fig. 1). In the paraelectric phase the ions are centrally located between the two stable positions [62]. The energy barrier separating the two equivalent polarization orientations scales with the volume of the microscopic domain, so a typical microdomain of characteristic length on the order of 10 nm may thermally fluctuate between the polarization states allowed by crystalline symmetry, analogous to a superparaelectric [63]. These thermal fluctuations are believed to freeze out as competing interactions (ferroelectric and antiferroelectric ordering) frustrate the fluctuations, leading to an enhanced polarization grating lifetime.

The glassy nature of SBN has important implications for the lifetime of domain holograms. Although it is conventional to apply Gibb's free energy equilibrium arguments to evaluate the stability of domains, the stable glassy domain configuration may not reach the equilibrium configuration within a finite time. A nonequilibrium domain grating may be stable below the freezing temperature because the glassy phase is nonergodic. At these low temperatures the viscosity of the dipoles increases dramatically and freezes in the polarization modulation. Figure 18a illustrates the thermally activated erasure of the domain grating at 35 °C. The temporal response follows a $t^{-\beta}$ law, characteristic of the decay of the remnant magnetization in spin glasses [64]. β is a measure of the polarization viscosity. The decay of the polarization grating in these experiments [61] was monitored for 10^4 to 7×10^5 seconds. The lifetime is arbitrarily defined as the time during which the polarization modulation amplitude decays to 10% of its value at t = 1 second, to avoid the unphysical singularity of the power law at t = 0. The lifetime of the grating at 20 °C extends well beyond measurable times, yet the onset of freezing is apparent from the absence of grating decay after 4000 seconds. By fitting a $t^{-\beta}$ curve, the lifetime was estimated to be 10^{15} seconds at 20 °C. The temperature dependence of the hologram lifetime is summarized in Fig. 18b, with the extrapolated data point indicated by a circle. The functional fit (i) is the Arrhenius law for the characteristic time τ for thermal excitation out of a potential well of depth E_a:

$$\frac{1}{\tau}=\frac{1}{\tau_o}\exp\left(\frac{-E_a}{k_B T}\right) . \tag{13}$$

For a system that exhibits a freezing temperature T_f, the Vogel-Fulcher law [65, 66] is a more accurate empirical relationship between the decay time and the temperature:

$$\frac{1}{\tau}=\frac{1}{\tau_o}\exp\left(\frac{-E_a'}{k_B(T - T_f)}\right) . \tag{14}$$

In a system undergoing freezing, the activation energy $E_a = E_a' \, T/(T - T_f)$ diverges at the freezing temperature. The freezing behavior is apparent in this SBN:75 sample at temperatures below 30 °C, where the Arrhenius fit is no longer appropriate. Fit (ii) in Fig. 18b illustrates the Vogel-Fulcher fit to the hologram lifetime data, with T_f = 19 °C, E_a' = 5 meV and τ_o = 1 sec^{-1}, and displays excellent agreement with the data. This indicates that polarization gratings can be permanently frozen-in.

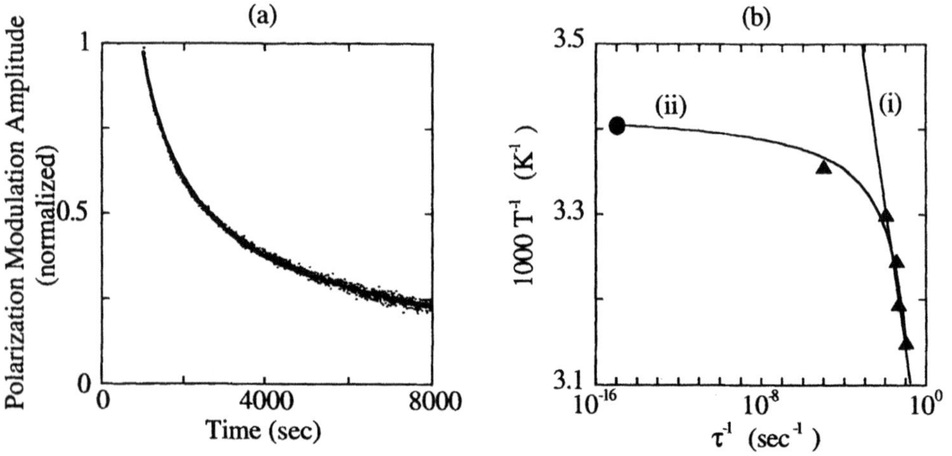

Fig. 18 (a) Power law decay of remnant polarization modulation amplitude at 35 °C. (b) Lifetime of remnant polarization modulation amplitude for different temperatures. Curve fit (i) describes an Arrhenius and (ii) a Vogel-Fulcher thermal activation process. Source: Reference [61].

The magnitude of the remnant polarization modulation exhibits a temperature resonance for gratings recorded using a short exposure at 30 to 37 °C (Fig. 19) and subsequently cooled to room temperature during a dark development stage of 15 minutes. The dynamic grating response shows a similar temperature dependence. The weak response at low temperatures is due to the large viscosity of the glassy polarization. On the other hand, at temperatures above 37 °C, the static electric susceptibility diverges according to the Curie-Weiss law

$$\chi(T) = \beta\,(T_c - T)^{-1}, \quad T < T_c \ , \tag{15}$$

the linear electrooptic coefficients decrease as the spontaneous polarization decreases, the domain grating is thermal scrambled, and the dark conductivity increases. These four effects conspire to reduce the diffraction efficiency of the domain grating. Consequently, a two-stage fixing process, in which the domain grating generation is thermally assisted at elevated temperatures (30 to 37 °C) and then is frozen-in at slightly lower temperatures (20 to 25 °C), enhances both the grating modulation and lifetime.

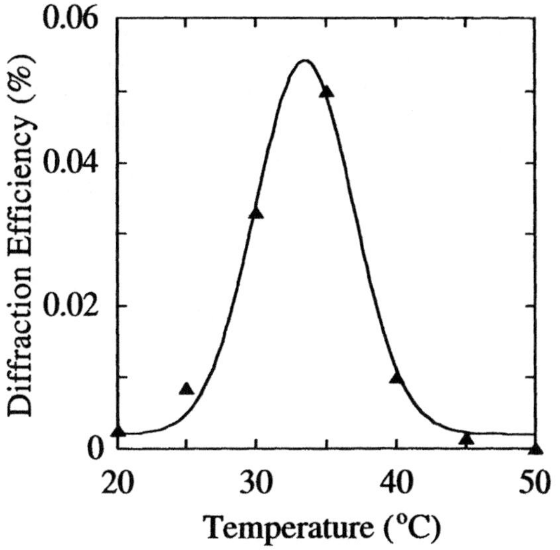

Fig. 19 Resonance in the diffraction efficiency of remnant polarization grating in for holograms recorded between 30 and 37 °C.

The mechanism of optically induced domain reversal in glassy ferroelectrics is believed to be closely related to the macro-microdomain transformation under depoling electric fields [67]. A thermodynamic description of this transformation is rooted in the theory of spin glasses. Consider the distinction between the field-cooled and zero-field-cooled state [64]. By cooling the crystal from above the Curie temperature to the freezing temperature T_f in the absence of an applied field, the ferroelectric is in a zero-field-cooled state in which long range ferroelectric order is absent. However, by cooling under an applied poling field, long range order can be frozen-in below T_f. This is conventionally called the field-cooled state. This is the state the crystal is prepared in before recording holograms. A depoling electric field **E** applied parallel to the c axis effectively lowers the local freezing temperature through the deAlmedia-Thouless relationship [67, 68]

$$T_f(E) = T_f(0)\left[1-\left(\frac{E}{A}\right)^{2/3}\right] \tag{16}$$

where A is a constant defined as $A = k_BT/p_s$ and p_S is the freezing dipole moment. The photogenerated space-charge field is the source of this applied field. The periodic field spatially modulates the freezing temperature. If the temperature of the crystal is slightly above the unperturbed freezing temperature, then as the freezing point is perturbed downward, the relaxation rate of the glassy polarization to the un-poled state increases dramatically. Therefore, in regions where the space-charge field is parallel to the c axis, the long range order (or frozen-in polarization) will "thaw out" and reduce the local polarization. In principle, large negative space-charge fields (>1 kV cm^{-1} for SBN:75) not only lower the freezing temperature as in Eq. 17, but also locally transform these microdomains into new inverted macrodomains (with dimensions typically greater than microns). An idealized model of the microscopic domain structure for the grating vector parallel to the c axis is illustrated in Fig. 20. This model is based on (i) observations of electrically erasable remnant volume gratings [1, 61, 69], (ii) the decrease of the spatially averaged electrooptic coefficient r_c by an order of magnitude and the resulting decrease of photorefractive beam fanning following grating formation, (iii) the significant enhancement of the second harmonic while writing domain gratings for quasi-phase matching [2] and (iv) the manifestation of freezing behavior in the lifetime of polarization gratings [61].

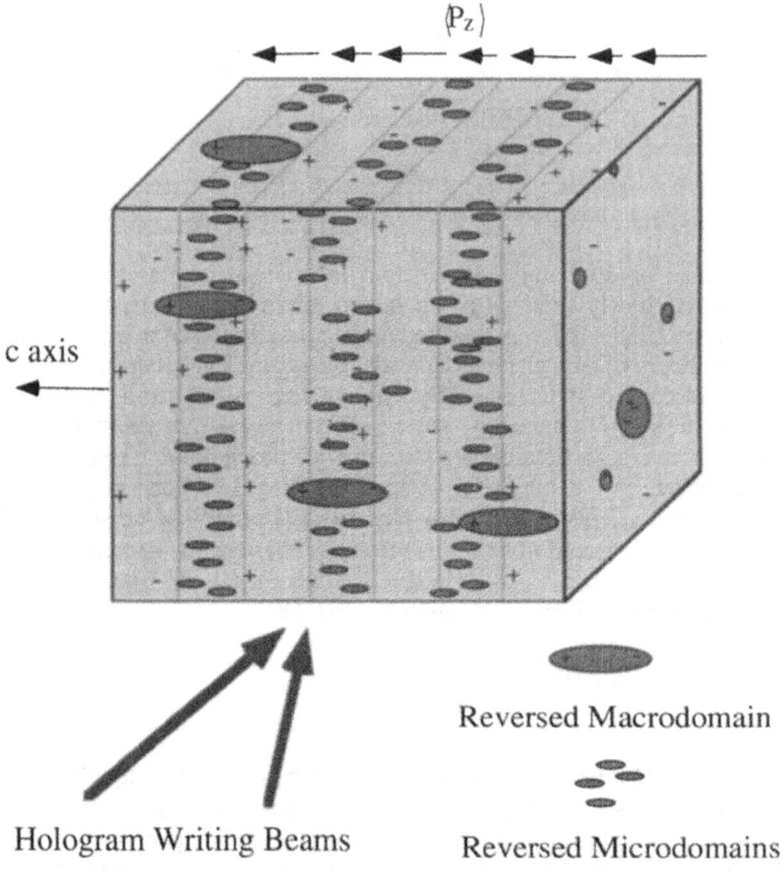

Fig. 20 Model of glassy domain grating microstructure in the geometry of Fig. 3a.

9. APPLICATIONS OF FERROELECTRIC DOMAIN GRATINGS

9.1 Selective Page Addressable Fixing

A primary application of domain fixing is holographic data storage. A potential benefit of polarization fixing are its complementary properties of updatability and long term memory. Taking advantage of these properties, a technique has been implemented to individually fix holograms that may be individually overwritten at a later time [1]. This rewrite property is desirable in applications such as holographic random access memories or reconfigurable optical interconnects, yet it has not been demonstrated using conventional thermal fixing techniques in $LiNbO_3$. For instance, three angle-multiplexed holograms have been stored in the image plane and individually fixed for 15 minutes with incident intensities of 4 W cm^{-2}. The third page was then overwritten by a fourth page for 20 minutes. The 3 original addresses were then read out (Fig. 21). The original 2 pages remained fixed, while the third page was replaced by the fourth. The poor image quality arises from index inhomogeneities caused by growth striations and scattering from domain interfaces. While the fixing process of the latter holograms degrades all previously fixed gratings sharing the same volume, proper scheduling of exposures can be implemented to optimize the overwriting process and reduce crosstalk.

9.2 Quasi-phase Matching

In addition to holographic storage applications, domain gratings can be used for optical frequency conversion. Polarization gratings spatially modulate the nonlinear susceptibility of the crystal and offer the potential of significantly increasing the conversion efficiency of nonlinear wave mixing processes. Phase matching, which is usually a prerequisite for efficient conversion, can only be achieved in a small subset of nonlinear materials. The obstacle is overcome by using quasi-phase matching (QPM) [70, 71] to periodically compensate for the phase mismatch between the fundamental and second harmonic waves caused by the dispersion of the index of refraction. QPM can be achieved by periodically poling a ferroelectric crystal so the nonlinear coefficients for second harmonic generation (SGH) are spatially modulated with a period equal to twice the coherence length. For collinear beams, the coherence length is $\ell_c = \lambda/4\,(n^{2\omega}-n^{\omega})$, where λ is wavelength in vacuum of the fundamental, n^{ω} is the index of refraction at the fundamental wavelength and $n^{2\omega}$ is the index at the second harmonic. In the absence of periodic domain inversion, the coherence length is the maximum effective crystal length that contributes to the second harmonic power. Coherence lengths exceeding a few mm are needed in practice for efficient

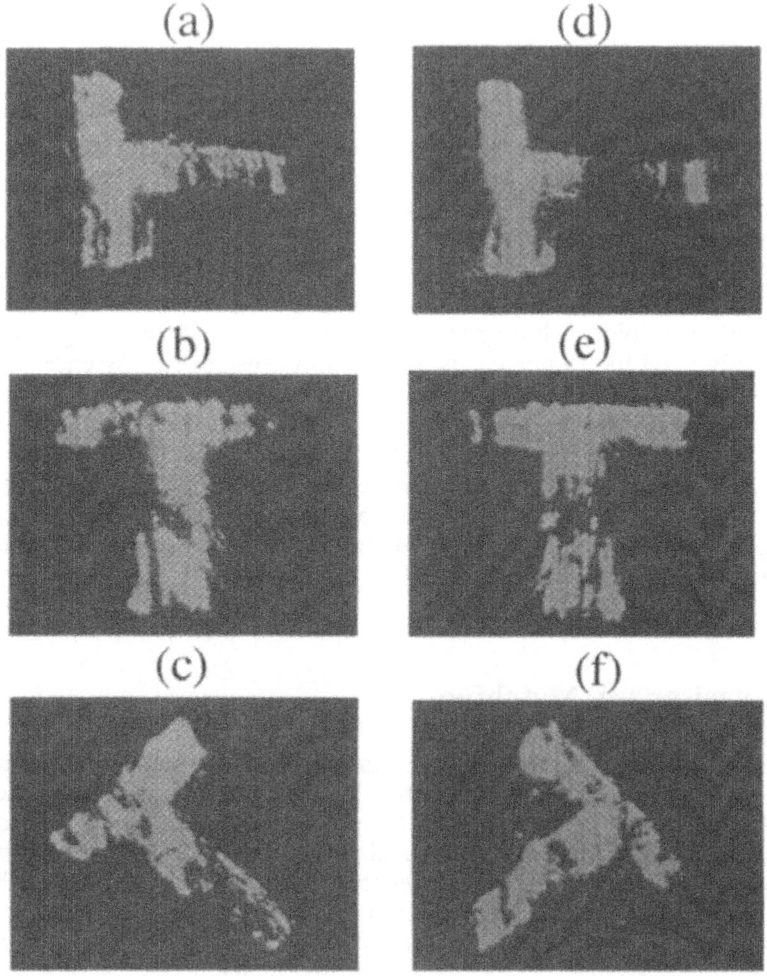

Fig. 21 Selective, page addressable fixing of 3 pages of angle multiplexed data, each page spaced by $\Delta\theta = 0.01$ rad. Source: Reference [1].

conversion. In ferroelectric crystals such as $LiNbO_3$ and SBN:75, the coherence length is on the order of microns for second harmonic generation in the visible; consequently, the non-phase matched conversion is negligibly small despite the relatively large nonlinear coefficients.

Quasi-phase matching has been achieved in SBN using optically induced domain holograms, which produce both broadband and narrowband enhancements of the conversion efficiency. Permanent enhancements using remnant domain gratings in the configuration of Fig. 8, under applied electric fields, have been demonstrated [56, 72, 73]. In addition, tunable quasi-phase matching has been achieved using dynamic polarization gratings [2, 61]. In these latter experiments, photorefractive space-charge fields alone generate the dynamic domain grating. This discussion will focus on the second phenomenon.

The experimental results described here were obtained using an SBN:75 crystal for doubling fundamental wavelengths of 880 to 990 nm. Figure 9 illustrates the experimental setup for writing domain gratings and simultaneously generating the second harmonic. The orientation of the crystalline axes of the SBN:75 crystal are chosen so that both the grating vector and polarization of the fundamental beam have a significant projection along the c axis. The space-charge field is used to invert domains, so a component of the grating vector along the poling axis is required. Similarly, the fundamental must have a significant projection of its electric field along the c axis to produce a large induced polarization oscillating at the second harmonic. The magnitude of the polarization at the second harmonic perpendicular to $\mathbf{k}^{2\omega}$ (the second harmonic wave vector) is:

$$P^{2\omega} = 2\left(2\, d_{15} \sin^2\theta \cos\theta + d_{33} \cos^3\theta + d_{31} \sin^2\theta \cos\theta\right) E^{\omega} E^{\omega}, \tag{17}$$

where θ is the angle of both $\mathbf{P}^{2\omega}$ and $\mathbf{E}^{\omega}$ relative to the c axis. Although $\mathbf{P}^{2\omega}$ attains a maximum value at $\theta = 0$ degrees, this orientation prevents the space-charge field from having a significant component parallel to the c axis. However, $\mathbf{P}^{2\omega}$ remains significant for θ as large as 45 degrees. $\theta = 45$ degrees is a convenient tradeoff. By symmetry arguments, inverting the orientation of a ferroelectric domain changes the sign of all three nonlinear coefficients. Thus the effective nonlinear coefficient $d_{eff} = (2\, d_{15} + d_{33} + d_{31})/2^{3/2}$ is modulated. The domain grating vector $k_g = \pi/\ell_c$ is oriented parallel to $\mathbf{k}^{\omega}$ and $\mathbf{k}^{2\omega}$ to maximize the overlap of the fundamental and second harmonic field profiles.

Wavelength tunability is inherent in this technique. For instance, the QPM peak can be shifted in wavelength by writing polarization gratings with different periods. Figure 22a illustrates a typical QPM peak centered on 457.25 nm. A FWHM of 0.2 nm for a crystal of thickness 4.25 mm is predicted, in close agreement with the measured FWHM of 0.175 nm. This observation confirms that the grating is uniform over the entire 4.25 mm propagation distance.

The spectral response of the SHG can be tailored by writing an ensemble of gratings with different periods, in the same or different locations within the volume. Applying the fixing techniques discussed in section 8, these gratings can be rendered permanent. Figure 22b illustrates a series of QPM peaks written sequentially in the same volume, with the strongest enhancement peak corresponding to the last grating written. The spectral response is designed by writing several gratings (either simultaneously or sequentially) with different periods. An enhancement of a factor of 1000 above the background has been achieved.

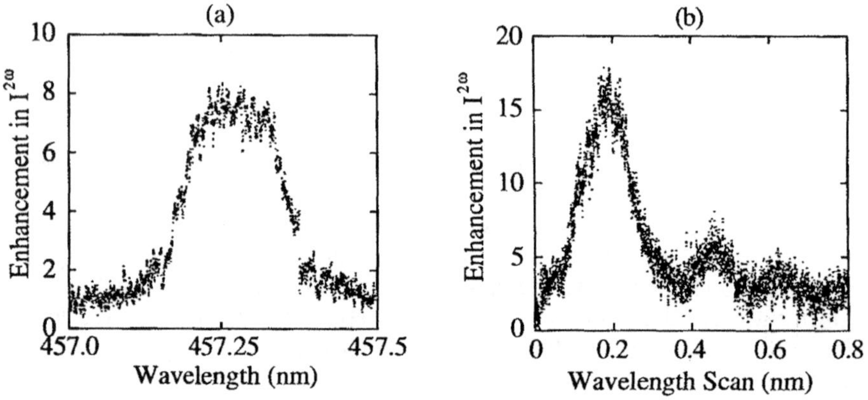

Fig. 22 (a) Typical QPM spectral response peak with a FWHM of 0.175 nm. (b) Multiple QPM peaks by writing several gratings sequentially. Source: Reference [2].

The second harmonic power grows rapidly while writing the phase matching volume hologram. Figure 23 illustrates the time evolution of second harmonic power while writing a QPM grating with a total intensity of 1 W cm^{-2}. The dynamic polarization grating (described by the third term of Eq. 3) tracks the free space-charge grating (described by the second term of Eq. 3) in time. The transients in the peak enhancement are attributed to the decay and buildup of the grating as the optical fringes shift due to vibration, air currents and temperature changes. The buildup time constant of the QPM peak is approximately inversely proportional to intensity and is consistent with the photorefractive grating buildup times for SBN:75. At 1 W cm^{-2} the buildup time is 0.5 seconds, and at 0.33 W cm^{-2} it is 1.5 seconds. The small permanent enhancement Δ following optical erasure of the space-charge field (Fig. 23) is attributed to the remnant polarization grating. This remnance is increased by applying one of the domain fixing techniques described earlier.

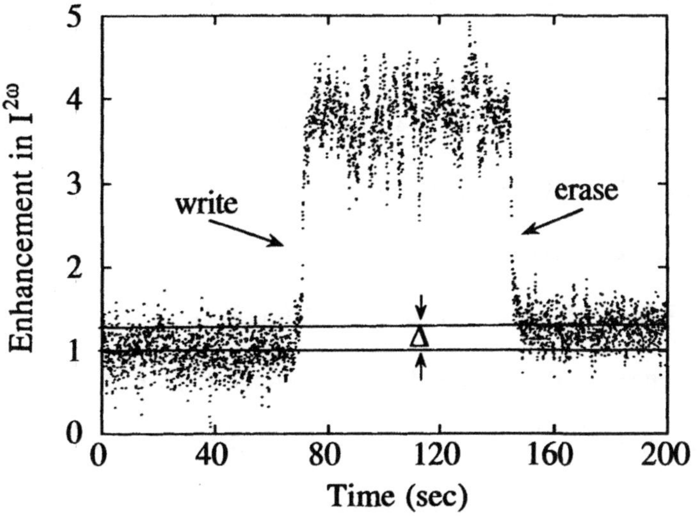

Fig. 23 Time evolution of second harmonic enhancement near the center of QPM spectral response peak. The rapid response indicates the presence of a dynamic domain grating. Source: Reference [2].

10. Conclusion

Ferroelectrics display a wealth of phenomena that can be utilized for fixing permanent phase gratings with potential applications such as holographic data storage and quasi-phase matching. This chapter describes fixing techniques based on thermally activated drift of ions as in $LiNbO_3$ and polarization hysteresis in SBN:75 and $BaTiO_3$, focusing primarily on the phenomenology and experimental results. The weight of the chapter has rested on polarization fixing in SBN; however, thermal fixing in $LiNbO_3$ is at a much more technologically advanced stage in terms of fixed hologram image quality, resolution, and diffraction efficiency. These fixing techniques may also find related applications in the field of acoustics and general nonlinear mixing processes. Extension of these techniques to photorefractive thin film materials is also of great fundamental and practical importance.

References

[1] A. S. Kewitsch, M. Segev, A. Yariv, R. R. Neurgaonkar, *Opt. Lett.* **18**, 1262 (1993).

[2] A. S. Kewitsch, M. Segev, A. Yariv, G. J. Salamo, T. W. Towe, E. J. Sharp, R. R. Neurgaonkar, *Appl. Phys. Lett.* **64**, 1023 (1994).

[3] J. P. Herriau, J. P. Huignard, *Appl. Phys. Lett.* **49**, 1140 (1986).

[4] S. Redfield, L. Hesselink, *Opt. Lett.* **13**, 880 (1988).

[5] D. Psaltis, F. Mok, H. S. Li, *Opt. Lett.* **19**, 210 (1994).

[6] D. von der Linde, A. M. Glass, K. F. Rodgers, *Appl. Phys. Lett.* **26**, 22 (1975).

[7] M. Carrascosa, F. Agulló-López, *J. Opt. Soc. Am. B* **7**, 2317 (1990).

[8] P. Hertel, K. H. Ringhofer, R. Sommerfeldt, *Phys. Stat. Sol. (a)* **104**, 855 (1987).

[9] R. Matull, R. A. Rupp, *J. Phys. D: Appl. Phys.* **21**, 1556 (1988).

[10] B. I. Sturman, V. M. Fridkin, *The Photovoltaic and Photorefractive Effects in Noncentrosymmetric Materials*. G. W. Taylor, Eds., Ferroelectricity and Related Phenomena (Gordon and Breach, Philadelphia, 1992), vol. 8.

[11] S. H. Wemple, M. Di Domenico, *Appl. Phys. Lett.* **12**, 209 (1968).

[12] M. E. Lines, A. M. Glass, *Principles and Applications of Ferroelectrics and Related Materials*. (Clarendon Press, Oxford, 1977).

[13] L. Bursill, P. Lin, *Philos. Mag. B* **54**, 157 (1986).

[14] A. S. Bhalla, R. Guo, L. E. Cross, G. Burns, F. H. Dacol, R. R. Neurgaonkar, *J. Appl. Phys.* **71**, 5591 (1992).

[15] W. H. Huang, D. Viehland, R. R. Neurgaonkar, *J. Appl. Phys.* **76**, 490 (1994).

[16] V. M. Fridkin, *Photoferroelectrics*. M. Cardona, P. Fuldeand H.-J. Queisser, Eds., Springer Series in Solid State Sciences (Springer-Verlag, Berlin, 1979), vol. 9.

[17] F. Kahmann, R. Pankrath, R. A. Rupp, *Opt. Comm.* **107**, 6 (1994).

[18] H. Barkhausen, *Physikalische Zeitschrift* **20**, 401 (1919).

[19] V. M. Rudyak, *Soviet Physics Uspekhi* **13**, 461 (1971).

[20] A. A. Bogomolov, V. V. Ivanov, V. M. Rudyak, *Sov. Phys.-Cryst.* **14**, 894 (1970).

[21] A. G. Chynoweth, *Phys. Rev.* **110**, 1316 (1958).

[22] T. R. Bolk, A. A. Grekov, N. A. Kosonogov, A. I. Rodin, V. M. Fridkin, *Sov. Phys.-Cryst.* **16**, 198 (1971).

[23] V. M. Fridkin, I. I. Groshik, V. A. Lakhovizkaya, M. P. Mikhailov, V. N. Nosov, *Appl. Phys. Lett.* **10**, 354 (1967).

[24] M. DiDomenico Jr., S. H. Wemple, *J. Appl. Phys.* **40**, 720 (1969).

[25] J. C. Phillips, *Bonds and Bands in Semiconductors*. A. M. Alperand A. S. Nowick, Eds., Materials Science and Technology (Academic Press, New York, 1973).

[26] C. R. Jeggo, G. D. Boyd, *J. Appl. Phys.* **41**, 2741 (1970).

[27] C. Shih, A. Yariv, *Phys. Rev. Lett.* **44**, 281 (1980).

[28] C. Shih, A. Yariv, *J. Phys. C: Solid State Phys.* **15**, 825 (1982).

[29] R. Hofmeister, A. Yariv, S. Yagi, A. Agranat, *Phys. Rev. Lett.* **69**, 1459 (1992).

[30] J. J. Amodei, D. L. Staebler, *Appl. Phys. Lett.* **18**, 540 (1971).

[31] G. Montemezzani, P. Günter, J. Opt. Soc. Am. B**7**, 2323 (1990).

[32] E. Krätzig, R. Orlowski, *Appl. Phys.* **15**, 133 (1978).

[33] J. J. Amodei, D. L. Staebler, W. Stephens, Appl. Phys. Lett. **18**, 507 (1971)

[34] S. W. McCahon, D. Rytz, G. C. Valley, M. B. Klein, B. A. Wechsler, *Appl. Opt.* **28**, 1967 (1989).

[35] R. Müller, L. Arizmendi, M. Carrascosa, J. M. Cabrera, *Appl. Phys. Lett.* **60**, 3212 (1992).

[36] W. Bollmann, H. J. Stöhr, *Phys. Status Solidi (a)* **39**, 477 (1977).

[37] D. L. Staebler, J. J. Amodei, *Ferroelectrics* **3**, 107 (1972).

[38] H. Vormann, G. Weber, S. Kapphan, E. Krätzig, *Solid State Commun.* **40**, 543 (1981).

[39] D. L. Staebler, W. J. Burke, W. Phillips, J. J. Amodei, *Appl. Phys. Lett.* **26**, 182 (1975).

[40] R. Müller, M. T. Santos, L. Arizmendi, J. M. Cabrera, *J. Phys. D: Appl. Phys.* **27**, 241 (1994).

[41] V. Leyva, G. A. Rakuljic, B. O'Conner, *Appl. Phys. Lett.* **65**, 1079, (1994).

[42] A. Ashkin, G. D. Boyd, J. M. Dziedzic, R. G. Smith, A. A. Ballman, J. J. Levinstein, K. Nassau, *Appl. Phys. Lett.* **9**, 72 (1966).

[43] V. I. Kovalevich, L. A. Shuvalov, T. R. Volk, *Phys. Stat. Sol. (a)* **45**, 249 (1978).

[44] V. V. Lemeshko, V. V. Obukhovskii, *Sov. Phys. Solid State* **30**, 933 (1988).

[45] V. V. Lemeshko, V. V. Obukhovskii, A. V. Stoyanov, *Sov. Phys. Solid State* **34**, 977 (1992).

[46] N. A. Abramov, V. V. Voronov, Y. S. Kuzminov, *Ferroelectrics* **22**, 649 (1978).

[47] F. Micheron, J. C. Trotier, *Ferroelectrics* **8**, 441 (1974).

[48] F. Micheron, G. Bismuth, *Appl. Phys. Lett.* **20**, 79 (1972).

[49] F. Micheron, G. Bismuth, *Appl. Phys. Lett.* **23**, 71 (1973).

[50] J. B. Thaxter, M. Kestigian, *Appl. Opt.* **13**, 913 (1974).

[51] Y. Qiao, S. Orlov, D. Psaltis, R. R. Neurgaonkar, *Opt. Lett.* **18**, 1004 (1993).

[52] R. S. Cudney, J. Fousek, M. Zgonik, P. Günter, M. H. Garrett, D. Rytz, *Appl. Phys. Lett.* **63**, 3399 (1993).

[53] R. S. Cudney, J. Fousek, M. Zgonik, P G¨nter, M. H. Garrett, D. Rytz, Phys. Rev. Lett. **73**, 3883 (1994)

[54] W. D. Smith, C. E. Land, *Appl. Phys. Lett.* **20**, 169 (1972).

[55] F. Kahmann, R. Matull, R. A. Rupp, J. Seglins, *Europhys. Lett.* **13**, 405 (1990).

[56] M. Horowitz, A. Bekker, B. Fischer, *Appl. Phys. Lett.* **62**, 2619 (1993).

[57] M. Horowitz, A. Bekker, B. Fischer, *Opt. Lett.* **18**, 1964 (1993).

[58] P. Günter, J.-P. Huignard, Eds., *Photorefractive Materials and Their Applications I*, vol. 61 (Springer-Verlag, Berlin, 1989).

[59] A. S. Kewitsch, A. Saito, A. Yariv, M. Segev, R. R. Neurgaonkar, *to be published* (1994).

[60] W. Clark III, G. Wood, M. Miller, E. Sharp, G. Salamo, B. Monson, R. R. Neurgaonkar, *Appl. Opt.* **29**, 1249 (1990).

[61] A. S. Kewitsch, M. Segev, A. Yariv, G. J. Salamo, T. W. Towe, E. J. Sharp, R. R. Neurgaonkar, *Phys. Rev. Lett.* **73**, 1174 (1994).

[62] P. B. Jamieson, S. C. Abrahams, J. L. Bernstein, *J. Chem. Phys.* **48**, 5048 (1968).

[63] L. E. Cross, *Ferroelectrics* **76**, 241 (1987).

[64] K. Binder, A. P. Young, *Rev. Mod. Phys.* **58**, 801 (1986).

[65] H. Vogel, *Phys. Z.* **22**, 645 (1921).

[66] G. J. Fulcher, *J. Am. Cer. Soc.* **8**, 339 (1925).

[67] D. D. Viehland, Dissertation, The Pennsylvania State University (1991).

[68] J. R. de Almedia, D. J. Thouless, *J. Phys. A.* **11**, 983 (1978).

[69] A. S. Kewitsch, M. Segev, A. Yariv, R. R. Neurgaonkar, *Jap. J. Appl. Phys.* **32**, 5445 (1993).

[70] J. A. Armstrong, N. Bloembergen, J. Ducuing, P. S. Pershan, *Phys. Rev.* **127**, 1918 (1962).

[71] S. Somekh, A. Yariv, *Opt. Comm.* **6**, 301 (1972).

[72] B. Fischer, M. Horowitz, Appl. Phys. Lett. **64**, 1756 (1994).

[73] M. Horowitz, A. Bekker, B. Fischer, Appl. Phys. Lett. 65, 679 (1994).

4.

PHOTOREFRACTIVE SPATIAL SOLITONS

Mordechai Segev[(a)], Bruno Crosignani[(b)]
Gregory Salamo[(c)], Galen Duree[(c)]
Paolo Di Porto[(b)] and Amnon Yariv[(d)].
(a) Electrical Engineering Department, Princeton University, Princeton NJ 08544
(b) Dipartimento di Fisica, Universita' dell'Aquila, L'Aquila, Italy.
(c) Physics Department, University of Arkansas, Fayetteville, AR 72701.
(d) Department of Applied Physics, California Institute of Technology,
Pasadena, CA 91125.

1. INTRODUCTION

Self-trapping of light beams in nonlinear self-focusing Kerr media has been explored intensively during the last three decades. The self trapped beams, often in the form of spatial solitons, evolve from nonlinear changes in the refractive index (δn) of the material proportional to the intensity distribution of the light, when the confining effect of the refractive index exactly compensates for the defocusing effect of diffraction. Motivated by experimental results of self-focusing of light beams [1] in the presence of the optical Kerr effect, a theoretical solution for the resulting spatial solitons was found [2,3]. This was followed by demonstrations of self-trapping in vapors [4], liquids [5], and finally, in the last few years, spatial solitons in solid (glass) media [6]. However, the index changes needed for optical-Kerr spatial solitons require high intensities, often exceeding 1 MWatt/cm^2. Several other materials, such as slightly-absorbing liquids, exhibit Kerr-like thermal nonlinearities at much lower power densities (Watts/cm^2), typically with a negative δn. These materials enabled the observations of dark spatial solitons [7] (narrow dark "notches") and optical vortex solitons [8,9] (dark solitons were also observed using optical-Kerr nonlinearities in ZnSe [10] at GWatt/cm^2 power densities). The most distinct properties of all these Kerr-like solitons are (i) a strong intensity dependence and the existence of a power threshold, and (ii) stability of bright solitons only in a single transverse dimension (consequently, bright Kerr-type solitons can be observed only in a planar waveguide, otherwise the soliton becomes unstable and cannot be observed).

Our study of self-trapping effects of optical beams in photorefractive (photorefractive) media [11] has predicted spatial solitons that stem from optical nonlinearities which are very different from the traditional Kerr effects. Until now, three different types of photorefractive solitons have been investigated. The first type which has been studied results from the nonlocal nature of the photorefractive effect, and manifested in the dependence of the perturbation in the refractive index on the transverse derivatives of the light intensity distribution [11,12]. The intensity profile of the beam modulates the refractive index via the photorefractive effect, which results in self-scattering (multiple two-wave mixing) of its spatial frequency components, and an exact compensation for the diffraction effects. These solitons can be generated even at very moderate power densities, because the efficiency of the photorefractive effect is independent of the absolute light intensity (whenever the intensity is much larger than the dark irradiance, which can be less than 100 μWatts/cm^2 in some photorefractive materials). Recent experimental observations [13,14] have demonstrated the existence of one and two dimensional bright solitons, planar dark solitons and vortex solitons [15]. These photorefractive solitons of the first (nonlocal) type exist when an external voltage is applied to the photorefractive material, after the index gratings have been formed, but before the external field is screened by the background conductivity. Solitons of this type are transient by nature and we refer to their time-window of existence as quasi-steady-state. Their most distinct properties are (i) the independence of the absolute light intensity [11-13] (for intensities much larger than the dark irradiance) and (ii) the capability of trapping in two transverse dimensions [13-17].

The second type of photorefractive soliton, which we call the screening-soliton [18], appears in the steady-state after the external field is nonuniformly screened, because of the transverse, nonuniform intensity distribution. This effect is local and results in an index perturbation that is inversely proportional to the sum of the optical and the dark irradiances. Its most distinct properties are (i) dependence of the soliton size on the ratio between the optical and the dark irradiances (the narrowest size is obtained for ~1000:1 ratios) and (ii) existence of bright solitons for a negative perturbation in the index while dark solitons require a positive perturbation (this implies that the polarity of the applied field is opposite to the polarity required to generate photorefractive solitons of the first type). Although a definite observation of screening solitons has not yet been reported, some experiments [19,20] have shown self-focusing and self-defocusing caused by these screening effects.

The third type of photorefractive soliton is present in materials that are both photorefractive and photovoltaic. These photovoltaic solitons [21] stem from photovoltaic currents that generate (in steady-state) an index perturbation analogous to the nonlinearity in a saturable absorber (sometimes called thresholding nonlinearity) [22,23], which is a local effect as well. Their most distinct property is the dependence on the ratio between the optical and the dark irradiances, the narrowest solitons being obtained when this ratio is between 1 and 2. A recent experimental observation [24] has demonstrated photovoltaic dark solitons in bulk $LiNbO_3$.

In this text we provide a detailed overview of the current theoretical understanding of photorefractive solitons of the first (nonlocal) type, along with an updated description of experimental results. We discuss the properties of the photorefractive solitons, and specify the material requirements and the design considerations for experimental realizations. We do not discuss here (beyond this introductory part) photorefractive solitons of the second and the third types (screening and photovoltaic solitons), simply because their experimental studies are still in a preliminary stage.

2. BEAM PROPAGATION IN PHOTOREFRACTIVE MEDIA

Spatial solitons are nondiffracting solutions of the nonlinear wave equation that describes beam propagation in photorefractive media. To account for the nonlinear effects, it is necessary to find the explicit dependence of the refractive index on the intensity distribution of the light. The general procedure includes solving the rate equation for the excitation of charge carriers from (deep) photorefractive traps, Gauss' Law, the continuity equation and Pockel's relation between the space-charge-field and the perturbation in the refractive index through the (linear) electro-optic effect. These equations were linearized and solved [25,26] for a single pair of plane waves that form a single interference grating (and, consequently, a single index grating) of a low visibility (low modulation depth).

As an intuitive approach to describe beam propagation in photorefractive media, we extended this existing single-grating theory and formulated the interaction between all possible pairs of plane-wave components that compose the optical beam. In this manner, our general formalism accounts for the transverse beam profile simply by decomposing it into a set of plane-wave components and treating them individually. This phenomenological model of nonlinear beam propagation has one major drawback: it cannot account for large modulation depth, and therefore it cannot account for the margins of the beam. Nevertheless, it is useful primarily because one can isolate the quasi-steady-state time frame [26], where the photorefractive index-gratings have already formed but before the externally-applied field is screened [11-13] by the background conductivity.

We start by deriving the nonlinear wave equation which describes the propagation of a monochromatic optical beam of a given frequency (ω) and polarization, traveling in the positive direction of an arbitrary axis z. We assume the absence of nonlinear interaction between orthogonal polarizations (anisotropic scattering [27]), so that our problem can be reduced to a scalar formulation. Light propagation in nonlinear media can be conveniently described using coupled-mode theory [28], applied to the case of unbounded media for which an appropriate set of spatial modes is the continuum of plane waves [29]. The electric field associated with the light beam propagating primarily along the z direction is written as

$$E(\mathbf{r},z,t) = \frac{1}{2}\left[e^{i(kz-\omega t)} \int E(\mathbf{q},\mathbf{r})\, e^{i(\beta_q - k)z} f(\mathbf{q},z)\, d\mathbf{q} + c.c.\right] \quad (1)$$

$$\equiv \frac{1}{2}\left[\, A(\mathbf{r},z)\, e^{i(kz-\omega t)} + c.c.\right]$$

where

$$E(\mathbf{q},\mathbf{r}) = \frac{1}{2\pi}\sqrt{\frac{\mu_0}{\varepsilon_0 n_1}}\; e^{i\mathbf{q}\cdot\mathbf{r}} \quad (2)$$

with $r \equiv (x,y)$, $k = \omega n_1/c$ is the light wavenumber, n_1 is the unperturbed index of refraction in the medium, and $f(\mathbf{q},z)$ is the spatial frequency (angular) distribution of the complex amplitude $A(\mathbf{r},z)$. A spatial mode (plane-wave component) is characterized by the projections of its wave vector ($\mathbf{q}$ and β_q) on the transverse ($\mathbf{r}$) and longitudinal (z) directions respectively, with $\beta_q = (k^2 - q^2)^{1/2}$ (where q is restricted to $0 \le |q| \le k$ and, in the paraxial approximation, $(k - \beta_q) << k$). Neglecting absorption, and under rather general conditions [29], it is easy to show that in the presence of a refractive index distribution $n(\mathbf{r},z) = n_1 + \delta n(\mathbf{r},t)$, $A(\mathbf{r},t)$ obeys the equation

$$\left(\frac{\partial}{\partial z} - \frac{i}{2k}\nabla^2_{\mathbf{r}}\right) A(\mathbf{r},z) = \frac{ik}{n_1}\,\delta n(\mathbf{r},z)\, A(\mathbf{r},z) \quad (3)$$

The nonlinear term $\delta n(\mathbf{r},z)$ is obtained by considering the mixing processes between all possible pairs of plane-wave components of the beam (which are all of the same frequency ω). When only one pair of such plane waves (spatial modes) $\mathbf{q}_1$ and $\mathbf{q}_2$, of field amplitudes $a_1(z)$ and $a_2(z)$, is present in the medium, it induces an index grating $\delta n(\mathbf{r},z)$ which is proportional to the time-averaged (over the characteristic photorefractive response-time of the medium) interference pattern between the waves. The proportionality coefficient is a complex factor $\widetilde{\delta n}(\mathbf{q}_1,\mathbf{q}_2)$ which represents the photorefractive coupling coefficient between the two plane waves, given the material properties (the orientation of the photorefractive crystalline medium, its traps density P_d, its refractive index n_1, the DC dielectric constant ε_r), the externally-applied electric field E_o, and the polarization of the waves. In this simple case, $\delta n(\mathbf{r},z)$ is given by [30]

$$\delta n(\mathbf{r},z) = \frac{1}{I_0}\left\{ a_1(z)\, e^{i(\mathbf{q}_1\cdot\mathbf{r} + \beta_{q1}z)}\, a_2^*(z)\, e^{-i(\mathbf{q}_2\cdot\mathbf{r} + \beta_{q2}z)}\, \widetilde{\delta n}(\mathbf{q}_1,\mathbf{q}_2) + c.\,c. \right\} \quad (4)$$

where $I_o=|a_1|^2+|a_2|^2$ is the absolute light intensity ($I_o >> |a_1 a_2^*|$). Because the perturbation $\delta n(\mathbf{r},z)$ in the refractive index is real (no absorption) we get $\widetilde{\delta n}(\mathbf{q}_1,\mathbf{q}_2) = \widetilde{\delta n}^*(\mathbf{q}_2,\mathbf{q}_1)$. This light induced index change may be viewed as the sum of two grating components, that is, the real and imaginary parts of $\widetilde{\delta n}(\mathbf{q}_1,\mathbf{q}_2)$, which correspond to index gratings that are in phase and shifted by $\pi/2$, respectively, with respect to the light interference grating $a_1 a_2^*/I_o$. The $\pi/2$ shifted gratings are responsible for (light) power transfer between the two waves, a phenomenon which can manifest itself as optical amplification [31], photorefractive oscillation [30,32], and stimulated noise scattering (fanning [33]). The in phase gratings are responsible for nonlinear phase-coupling. The real/imaginary parts of $\widetilde{\delta n}(\mathbf{q}_1,\mathbf{q}_2)$ are even/odd, respectively, under the exchange $(\mathbf{q}_1,\mathbf{q}_2) \rightarrow (\mathbf{q}_2,\mathbf{q}_1)$. The grating coefficient $\widetilde{\delta n}(\mathbf{q}_1,\mathbf{q}_2)$, when multiplied by "i", is often called γ, and yields an intensity coupling coefficient $\Gamma=2\mathrm{Re}\{\gamma\}= -4\pi\, \mathrm{Im}\{\ \widetilde{\delta n}(\mathbf{q}_1,\mathbf{q}_2)\ \}/\lambda$. The coefficient $\widetilde{\delta n}(\mathbf{q}_1,\mathbf{q}_2)$ is given by [31]:

$$\widetilde{\delta n}(\mathbf{q}_1,\mathbf{q}_2) = \frac{-\omega}{2c}\, n_1^3\, r_{eff}(\mathbf{q}_1,\mathbf{q}_2)\, E_m(\mathbf{q}_1,\mathbf{q}_2)\, (\mathbf{e}_1 \cdot \mathbf{e}_2^*) \tag{5}$$

where $\mathbf{e}_1,\mathbf{e}_2$ are the polarization vectors of the two interfering plane waves, $r_{eff}(\mathbf{q}_1,\mathbf{q}_2)$ is an effective electro-optic coefficient (which depends on the orientation of the photorefractive crystalline medium), and $E_m(\mathbf{q}_1,\mathbf{q}_2)$ the coefficient of the induced space charge field

$$E_m(\mathbf{q}_1,\mathbf{q}_2) = E_p(\mathbf{q}_1,\mathbf{q}_2)\, \frac{E_d(\mathbf{q}_1,\mathbf{q}_2) - iE_0}{E_0 + i[E_d(\mathbf{q}_1,\mathbf{q}_2) + E_p(\mathbf{q}_1,\mathbf{q}_2)]} \tag{6}$$

where E_0 is the externally applied electric field, $E_p= eP_d/(\varepsilon_0\varepsilon_r K_g)$ the limiting space charge field, and $E_d=k_B T K_g/e$ the diffusion field (e is the magnitude of the electron charge, T the temperature and k_B Boltzman constant). Both E_d and E_p depend on the magnitude and direction of the interference grating wave vector $\mathbf{K}_g=\mathbf{k}_1-\mathbf{k}_2=(\mathbf{q}_1-\mathbf{q}_2, \beta_{q1}-\beta_{q2})$ between the two plane waves. In some materials, E_p has also a small angular dependence through ε_r [26, 33]. Because r_{eff} is always real, the phase shift of $\widetilde{\delta n}(\mathbf{q}_1,\mathbf{q}_2)$ with respect to the interference grating $\mathbf{K_g}$ is dictated solely by E_m, which can be split into its real and imaginary parts, $E_m=-(E_1+iE_2)$, as

$$E_1 = \frac{E_0}{(\frac{E_0}{E_p})^2 + (\frac{E_d}{E_p} + 1)^2} \quad \text{and} \quad E_2 = \frac{E_p\,(E_0^2+E_d^2+E_dE_p)}{E_0^2 + (E_d+E_p)^2} \tag{7}$$

It is important to note that in the planar (two-dimensional) cases, where $\beta_{q1}-\beta_{q2} << |\mathbf{q}_1 - \mathbf{q}_2|$, $\mathbf{K}_g$ is odd under both exchanges $(\mathbf{q}_1,\mathbf{q}_2) \rightarrow (\mathbf{q}_2,\mathbf{q}_1)$ and $(\mathbf{q}_1,\mathbf{q}_2) \rightarrow (-\mathbf{q}_1,-\mathbf{q}_2)$, E_2 behaves similarly, while E_1 is always even. These symmetry properties imply design considerations for a proper choice of the photorefractive crystalline medium and the direction of propagation, polarization and application of external field, and will be addressed later.

When more than two plane waves are present, $\delta n(\mathbf{r},z)$ involves a summation over all the possible interacting pairs of plane waves. For a given light beam $A(\mathbf{r},z)$, which consists of a continuous spatial-frequency spectrum of plane waves $f(\mathbf{q},z)$, this summation takes the integral form:

$$\delta n(\mathbf{r},z) = \frac{1}{|A(\mathbf{r},z)|^2} \int d\mathbf{q}_1 \int d\mathbf{q}_2$$
$$f(\mathbf{q}_1, z)\, f^*(\mathbf{q}_2, z)\, E(\mathbf{q}_i, \mathbf{r})\, E^*(\mathbf{q}_2, \mathbf{r}) \exp\left(i\,(\beta_{q1} - \beta_{q2})\, z\right) \delta\hat{n}(\mathbf{q}_1,\mathbf{q}_2) \tag{8}$$

Because the photorefractive nonlinearity is independent of the absolute light intensity, δn is normalized by the factor $|A(\mathbf{r},z)|^2$. In addition, a constant factor representing the dark irradiance [34] should be added to the light intensity in the denominator of Eq. (8) (their sum represents the total, dark and photo-induced, current). This expression for δn is based on the assumption that the induced gratings are of low modulation depth, and hence can be linearly superposed. This assumption, however, is not a major restriction when a beam is considered, provided that it has a small angular divergence. This representation of the photorefractive light induced index change fails either in the near proximity of dark spots or at the margins of the beam, where the local light intensity $|A(\mathbf{r},z)|^2$ is of the order of the dark irradiance; there, high modulation depth effects become significant [35] and the linearization assumption is no longer valid. We restrict our analysis to regions where the light intensity is much higher than the dark irradiance, and thus neglect the additional constant in the denominator of Eq. (8).

The stead-state solution used in Eqs. (4) and (8) does <u>not</u> include screening effects [18]. The assumption of low visibility interference between each pair of plane-wave components precludes the analysis of the margins of the beam and, consequently, the screening effects of the external field are eliminated from the theoretical analysis. In practice, however, screening effects are present [36] and eventually the external field is partially screened. The response times of these two events differ fundamentally, because the formation time of the photorefractive grating is dependent on the recombination rate into the deep traps and the screening time depends only on the dielectric relaxation time [37]. In some photorefractive materials (all ferroelectric photorefractive crystals) the grating formation time is much smaller than the dielectric relaxation time, and between these time frames exists a time-window where the photorefractive gratings have already formed but the external field has not been screened yet. These materials are appropriate for realization of photorefractive solitons of the first type [13-17]. In other photorefractive materials (such as photorefractive semiconductors) the grating formation time exceeds the dielectric relaxation time, and the necessary time-window for observation of photorefractive solitons of the first type does not exist. These materials are, however, still appropriate for observation of screening-solitons [18].

This model of beam propagation in photorefractive media has proven effective in the interpretation of a variety of wave-mixing processes and in extracting predictions [33,38-43] that have been verified experimentally [44]. For a detailed analysis, which includes the temporal evolution, we refer the reader to the appendix of Ref. [43]. The major advantage of using this model is the steady-state treatment of δn, which is justified when a quasi-steady-state is reached between the characteristic formation and the screening time. A full treatment that describes the temporal evolution and includes screening will be given elsewhere [45].

3. THE PHOTOREFRACTIVE SOLITON EQUATION

We define a self-trapped optical beam (spatial soliton) as a waveform which does not change its shape during propagation, along the optical axis z, that is

$$A(\mathbf{r},z) = U(\mathbf{r})\, e^{i\gamma z} \tag{9}$$

where $U(\mathbf{r})$, the transverse amplitude, and γ, the characteristic soliton propagation constant, may be complex (allowing for amplification or decay). Because diffraction is basically a symmetric process, the simplest way to compensate for it, and generate a spatial soliton in a photorefractive medium, is through a symmetric nonlinear process. Hence we require $\delta n(\mathbf{r},z)=\delta n(-\mathbf{r},z)$ for all $\mathbf{r}$ and z. The substitution of this relation in Eq. (3) implies

$$U(\mathbf{r}) = \pm U(-\mathbf{r}) \quad . \tag{10}$$

First we discuss bright solitons, and therefore consider the simplest case of symmetric solutions, which require $\widetilde{\delta n}(\mathbf{q}_1,\mathbf{q}_2)=\widetilde{\delta n}(-\mathbf{q}_1,-\mathbf{q}_2)$ (note that an anti-symmetric nonlinearity $[\delta n(\mathbf{r},z)=-\delta n(-\mathbf{r},z)]$ does not satisfy Eq. (3) with a symmetric soliton waveform, and hence does not provide any additional solutions). In the most general case, $\widetilde{\delta n}(\mathbf{q}_1,\mathbf{q}_2)$ can be expressed as:

$$\widetilde{\delta n}(\mathbf{q}_1, \mathbf{q}_2) = \int\int g(\boldsymbol{\rho}, \boldsymbol{\rho}')\, e^{-i(\mathbf{q}_1\cdot\boldsymbol{\rho} + \mathbf{q}_2\cdot\boldsymbol{\rho}')} d\boldsymbol{\rho}\, d\boldsymbol{\rho}' \tag{11}$$

so that, from Eqs. (1) and (8), we get

$$\delta n(\mathbf{r}, z) = \frac{1}{|A(\mathbf{r}, z)|^2} \int\int A(\mathbf{r} - \rho, z)\, A^*(\mathbf{r} + \rho', z)\, g(\rho,\rho')\, d\rho\, d\rho' \tag{12}$$

Note the explicit nonlocal nature of the photorefractive effect which is brought out by numerator in Eq. (12), and the local behavior of the "damping factor" $|A(\mathbf{r},z)|^2$ in the denominator. The last argument was examined carefully and verified experimentally by Mamaev and Shkunov [34]. By inserting Eq. (12) into Eq.(3), the equation of evolution of A(r,z) reads

$$\left(\frac{\partial}{\partial z} - \frac{i}{2k} \nabla_{\mathbf{r}}^2\right) A(\mathbf{r}, z) = \frac{ik}{n_1} \frac{1}{A^*(\mathbf{r}, z)} \int\int A(\mathbf{r} - \rho, z)\, A^*(\mathbf{r} + \rho', z)\, g(\rho, \rho')\, d\rho\, d\rho' \ . \tag{13}$$

In particular, if we look for soliton solutions, we substitute equation (9) in equation (13) and get

$$\left(\gamma - \frac{1}{2k} \nabla_{\mathbf{r}}^2\right) U(\mathbf{r}) = \frac{k}{n_1} \frac{1}{U^*(\mathbf{r})} \int\int U(\mathbf{r} - \rho)\, U^*(\mathbf{r} + \rho')\, g(\rho, \rho')\, d\rho\, d\rho' \tag{14}$$

The integro-differential equation (14) can be transformed into an ordinary differential equation by a Taylor expansion of U(r-ρ) near ρ=0:

$$U(\mathbf{r} - \rho) = U(\mathbf{r}) - \nabla_{\mathbf{r}} U(\mathbf{r}) \cdot \rho + \frac{1}{2} [\nabla_{\mathbf{r}} \nabla_{\mathbf{r}} U(\mathbf{r})] : \rho\, \rho + \ldots. \tag{15}$$

along with an analogous expansion for $U^*(\mathbf{r}+\rho')$ around ρ'=0, and inserting it into its right-hand side.

The smallness parameter in the Taylor expansion (15) is given by ε=d/l, where d is the typical scale of nonlocality (dictated by the form of $\widetilde{\delta n}(\mathbf{q}_1,\mathbf{q}_2)$), and l the characteristic transverse dimension of the beam. To truncate the expansion at a given order, one has to assume ε<1. It is worthwhile to note that, because of the invariance of δn(r,z) under the exchange $\mathbf{r} \rightarrow -\mathbf{r}$, the odd-order terms of Eq. (15) do not contribute to the right hand side of Eq. (14).

3.1 The Two-Dimensional Case

Any photorefractive material is a non-centrosymmetric crystalline medium, which does not have a cylindrical symmetry; as a result the propagation equation is very complicated in the general three-dimensional case. We therefore restrict this initial treatment to one transverse dimension (x) and the propagation direction z. Dimensionality effects are discussed in a separate section later. In the two-dimensional case U(**r**) = U(x), and we obtain, by truncating the Taylor expansion after the second term,

$$\left(\frac{\partial}{\partial z} - \frac{i}{2k}\frac{\partial^2}{\partial x^2}\right) A(x,z) = \frac{-ik}{n_1} I_{11} \frac{1}{A^*}\left|\frac{\partial A}{\partial x}\right|^2 + \frac{ik}{2n_1} I_{20} \frac{\partial^2 A}{\partial x^2} + \tag{16}$$

$$\frac{ik}{2n_1} I_{02} \frac{A}{A^*}\frac{\partial^2 A^*}{\partial x^2} + \frac{ik}{4n_1} I_{22} \frac{1}{A^*}\left|\frac{\partial^2 A}{\partial x^2}\right|^2$$

with

$$I_{mn} = \int d\rho \int d\rho' \, g(\rho,\rho')\, \rho^m \rho'^n = e^{i(m+n)\pi/2} \frac{\partial^m}{\partial q_1^m}\frac{\partial^n}{\partial q_2^n} \widetilde{\delta n}(q_1,q_2)\Big|_{q_1=q_2=0} \tag{17}$$

Note that, although it appears that the index perturbation depends on the complex amplitude A (Eq. 16), one can readily express it in terms of $I=|A|^2$ only (i.e., the phase of A does appear in δn).

We expand

$$\widetilde{\delta n}(q_1,q_2) = \sum_{n=0}^{\infty}\sum_{m=0}^{\infty} a_{mn}\, q_1^m\, q_2^n \tag{18}$$

and obtain

$$I_{mn} = m!\, n!\, a_{mn}\, e^{i(m+n)\pi/2} \tag{19}$$

Under the symmetry condition $\widehat{\delta n}(q_1,q_2)=\widehat{\delta n}(-q_1,-q_2)$ it follows that $a_{mn}=0$ if m+n is odd. Using the relation $\widehat{\delta n}(q_1,q_2)=\widehat{\delta n}^*(q_2,q_1)$ (from the reality of the perturbation in the refractive index $\delta n(x,z)$), we obtain $a_{mn}=a^*_{nm}$ and hence $I_{mn}=I^*_{nm}\, e^{i(m+n)\pi/2}$.

First let us discuss the case where the phase does not change in the transverse direction, i.e. U(x) is real, while the propagation constant is allowed to be complex: $\gamma=\gamma_1+i\gamma_2$. By writing the real and imaginary parts of Eq. (16) we get, after setting $I_{mn}=I^{(re)}_{mn}+i\,I^{(im)}_{mn}$,

$$\gamma_1\,U=\frac{-k}{2n_1}\,I^{(im)}_{20}\,\frac{d^2U}{dx^2} \tag{20}$$

and

$$\gamma_2\,U-\frac{1}{2k}\frac{d^2U}{dx^2}=\frac{-k}{n_1}\,I_{11}\,\frac{(dU/dx)^2}{U}+\frac{k}{2n_1}\,(I^{(re)}_{20}+I^{(re)}_{02})\,\frac{d^2U}{dx^2}+\frac{k}{4n_1}\,I_{22}\,\frac{1}{U}\,(\frac{d^2U}{dx^2})^2\ . \tag{21}$$

From the symmetry conditions of I_{mn}, we get $I^{(im)}_{20}=-\,I^{(im)}_{02}$, $I^{(re)}_{20}=I^{(re)}_{02}$, while I_{mm} is real for all m. Thus Eq. (20) yields $\gamma_1 U=0$, so $\gamma_1=0$ for every nonzero solution, while Eq. (21) becomes

$$\gamma_2\,U-\frac{1}{2k}\frac{d^2U}{dx^2}=\frac{-k}{n_1}\,I_{11}\,\frac{(dU/dx)^2}{U}+\frac{k}{n_1}\,I^{(re)}_{20}\,\frac{d^2U}{dx^2}+\frac{k}{4n_1}\,I_{22}\,\frac{1}{U}\,(\frac{d^2U}{dx^2})^2\ . \tag{22}$$

Thus, whether $\widehat{\delta n}(q_1,q_2)$ is real or complex, the equation governing the evolution of U is

$$\gamma U-\frac{1}{2k}\,U''=\frac{-k}{n_1}\,I_{11}\,\frac{U'^2}{U}+\frac{k}{n_1}\,I^{(re)}_{20}\,U''+\frac{k}{4n_1}\,I_{22}\,\frac{U''^2}{U} \tag{23}$$

where $\gamma=\gamma_2$ (real), and U',U" indicate the first and the second derivatives of U with respect to x.

At this point, it is necessary to recall the general expression for $\widehat{\delta n}(q_1,q_2)$ (see Eqs. 5-7) and apply it to the two dimensional case. The interference grating vector obeys the relation $\mathbf{K_g}(q_1,q_2)\equiv(q_1-q_2)\,\hat{x}$, so that $E_2(q_1,q_2)=-E_2(-q_1,-q_2)$ while $E_1(q_1,q_2)=E_1(-q_1,-q_2)$. Because the coefficient $\widehat{\delta n}(q_1,q_2)$ is proportional to the product $(E_1+iE_2)^{*}r_{eff}$ (see Eq. (5)), the symmetry requirement $\widehat{\delta n}(q_1,q_2)=\widehat{\delta n}(-q_1,-q_2)$ can be satisfied either by the in phase perturbation in the refractive index (generated by E_1, and responsible for the real parts of I_{mn}), or by the $\pi/2$ phase shifted gratings (generated by E_2 and responsible for the imaginary parts of I_{mn}), but not by both types of gratings. For the solutions where U is real, it follows from Eq. (23) that the coefficients I_{mn} (up to the second order) must be real. Therefore we require in the design considerations $r_{eff}(q_1,q_2)=r_{eff}(-q_1,-q_2)$ and it immediately follows that all the imaginary parts of I_{mn} have to be zero, because they do not satisfy this requirement.

By keeping the first significant term, Eq. (23) becomes

$$\gamma - a\frac{U''}{U} + b\left(\frac{U'}{U}\right)^2 = 0 \qquad (24)$$

where $a=\frac{1}{2k}+\frac{k}{n_1}I_{20}^{(re)}$ and $b=\frac{k}{n_1}I_{11}$. Equation (24) is solved using the change of the dependent variable $y=U'/U$, thus obtaining two types of solutions [12].

The first class of solutions (subject to the boundary condition $U'(0)/U(0)=y(0)=0$) is

$$U(x) = U_0\,[\mathrm{sech}(\alpha x)]^D \qquad (25)$$

where $D=a/(b-a)$ and $\alpha=-[\gamma(a-b)/a^2]^{1/2}$.

If we require U to go to zero when x goes to infinity, then D has to be positive, that is, after setting $a=\eta b/2$,

$$D=\frac{a}{b-a}=\frac{\eta}{2-\eta}>0 \qquad (26)$$

which in turn implies $0<\eta<2$. In particular, if we assume D=1, that is a=b/2 or, equivalently, $\eta=1$, we have the hyperbolic-secant solution $U(x)=U_0\,\mathrm{sech}(\alpha x)$.

The soliton propagation constant may be written as

$$\gamma=\frac{a^2}{a-b}\alpha^2 \qquad (27)$$

so that Eq. (25) represents a general class of solutions of Eq. (24) for <u>any</u> α and U_0, provided D is given by Eq. (26) and the eigenvalue γ being determined by Eq. (27). In practice, the condition $\varepsilon<<1$ required for truncating the Taylor expansion of the field amplitude (see after Eq. (15)) restricts the choice of a to value such that $\alpha d<<1$. More precisely, to put this intuitive criterion on a firmer ground, we generalize Eq. (24) to include the term containing U''^2 (see Eq. (21)), thus getting

$$\gamma + b\left(\frac{U'}{U}\right)^2 - a\frac{U''}{U} + c\left(\frac{U''}{U}\right)^2 = 0 \quad , \tag{28}$$

where $c=(k/4n_1)I_{22}$. If we now assume, as suggested by Eq. (25), U(x) to possess an asymptotic behavior of the kind $\exp(-\alpha D|x|)$, Eq. (28) furnishes

$$\gamma = \gamma\left(1 - \frac{c\,D^2\,\alpha^2}{a-b}\right) \ldots\ldots , \tag{29}$$

so that the smallness parameter of our expansion is given by $\varepsilon'=cD^2\alpha^2/(a-b)$.

The second class of solutions of Eq. (24) is given, as can be checked by direct substitution, by

$$U(x) = U_0\, e^{-\alpha^2 x^2} \tag{30}$$

for any U_0 and α, provided that a=b, the eigenvalue γ being furnished in this case by the relation $\gamma=-2b\alpha^2$. This solution cannot be obtained in the frame of the derivation leading to the hyperbolic-secant type pulse (see discussion in Ref. [12]).

The above considerations point to the existence of a lower bound on the material nonlinearity (corresponding to $\eta=0$). Below it, diffraction overcomes self-focusing for all possible waveforms. This implies a lower bound on the product $E_0\, r_{eff}$ (the external field times the material Pockel's coefficient), that is the so-called "soliton threshold". It is derived from the requirement D>0 (or, alternatively, $\eta<2$) that is necessary to guarantee that the boundary condition as $|x|\rightarrow\infty$ is satisfied. More specifically, the soliton threshold is calculated from the condition

$$\frac{1}{2k} + \frac{k}{n_1} I_{20}^{(re)} < \frac{k}{n_1} I_{11} \quad . \tag{31}$$

Interestingly, the condition D>0 also implies an upper bound on the product $E_0 r_{eff}$, above which the nonlinear focusing effects exceed the diffraction effects and a photorefractive soliton cannot be observed. However there may exist higher order soliton solutions for large fields.

The soliton threshold condition furnished by Eq. (31) practically expresses the known guidance condition [46]

$$V = kr\sqrt{2n_1\,\Delta n} \geq 2 \tag{32}$$

for an optical waveguide of radius r. As an example, for a Gaussian photorefractive soliton (see Eq. (30)) it is easy to show, after assuming $r \approx 1/\alpha$, that Eq. (32) corresponds to

$$\frac{1}{2n_1^2 k} + \frac{k}{n_1} I_{20}^{(re)} < \frac{k}{n_1} I_{11} \tag{33}$$

which is practically equivalent to Eq. (31).

It is well known [26,30] that the efficiency of the photorefractive effect is independent of the absolute light intensity. This property is manifested here through the normalization by $|A(\mathbf{r},z)|^2$ in Eq. (8). These photorefractive spatial solitons are therefore independent of their absolute intensity (U_0 is arbitrary in Eqs. (25),(30)). The regime of intensity-independence starts when $|U_0|^2$ is much larger than the dark irradiance [15,26]. The inclusion of material loss (or gain) in this model results in a beam which maintains its transverse profile even as the total light intensity increases (gain) or attenuates (loss) with propagation. In the absorption case, for example, a linear term $\sigma A(\mathbf{r},z)$ is added to the left hand side of Eq. (13), and the soliton propagation constant γ becomes complex. Equation (14) is obtained by taking the imaginary part of γ to be equal to $-\sigma$, and replacing γ by its real part γ_r. The resulting solution is $U(r)\,e^{i\gamma_r z - \sigma z}$, and the transverse structure remains unchanged.

4. DESIGN CONSIDERATIONS

Several considerations have to be taken into account when designing an experiment in order to observe photorefractive spatial solitons.

The first design consideration is the necessity of a real coupling coefficient $\widetilde{\delta n}(\mathbf{q}_1,\mathbf{q}_2)$ for all $\mathbf{q}_1$ and $\mathbf{q}_2$. This can be achieved by applying an external DC field E_0 along the general direction of the grating vectors K_g, i.e. perpendicularly to the optical axis z. In this way, we introduce a drift transport mechanism which redistributes the optically excited charges and the induced index grating is in phase with the interference grating, for each pair of plane wave components. The interacting pairs of plane waves are not capable of exchanging energy, but their phase is altered by the nonlinear coupling effects. Therefore the initial amplitudes of the waves remain constant, and their diffraction effects (which are caused by the accumulation of linear phases) can be compensated by the nonlinear phase coupling.

The second design consideration is the necessity to eliminate the imaginary part of $\widehat{\delta n}(\mathbf{q}_1,\mathbf{q}_2)$ for all $\mathbf{q}_1$ and $\mathbf{q}_2$, which in turn avoids the energy transfer process and keeps the wave amplitudes fixed. In the "conventional" photorefractive crystals, this process is caused by the unavoidable diffusion transport mechanism, which results in a $\pi/2$ phase shift between the interference pattern and the index gratings. Although it can not be completely extinguished, the magnitude of this effect can be greatly reduced by a proper choice of the amplitude of the externally applied electric field E_0. Because the paraxial approximation is assumed in our model, the spatial spectrum f(q,z) of the photorefractive soliton is restricted to a relatively small angular deviation, and the resulting interference gratings (and the index gratings as well) possess large periods. Consequently the limiting space charge field E_p is relatively large and the diffusion field E_d is small. Application of an appropriate external field, such that $|E_d| << |E_0| << |E_p|$ allows us to neglect the imaginary part of the coupling coefficient, so that $E_m(q_1,q_2) \cong -E_1(q_1,q_2)$. The second alternative to satisfy the above design requirement is to take advantage of the newly discovered effects in the paraelectric crystals [47,48], in the absence of externally applied electric fields.

Note that by reversing the polarity of E_0, the linear diffraction and the nonlinear phase add up and the diffraction effect is enhanced. This effect is a manifestation of the self-defocusing nature of a photorefractive nonlinearity under the influence of a negative field. In quasi-steady-state it has the capability of self-trapping dark spatial solitons and vortex solitons [15], as discussed later.

A third design consideration is associated with the symmetry requirement $\widehat{\delta n}(q_1,q_2)=\widehat{\delta n}(-q_1,-q_2)$. As shown in the previous section, to satisfy this requirement for the real part of $\widehat{\delta n}(q_1,q_2)$, one needs to choose the crystalline orientation and beam polarization in such a way that $r_{eff}(q_1,q_2) = r_{eff}(-q_1,-q_2)$.

Another difficulty that may arise in an experimental realization of photorefractive solitons is the interaction with the fanning noise, which can alter the beam's transverse structure and degrade the soliton. Fanning is present in all the efficient photorefractive crystals which support the energy coupling process [and hence have a nonzero imaginary part of $\widehat{\delta n}(q_1,q_2)$]. Scattered light is amplified by this process, at the expense of the input beam [33,41,49]. Eliminating the imaginary part of the coupling coefficient prevents the evolution of the fanning effect and avoids the problem. Should residual $\pi/2$ phase-shifted gratings exist (Eq. (43)), the choice of a transversely narrow soliton waveform reduces the cross-section of interaction with the noise, and the fanning all but disappears (in Ref. 33 it has been demonstrated that a choice of a small spot-size Gaussian beam input suppresses the fanning completely, even in the presence of a very large imaginary part of $\widehat{\delta n}(q_1,q_2)$).

Finally, the most crucial design consideration is the choice of a photorefractive medium that possesses the necessary time window to enable the observation of these nonlocal solitons (of the first type). Because these solitons are transient by nature and observed in a quasi-steady-state (after the gratings have formed but before the external field is screened), one must require, in the first place, that the quasi-steady time window does exist. This requirement implies that the formation time of the photorefractive index perturbation is much shorter than the dielectric relaxation time of the medium. This requirement is easily satisfied by all ferroelectric photorefractive crystals, but in photorefractive semiconductors (such as GaAs and InP) the dielectric relaxation time is shorter than the grating formation time [37]. Photorefractive solitons of the first type cannot, therefore, be observed in semiconductor crystals.

4.1 Example: A Spatial Soliton in SBN

As an example, we analyze the design considerations for experimental observation of a photorefractive spatial soliton in Strontium Barium Niobate (SBN). The symmetry requirement $r_{eff}(q_1,q_2) = r_{eff}(-q_1,-q_2)$ is satisfied by the choice of an optical field linearly polarized along x, once the x direction is adjusted to coincide with the crystalline c axis (the relatively small Pockel coefficients, other than r_{33}, are neglected) which yields

$$\widehat{\delta n}(q_1,q_2) \equiv \frac{n_1^3}{2} r_{33} E_0 \frac{1}{1+\left(\frac{E_0\varepsilon_0\varepsilon_r}{eP_d}\right)^2 (q_1-q_2)^2} \equiv \frac{B}{1+d^2(q_1-q_2)^2} \qquad (34)$$

where the parameters e, P_d, ε, and ε_r were defined in a previous section, and E_m is approximated (in the limit of $|E_d| << |E_0| << |E_p|$) as $E_m(q_1,q_2) \equiv -E_1(q_1,q_2) \equiv -E_0/[(E_0/E_p)^2+1]$. The extension of the region of influence of the nonlocal photorefractive effect is therefore associated with the parameter d appearing in Eq. (34).

The coefficients I_{11} and $I_{20}^{(re)}$ can be found, according to their definitions in Eqs. (18)-(19), to be $I_{11}=-I_{20}^{(re)}=-2Bd^2$. The third-order term in the Taylor expansion of equations (15)-(17) is identically zero, and the fourth order term corresponds to the coefficient $I_{22} = -I_{40}^{(re)}=-I_{04}^{(re)} = 2Bd^4$.

The soliton threshold is manifested by a positive value of D in Eq. (25). This condition implies a negative value on B (the polarity of the externally applied electric field E_0 should be taken as such that the resultant index perturbation is positive, for observation of bright solitons; in this respect, the important quantity is the absolute value of E_0). The soliton threshold corresponds to

$$\frac{n_1}{8k^2d^2} < |B| < \frac{n_1}{4k^2d^2} \qquad (35)$$

and therefore $1/(4n_1{}^2k^2d^2) < r_{33}|E_0| < 1/(2n_1{}^2k^2d^2)$. Recalling that $d=E_0\varepsilon_r\varepsilon_0/eP_d$ and $k=2\pi n_1/\lambda$ yields the condition

$$\left(\frac{\lambda^2 e^2 P_d^2}{16\pi^2 r_{33} n_1^4 \varepsilon_0^2 \varepsilon_r^2}\right)^{1/3} < |E_0| < \left(\frac{\lambda^2 e^2 P_d^2}{8\pi^2 r_{33} n_1^4 \varepsilon_0^2 \varepsilon_r^2}\right)^{1/3} \qquad (36)$$

By taking $r_{33} = 224\times10^{-12}$ [m/V], $\lambda \approx 0.45$ μm, $\varepsilon_r \approx 1100$, $n_1 = 2.35$ and $P_d=5\times10^{15}$ cm^{-3}, the magnitude of the electric field that can induce self-trapping of a light beam in the SBN crystal is between the values 1.9 and 2.4 [kV/cm]. In practice, the external fields required to trap a bright soliton in SBN were found to be between 200 and 500 [V/cm] [13-17]; the discrepancy is attributed to the approximations implicit in the model and in Eq. (34). The existence of a bounded range for the trapping field has been confirmed by all experimental observations.

5. EXPERIMENTAL OBSERVATION OF BRIGHT SOLITONS

In this section, the recent experimental observations [13,14,16,17] of photorefractive spatial solitons are described. The basic apparatus consisted of a CW, 457 nm argon-ion laser and a 5 mm x 5 mm x 6 mm Strontium Barium Niobate (SBN) crystal with 0.01% (by weight) Rhodium dopant. A schematic diagram of the apparatus is shown in Fig. 1. The optical beam was focused by a lens and directed onto the input face of the SBN crystal, which was placed well beyond the Rayleigh range from the beam waist. The beam diameter at the crystal entrance face was 71 μm and 81 μm in the vertical and horizontal directions, respectively. Along with the configuration proposed in Ref. 13, the beam propagated along the crystalline **a** axis and its polarization was initially chosen to be along the **c** axis (extraordinary polarization).

The beam diameter throughout the crystal was measured using an imaging system consisting of an imaging lens and a two-dimensional detector array. Unlike previous studies of propagation of spatial solitons in other nonlinear media [4-10] that included only input and output cross-sections of the soliton beam, these studies examined the actual beam profile at various planes inside the nonlinear medium.

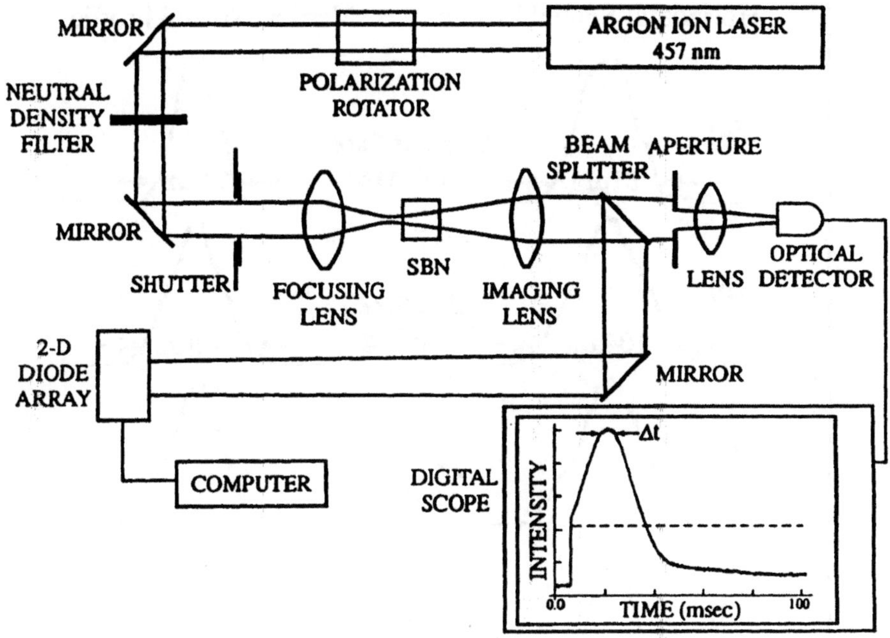

Figure 1. The experimental apparatus used to observe photorefractive solitons.

The method of imaging different planes of the soliton beam deserves special attention and will be analyzed elsewhere. As an intuitive argument, note that, because the material resembles a graded-index waveguide, at least 100x100 transverse modes can be resolved for a propagation distance of about 1 cm before imaging becomes impossible because of modal-dispersion. The imaging system imaged the beam spot at the entrance face (with some magnification) onto the detector array. As the imaging lens and the detector array were moved away from the crystal, different cross sections of the Gaussian beam were imaged onto the detector array. In this manner, the beam diameter at different locations throughout the crystal was monitored. The imaging system was calibrated using known objects (thin apertures) at the input and output faces, and imaging them onto the detector array. Using the known value of the reference apertures, the magnification was determined and the positions of the exit and entrance faces of the crystal were located. Using this information, the cross sections of the incident beam on the entrance and exit faces were determined and are shown in Figs. 2(a) and 2(b). An important restriction on this method

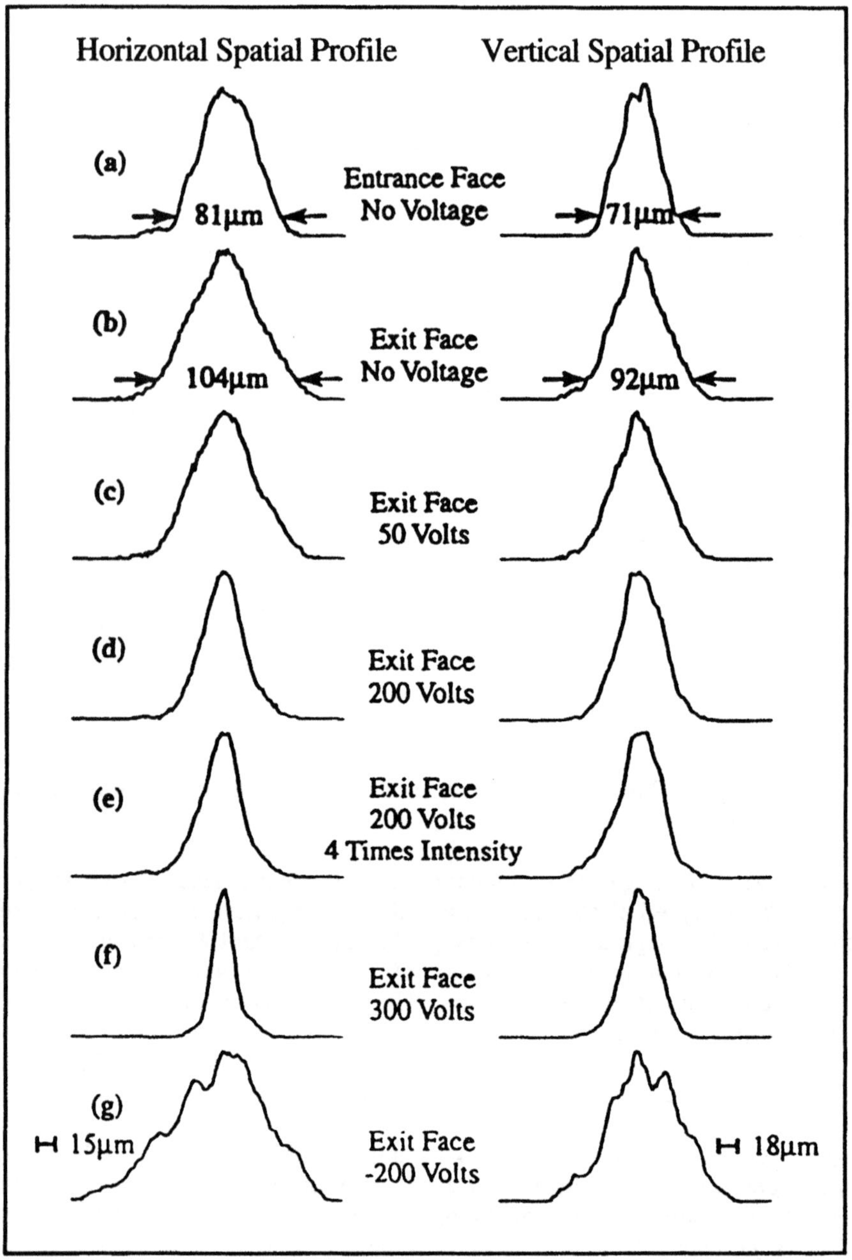

Figure 2. The horizontal and the vertical beam profiles, for various conditions (applied voltages and input powers).

stems from the necessary requirement of resemblance to a graded-index waveguide. One cannot, for example, use this method (as it is) when two solitons are launched, because imaging through a pair of coupled graded-index waveguides is a more complicated issue.

At relatively low applied voltages ($E_0 < 1$ kVolt/cm) and at relatively large beam diameters ($2w \geq 20$ μm) the dominant space charge field is E_1 rather than E_2. We have therefore chosen to work with crystals of proper parameters that enable low voltage observations of solitons. This solves the potential problem of soliton-beam breakdown due to the asymmetric energy-exchange interaction, because in this regime the gain-length product (coupling coefficient times length) is rather small even for spatial frequencies that exceed the beam bandwidth, and fanning is suppressed [33,49]. An important observation is that the propagating spatial profile of the soliton beam remained constant during the period of observation, which may be therefore referred to as quasi-steady-state. This period was determined by monitoring the intensity of the incident beam after passing through the crystal and an exit aperture which was adjusted to be proper size as if diffraction did not occur in the crystal (but did occur, of course, in the air between the exit face and the aperture). The transmission data indicated that self focusing took place before any fanning could be observed. The insert in Fig. 1 shows the total energy transmitted through the aperture as a function of time. Without the applied voltage, no self-focusing effects were observed (dashed curve). After applying the external voltage, self-focusing effects occur (solid curve): the transmitted energy first increased, then reached a quasi-steady-state, and finally diminished (for large times) because of screening of the applied field. For much higher voltages (as discussed below) the disappearance of the self-focusing effects was caused by the combined effects of fanning (very large E_0 enhances E_2 and, at a certain point, energy-exchange dominates) and screening. The quasi-steady-state region provided a window in time when diffraction effects (or solitons) could be observed with very little fanning noise present.

The most convenient method for observing photorefractive solitons with a specific input beam is to scan the applied voltage over the calculated range. In the experiments described in Ref. 13, for a voltage of 50 V (or electric fields of 100 V/cm) applied along the **c** axis, the spatial profile and beam diameter at the exit face of the crystal indicated that self focusing effects were insufficient to compensate for diffraction (Fig. 2c), although it was evident (by comparison to Figs. 2a and 2b) that the beam divergence was reduced. When the voltage was increased to 300 V (field of 600 V/cm) the beam converged as it propagated through the crystal (Fig. 2f). The photorefractive focusing produced with 50 V was insufficient to compensate for diffraction, while 300 V overcompensated. At 200 V (or 400 V/cm) a spatial soliton was formed. This is shown both in Figs. 2d and 3. The latter shows the horizontal and vertical profiles of the beam as it propagated through the crystal. Both profiles were observed to remain nearly constant, regardless of the material inhomogeneities (or structural perturbations in the index, as may be observed in the profiles of Figs. 2a and 2b). The temporal window, or steady-state regime, for this observation was at least 130 msec (in later experiments [15] this window was extended to 10 seconds).

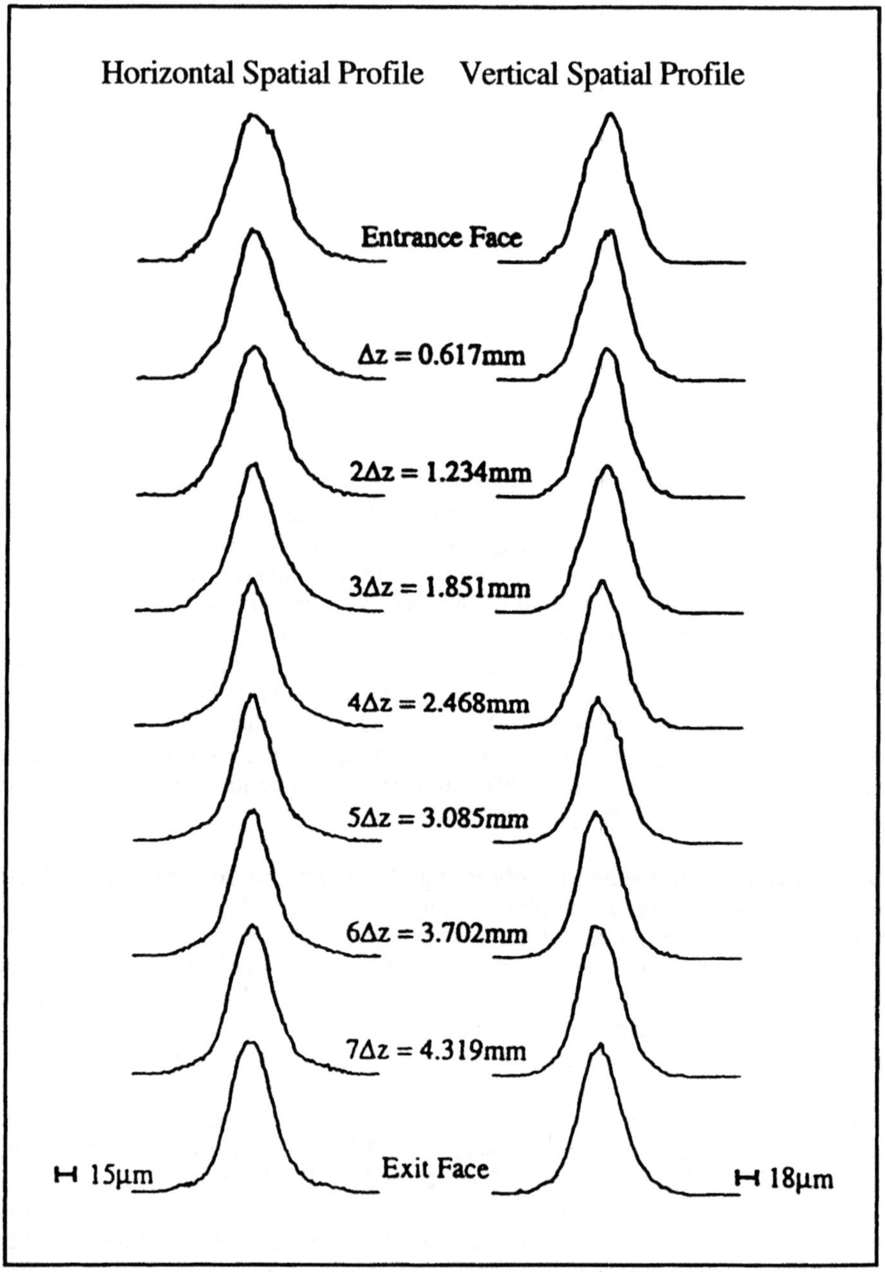

Figure 3. Spatial beam profiles inside the crystal with 200 V.

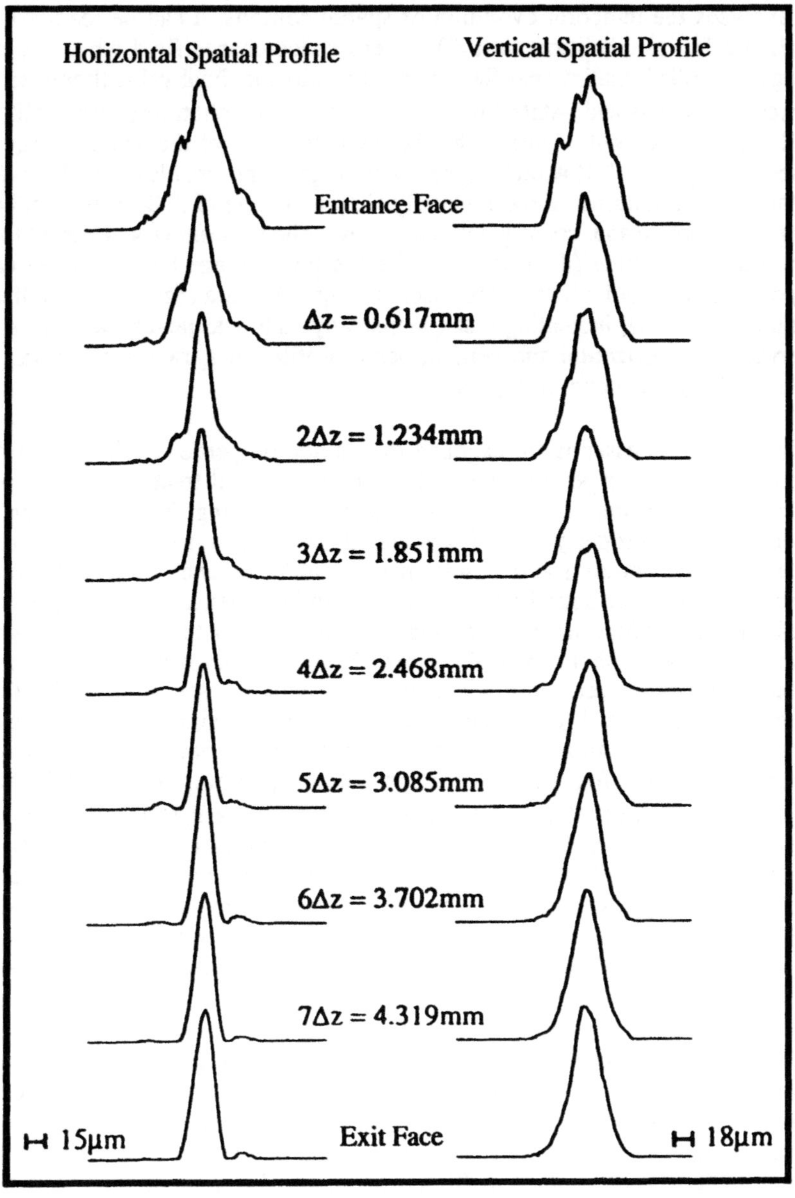

Figure 4. Spatial beam profiles inside the crystal with 3000 V.

At large voltages that exceed the upper limit on the space charge field, it has been predicted [12] that solitons cannot form because the nonlinear self-focusing effects greatly exceed diffraction. This is true in a quasi-steady-state regime. However, as one analyses the temporal evolution of spatial solitons, it can be readily observed that if the initial condition (at t=0) is zero space-charge field when the external voltage is applied, the internal field increases to its maximal value (bounded by E_p), reaches the quasi-steady-state (that cannot support the photorefractive soliton), and finally is screened out. During the rise (and the fall) of the space charge field it passes through the "allowed" region of voltages, and enables the observation of photorefractive solitons. Experimentally [13], a voltage of 3000 V (or field of 6000 V/cm) was applied and, consequently, a narrow solitary beam was observed for a very short window in time (4 msec). Unlike the low voltage (steady-state) case, the temporal curve (equivalent to the insert in Fig. 1) had a significant gradient even within this window, indicating that quasi-steady-state was never reached. A typical observation of horizontal and vertical beam profiles of a transient soliton (at 6000 V/cm in SBN) are shown in Fig. 4.

At this point it is essential to examine the unique properties of the photorefractive soliton (of the first type) as predicted by the theoretical treatment of the previous sections. The existence of a voltage range for trapping the soliton was, in this respect, the first property to be investigated. A second, fundamentally important property of the photorefractive soliton is its independence on the absolute light intensity. The lower bound on this intensity-independence regime is given by the dark irradiance, while the upper bound is determined by the intensity at which the number density of photo-excited charges becomes comparable to the density of ionized donors (a lower, more practical, upper bound is set by thermal effects that become significant at higher intensities and appear at similar time frames as the solitons). The prediction of intensity independence has been tested by varying the input intensity over four orders of magnitude [15] (powers of 0.5-785 μWatts or average intensities of 0.05-78.5 Watts/cm^2; in this respect, 0.05 μWatts is also the lowest power with which a bright spatial soliton has ever been observed). In the experiments of Ref. 15, however, the range of intensity variation was limited by the response of the detector array , i.e., by external background noise. Figures 2d and 2e show that the soliton profile and the trapping voltage did not change under an intensity variation of a factor of 4. Similar figures with a factor of 100 are presented in the section on dark solitons.

As conclusive evidence that the observed soliton indeed had a photorefractive origin, the writing process of the index gratings (that eventually lead to the soliton formation) was intentionally interrupted. The recorded index perturbation remained almost unchanged for days in the dark (the dark decay mechanism is caused by dark current, which is very small in SBN), but could be immediately erased with a spatially uniform, incoherent, light beam of average intensity higher than the soliton peak intensity [13].

Another property of the photorefractive soliton is the sensitivity to the polarization of the input light beam. When the polarization was varied (from extraordinary to ordinary), the voltage required to support the soliton increased, in agreement with Eq. (36), as the electro-optic coefficient changed from r_{33} to r_{13} [the increase was proportional to $(r_{33}/r_{13})^{1/3}$]. A related issue is the polarity of the external field ($\mathbf{E_0}$) with respect to the **c** axis. As predicted by the theory [11,12], when the polarity is reversed, no bright soliton could be observed. Moreover, diffraction effects were enhanced and index inhomogeneities completely distorted the beam (Fig. 2g). Intuitively, the self-generated positive lens that was responsible for the self-focusing effects and for the bright soliton turned into a negative lens (of the same curvature) which increased the diffraction. As described in the next sections, this regime of negative index perturbations gives rise to planar dark solitons and to vortex solitons [15].

The transverse phase of the soliton beam is another interesting issue. In the theoretical model it was assumed, rather than proven, that the transverse phase is uniform. One can show, however, by assuming a soliton form (in Eq. (9)) with a non-uniform transverse phase, and substituting in Eq. (16), that the only solutions that do not lead to a contradiction are those that possess a uniform transverse phase. This was also verified experimentally by examining the diffraction (propagation) of the light beam exiting the crystal. The beam spot at the exit face became the "minimal waist" of the nearly Gaussian beam propagating after the crystal, verifying the uniform transverse phase, regardless of the phase at the input plane [13,16].

The width (transverse cross-section) of the photorefractive soliton is an intriguing issue: while the width of Kerr-like solitons is uniquely determined by the optical power, photorefractive solitons are independent of it. One might expect that because the photorefractive nonlinearity is driven by the external field, the soliton width should depend on it. However the theory [11,12] predicts that the solitons should exist for a well-defined range of external fields (discussed above), and that <u>within this range</u> the soliton width will be <u>independent</u> of this external field. This property is a consequence of the optical nonlinear properties of the medium: the perturbation in the refractive index is <u>not</u> proportional to the external voltage, but rather to the light-induced space-charge field, which is a complicated function of the beam profile. The theoretical consequence of this property is that the soliton width is determined by the parameter a in equations (25) and (30). This parameter is not uniquely determined by a or b, but is proportional to the square root of the soliton propagation constant γ (see Eq. (27) and the paragraph below Eq. (30)). The immediate implication is that, within the rage of existence, α is independent of a and b, and consequently the soliton width is independent of the external field.

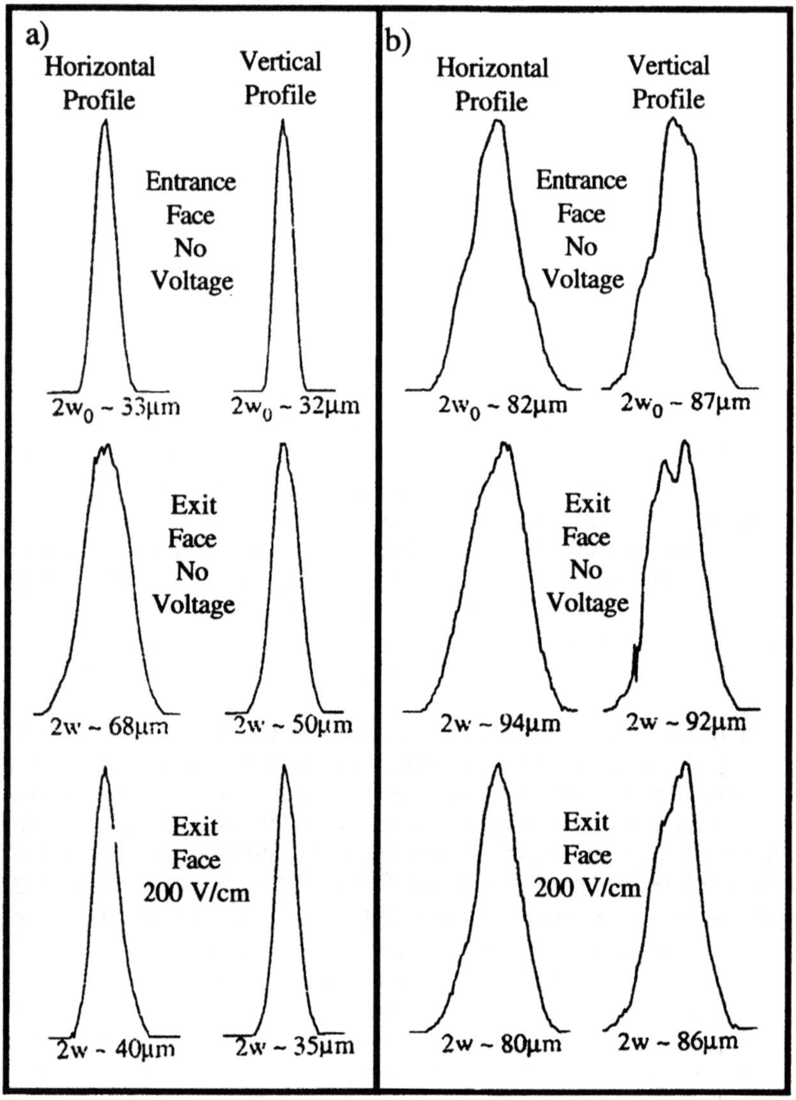

Figure 5. Horizontal and vertical profiles of input (upper), diffracted output (middle) and self-trapped (lower) beams of 35 (a) and 85 (b) μm waists, which are located at the input plane.

The experimental verification of the predictions on the behavior of the soliton width is described in Ref. [16]. The external field was held at 200 V/cm (the allowed region in that specific experiment ranged from 200 to 400 V/cm), and the propagation of circular-input spatial solitons of different sizes was examined by varying the boundary conditions only, i.e., the diameter of the input beam. To study this issue alone and exclude any evolutionary process, through which an arbitrary input evolves into a soliton, we launched beams with a uniform transverse phase that did not require evolution (by launching a Gaussian beam whose waist was always located at the input face of the crystal). The variations in the cross-sections of the input beam were obtained by using a set of different lenses, and placing, in each measurement, the waist generated by each lens at the entrance face of the crystal. Keeping the external voltage fixed, we were able to trap beams of cross-sections between 30 and 90 μm, in a 6 mm long crystal. Narrower solitons required higher fields, while for broader beams, diffraction is very small for this crystal length, and the comparison between self-trapping and diffraction yields an insignificant difference. A typical observation is presented in Fig. 5, where the input, diffracted output and self-trapped output are shown for input beam waists of ~35 μm (Fig. 5a) and ~85 μm (Fig. 5b), respectively. These observations clearly indicate that the diameter of the soliton beam is not unique for a specific (allowed) value of external field, but rather depends on the diameter of the input beam, and evolves into the soliton that is the closest to the boundary condition.

5.1 Soliton Dimensionality

The theoretical model described in the previous sections is two dimensional, i.e. includes one transverse and one longitudinal dimension. The experimental observations, however, have shown that self-trapping is obtained, in some cases, also in two transverse dimensions. Extending the theory to include the additional dimension is inherently a complicated task, because the photorefractive materials are crystalline media, and therefore do not possess a cylindrical symmetry. In the absence of a full three dimensional treatment, several issues have been addressed experimentally from a heuristic point of view only.

Experimental observations [13, 16] showed that the trapping effects were inherently asymmetric with respect to the two transverse dimensions. Assuming the anisotropy of the photorefractive medium to be negligible, the only remaining asymmetry between the transverse dimensions can be associated with the direction of the external field. With this in mind, we recall that the influence of the external field is maximal for gratings whose **K**-vectors are parallel (or anti-parallel) to its direction, and vanishes for gratings that are perpendicular to it [26]. To isolate the asymmetric behavior, we have studied the behavior of a beam that is narrow in one transverse dimension but very wide (virtually uniform) along the other direction. Consequently,

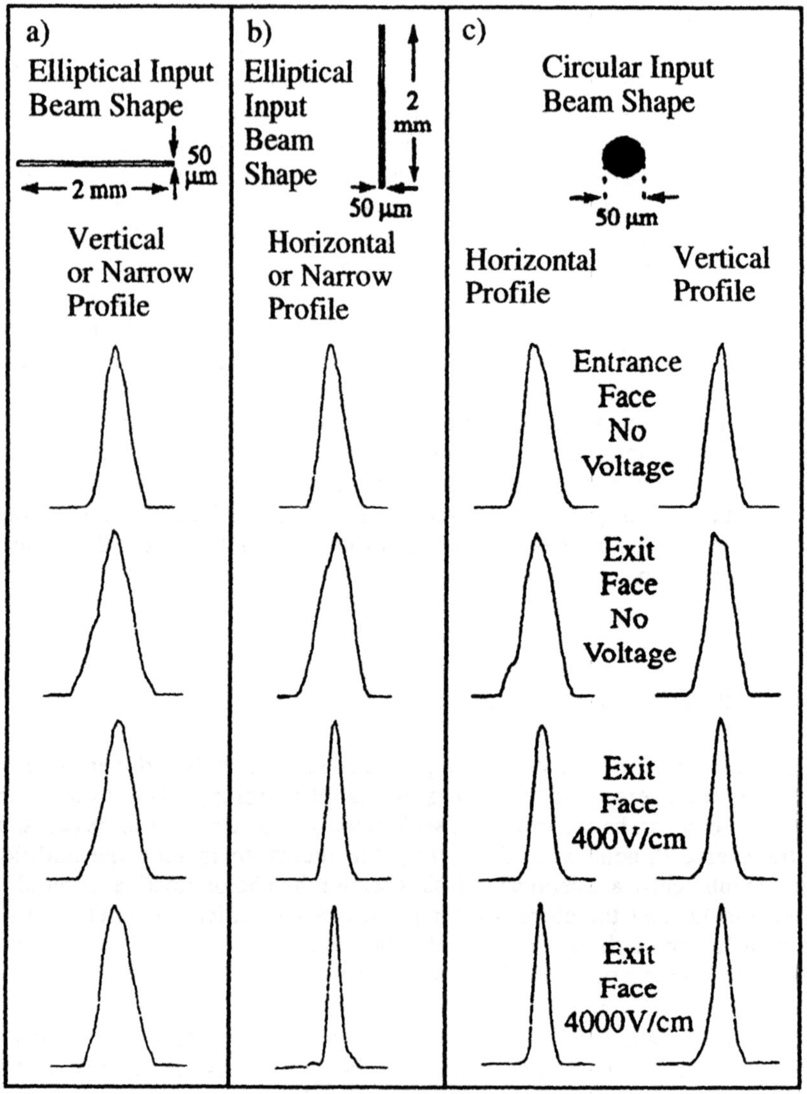

Figure 6. The narrow profiles of an asymmetric beam focused by a cylindrical lens and elongated parallel (a) and perpendicular (b) to the direction of the external field, and the profiles of a circular input beam (c).

we expected to observe a fundamental difference between self-trapping of beams that are uniform in the direction of the applied field and those that are uniform in the direction perpendicular to it. This qualitative experiment was performed using the setup shown in Fig. 1. To simplify the description of the experiment, let us identify the directions parallel and perpendicular to the direction of the external voltage with the transverse x and y axes, respectively, and recall that they correspond to the crystalline **c** and **a** axes, respectively. We employed a cylindrical lens to launch beams that are uniform in one transverse dimension and narrow in the other.

The experimental observations are shown in Fig. 6. In all the experiments the direction of the external field was kept parallel to the x (**c**) axis and the beam was always extraordinarily (**c**) polarized. First, the trapping effects for an input beam uniform in the x direction (roughly 2 mm long) and narrow in the y direction (~ 50 μm wide) were examined. The grating **K**-vectors (which were in the directions of the gradients of the optical intensity) were perpendicular to the direction of the external field, and therefore the external voltage could not influence the beam. In agreement with this intuitive picture, Fig. 6a shows the narrow (y) cross-sections of the input and output beam. As clearly seen in Fig. 6a, self-trapping was observed neither in the "ordinary" trapping fields, nor even at very high fields (~4 kVolt/cm). Thus, this observation indicated that a beam that is uniform in the direction parallel to the external field <u>cannot</u> be trapped. Next, the cylindrical lens was rotated and an input beam uniform in the y direction and narrow in the x direction was obtained (Fig. 6b). Unlike the first case, the beam could now be easily trapped by fields as low as 400 V/cm, and also exhibited transient trapping at high voltages (~ 4 kV/cm) in a manner similar to the one described in the previous section [13]. The corresponding input and output narrow cross-sections (x) of the beam are shown in Fig. 6b. Finally, the cylindrical lens was replaced by a spherical lens and a circular beam with symmetric cross-sections (~ 50 μm) in both x and y, was launched. Trapping was present in both x and y directions, at fields identical to those of Fig. 6b, as shown in Fig. 6c. This implies that the trapping in the y direction relied on tilted gratings, that is, on non-uniform intensity distribution of the input beam which had a significant projection on the x direction. Thus, a beam that is narrow in the direction parallel to the external field <u>can</u> be trapped either in <u>one or</u> in <u>two</u> directions, at exactly the same external field. These observations emphasize the asymmetry between the transverse dimensions, as imposed by the vector nature of the external field.

6. PLANAR DARK SOLITONS AND VORTEX SOLITONS

A planar dark soliton is a nondiffracting notch placed on a background of uniform illumination. In Kerr-like materials [3,7] the dark solitons are hyperbolic tangent functions (of the transverse coordinate x), which undergo a π phase jump at the origin (x=0). Kerr-type dark solitons can be realized only in defocusing media, where the perturbation in the refractive index is negative and proportional to the intensity of the light beam, the soliton size decreasing with increasing intensity.

As expected from the behavior of the bright photorefractive solitons, the photorefractive dark solitons are essentially different from the Kerr-like dark solitons. The theoretical investigations of photorefractive solitons of the second (screening solitons [18]) and the third (photovoltaic soliton [21]) types predicted that dark screening solitons exist for positive perturbations in the index only, and that the narrowest photovoltaic soliton can be obtained for a light intensity on the order of the dark irradiance. Photorefractive solitons of the first type, however, were expected to possess the properties of their bright counterparts. From the behavior of the bright beam when the polarity of the voltage was negative (Fig. 2g) one could immediately infer that in this case the material acted as a self-defocusing medium, and therefore dark solitons were anticipated.

The dark solitons were observed [15] using the experimental setup shown in Fig. 1. As in the previous section, we identify the directions parallel and perpendicular to the direction of the external voltage with the transverse x and y axes, respectively, and recall that they correspond to the crystalline **c** and **a** axes, respectively. In this experiment the beam was extraordinarily (x) polarized. To observe the planar dark solitons, we launched a dark notch with the necessary π phase jump in its center by inserting a thin glass slide in one half of the beam. The phase was adjusted until the notch was obtained by tilting the glass slide. The beam was then focused and entered the photorefractive crystal with a notch size of 21 μm (FWHM). The output beam was split and monitored by a photodetector (to locate the quasi-steady-state time window), by a CCD camera (after an imaging system), and a portion of it was recombined with a portion of the original beam for an interferometric measurement.

The experimental observations are shown in Figs. 7-9. Figure 7 shows the transverse profiles of the beam as it propagated through the crystal. In the left column the beam simply diffracts (zero applied field): both the bright lobes and the dark notch increase their sizes in the same manner (from 21 μm at the input face to 35 μm at the exit face, FWHM of the notch). When a <u>negative</u> voltage was applied ($V/d \approx -400$ V/cm), the dark notch maintained its size and shape, while the bright lobes increased (doubled) their diffraction (right column). This is a good indication that the photorefractive crystal behaves (in the quasi-steady-state) as a self-defocusing medium when a negative voltage is applied. Figure 8 displays the results of the interferometric measurement after the beam had passed through the crystal. It is evident that when the slide was missing (Fig. 8a) or adjusted with no phase jump across the beam (Fig. 8b), the interference fringes were continuous. When the slide was adjusted for a phase jump, the discontinuity of the fringes at the center of the notch was observed both for normal diffraction (Fig. 8c) and when the proper negative voltage was applied (Fig. 8d).

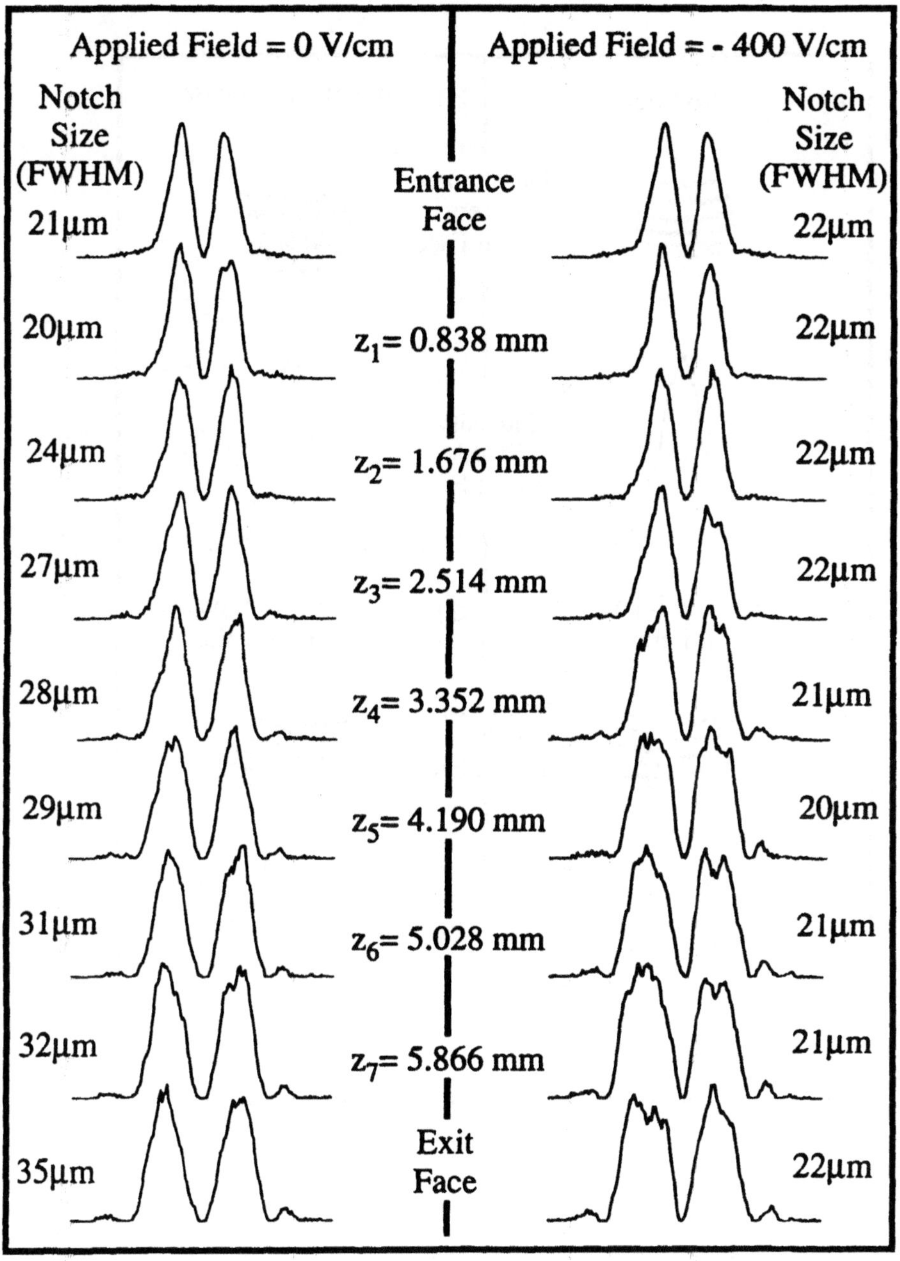

Figure 7. Transverse profiles of the diffracting beam (left column) and the dark soliton (right column) as they propagate through the crystal. The size of the dark notch (FWHM) is given for each plane.

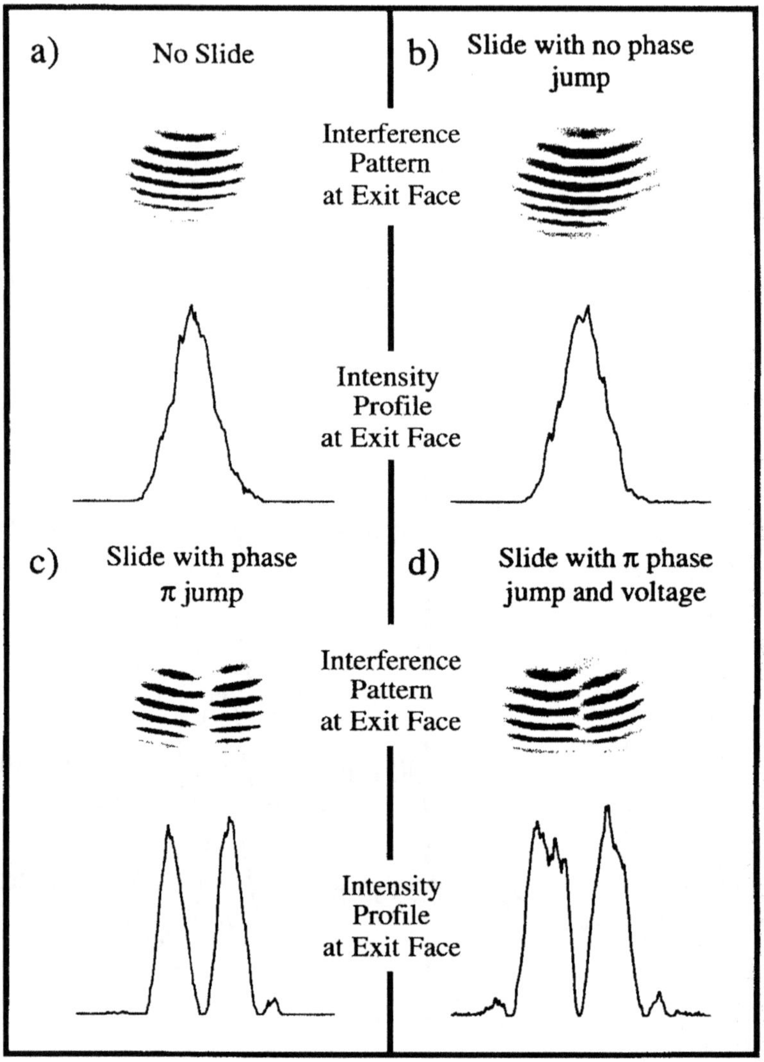

Figure 8. Transverse beam profiles at the output plane of the crystal, and their corresponding interferograms that manifest the continuous phase for a Gaussian input beam (a and b) and the π phase jump at the origin for the diffracted notch (c) and the dark soliton (d).

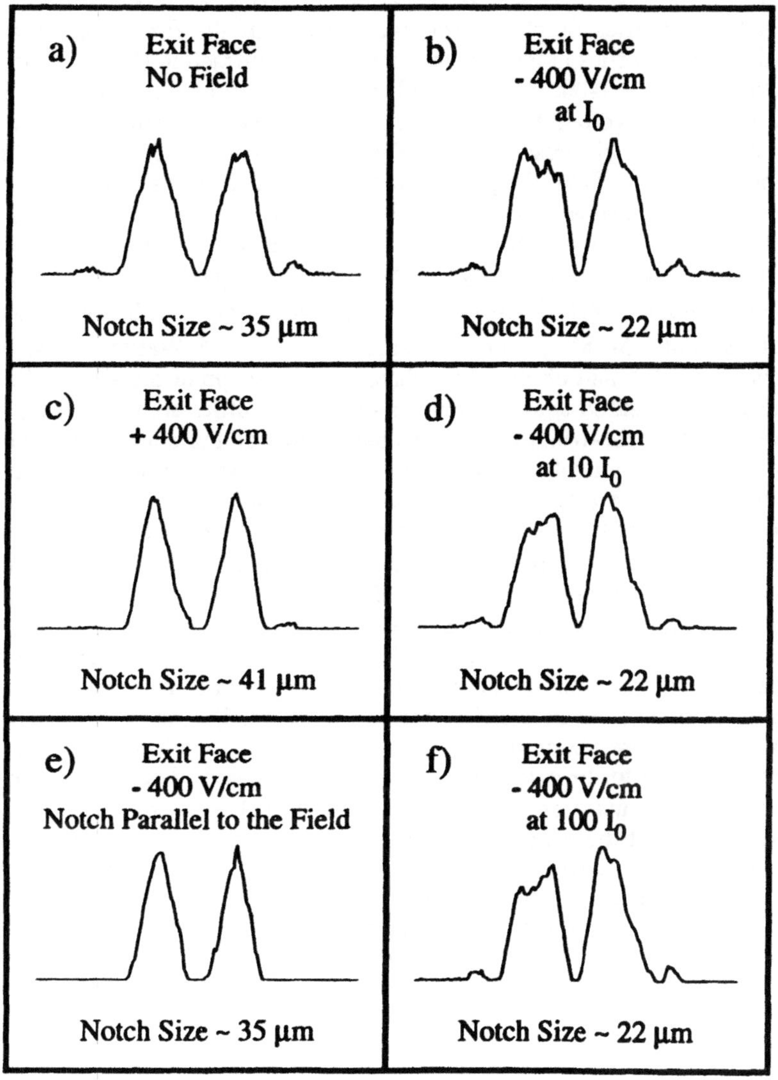

Figure 9. Transverse beam profiles at the output plane of the crystal for normal diffraction (a), for a dark soliton at intensities of 0.3, 3 and 30 Watts/cm^2 (optical powers of 3, 30 and 300 μWatts; b,d,f), for the case of self-focussing (at positive voltages; c) and for a $\pi/2$ rotated notch that cannot be self-trapped (e).

Having demonstrated the existence of planar dark solitons, we proceeded to test the characteristic properties of photorefractive solitons of the first type. Figure 9 displays the beam profiles at the exit face for normal diffraction (Fig. 9a) and for a dark soliton (at $V/d \approx -400$; Fig. 9b). When the polarity of the applied field was reversed (to positive voltage) the notch increased its size significantly, while the bright lobes reduced their cross-sections (Fig. 9c). This is an indication that for the positive polarity of the field, the photorefractive crystal behaves as a self-focusing medium (which is capable of trapping a bright soliton but not a dark one). An important signature of the nonlocal photorefractive solitons is their independence of the absolute light intensity. To verify this we varied the input power over 2-3 orders of magnitude (3-300 μWatt, intensities of 0.3-30 Watt/cm^2, for the experiment on dark solitons, and 0.5-785 μWatt, intensities of 0.05-78.5 Watt/cm^2, for bright solitons experiments), and observed no change in the shape or the size of the dark soliton. Figures 9d and 9f show typical soliton beam-profiles at 30 and 300 μWatt, respectively. The time-window for the observation of the solitons is inversely proportional to their optical intensity which, because the soliton shape remained constant over the entire range of experiments, is proportional to the optical power. The lower limit on this intensity was dictated by detector limitations (rather than by the dark irradiance, which was estimated to be much smaller than 10 mWatt/cm^2). The upper limit was determined by thermal effects which induced nonuniformities in the index. Finally, the input notch was rotated, so that its cross-section was in the y direction (perpendicular to the external field). As clearly seen in Fig. 9e, self-trapping effects of the notch could not be observed even at higher fields (-1000 Volt/cm; much higher negative fields depoled the crystal). As in the case of the bright soliton, this observation indicates that a beam (or a notch) that is uniform in the direction parallel to the external field <u>cannot</u> be trapped. For comparison with previous theoretical analyses [3] and observations [7] of spatial Kerr-like dark solitons, we point out the major differences among the properties of Kerr-like and photorefractive solitons: (i) independence of photorefractive solitons on the light intensity, including the absence of a threshold, (ii) dependence of photorefractive solitons on the direction of trapping (a manifestation of the tensorial nature of the photorefractive effect), (iii) possibility of switching from a regime of dark photorefractive solitons to one of bright photorefractive solitons by reversing the polarity of the applied field and (iv) stability of spatial photorefractive solitons (both dark and bright) in two transverse dimensions.

Next, we describe the observation of photorefractive vortex solitons. Optical vortices [8,9] are beams possessing a uniform amplitude and a phase that varies as $\exp(im\theta)$ (referring to a polar, transverse, coordinate system r and θ) where m is an integer (called the topological charge). Vortices on Gaussian beams [8] exhibit also a radial Gaussian dependence, and are nearly identical to the familiar "donut mode" [50] of a laser that is often observed when its intra-cavity aperture is large. This kind

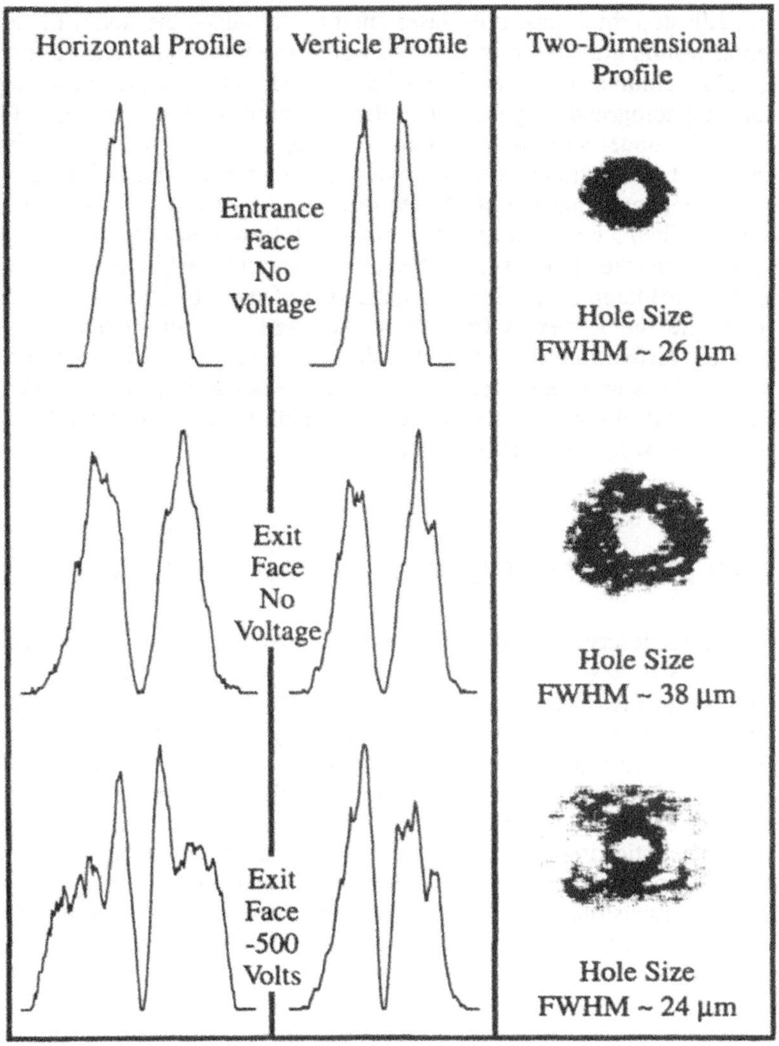

Figure 10. Two-dimensional transverse profiles and photographs of the input (upper) and diffracting optical vortices (middle) and of the photorefractive vortex soliton (lower).

of mode can be represented as a Laguerre-Gaussian mode with p=1 and azimuthal dependence exp(imθ) where $m \geq p$ [50]. We used the experimental setup shown in Fig. 1, and adjusted the intra-cavity laser aperture to launch the donut mode. The experimental results are shown in Fig. 10, where the horizontal (left column) and vertical (middle column) cross-sections of the cylindrical beam are shown together with the actual photographs (right column; darker regions represent regions of higher optical intensity), under various conditions. The upper and the middle sets show the beam profiles at the entrance and the exit faces of the crystal, with no applied voltage. It is apparent that the bright regions and the dark regions diffract in the same manner. When a negative voltage was applied (lower profiles) the central two-dimensional region (the optical vortex) became self-trapped, while the bright regions increased their diffraction significantly (indicative of the self-defocusing behavior). Note that, while the vortex is trapped in the same manner in both transverse dimensions, the diffraction of the bright regions is more enhanced in the horizontal direction. Comparison to observations of optical vortex solitons in Kerr media [9] reveal that all the differences in the properties of both types, as discussed above for dark solitons, persist for the vortex solitons.

7. STABILITY OF PHOTOREFRACTIVE SOLITONS

The observations described in previous sections revealed that the photorefractive soliton is stable and may be observed despite index inhomogeneities that are always present in photorefractive materials. Furthermore, we observed that the photorefractive soliton can evolve from an arbitrary input waveform (see Fig. 4). Although this evolution property of the solitons may be viewed as another manifestation of their stability, we note that the evolution stage encompasses other properties (such as modulation instability). We therefore isolate stability alone and study its characteristics in this section. Because our theoretical analysis is restricted to two (transverse x and longitudinal z) dimensions and cannot explain trapping in two transverse dimensions, it is desirable to investigate stability to perturbations in the soliton waveform that may occur in one dimension only. Fortunately, in most ferroelectric photorefractive crystals inhomogeneities consist primarily of index variations parallel to the crystalline c axis, which is also the growth direction as well as our transverse axis x. This enabled an experimental study, consistent with the theoretical formulation, of the soliton stability to perturbations in one transverse dimension [17].

We start by recalling the two-dimensional version of Eq. (3) (the paraxial nonlinear wave equation), that is

$$\left(\frac{\partial}{\partial z} - \frac{i}{2k}\frac{\partial^2}{\partial x^2}\right) A(x,z) = \frac{ik}{n_1}\,\delta n(x,z)\, A(x,z) \qquad (37)$$

and assume the presence of some material index-perturbation giving rise to a deviation from these ideal solutions. We examine whether the consequent perturbation in the field amplitude grows or is confined to a small propagation distance. If the photorefractive light-induced index perturbation is real (excluding absorption gratings but permitting uniform absorption), i. e. $\delta n(x,z) = \delta n^*(x,z)$, one obtains, by multiplying Eq. (37) by A^* and adding it to its complex conjugate equation,

$$\frac{\partial}{\partial z}(AA^*) + \frac{i}{2k}\left(A\frac{\partial^2 A^*}{\partial x^2} - A^*\frac{\partial^2 A}{\partial x^2}\right) = 0 \tag{38}$$

which is equivalent to conservation of energy. Introducing a deviation $U^{(1)}(x,z)$ from the soliton solution $U^{(0)}(x)\,e^{i\gamma z}$ (where $U^{(0)}$ is either the $U_0[\mathrm{sech}(\alpha x)]^D$ solution of Eq. (25), or the Gaussian solution of Eq. (30)), we write

$$A(x,z) = U^{(0)}(x)\,e^{i\gamma z} + U^{(1)}(x,z) \tag{39}$$

where $|U^{(0)}|^2 >> |U^{(1)}|^2$ for all x and z. Substituting A(x,z) into Eq. 38, and neglecting small terms of the order of $|U^{(1)}|^2$, yields

$$\left[U^{(0)}(U_z^{(1)*}+i\gamma U^{(1)*}) + \frac{i}{2k}(U^{(0)}U_{xx}^{(1)*} - U^{(1)*}U_{xx}^{(0)})\right]e^{i\gamma z} +$$

$$\left[U^{(0)}(U_z^{(1)}-i\gamma U^{(1)}) + \frac{i}{2k}(U^{(1)}U_{xx}^{(0)} - U^{(0)}U_{xx}^{(1)})\right]e^{-i\gamma z} = 0 \tag{40}$$

where the subscripts stand for partial x or z derivatives. Grouping terms with identical exponential arguments, which corresponds to considering distances $z >> 1/\gamma$ (in practice, z of the order of 100 λ or larger), we get

$$U_z^{(1)} - i\gamma U^{(1)} + (i/2k)\left[(U^{(1)}\,U_{xx}^{(0)}/\,U^{(0)}) - U_{xx}^{(1)}\right] = 0 \tag{41}$$

Let us now assume that the perturbation $U^{(1)}$ takes place on a scale ℓ_p much smaller than the soliton size $\ell \sim 1/D\alpha$ ($\ell >> \ell_p >> \lambda$) (for the Gaussian solution, $\ell \sim 1/\alpha$). Because $U_{xx}^{(0)}/U^{(0)} \approx 1/\ell^2$, $U_{xx}^{(1)}/U^{(1)} \approx 1/\ell_p^2$ and recalling that $\gamma = D/(\ell^2 k) \approx 1/(\ell^2 k)$, the second and the third terms on the left-hand-side of Eq. (41) may be neglected with respect to the first and the fourth terms, and it becomes

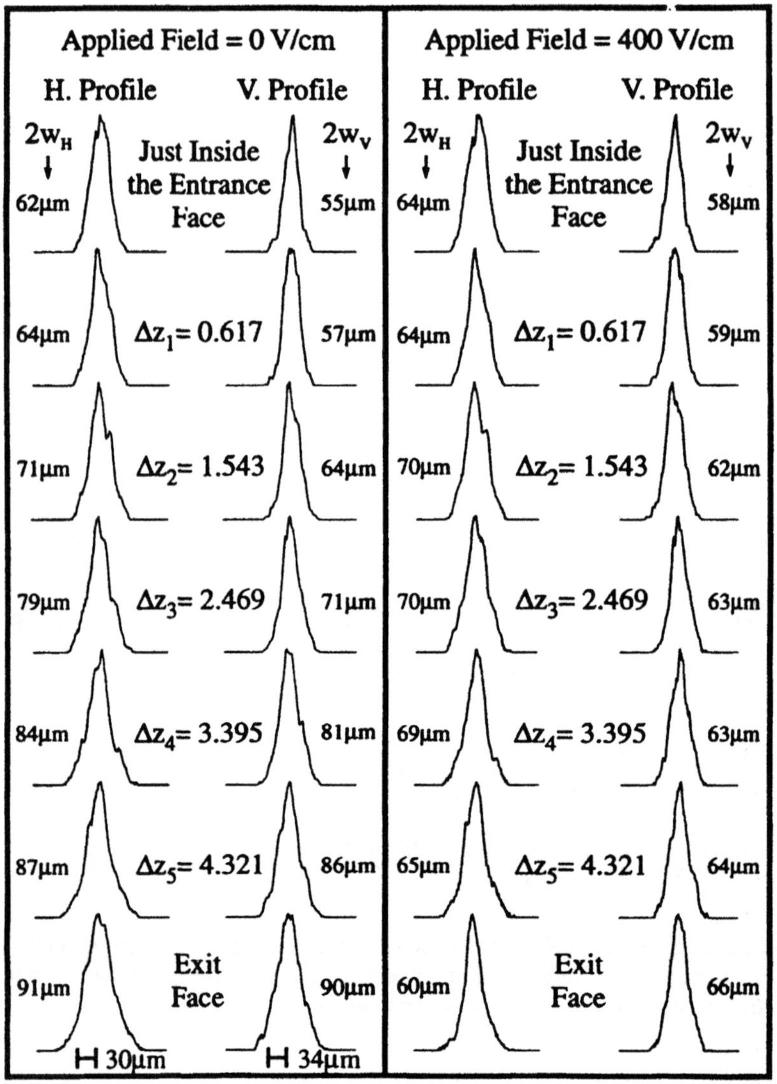

Figure 11. Horizontal and vertical cross-sections of the propagating diffracted (zero field, left column) and soliton beam (right column) at various planes (inside the photorefractive crystal) in the presence of small index perturbations. The horizontal and vertical beam spot-sizes (ω_H and ω_V) are calculated according to a Gaussian fit.

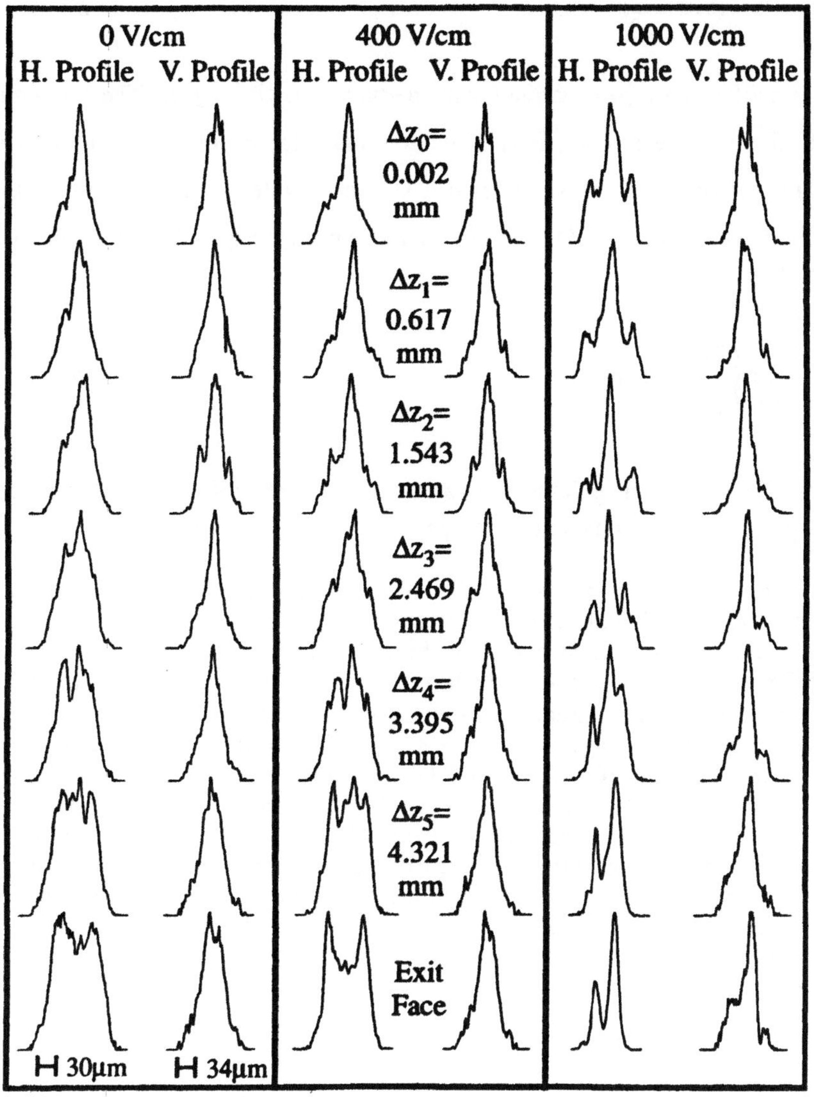

Figure 12. Horizontal and vertical cross-sections of the propagating beam with zero (left column), 400 V/cm (middle column) and 1000 V/cm (right column) external fields at different planes (inside the photorefractive crystal) in the presence of large index perturbations.

$$U_z^{(1)} - (i/2k)\, U_{xx}^{(1)} = 0 \ , \tag{42}$$

which is a diffusion-type equation for the perturbation $U^{(1)}$ only. The longitudinal scale of $U^{(1)}$ may be as small as 10 λ, its minimal value being restricted only by the paraxial approximation. This implies that $U^{(1)}$ propagates (almost) independently of the soliton solution $U^{(0)}$. On the other hand, because Eq. (42) entails the energy conservation relation

$$\frac{d}{dz}\int_{-\infty}^{\infty}\left|U^{(1)}\right|^2 dx = 0 \tag{43}$$

$U^{(1)}$ remains small in magnitude, separated from $U^{(0)}$ and diffuses (by diffraction) as it propagates away from its point of origin.

We now present experimental results that illustrate the stability of the photorefractive solitons. Figures 11 and 12 show transverse cross-sections (normalized to the maximal amplitudes in each plane) of the propagating beams at various external fields. The minimal waist $2\omega_0$ of the beam in Figs. 11 and 12 was ≈30 μm, located about 2 mm before the entrance face of the crystal (at z = -2mm). The confocal parameter of the beam was $z_0 \approx 3.1$ mm, so that diffraction effects became apparent mainly at the last three planes. In Fig. 11, the input beam was launched through a region of small perturbations and we examined its behavior at zero field (normal diffraction, left column), and with an external field of 400 V/cm (soliton, right column), as it propagated through the medium (z = 5mm indicates the exit face). This figure shows that, despite significant index inhomogeneities (apparent mainly at the third and fifth planes from the top in Fig. 11) the photorefractive soliton was rather stable and smooth, and the perturbations in the wave-forms diffused with increasing distance from the plane in which they were generated. The soliton size ℓ (FWHM) was ≈30 μm and we estimated the perturbation size ℓ_p to be ≈3 μm.

The soliton stability breaks down when the scales of the soliton and the perturbation become comparable, i.e., $\ell \approx \ell_p$. This situation is present in Fig. 12, which illustrates a case in which the input beam was launched directly into a region of a growth striation in the same crystal (the striations planes are perpendicular to the crystalline c axis, which is the transverse x axis in our experiment). The left column shows the propagation of the input beam with zero external field at different planes. The middle column shows the propagation with the external field that trapped the beam in the case of Fig. 11. The situation is even more dramatic in the right column where the external field exceeded the trapping field. The beam broke down into two filaments, each trying to form its own non-diffracting beam, but failing to do so because of their mutual, transversely irregular, interaction (this experiment is also a good indication that soliton beams which are sufficiently close to each other do interact). The soliton size ℓ was again ≈30 μm but ℓ_p was now estimated to be ≈10 μm. We estimated the size and dimensionality of the index perturbation which destroyed the soliton by varying the location of the input beam and scanning it in both transverse directions. The x region over which a soliton was not able to form was roughly ≈ 10 μm wide, which corresponds to $\ell/\ell_p \approx 1/3$. Translation in the y direction, on the other hand, did not change significantly the region of self-trapping, a property which is consistent with the assumption that the dominant index perturbations are along the x direction and more or less uniform in the y direction.

8. DISCUSSION

In this chapter we have presented an updated overview of the current state of research on photorefractive spatial solitons. We have concentrated on the nonlocal solitons (photorefractive solitons of the first type) and described the existing theoretical model and the experimental observations. There remains, however, a wide range of open questions and various directions for further extensive research. Some of these are related to collisions between solitons and asymmetric solitons (that are launched in a direction not parallel to a principal crystalline axis), dark solitons and inclusion of nonuniform losses. Other questions are brought up by the physical complexity of our problem: we have experimentally found that the photorefractive soliton exists in two transverse dimensions, but we have restricted our analysis to one transverse dimension only. Because the nonlinear medium in our case is always a crystal, which does not have cylindrical symmetry, the extension of our model to three dimensions is not an easy task. The current effort is directed towards modeling a full three-dimensional analysis that includes the necessary temporal evolution and accounts for screening. We believe, however, that the richness of nonlinear effects in photorefractive crystals entails a large variety of soliton effects yet to be discovered.

We gratefully acknowledge the collaboration we had in part of the research on the nonlocal photorefractive solitons with Ed Sharp (of the Army Research Laboratory) and the never-ending supply of high quality photorefractive crystals we had from R^2 Neurgaonkar of Rockwell International Science Center. We would also like to acknowledge the extensive and useful collaborations with George C. Valley (of Hughes Research Laboratories) on the photorefractive screening-solitons and on photovoltaic solitons, and with Matthew Bashaw, Martin Fejer and Minoru Taya (of Stanford University) on the photovoltaic solitons.

References

[1] M. Hercher, J. Opt. Soc. Am. **54**, 563 (1964).

[2] R. Y. Chiao, E. Garmire and C. H. Townes, Phys. Rev. Lett. **13**, 479 (1964).

[3] V. E. Zakharov and A. B. Shabat, Sov. Phys. JETP **34**, 62 (1972) and Sov. Phys. JETP **37**, 823 (1973).

[4] J. E. Bjorkholm and A. Ashkin, Phys. Rev. Lett. **32**, 129 (1974).

[5] A. Barthelemy, S. Maneuf and C. Froehly, Opt. Comm. **55**, 201 (1985).

[6] J. S. Aitchinson, A.M. Weiner, Y. Silberberg, M. K. Oliver, J. L. Jackel, D. E. Leaird, E. M. Vogel and P. W. Smith, Opt. Lett. **15**, 471 (1990)

[7] G. A. Swartzlander, D. R. Andersen, J. J. Regan, H. Yin and A. E. Kaplan, Phys. Rev. Lett. **66**, 1583 (1991).

[8] G. S. McDonald, K. S. Syed and W. J. Firth, Opt. Comm. **94**, 469 (1992).

[9] G. A. Swartzlander and C. T. Law, Phys. Rev. Lett. **69**, 2503 (1992).

[10] S. R. Skinner, G. R. Allan, D. A. Andersen and A. L. Smirl, IEEE J. Quant. Elect. **QE-27**, 2211 (1991).

[11] M. Segev, B. Crosignani, A. Yariv and B. Fischer, Phys. Rev. Lett. **68**, 923 (1992).

[12] B. Crosignani, M. Segev, D. Engin, P. DiPorto, A. Yariv and G. Salamo, J. Opt. Soc. Am. **B-10**, 449 (1993).

[13] G. Duree, J. L. Shultz, G. Salamo, M. Segev, A. Yariv, B. Crosignani, P. DiPorto, E. Sharp and R^2 Neurgaonkar, Phys. Rev. Lett. **71**, 533 (1993).

[14] M. Segev, A. Yariv, G. Salamo, G. Duree, J. Shultz, B. Crosignani, P. DiPorto and E. Sharp, Opt. & Phot. News **4**, 8 (1993).

[15] G. Duree, M. Morin, G. Salamo, M. Segev, A. Yariv, B. Crosignani and P. DiPorto, submitted to Phys. Rev. Lett., June 1994.

[16] G. Duree, G. Salamo, M. Segev, A. Yariv, B. Crosignani, P. DiPorto and E. Sharp, Opt. Lett. 19, 1195 (1994).

[17] M. Segev, A. Yariv, B. Crosignani, P. DiPorto, G. Duree, G. Salamo and E. Sharp, tOpt. Lett. 19, 1296 (1994).

[18] M. Segev, G. C. Valley, B. Crosignani, P. DiPorto and A. Yariv, to be published Phys. Rev. Lett. (1994).

[19] M. D. Iturbe-Castillo, P. A. Marquez-Aguilar, J. J. Sanchez-Mondragon, S. Stepanov and V. Vysloukh, Appl. Phys. Lett. **64**, 408 (1994). We point out that although this paper has claimed the observation of solitons, no physical evidence for a soliton beam (that should have propagated throughout the medium with no change of shape) was presented. Instead, the observations indicated self-focusing effects that **may** lead to solitons. The observations were attributed (by the authors) to Kerr-type solitons. It is **our** interpretation, however, that these observations are related to the screening effects and may lead to screening solitons.

[20] G. Salamo, University of Arkansas, Fayettville, Arkansas, private communication.

[21] G. C. Valley, M. Segev, B. Crosignani, A. Yariv, M. M. Fejer and M. Bashaw, to be published in Rapid Commun., Phys. Rev. A. (1994).

[22] S. Gatz and J. Herrmann, J. Opt. Soc. Am. B **8**, 2296 (1991).

[23] Y. Chen, Opt. Comm. **90**, 317, (1992).

[24] M. Taya, M. Bashaw, M. M. Fejer, M. Segev and G. Valley, submitted to Phys. Rev. Lett.

[25] V. L. Vinetskii and N. V. Kukhtarev, Sov. Phys. -Solid State **16**, 2414 (1975) and N. V. Kukhtarev, Sov. Tech. Phys. Lett. **2**, 438 (1976).

[26] For a comprehensive overview of a single-grating interaction see G. C. Valley and J. Lam, Chapter 3 in *Photorefractive Materials and Their Applications I*, ed. P. Gunter and J. P. Huignard, Springer-Verlag, (New York, 1988).

[27] D. A. Temple and C. Warde, J. Opt. Soc. Am. **B-3**, 337 (1986).

[28] S. Solimeno, B. Crosignani and P. Di Porto, *Guiding. Diffraction and Confinement of Optical Radiation*, Chapter 8 (Academic Press, Orlando, 1986).

[29] B. Crosignani and A. Yariv, J. Opt. Soc. Am. **A-1**, 1034 (1984).

[30] J. O. White, S.K. Kwong, M. Cronin-Golomb, B. Fischer and A. Yariv, Chapter 4, in *Photorefractive Materials and Their Applications II*, ,ed. by P. Gunther and J. P. Huignard, Springer-Verlag (New York, 1989).

[31] D. L. Staebler and J. J. Amodei, J. Appl. Phys. **43**, 1042 (1972).

[32] J. Feinberg and R. W. Hellwarth, Opt. Lett. **5**, 519 (1980); erratum- Opt. Lett. 6, 257 (1981).

[33] M. Segev, Y. Ophir and B. Fischer, Opt. Comm. **77**, 265 (1990).

[34] A. V. Mamaev and V. V. Shkunov, Sov. J. Quant. Elect. **19**, 1199 (1990).

[35] F. Vachss and L. Hesselink, J. Opt. Soc. Am. **A-5**, 690 (1988).

[36] G. Albanese, J. Kumar and W. H. Steier, Opt. Lett. **11**, 650 (1986).

[37] M. B. Klein, S. W. McCahon, T. F. Boggess and G. C. Valley, J. Opt. Soc. Am **B-5**, 2468 (1988).

[38] M. Segev, Y. Ophir and B. Fischer, Appl. Phys. Lett. **56**, 1086 (1990).

[39] M. Segev and A. Yariv, Opt. Lett. **16**, 1938 (1991).

[40] G. C. Valley, J. Opt. Soc. Am. **B-9**, 1440 (1992).

[41] M. Segev, D. Engin, A. Yariv and G. C. Valley, Opt. Lett. **18**, 956 (1993).

[42] M. Segev, D. Engin, A. Yariv and G. C. Valley, Opt. Lett. **18**, 1828 (1993).

[43] D. Engin, M. Segev, S. Orlov, A. Yariv and G. C. Valley, J. Opt. Soc. Am. **B11**, 1708 (994).

[44] S. Orlov, M. Segev. A. Yariv and G. C. Valley, Opt. Lett. **19**, 578 (1994).

[45] B. Crosignani, M. Segev, G. Salamo, G. Duree and P. DiPorta, submitted.

[46] D. Marcuse, *Theory of Dielectric Optical Waveguides*, Academic Press (New-York, 1974).

[47] A. Agranat, V. Leyva and A. Yariv, Opt. Lett. **14**, 1017 (1989).

[48] R. Hofmeister, A. Yariv, S. Yagi and A. Agranat, Phys. Rev. Lett. **69**, 1459 (1992).

[49] Because beam fanning is a result of energy-exchange interactions (generated by diffusion fields) between the beam and scattered noise, we use two separate effects to eliminate it: (i) for beam size that is much smaller than 100 μm, the cross-section for interaction with noise in directions that differ significantly from z is very small, and (ii) we operate at external voltages that generate space-charge-fields which are much larger than the diffusion field: $E_0 >> E_d \approx \frac{k_B T}{q} \frac{\partial I/\partial x}{I}$ ~8 [V/cm] in the experiments of [13-17].

[50] A. E. Siegman, *Lasers*, (University of Science Books, Mill Valley, California, 1986), Chap. 17, p. 689-690.

5.

PHOTOREFRACTIVE POLYMERS

S. M. Silence*, D. M. Burland, and W. E. Moerner
IBM Research Division, Almaden Research Center,
650 Harry Road, San Jose, CA 95120-6099
*current address: Xerox Corporation, Webster, NY 14580

1. INTRODUCTION

The photorefractive effect was first observed in $LiNbO_3$ more than 25 years ago[1]. Since that time a wide variety of inorganic systems have been shown to be photorefractive, including $BaTiO_3$, $Bi_{12}SiO_{20}$ (BSO), $Sr_xBa_{1-x}NbO_3$, InP:Fe, GaAs, and multiple-quantum-well materials[2]. These inorganic systems are discussed elsewhere in this volume. This chapter is concerned with a new class of photorefractive systems, photorefractive polymers, first unambiguously identified and characterized in 1990[3],[4]. Throughout the history of the photorefractive effect, a variety of applications have been proposed[5]-[11], including high density optical data storage, multiple image processing, phase conjugation devices, beam fanning intensity limiters, simulations of neural networks and associative memories, and programmable optical interconnects. However, the difficult crystal growth, sample preparation and material reproducibility required for inorganics have limited the widespread use of these materials. Photorefractive polymers have the potential to alleviate some of these problems. They are easily processable into thin films of high optical quality, can be modified easily by chemical doping, and are compatible with integrated circuit processing techniques, making them potentially useful in integrated optical device applications.

Many mechanisms exist by which the index of refraction of a material can be modified in response to an incident optical beam; photochromism, thermochromism, thermorefraction, generation of excited states, conventional third order nonlinear electronic processes related to the third order susceptibility $\chi^{(3)}$, etc.[12],[13]. All of these mechanisms, however, lack the key spatial nonlocality property that defines the photorefractive effect. This nonlocal character arises from the physical motion of charges in the material, usually over distances on the order of a few microns. This charge transport gives rise to a spatial phase shift between the incident light intensity pattern and the refractive index modulation. An important consequence of this phase shift is energy transfer between two light beams interfering in a photorefractive medium, called asymmetric two-beam coupling. If the coupling is sufficiently strong, the asymmetric two-beam coupling gain may exceed the absorption and reflection losses of the sample, resulting in optical amplification. Asymmetric two-beam coupling cannot occur with any purely local mechanism of index change (unless the material or the light intensity pattern is physically translated or ac electric fields are applied[14]). Many of the most interesting photorefractive applications[5]-[11] that have been proposed rely on this two-beam coupling gain. These applications include coherent image amplification, novelty filtering, self-phase conjugation, beam fanning limiters, and simulations of neural networks and associative memories. In this chapter, we consider as photorefractive materials only those systems in which nonlocal index of refraction modulation has been observed. Two-beam coupling or its experimental equivalent is thus required for observing such a nonlocal spatial phase shift.

There are several reasons for pursuing the development of organic photorefractive materials, and polymeric photorefractives in particular, in addition to ease of processing mentioned above. One motivation comes from a consideration of a figure-of-merit that permits one to compare the maximum refractive index change possible in different materials (assuming equal densities of trapped charges). This figure-of-merit is given by the expression $Q = n^3 r_e/\varepsilon_r$ where n is the optical index of refraction, r_e is an effective electro-optic coefficient, and ε_r is the dc dielectric constant relative to the permittivity of free space ε_o. Q approximately measures the ratio of the optical nonlinearity to the dielectric screening of the internal space charge distribution. Table 1 shows a comparison of the measured Q values for inorganics along with estimated possible values for Q in organic crystals and polymers. For inorganics, it is well-known that Q does not vary much from material to material. This is a result of the fact that the optical nonlinearity is driven chiefly by the large ionic polarizability so that the large electro-optic coefficient and large dielectric constant are inextricably linked.

For organic nonlinear materials, the nonlinearity is a molecular property arising from

Table 1: Figure of Merit $Q=n^3r/\varepsilon_r$ for Inorganics[a] and Organics

material	electro-optic coefficient r (pm/V)	refractive index n	dielectric constant ε_r	Q (pm/V)
$Bi_{12}SiO_{20}$	5	2.54	56	1.5
GaAs	1.43	3.4	12	4.7
$BaTiO_3$	1640	2.4	3600	6.3
$LiNbO_3$	31	2.2	32	10.3
$KNbO_3$	380	2.3	240	19.3
(Sr, Ba)Nb_2O_6	216	2.3	750	3.5
organic crystal (potential)	67	2.0	3.2	168
organic polymer (potential)	30	1.6	4.0	31
photorefractive polymer (example)[b]	3.1	1.7	7.0	2.2

a. Ref. [2]; b. Ref. [16].

the asymmetry of the electronic charge distributions in the ground and excited states of the individual molecules[15]. For this reason, in organics large electro-optic coefficients are not accompanied by large dc dielectric constants, thus a potential improvement in performance of up to a factor of ten is possible with organic photorefractives as was recognized several years ago[3].

In 1990, the first observation of the photorefractive effect in an organic system was described. The material used was a carefully grown electro-optic organic crystal, 2-cyclooctylamino-5-nitropyridine (COANP) doped with the charge transporting molecule 7,7,8,8-tetracyanoquinodimethane (TCNQ)[17],[18]. The growth of high optical quality organic crystals containing reasonably large dopant concentrations is a very difficult process because most dopants are expelled during the crystal preparation.

Consequently, the strength of the photorefractive effect observed in doped organic crystals is limited. Polymeric materials, on the other hand, can be doped with molecules of quite different sizes and at high concentrations with relative ease. The second-order nonlinearity of polymers containing nonlinear chromophores can be produced by electric field poling to orient the nonlinear molecules, whereas in crystals one must take the crystal structure as is. Only the relatively rare subset of crystals with noncentrosymmetric crystal structures will be electro-optic. Sutter, et al.[19] have recently reported observation of the photorefractive effect in an undoped noncentrosymmetric organic crystal of 4'-nitrobenzylidene-3-acetamino-4-methoxyaniline (MNBA). In this case the MNBA molecules are not only nonlinear but also participate in the photoconductivity.

The first polymeric material to be definitely shown to be photorefractive[3],[4] was composed of the electric-field-poled nonlinear epoxy polymer bisphenol-A-diglycidylether 4-nitro-1,2-phenylenediamine (bisA-NPDA; see Fig. 5a)[20] made photoconductive by doping with the hole transport agent diethylaminobenzaldehyde-diphenylhydrazone (DEH; see Fig. 5b). This material provided a proof-of-principle that the simultaneous requirements of optical nonlinearity, charge generation, transport, trapping, and absence of interfering photochromic effects could be combined in one polymeric material to produce photorefractivity.

2. SPECIAL PROPERTIES OF ORGANIC PHOTOREFRACTIVE POLYMERS

2.1 CHARGE TRANSPORT PROCESSES

Many of the physical processes occurring in organic materials, and necessary for the photorefractive effect, either do not occur in inorganic crystals or have very different characteristics. This is particularly true of transport properties in polymers where, unlike organic or inorganic crystals, both the quantum yield and carrier mobility are electric field dependent.

The quantum efficiency Φ of charge carrier (usually positive hole) generation has been shown to be highly electric field dependent. The mechanism of charge carrier generation in organic systems is presumed to involve the initial formation of a bound electron-hole pair upon absorption of a photon of suitable energy. Separation of this bound pair creates a free hole. Competing with the process of free hole generation is geminate recombination in which the electron and hole simply recombine. Geminate recombination is important in organic systems because of their low dielectric constants. The electron and hole are not electrostatically shielded from each other and, in the absence of an external electric field, recombination dominates over free carrier generation. With the application of an electric field, free carrier generation becomes increasingly important and the quantum efficiency exhibits a strong electric field dependence. This electric field dependence is shown for the photoconducting polymer poly(N-vinylcarbazole) (PVK; see Fig. 5c) doped with trinitrofluorenone (TNF; see Fig 5e)[21] in Fig. 1. One clearly sees a three order-of-magnitude variation in charge generation efficiency with electric field in this system. This electric-field-dependent quantum efficiency has been theoretically described by Onsager[22] and others[23]-[25] and is widely observed in organic photoconductors, both polymeric and crystalline[21],[26]-[28].

The mobility, defined as the charge carrier velocity per unit electric field, is also electric field dependent in polymeric systems. This is not the case for inorganic and most organic crystals where a well defined field-independent mobility μ is measured. In many molecularly doped organic polymers, the mobility of the photogenerated charges has been observed to increase as $\exp\sqrt{E}$ [29]-[33]. This behavior has been observed in several photorefractive polymers as well[34]-[36]. Fig. 2 shows the measured dependence of the (hole) mobility on electric field for the photorefractive polymer bisA-NPDA:DEH mentioned above.

The mobility is also strongly dependent upon the interparticle separation between the transport molecules[30]. This is a consequence of the hopping transport mechanism in polymeric photoconductors that relies on the overlap of electronic wavefunctions on neighboring transport molecules. This means that these transport molecules must be present in high concentration in order to form an efficient connected network for the hopping of carriers. The polymeric system must also contain polar molecules in order to be electro-optic. In many cases, addition of such polar dopant molecules or the use of polar host polymers has been shown to reduce the observed mobility[37] (doped poly(silane) materials provide a counterexample to this trend[37],[38]). Since the nonlinear optical chromophores required for the photorefractive effect are almost always polar, the need for a high concentration of such chromophores may reduce the carrier mobility in some cases.

In addition to charge generation and transport, a photorefractive system must also possess charge carrier traps that allow the formation of reasonably long-lived space-charge fields within the material. The density of traps and the lifetime of carriers in

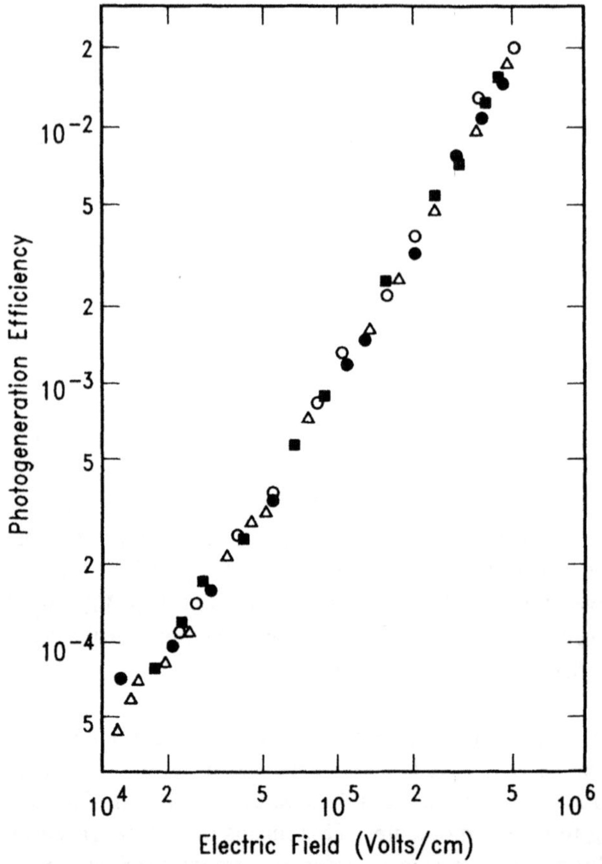

Fig. 1. The dependence of the quantum efficiency of carrier generation on the electric field for poly(N-vinylcarbazole):trinitrofluorenone. Reprinted from Ref. [21].

the traps controls the growth rate, resolution and maximum efficiency of the photorefractive gratings. The nature of the trapping sites in photorefractive materials is only rarely known in detail. For example, several decades of detailed studies on $BaTiO_3$ were necessary[39] in order to reach the present (still qualitative) understanding of the trapping centers, which are most likely oxygen vacancies and Fe^{n+} ions. It is generally agreed that the deep traps in the inorganic crystals are either vacancies, interstitials, or impurities with multiple valence states.

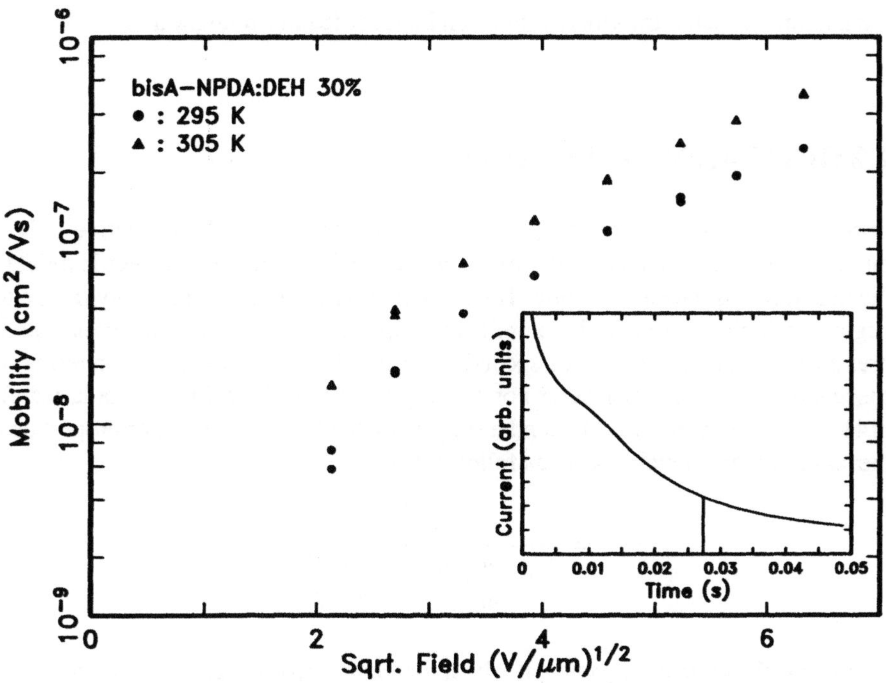

Fig. 2. Dependence of measured hole mobility on electric field for the photorefractive polymer bisA-NPDA:DEH. Reprinted from Ref. [34].

For all of the photorefractive polymers reported in this chapter, no trapping components were intentionally added. Because the photorefractive polymers are a relatively new class of photorefractive materials, it is not surprising that the nature of the traps is not well-known. Some general comments can be made however. Since transport usually occurs via a network of transport molecules in an amorphous host with configurational disorder, there should be a distribution of energies for carriers[23]. The tail of this carrier energy distribution provides a continuous distribution of shallow traps which empty thermally. Deeper traps could result from molecules or ionic impurities with several stable oxidation states. For some photorefractive polymers, the trap density has been estimated and found to be on the order of 10^{15} to 10^{16} cm^{-3}, values which are roughly comparable to those for inorganic crystals[40]. It is clear that further study and optimization of the trapping centers in photorefractive polymers using techniques like ESR and thermally stimulated currents are needed, especially because the trap density determines the maximum space-charge field that can be generated and the maximum spatial information that can be recorded in a hologram.

2.2 OPTICAL NONLINEARITY

In addition to the transport properties discussed above, a photorefractive polymer must also show a dependence of the optical index of refraction upon electric field, i.e., the material must be electro-optic. The second order nonlinearity of a polymer system begins with the second order nonlinear properties of the chromophores that are present in the polymers. Just as second order nonlinearity is described in terms of a susceptibility $\chi^{(2)}$, the molecular nonlinearity is described by the hyperpolarizability β[41]. The hyperpolarizability can be simply related to other molecular properties via the so-called "two level model" as follows[42]:

$$\beta = \frac{1}{\left(1-\frac{\omega^2}{\omega_{eg}^2}\right)\left(1-4\frac{\omega^2}{\omega_{eg}^2}\right)}\beta_0 \tag{1}$$

where ω is the optical frequency at which the photorefractive grating is being written and ω_{eg} is the frequency of the lowest energy electronic transition in the chromophore. The hyperpolarizability at zero frequency, β_0, is given by:

$$\beta_0 = \frac{3}{\hbar^2}\frac{\mu_{eg}^2(\mu_e-\mu_g)}{\omega_{eg}^2} \tag{2}$$

where μ_{eg} is the transition dipole moment of the lowest energy transition, μ_e is the dipole moment of the excited state and μ_g is the dipole moment of the ground state. Several different conventions are used to define β and β_0[43] which will affect the numerical factor in Eq. (2). From these expressions one can get some idea of what an efficient nonlinear chromophore must look like. One clearly wants a molecule where the change in dipole moment upon excitation is as large as possible. An electronic transition possessing such a large change is called a charge transfer transition. This kind of transition is illustrated in Fig. 3 in which the highest occupied molecular orbital (HOMO) and the lowest unoccupied molecular orbital (LUMO) for the nonlinear chromophore p-nitroaniline are shown. The lowest energy (HOMO-to-LUMO) optical transition involves a significant transfer of charge from the electron donor

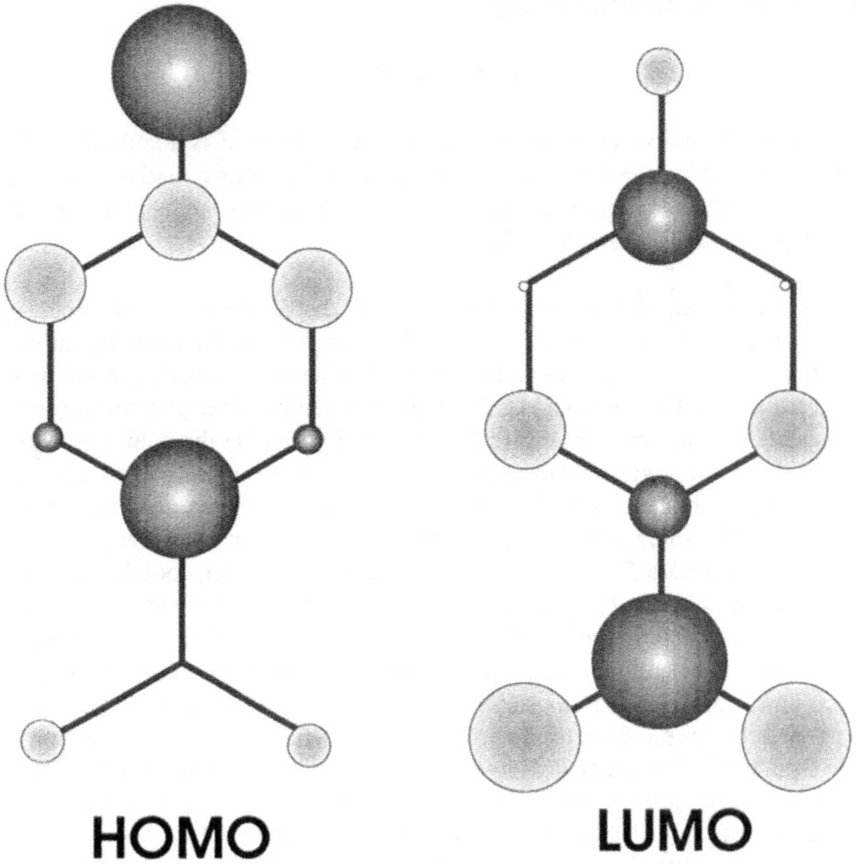

Fig. 3. The highest occupied molecular orbital (HOMO) and lowest unoccupied molecular orbital (LUMO) obtained from a semi-empirical calculation of the electronic structure of p-nitroaniline (PNA, H atoms not shown). The p_z atomic orbital coefficients are indicated by the diameter of the circles at the atoms. The dark or light shading indicates the sign of the coefficient. Reprinted from Ref. [44].

-NH_2 end of the molecule to the electron acceptor -NO_2 end. All nonlinear chromophores used to date have this charge transfer character. In addition, Eq. (1) shows how producing photorefractive gratings using light as close as possible in energy to the charge transfer transition of the chromophore can result in significant resonance enhancement. Of course this resonance enhancement is often counteracted by enhanced absorption and some tradeoff must be made.

The molecular hyperpolarizability is related to the macroscopic second-order susceptibility by the following relationship[45]:

$$\chi^{(2)} = NF\langle\beta\rangle \tag{3}$$

where N is the number of chromophores per unit volume, F is a collection of local field factors relating the externally applied field to the electric field actually experienced by the chromophore and $\langle\beta\rangle$ is an effective hyperpolarizability obtained by averaging over all chromophore orientations.

In a randomly oriented polymer system $\langle\beta\rangle = 0$ and the system is not electro-optic. Second-order nonlinearity must be induced in the polymer by applying an electric field to align the chromophores and remove the center of symmetry present in a random distribution. The alignment of the molecules can be either permanent (or quasi-permanent), in which case the molecules remain aligned after the field is removed for periods of months or years, or temporary, in which case the molecules remain aligned only as long as the field is applied to the material. The scheme used most often for achieving stable poling is to heat the material to a temperature near the polymer glass transition temperature T_g, apply an electric field to the material, cool the material to a temperature well below T_g, and then remove the field. At this lower temperature, the molecules are unable to easily relax back to a random distribution of orientations. Much work has been done on the poling process to maximize both the value of the electro-optic coefficient and its stability[46]; for example, chemical cross-linking of the polymer chains has been used during poling to achieve larger net alignment at the same time that T_g is increased. However, if the T_g of the material is sufficiently low such that the nonlinear optical chromophores can be oriented at room temperature, the value of the effective electro-optic coefficient r_e is a function of the applied field. The coefficient r_e is generally observed to be linearly proportional to the field because saturation of the orientation is seldom reached. This is the situation with many of the photorefractive polymers studied so far and is the origin of a novel enhancement effect[47] related to the ability of the nonlinear optical chromophores to orient at ambient temperatures, resulting in significantly increased photorefractive grating efficiencies (see Section 4.3). It is worth noting that a linear electro-optic coefficient depending upon applied field is formally equivalent to the system having an orientational $\chi^{(3)}$. The applied field converts this $\chi^{(3)}$ into an effective $\chi^{(2)}$, similar to the situation with the electronic $\chi^{(3)}$ in semiconductor materials.

2.3 DESIGN OF PHOTOREFRACTIVE POLYMERS

As discussed above, the necessary elements for a material to be considered as potentially photorefractive are photoinduced charge generation, charge transport, trapping, and electro-optical nonlinearity. Therefore a candidate photorefractive polymer can be made either by incorporating these properties directly into the polymer or by doping guest molecules into the polymer to produce these properties. It is usually desirable to incorporate at least some of the photorefractive properties into the polymeric host itself to minimize the amount of inert volume in the material.

The schematic drawings in Fig. 4 show the three different approaches to the design of photorefractive polymers that have been pursued to date. The first class, illustrated in Fig. 4(a), consists of a host polymer with nonlinear optical (NLO) moieties covalently bonded to the polymer backbone, either as side-chain or main-chain substituents. An alternative approach to providing the optically nonlinear host polymer already familiar from the work on inorganics[48]-[52] would be to utilize a third order nonlinear polymer with a significant electronic $\chi^{(3)}$ such as poly(*p*-phenylene vinylene) in a bias electric field to produce the required optical nonlinearity. The nonlinear host material is doped with small molecule charge transport agents (CT) in concentrations high enough to ensure bulk transport of the photogenerated charges by hopping from transport molecule to transport molecule, a well-known procedure in the electrophotography industry[53]. Charge generation and trapping may be provided by the nonlinear optical polymer itself. Alternatively, the photorefractive response can be optimized by including additional dopants in relatively low concentration to perform the functions of charge generation(CG) and wavelength sensitization or trapping. Various charge generating dyes and pigments (sensitizing agents[54]) are known in the fields of photography and electrophotography. These CG agents permit the sensitization of photorefractive polymers at a variety of wavelengths, a distinct advantage over inorganic systems. One particularly interesting new CG agent for photorefractive polymers is the fullerene C_{60}[55].

The second approach to producing a photorefractive polymer is shown in Fig. 4(b). In this case, the host material consists of a polymer containing charge transporting groups (CT) either as main-chain or side-chain substituents on the polymer backbone. The material is made optically nonlinear by the addition of a nonlinear optical chromophore in as high a concentration as possible to maximize the optical nonlinearity upon poling. As in the first class, additional dopants can be added to provide or improve charge generation (CG) and trapping.

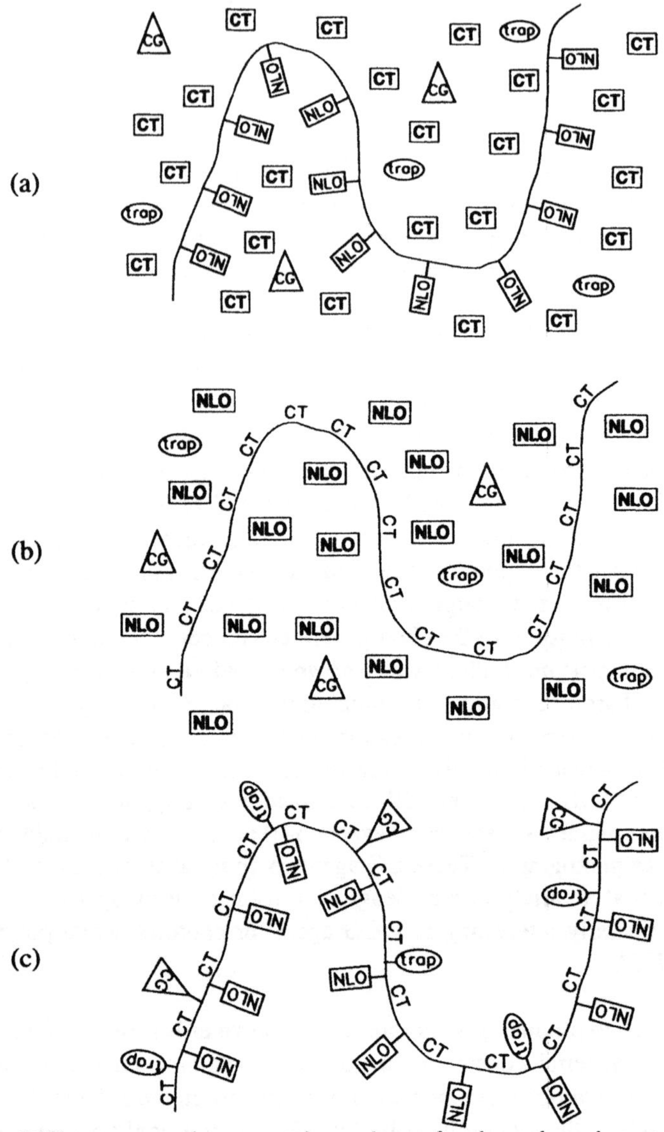

Fig. 4. Different approaches to creating a photorefractive polymeric material. (a) The nonlinear optical (NLO) property is incorporated into the polymer, while the functional groups responsible for charge generation (CG), charge transport (CT), and trapping are incorporated as guest molecules in the polymer matrix. (b) The CT occurs along the polymer chain and the remaining functionalities are incorporated as guests in the matrix. (c) The functional groups responsible for all of the necessary elements are attached to the polymer backbone. Reprinted from Ref. [56].

A third approach to producing a photorefractive polymer involves covalent attachment of all the required functionalities directly to the polymer backbone, referred to as fully functionalized polymers (Fig. 4(c)). To date, several potential materials of this type have been reported, but the performance has been inferior to that for the other two materials classes. In addition, materials in this class pose substantial chemical synthesis challenges. Specific examples of all three classes of materials will be discussed in Section 3. The structures of the most important polymers and dopants described in this chapter are shown in Fig. 5.

2.4 IMPORTANCE OF PROVING PHOTOREFRACTIVITY WITH TWO-BEAM COUPLING

In many cases, polymeric systems have been assumed to be photorefractive solely from measurements indicating that they are optically nonlinear, photoconducting and possess carrier traps. While these features are necessary for photorefractivity they are not sufficient to insure that the photorefractive effect dominates over competing mechanisms of grating formation. A polymer system has been described recently[57] that satisfies all of the necessary criteria for photorefractivity and which also exhibits an electric-field-dependent diffraction efficiency. However, detailed measurements (using methods described in Ref. [58]) reveal no phase-shift between the light intensity pattern and the Δn grating at all fields, clearly demonstrating that the material is not photorefractive. This nonphotorefractive system consists of the photoconducting polymer PVK doped with a nonlinear chromophore Lophine 1 (see Fig. 5f) and in which a small amount of TNF is added. PVK and TNF form a charge transfer complex absorbing in the red. The PVK:TNF complex acts as a charge generator in this wavelength region. Fig. 6 shows detailed measurements of the diffraction grating phase produced in this material as a function of electric field, clearly indicating that the system is not photorefractive despite its electric-field-dependent diffraction efficiency. The precise details of the mechanism responsible for the electric-field-dependent nonphotorefractive grating are not known. A plausible explanation has been given in terms of local electric-field-dependent photochemistry combined with quadratic electro-absorption[57]. Recently, dual gratings produced by the simultaneous presence of reversible photochemistry and photorefractivity have also been described[59].

To unambiguously distinguish between the photorefractive effect and other types of gratings which may occur in polymers, two-beam coupling measurements must be

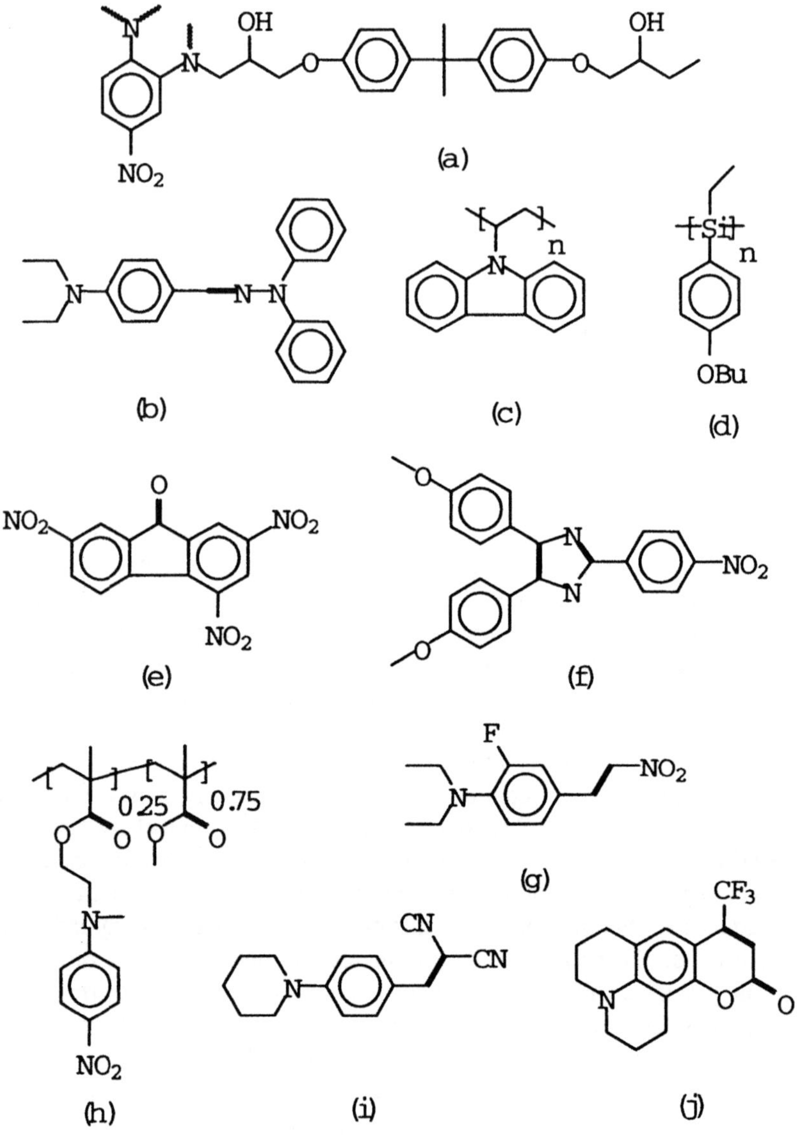

Fig. 5. Structures of materials described in this chapter. (a) bisA-NPDA, (b) DEH, (c) PVK, (d) PBPES, (e) TNF, (f) Lophine 1, (g) FDEANST, (h) PMMA-PNA, (i) PDCST, (j) C-153. For full names, see text.

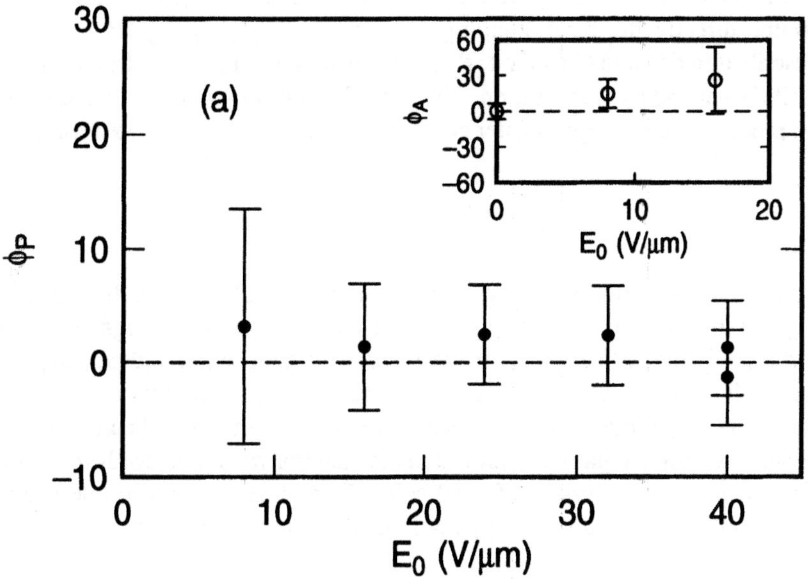

Fig. 6. Diffraction grating phase shift for PVK: TNF: Lophine 1 versus electric field. The phase shifts of the absorption (ϕ_A) and index (ϕ_P) gratings are consistent with 0° phase shift. Reprinted from Ref. [57].

performed. In a two-beam coupling experiment, the change in the transmitted intensity of either of the writing beams is recorded as the other writing beam is switched on and a grating is formed. Each writing beam is partially diffracted in the direction of the other writing beam by the grating. By monitoring the changes in intensity of both beams as the grating grows, the nature of the grating can be determined. This is done most easily when the intensities of the two beams are equal (before the sample). Symmetric loss (gain) in both of the writing beams is the signature of an increase (decrease) in the absorption coefficient which is in-phase with the intensity grating. Such in-phase absorption gratings can result from many processes[12], as we have just seen, such as photochemistry, excited state population, electronic $\chi^{(3)}$, and thermal effects. No change in the intensity of the writing beams can be produced by either an in-phase index-of-refraction grating or a 90° phase-shifted absorption grating. An asymmetric change in the intensities of the two writing beams, i.e., an increase of intensity of one of the writing beams and an approximately equal decrease in the intensity of the other writing beam can only occur from an index of refraction grating which is phase-shifted relative to the intensity grating. Thus, the asymmetric change of intensity of the two writing beams as the grating is formed is an unambiguous signature of the photorefractive effect.

From the increase in intensity of the transmitted signal having gain, it is possible to determine directly the beam-coupling ratio $\gamma_0 = P_{signal,\ with\ pump}/P_{signal,\ without\ pump}$. As described in detail in Chapter 1, P_{signal} is the measured power (after the sample) of the write beam under consideration. The normalized beam-coupling gain coefficient Γ (more precisely, the energy transfer coefficient) may then be calculated using[60]

$$\Gamma = \frac{1}{L}[ln(\gamma_0\beta_P) - ln(\beta_P + 1 - \gamma_0)] \tag{4}$$

where β_P is the ratio of the write powers before the sample (the P subscript is used to distinguish this variable from the molecular hyperpolarizability designated by β), and $L = d\cos\theta$ is the optical path for the beam with gain with incidence angle θ in a sample with thickness *d*. In this expression, the effects of obliquity, i.e. different path lengths, total absorption, and reflection losses for the two beams, have been neglected. Net internal gain is achieved in the material when Γ exceeds the absorption coefficient α of the material ($(\Gamma - \alpha) > 0$). To realize net gain in one beam for a practical application, beam ratios $\beta_P >> 1$ should be used (undepleted pump regime).

As a specific example of the application of Eq.(4) to a photorefractive polymer, Fig. 7 shows the normalized transmitted intensity $I/I(t=0)$ of both writing beams measured in a two-beam coupling experiment for the photorefractive polymer mixture PVK:PDCST:TNF (where PDCST is the nonlinear chromophore 4-piperidinobenzylidene-malonitrile; see Fig. 5i) at $E_0 = 32$ $V/\mu m$ with $\beta_P = 1$[61]. The figure shows the results of two separate experiments. In the first experiment, the intensity of beam 1 is monitored as beam 2 is switched on at $t = 0$ and switched off at $t = 120$ s (open circles). In the second experiment, the intensity of beam 2 is monitored as beam 1 is switched on at $t = 0$ and switched off at $t = 120$ s (filled circles). As the grating is written, the intensity of beam 1 increases and the intensity of beam 2 decreases by an approximately equal amount, indicating an index-of-refraction grating which is spatially phase-shifted relative to the light intensity grating. From the saturation value of $I/I(t=0) = \gamma_0 = 1.06$ for beam 1, a two-beam coupling gain coefficient of $\Gamma = 7.8$ cm^{-1} is calculated. This exceeds the measured absorption coefficient α of the material (2.3 cm^{-1}) and therefore net internal gain is observed in PVK:PDCST:TNF.

The contributions to the optical response due to absorption and refractive index modulation as well as the phase of each of the contributions can be determined by a grating translation technique developed by Sutter and Günter[18] and modified for the oblique geometry common for photorefractive polymers[58]. This can be accomplished either by changing the relative phase of the two writing beams, resulting in a shift of the interference pattern in the sample, or by translating the position of the sample while the interference pattern remains fixed. In either case, the grating translation rate must be significantly faster than the growth rate of the grating to be measured.

3. MATERIALS CLASSES

3.1 NONLINEAR OPTICAL POLYMER HOSTS

The three materials classes of photorefractive polymers examined to date are summarized in Fig. 4. In this Section we will describe a typical example from each of the classes. For a more detailed discussion of the entire range of photorefractive polymers

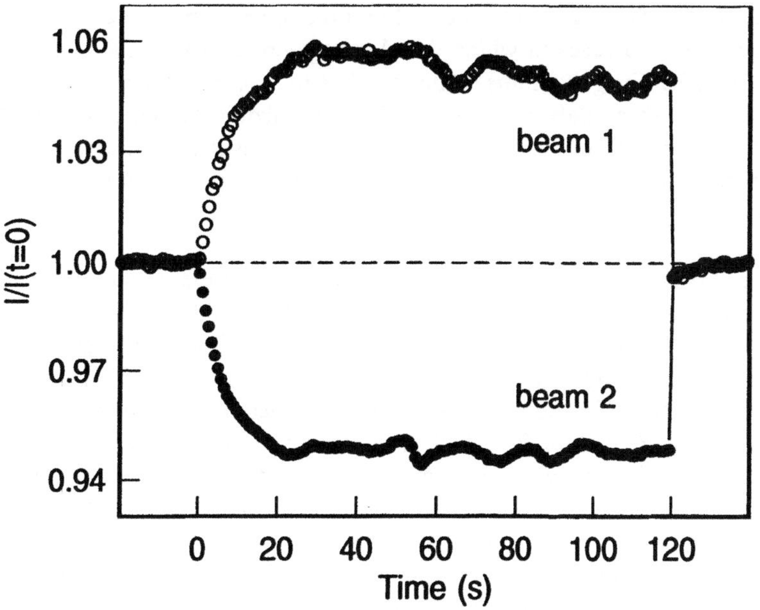

Fig. 7. Two-beam-coupling experiment for PVK:PDCST:TNF at 753 nm with E_0 = 32 V/μm and $\beta_P = 1$. The intensity of beam 1 is monitored as beam 2 is switched on at $t = 0$ and off at $t = 120$ s (open circles) and the intensity of beam 2 is monitored as beam 1 is turned on at $t = 0$ and off at $t = 120$ s (closed circles). The asymmetric energy transfer is proof of the photorefractive origin of the grating. Reprinted from Ref. [61].

identified trecently, the reader is referred to a separate review[56].

Photorefractive polymers in which the polymer itself is optically nonlinear (Fig. 4(a)) have been reported using three distinct types of nonlinear optical polymer hosts. The first type of host polymer is one in which the nonlinear optical chromophore is cross-linked into an epoxy material through multiple bonds between the epoxy chains and the chromophore. The second type of host is a copolymer of methyl methacrylate and an acrylate which has been derivatized with the nonlinear optical chromophore resulting in the chromophore appearing as a pendant group attached to the polymer backbone. Finally, the third type of host polymer is a linear epoxy, where the nonlinear optical chromophore is attached as a side group on an epoxy backbone.

It is worth noting that many other types of polymers have been (or are currently being) developed for purely nonlinear optical applications, such as electro-optic modulators and devices for generating frequency-doubled laser light[46]. These include polymers in which the nonlinear optical chromophore is incorporated into the polymer backbone, and cases in which the chromophore is attached at both ends to the polymer matrix. The goal of this related work is to maximize the second-order optical nonlinearity of these polymers, usually by maximizing the molecular nonlinearity, volume fraction, and the order parameter which measures the degree to which the chromophores are aligned. Many of these nonlinear optical polymers could also potentially serve as photorefractive polymer hosts.

As a specific example, we consider the polymer system that was the first to be definitely shown to be photorefractive, bisA-NPDA:DEH[3],[4]. Hole charge transport was provided by doping the nonlinear polymer host with the well-known[62] charge transport agent, diethylaminobenzaldehyde-diphenylhydrazone (DEH) at a concentration of 30 wt.%. No explicit traps or charge-generating agents were added to the bisA-NPDA:DEH mixture. The NPDA nonlinear chromophores, which are weakly absorbing at the 647 nm wavelength at which the experiments were performed, acted as the charge generators, and the traps were assumed to be naturally present in the polymer. Due to the partial cross-linking and low glass transition temperature T_g of the polymer (65 C; further depressed by the plasticizing effect of the high DEH doping level), the chromophores could be oriented in an electric field at room temperature, and all experiments were done in the presence of an applied field E_0.

The photorefractive effect was first established in bisA-NPDA:DEH by using four-wave mixing to investigate the dynamic behavior, field dependence, and polarization anisotropy η_p/η_s of gratings written in the sample[4], where the p and s subscripts

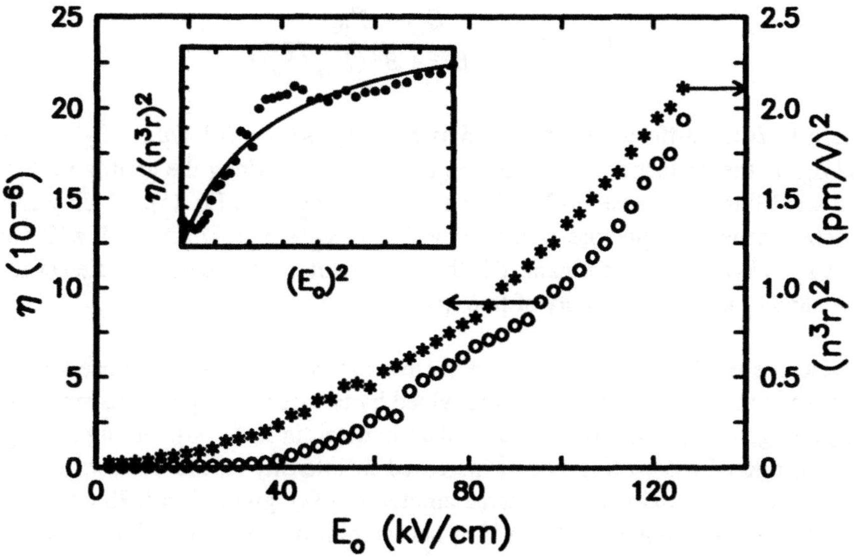

Fig. 8. Photorefractive and electro-optic response of bisA-NPDA:DEH. Right ordinate: square of the electro-optic response (n^3r_e, with n the index of refraction) for p-polarized 633 nm radiation obliquely incident on a 356 μm thick film. Left ordinate: Steady-state diffraction efficiency η_{ss} at 647 nm for the same sample as a function of the applied field. The inset shows the ratio of the latter to the former, which is proportional to E_{sc}^2. Reprinted from Ref. [4].

refer to measurements of the diffraction efficiency η using *p* or *s* polarized light. As with almost all of the photorefractive polymers reported to date, the sample investigated was composed of the polymeric material cast from solution and sandwiched between two glass plates coated with indium-tin-oxide (ITO), an optically transparent conductive material which allows an electric field to be applied across the sample. The thickness of the polymer layer was 350 μm. The steady-state diffraction efficiency η_{ss} is shown as a function of applied field E_0 for bisA-NPDA:DEH in Fig. 8 (left ordinate). As E_0 is increased η_{ss} increases from less than 10^{-6} (the detection limit) to a maximum value of 2×10^{-5} at $E_0 = 12.6$ V/μm. The square of the effective nonlinear coefficient of the sample n^3r_e is also shown as a function of applied field in the Figure (right ordinate).

The inset to Fig. 8 shows $\eta_{ss}/(n^3r_e)^2$, which should be proportional to the square of the space-charge field E_{sc} in the standard photorefractive theory, plotted as a function of $(E_0)^2$. An estimate of the trap density was obtained from the (one-carrier) photorefractive model[40], where the space-charge field is given by

$$E^2_{sc} = \frac{E^2_d + E^2_{0g}}{(1+B)^2 + B^2 (E_{0g}/E_d)^2} \tag{5}$$

where E_d *is* the diffusion field defined as $E_d = K_G k_B T/e = 0.1$ V/μm, $K_G = 3.9$ μm^{-1}, and E_{0g} is the projection of the electric field vector $\boldsymbol{E_0}$ along the grating vector $\mathbf{K_G}$. The solid line in the inset to Fig. 8 is a least-squares fit to Eq. (5), which yields the effective density of photorefractive traps $N_{pr} = K_G^2 \varepsilon_r \varepsilon_o k_B T/Be^2 = 1.9 \times 10^{15}$ cm^{-3}. Here k_B is Boltzmann's constant, T is the temperature, ε_r is the relative dielectric constant, and ε_0 is the permittivity of free space.

Conclusive evidence for the photorefractive nature of the grating observed in the four-wave mixing experiments was provided by two-beam coupling measurements of the grating spatial phase shift, as described in detail in Ref. [58]. During grating formation, one of the writing beams increased in intensity as the other writing beam decreased by approximately the same amount (0.6%), giving $\Gamma = 0.33$ cm^{-1}. This is considerably less than $\alpha = 10$ cm^{-1}, so net internal two-beam coupling gain was not observed in this system.

A similar system (nonlinear epoxy doped with DEH) with a more nonlinear chromophore (4,4'-nitroaminostilbene) has recently been shown to have a two-beam coupling gain $\Gamma > 50$ cm^{-1} at 650 nm[63]. The absorption coefficient was also high at this wavelength (200 cm^{-1}) so net gain was again not observed.

3.2 CHARGE-TRANSPORTING POLYMER HOSTS

Unlike the case for nonlinear polymer hosts discussed above, the number of polymers which inherently transport charge is quite small. To date, only two charge-transporting polymers have been used as hosts for photorefractive polymer mixtures. The first and much more thoroughly investigated of these polymers is poly(N-vinylcarbazole) (PVK), and the second is a poly(silane) derivative. Other charge-transporting polymers that should be investigated as hosts are poly(*p*-phenylene vinylene), poly(thienylene vinylene), and similar materials[64],[65] showing photoconductive properties yet also having a region of optical transparency.

The first reported[66] photorefractive polymer based on PVK consisted of PVK doped with the nonlinear optical chromophore 4-N,N'-diethylamino-(E)-β-nitrostyrene (DEANST, 32 wt.%) and the fullerene C_{60} (1 wt.%) as a charge generating agent. A second, improved PVK-based system was subsequently discovered consisting of PVK doped with the nonlinear optical chromophore 3-fluoro-4-N,N diethylamino-β-nitrostyrene (FDEANST, 33 wt.%; see Fig. 5g), with 1.3 wt.% TNF added as a long-wavelength photosensitizer[67]. FDEANST has reduced absorption in the red relative to the unfluorinated analog DEANST, and therefore absorption by the chromophore in the red and infrared region of the spectrum, where the well-known PVK:TNF charge-transfer complex[21],[68] absorbs, is eliminated. This feature is critical to the improved performance of the system over the unfluorinated analog. The glass-transition temperature T_g of the 125 μm thick films was approximately 40°C. Therefore these films could be poled at room temperature by the application of an electric field, and depoling occurred automatically when the electric field was removed. The PVK:FDEANST:TNF material was significant because it was the first one in which a > 1% diffraction efficiency and net two beam coupling gain were achieved.

The field dependence of the gratings observed in PVK:FDEANST:TNF was examined in p-polarized four-wave mixing experiments. For $\lambda = 647$ nm and 676 nm the grating growth times are comparable to the fastest nonlinear optical-polymer-based photorefractive polymers[55],[69] and η saturates after approximately one second. For $\lambda = 753$ nm where the optical density of the material and the power density in the write beams are smaller, it takes 20 to 30 times longer for the diffraction efficiency to reach saturation. In the absence of an externally-applied electric field, η is below the detection limit of 10^{-8} at both 753 and 676 nm. At 647 nm, a weak grating is observed to grow on the time scale of several minutes in zero field. Two-beam coupling measurements reveal this grating to be a nonphotorefractive 0° phase-shifted index grating[58]. The grating may be due to photochromic processes initiated by absorption in the nonlinear optical chromophore. This absorption accounts for approximately 10% of the total absorption at 647 nm but is negligible at the longer wavelengths.

As the applied field increases, the steady-state diffraction efficiency η_{ss} increases proportional to $E_0^{(3.6\pm0.3)}$ because of the increase in $n^3 r_e$ and in the space-charge field. At $\lambda = 647$ nm, η_{ss} reaches a value of 1.2% at $E_0 = 40$ V/μm. This value of η_{ss} is one to two orders of magnitude larger than the highest diffraction efficiency reported previous to this work for a photorefractive polymer[70],[71], if equal sample lengths are assumed. Recently, an efficiency of 5% was reported in a similar PVK-based material doped with a different optically nonlinear azo dye molecule at $E_0 = 40$V/μm[72]. This system also included N-ethylcarbazole as a supplementary charge transporting and plasticizing agent to further reduce the glass transition temperature and therefore realize additional orientational enhancement (see Section 4.3).

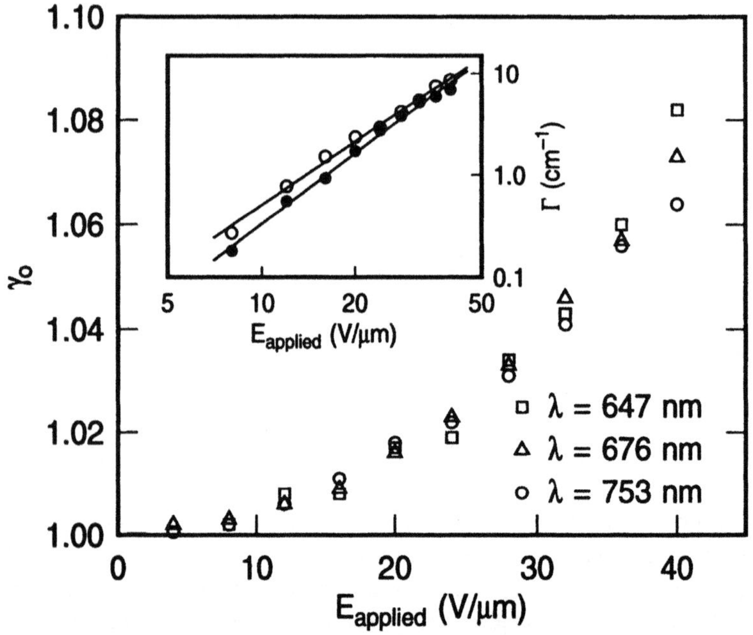

Fig. 9. γ_0 versus E_0 for PVK:FDEANST:TNF at the wavelengths shown with $\beta_P = 1$. Inset: Γ vs. E_0 for 753 nm and $\beta_P = 1$ (open circles) and $\beta_P = 100$ (closed circles). Absorption coefficient $\alpha = 1.4\ \text{cm}^{-1}$ at 753 nm. Reprinted from Ref. [67].

Two-beam coupling experiments verified that the field-dependent gratings observed in the four-wave mixing experiments were indeed photorefractive in origin. The beam-coupling ratio γ_0 for PVK:FDEANST:TNF is shown in Fig. 9 as a function of E_0 at a beam-power ratio $\beta_P = 1$ for 647 nm, 676 nm and 753 nm. The normalized beam-coupling gain coefficient Γ as defined by Eq. (4) is shown in the inset of Fig. 9, for $\lambda = 753$ nm, both for $\beta_P = 1$ and for $\beta_P = 100$. It was experimentally verified that Γ is essentially independent of β_P over the range 1 to 10^4 for the PVK:FDEANST:TNF system. For $\lambda = 753$ nm with a field of 40 V/μm across the sample, Γ reached a value of 8.6 cm^{-1}, which exceeded the material's absorption coefficient α of 1.4 cm^{-1} by more than a factor of six. This made PVK:FDEANST:TNF the first organic photorefractive material for which net internal gain was observed[67]. Although Γ was somewhat larger for $\lambda = 647$ nm (11.0 cm^{-1}), and 676 nm (9.9 cm^{-1}), the absorption coefficient of the material at these wavelengths was greater, 32 and 17 cm^{-1}, respectively, and net gain was not observed.

3.3 FULLY-FUNCTIONALIZED POLYMERS

The last class of polymers that have been considered as candidate photorefractive materials are composed of fully-functionalized polymers (Fig. 4(c)). Polymers which have both the nonlinear optical chromophore and the charge-transporting molecule attached to the polymer backbone could in principle have several advantages over those in which any (or all) of these components is present as a guest molecule. The first potential advantage is the stability of such single component systems against phase separation and sublimation during sample preparation. The second is stability of the electro-optic coefficient. Guest-host polymer systems are generally observed to have lower glass transition temperatures than single component systems, and therefore are less orientationally stable after being poled at high temperature. This advantage becomes less important (and potentially even a disadvantage) when the role of the orientational enhancement mechanism is considered (see Section 4.3).

There also exist several potential disadvantages of this approach. The first is the possibility of interrupting conduction pathways in such systems. In conducting polymers charge transport occurs along the chain, either directly in the backbone (as for polysilanes or PPV) or in the molecules connected to the chain as side groups (as for PVK). In systems in which the charge transport molecules are present as guests, some aggregation of these molecules undoubtedly occurs, which ensures contact between the transport molecules as long as the concentration is above a percolation threshold. However, in single component polymers the charge-transport molecules are attached to the chain in random locations with spacings which may be nonoptimal due to dilution by the comonomers, and as a result charge transport may be less efficient. Other disadvantages are the synthetic challenge in making fully-functionalized polymers, and the lack of flexibility; i.e., new syntheses are needed to make even simple modifications to the polymer.

The first example of a fully-functionalized polymer which was tested for photorefractivity was a methacrylate-based material containing both carbazole and tricyanovinylcarbazole groups linked to the polymer by an alkylene spacer[73]. Thin films of this polymer (1.4 - 3.3 μm) were poled at high temperature (100°C, 100 V/μm) and found to have both a high electro-optic coefficient (r_{33} = 6.1 pm/V) and high photoconductivity at 514 nm, near the absorption peak of the charge-transfer band. Gratings were written in these thin films at wavelengths between 550 and 600 nm[74]. However, two-beam coupling measurements revealed them to be nonphotorefractive, photochemically induced absorption gratings.

A recent report[75] describes photorefractive measurements on a previously discussed fully-functionalized system[76] in which the nonlinear optical chromophore, charge transporting molecule, and a charge generator were all attached to the same polymer backbone. The nonlinear optical chromophore was provided by a monomer with an amino-methylsulfonyl-stilbene functionality, the hole transport was provided by a similar monomer with a diphenylhydrazone related to DEH, and the charge-generator was provided by a third monomer to which one of two different derivatives of the dye Foron brilliant blue was attached. One advantage of this system is that in principle the relative concentrations of the various components (and therefore the volume fraction of the three functionalities) can be independently adjusted. The photoconductive response time at 632 nm was observed to be on the order of 100 ms at an intensity of 0.4 W/cm^2, from which a photocurrent quantum yield of 2.6 x 10^{-3} was reported. However, even though the absorption at this wavelength was appreciable, no values of optical density or absorption coefficient were presented. In a thin film 0.8 μm in thickness, values of r_{33} on the order of 12 pm/V were observed at 632 nm, with only a 5% reduction in this value 3 months after poling.

To test for the presence of photorefractive grating formation, a waveguide geometry was used with poling normal to the waveguide plane[75]. Two-beam coupling studies (in an external field of unspecified magnitude) were performed at 632 nm with a 3 mm interaction length, and gain coefficients up to 2.3 cm^{-1} were reported as a function of grating wavevector. Again, without knowledge of the absorption coefficient, the presence or absence of net gain could not be assessed. The authors did not observe a completely asymmetrical energy transfer between the two beams indicating the simultaneous presence of a photorefractive effect and a photochromic effect (absorption grating). These promising materials should be studied in the future at wavelengths with reduced absorption (or with improved charge generators) in order to try to minimize the photochromic process.

4. THEORETICAL ISSUES FOR PHOTOREFRACTIVE POLYMERS

4.1 ROLE OF CARRIER MOBILITY AND PHOTOCHARGE GENERATION EFFICIENCY

One possible limitation of photorefractive polymers that is often mentioned is the relatively slow response speed, especially when compared to semiconductor materials.

The speed of formation of a space-charge grating is proportional to the photoconductivity which in turn depends upon the quantum yield of photogeneration Φ, the carrier lifetime before trapping, and the mobility of the carriers μ. To increase the photoconductivity and hence the speed, it is often assumed that a material with high mobility is required. This section shows by analysis of the photogeneration process that mobility limitations only occur when the photocharge generation process itself is sufficiently efficient.

The temporal response of a photorefractive polymer can be assessed by using a simple model originally presented by Glass et al.[77],[78] to derive the maximum achievable sensitivity for a photorefractive material, where sensitivity is defined as the photorefractive response (usually index of refraction change) per unit energy required to produce this response. The model was recast by Yeh to estimate the fundamental limit on the speed of the photorefractive effect in inorganic crystals given a required beam-coupling gain[79],[80]. For our purposes, the analysis can be further rewritten[38] as an upper limit on the growth rate of the two-beam coupling gain coefficient, given by

$$\left(\frac{\Gamma}{t}\right)_{max} = \left(\frac{h\nu}{e}\right)\left(\frac{\Lambda_G}{\lambda}\right)\left(\frac{\pi}{2}\right)\left(\frac{n^3 r_e}{\varepsilon_0 \varepsilon_r}\right)(\alpha\Phi) I \qquad (6)$$

where the grating wavelength $\Lambda_G = 2\pi/K_G$, and all other parameters have been previously defined. It is understood that the left-hand-side of this expression can be replaced by a suitable derivative. This model assumes that the rate-limiting step in the formation of the photorefractive space-charge field at a given wavevector is charge generation; once a charge is generated, the mobility is assumed sufficiently high and the trapping sufficiently efficient that the charge contributes immediately to the photorefractive grating. More precisely, it is assumed that the transport length is optimal, that is, equal to Λ_G . Any decrease in the rate of grating formation due to inefficient transport and trapping should cause the measured Γ/t to fall below the value predicted by the right-hand side of Eq. (6). The reader may note the appearance of the figure of merit, Q, from section 1, in this expression.

For the photorefractive polymer PMMA-PNA:DEH (PMMA = polymethylmethacrylate, PNA = p-nitroaniline; see Fig. 5h), both with and without a sensitizing agent, the previously reported[55] values for all of the parameters in Eq. (6) are $n^3 r_e = 0.5$ pm/V at $E_0 = 11.4$ V/µm, $\varepsilon_r \approx 9$, $\lambda = 647$ nm, and $\Lambda_G = 1.6$ µm. These values fix a linear relation between Γ/t and $\alpha\Phi$, which is shown in Fig. 10 (dashed line). For the actual material (which has no explicit sensitizer), the value observed for Γ/t falls well below

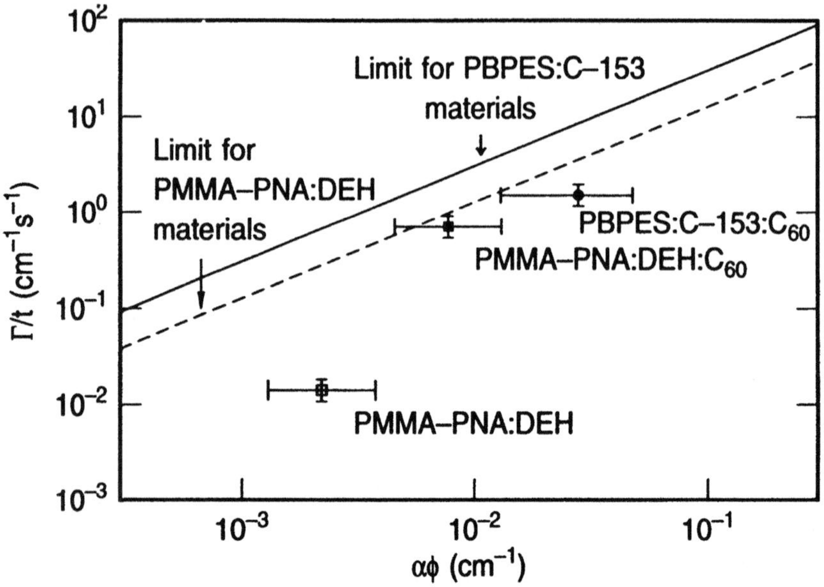

Fig. 10. The maximum observed value of Γ/t shown as a function of $\alpha\phi$ for PMMA-PNA:DEH (open square), PMMA-PNA:DEH:C_{60} (filled square), and PBPES:C-153:C_{60} (circle). Also shown are the theoretical limits based on the charge-generation-limited model for the PMMA-PNA:DEH materials (dashed line) and PBPES:C-153:C_{60} (solid line). Reprinted from Ref. [38].

the fundamental limit for the measured values of $\alpha = 3.7\ cm^{-1}$ and $\Phi = 6x10^{-4}$ (open square). For the same polymer doped with 0.1 wt.% C_{60}[55], the maximum value of Γ/t increases to a value near the predicted limit for measured values of $\alpha = 6.4\ cm^{-1}$ and $\Phi = 1.2 \times 10^{-3}$ (filled square). This is due to both an increase in the value of Γ and a decrease in the response time t. It has been suggested previously that the addition of C_{60} increases the concentration of traps in this photorefractive polymer[55]. Since the addition of 0.2 wt.% of C_{60} should have negligible effect on the mobility, it is most likely the improvement in trapping that raises the performance of this material to near the generation limit. Without increasing the value of $\Phi\alpha$, further increases in the mobility or more efficient charge transport should not improve the gain growth rate by a substantial amount in this polymer.

Further evidence that Φ is the limiting factor in the gain growth rate is given by the poly(silane)-based material PBPES:C-153:C_{60}, (PBPES = poly(4-n-butoxyphenylethylsilane), see Fig. 5d; C-153 = coumarin 153, see Fig. 5j), a sample with much higher mobility (and different $\Phi\alpha$)[38]. In this material, the mobility at high fields reaches

10^{-3} cm^2/Vs, approximately two to three orders of magnitude higher than other photorefractive polymers. Using the values $n^3r_e = 0.8$ pm/V, $\varepsilon_r \approx 6$, $\lambda = 647$ nm, and $\Lambda_G =$ 1.6 µm, the generation limit for this material is shown on Fig. 10 (solid line). An increase in μ by two orders of magnitude over that of PMMA-PNA:DEH:C_{60} has only a small effect on the measured Γ/t, and the experimentally determined value of Γ/t falls further below the line for the PBPES:C-153:C_{60} sample than for the PMMA-PNA:DEH:C_{60} sample. Optimization of the traps should lead to faster performance much closer to the generation limit in this material. Since the total absorption of the material will likely be bounded by thermal and photochemical effects occurring at large α, the route to even faster organic photorefractive polymers lies in increasing the quantum efficiency of charge generation, either through higher applied fields or intrinsically more efficient sensitizers.

4.2 THEORIES FOR THE SPACE-CHARGE FIELD

4.2.1 Modifications of Kukhtarev Theory to Include Geminate Recombination

The original rate equation theory of Kukhtarev[81] was intended to describe the generation of a space-charge field due to nonuniform illumination of an inorganic photorefractive material. The approach utilized a standard current equation, a continuity equation (for a single mobile charged species), Poisson's equation to relate the internal field to the charge density, and a rate equation for the single charged species. While this model has formed a very useful solid foundation for the interpretation of the grating experiments on inorganics[82], it contains several shortcomings when organic materials are considered. One limitation is related to the assumption of a field-independent photocarrier generation quantum yield Φ. In organic materials, quantum yields of charge generation are known to be highly field-dependent[21] as discussed in Section 1.

The first theoretical effort to include field-dependent generation was presented by Twarowski[83]. He used an expansion of the Onsager field-dependent charge generation probability[22] to first-order in E to modify Φ in the Kukhtarev equations. Due to this linear expansion of the complete Onsager field dependence, the specific results are only valid for low applied electric fields.

The modified Kukhtarev equations were solved[83] in the linearized regime under the assumption of small optical intensity modulation. While the saturation values for the internal space-charge field were not greatly affected by geminate recombination, in materials with low dielectric constant at low field where the quantum yield is small, the response times for grating formation and optical erasure were predicted to be long. For example, in zero field, the grating erasure time for choices of parameters suitable for organics would be on the order of 10^5 s at 1 W/cm^2. This is one reason why large fields have been necessary for the observation of useful photorefractive gratings in polymeric materials on reasonable time scales.

4.2.2 Rate Equation Models

The early work to include the effects of geminate recombination, though instructive, still did not include: (1) the full field dependence of the photogeneration quantum yield[84], (2) the effects of field-dependent mobility, (3) the possibility that in a molecularly doped polymer the diffusion coefficient for carriers might not be related to the carrier mobility by the usual Einstein relation[85], and (4) the possibility that the recombination and trapping rates might be mobility dependent (Langevin recombination[86]). Several approaches are possible to attack the formidable problem of formulation of more complete microscopic theories for the space-charge field. At this early stage, it is preferable to consider only the rate equation approach.

The relevant physical processes of photogeneration, recombination, trapping, and thermal detrapping can be treated in a fashion similar to that for the inorganics, with special modifications as required to account for the different charge generation and conduction mechanisms. The photogeneration process active in the illuminated regions may be described by

$$G + h\nu \rightarrow G^- + hole \qquad (7)$$

where G designates a (neutral) generator molecule. The trapping process can be described by

$$Tr + hole \rightarrow Tr^+ \qquad (8)$$

where Tr designates an unoccupied (neutral) trapping site. From these kinetic equations, a set of differential equations similar to the Kukhtarev equations can be written (see Ref. [56]). This set of equations is too complicated to yield closed-form solutions, so two approaches are available: numerical solution and linearization.

By considering several limiting cases, Schildkraut et al. have illustrated both approaches to the solution of these equations. In the first study[87], numerical simulation was utilized to calculate the space-charge field produced by photogeneration, transport, and trapping of mobile holes. The photogeneration quantum yield was assumed to follow $\Phi(E) = E^p$, where p is unity; this behavior is consistent with the behavior of some organic materials in the field range well below saturation. The field dependence of the mobility was taken as $\mu = exp[C(E^{1/2} - 1)]$ with C an experimentally determined constant. Examples with this mobility dependence are common in the literature on photoconducting polymers[29]-[33]. Both recombination and trapping rates were assumed to follow the Langevin form. The Langevin model is appropriate to the case where a mobile carrier and the generation site are oppositely charged so that a strong Coulomb attraction between them controls the recombination probability when they are within a sphere of radius equal to the Coulomb radius. This is a reasonable assumption for the generation process of Eq. (7) when the generator G is initially uncharged (the usual case). However, Langevin trapping may not be appropriate when the trapping center Tr is initially neutral.

Under these assumptions, numerical simulations were presented[87] for the cases of: (1) no traps present, and (2) hole trapping with no thermal detrapping. In case (1), a space-charge field is still produced because on average the holes drift away from the generator sites in the large external drift field. In case (2), a larger space-charge field is produced as expected. Fig. 11 shows the spatial distribution of the light intensity pattern (heavy solid line of large amplitude) and the calculated amplitude of the space-charge field at reduced times 0.1 (thin solid line), 1.0 (light dashed line), and 10.0 (heavy dashed line). The spatial phase-shift between the light pattern and the peaks of the space-charge field is clearly evident.

In a second study of the rate equation model[88], analytical solutions were presented for the zero-order and first-order Fourier components of the space-charge field. The authors considered two limiting cases: (a) once a charge is trapped, it cannot be released (deep traps) and (b) no hole traps are present. The Einstein relation between the mobility and the diffusion coefficient was assumed in order to remove one unknown physical quantity. The assumption of Langevin trapping described above was relaxed, as the hole traps were not required to be charged.

While these calculations provide an important first step toward a model framework for the understanding of photorefractive polymers, these model results must be tested on actual materials. Moreover, several additional physical effects should be included in future work: (i) the presence of shallow traps as well as deep traps (the evidence for which is the sublinear intensity dependence[89] of the growth rate described for

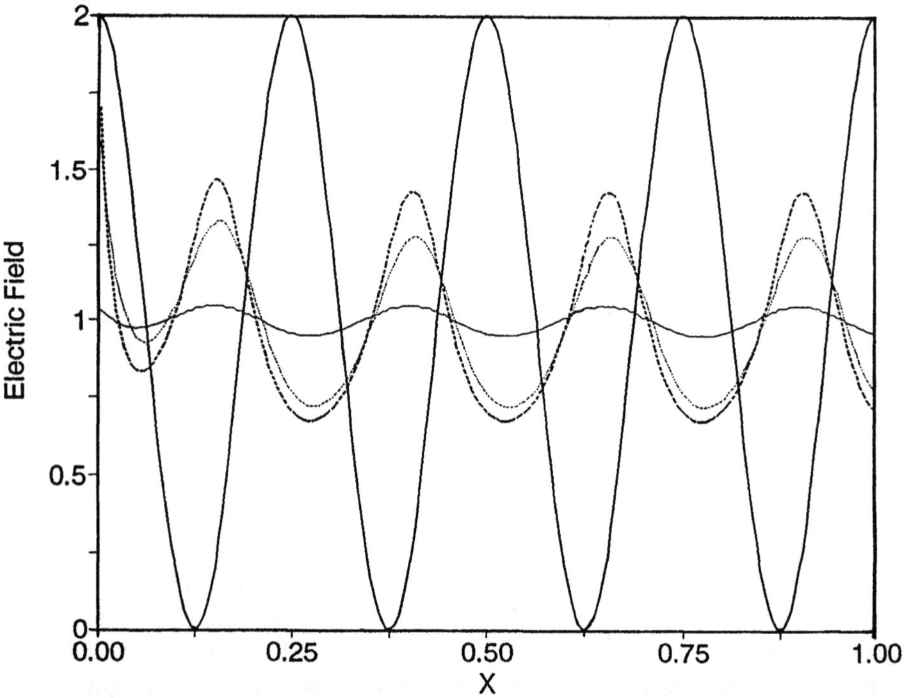

Fig. 11. Simulation of internal electric field *versus* position (normalized to sample length) in a dc field of unit value. The heavy solid line shows the light intensity profile for m = 1. The internal electric field is shown by the solid line of small amplitude at reduced time t = 0.1, the light dashed line at t = 1.0, and the heavy dashed line at t = 10.0. Reprinted from Ref. [87].

numerous systems), (ii) a more complete field-dependence of the photogeneration efficiency especially when sensitizers like C_{60} are used[90], and (iii) trapping mechanisms which do not follow the Langevin form.

4.3 ORIENTATIONAL ENHANCEMENT OF THE PHOTOREFRACTIVE EFFECT

In the standard model of the photorefractive effect, the modulation in the index of refraction by the internal space-charge field is a result of the electro-optic effect. Because the electro-optic coefficient is generally assumed to be constant (independent of position and field) in this model, it is useful to call this the "simple electro-optic" model for the photorefractive effect. Recently, a new mechanism for a significant enhancement compared to the simple electro-optic case has been reported for polymers[47] in which the nonlinear optical chromophores are orientationally mobile in the polymer matrix at the temperature of operation. The model system for this effect was PVK:FDEANST:TNF. The enhancement relies on the ability of the nonlinear optical chromophores to be aligned not only by the externally applied electric field, but also *in situ* by the sinusoidally varying space-charge field during grating formation. The resulting spatially periodic poling of the sample leads to a modulation of the birefringence of the material and to a modulation of the electro-optic response, the combination of which contributes favorably to the diffracted fields in the proper polarization.

Fig. 12 illustrates the effect schematically. The interfering light beams are assumed to produce a sinusoidally varying space-charge field by the usual mechanisms of drift and diffusion. The externally applied electric field E_0 will add to this field to produce a total field E_T. Since the nonlinear optical chromophores have orientational mobility due to the low T_g or to their relatively small molecular size, a spatially periodic orientational pattern is produced as the electric field orients the molecules by virtue of their ground-state dipole moment. In the two center panels of Fig. 12, this effect is illustrated with arrows representing a locally averaged value of the molecular (ground state) dipole moment. The two writing beams are assumed to enter the sample from the right and left sides of the Figure. If E_{sc} and E_0 lie in the same direction, as shown, only the magnitude of the average local dipole moment will be periodic. In the more general case, where there is a component of E_0 orthogonal to E_{sc}, the direction of the average local dipole moment will be periodic as well. Ref. [47] treats the general case of oblique incidence in detail.

The effect of the periodic orientation is to produce a spatially modulated birefringence and a spatially modulated electro-optic coefficient, both of which can contribute to the scattered light field. For these two effects, the nonlinear response of the material is quadratic in the total local electric field; hence the orientational enhancement effect may be regarded formally as a $\chi^{(3)}$ process. However, this orientational $\chi^{(3)}$ arises from different physical processes than other $\chi^{(3)}$ mechanisms previously reported for a variety of semiconductor and quantum well systems[48]-[50] based on Franz-Keldysh and band-edge effects and from the quadratic response in paraelectric crystals like potassium tantalate niobate[51],[52]. As Fig. 12 illustrates (right side), if no dc electric bias field is present, the resulting modulation of the index of refraction is at twice the wavevector K_G of the intensity pattern. When a bias field is present (left side of Fig. 12), the modulation appears at K_G as well, and can contribute strongly to

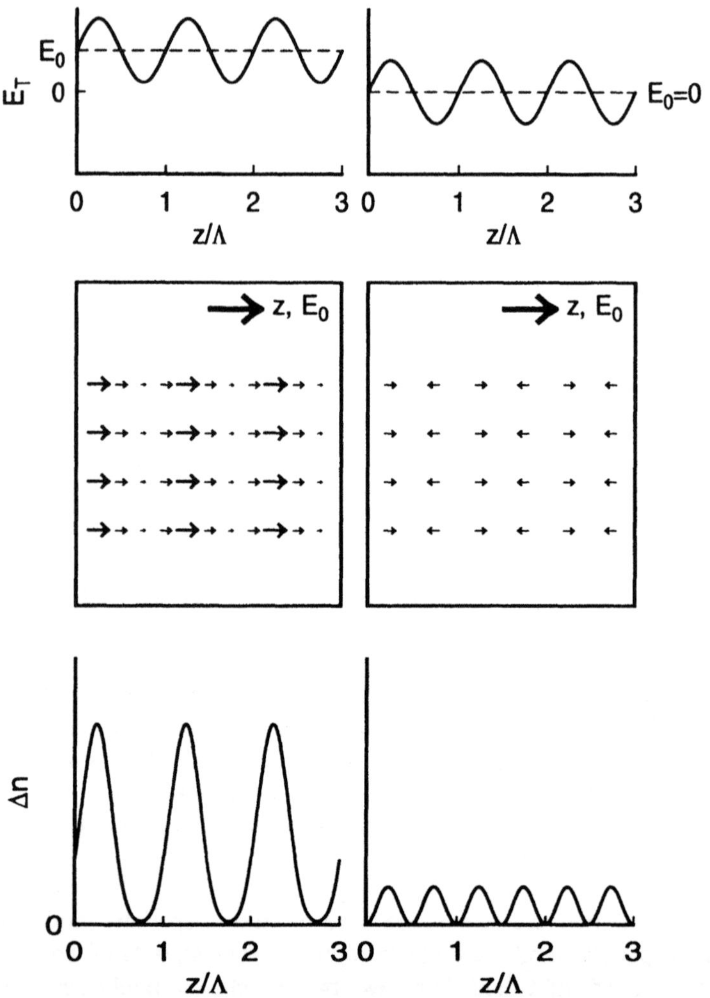

Fig. 12. Schematic illustration of the periodic poling responsible for the orientational enhancement effect. For simplicity, the background bias field E_0 and the space charge field are both directed along z in this figure, which would occur if the writing beams entered the sample from either side (symmetrically). The magnitude of the total field is shown at the top, for $E_0 \neq 0$ (left) and $E_0 = 0$ (right). The resulting average alignment of the chromophores is shown in the center, and the resulting refractive index grating is shown in the bottom panels. Note that for $E_0 = 0$, there is no component of the Δn grating at $1K_G$, while for $E_0 \neq 0$ a modulation at $1K_G$ appears. When the grating is written in an oblique geometry, there is a (uniform) component of the dc electric field perpendicular to z as well, resulting in a modulated direction as well as magnitude of the periodic poling field. Reprinted from Ref. [56].

the first-order Bragg-diffracted beam. The orientational enhancement effect will be operative for any photorefractive polymer in which the nonlinear chromophores have the ability to reorient appreciably in response to the local electric field; in contrast, it will not be important for permanently poled photorefractive polymeric systems nor for photorefractive crystals, either inorganic or organic.

The discovery of the orientational enhancement mechanism was motivated by the observation that, for the measured diffraction efficiency and electro-optic coefficient, the calculated space-charge field amplitude greatly exceeded the external applied field in PVK:FDEANST:TNF, a situation which simply cannot physically occur. The full theory[47] leads to three predictions: (1) an enhanced diffraction efficiency arising from the combined influence of both the modulated birefringence and electro-optic coefficients, (2) a change in the polarization anisotropy, and (3) presence of gratings with wavevectors of $2K_G$, $3K_G$, etc., even with a purely sinusoidal space-charge field. All three predictions were verified for PVK:FDEANST:TNF[47].

The modulation of poling produces a contribution to the diffraction efficiency even if the modulation of r is ignored and only the modulation of birefringence is present. Hence, a hypothetical chromophore with zero hyperpolarizability but large ground-state dipole moment and large birefringence would be potentially useful in the formulation of photorefractive polymers. In addition, if the spatially-modulated orientation were frozen in by cooling the sample before any decay of the space-charge field, the eventual decay of the space-charge field would not cause the diffracted signal from the stored orientation to disappear. This fixing phenomenon could allow non-destructive readout, as the long-term storage of information occurs via storage of a sinusoidal poling of the chromophores.

5. APPLICATIONS FOR PHOTOREFRACTIVE POLYMERS

Photorefractive polymers are of particular interest in a number of device applications for several reasons: (1) due to their potential for an order of magnitude gain over inorganic crystals in photorefractive sensitivity, (ii) due to their fabrication capabilities and ease of modification by doping, (iii) due to their low density and hence low device weight, and (iv) due to their potential low manufacturing cost. Even though photorefractive polymers have not yet been fully optimized and much materials research remains to be done, particularly in the area of chemical and physical fatigue, it is possible to discuss a few potential applications. Detailed descriptions of applications of photorefractive materials in general are included in other chapters in this volume.

In most applications, longer interaction lengths than just a few hundred microns are necessary either to increase the diffraction efficiency, the angular selectivity, or the ΓL product. There are two possible solutions to this problem that may be implemented easily with photorefractive polymers: multiple-layer geometries and rod-like geometries. A multi-layer sample would be made out of several layers of photorefractive polymer with dielectric spacer layers and optional bias electrodes as required. The holograms recorded in such a sample would have the properties of stratified volume holograms[91], and first results in such a sample geometry with a photorefractive polymer have recently been reported[92]. In the inorganics, rod-shaped crystals with length on the order of 1 cm and diameters on the order of 1 mm have been shown to provide enhanced angular selectivity for holographic recording due to multiple reflections[93]. Rod-shaped structures could easily be made from photorefractive polymers using extrusion, with the additional requirement that the final geometry should allow for the application of bias electric fields normal to the axis of the rods.

5.1 HOLOGRAPHIC OPTICAL STORAGE

One application for which photorefractive materials have been considered for many years is holographic optical storage[94]. Photorefractive polymers could provide a low cost holographic recording medium easily fabricated into large area formats such as optical disks. Here the important properties include recording sensitivity, recording response time, data retention time, nondestructive reading, holographic diffraction efficiency, angular selectivity, and erasability. Recording sensitivity, response time, and diffraction efficiency need be optimized for only one particular laser wavelength, preferably in the range accessible with diode lasers. Recording response times as long as 10-100 ms might be adequate when megabit pages are recorded in parallel. Desired data retention times under dark conditions should be as long as several years. To achieve long dark storage times, photorefractive polymers must be produced with an increased concentration of deep traps to immobilize the photoproduced carriers and a reduced concentration of shallow traps.

The dual requirements of non-destructive reading and facile erasability are somewhat in conflict for standard isotropic, single-photon charge generation mechanisms. Again due to the ease of modification of the individual functional components of generator, transporting species, and trap, it should be possible to achieve polarization selective or two-photon-activated charge generation in photorefractive polymers as has been done with other organic materials[95]-[98]. Such materials could in principle be gated between a light-insensitive state for nondestructive reading and a light-sensitive state for erasing by flood illumination.

In recent work, quasi-nondestructive reading (no decay of diffraction efficiency during continuous readout for 24 hours) has been observed for a new class of photorefractive polymers based on dual-function dopants[99]. A dual-function dopant is one in which the nonlinear chromophore is itself a charge transporting agent, thus, the host polymer (poly(methyl methacrylate)) acts only as a binder. The mechanism proposed for the extended reading time at low intensity involved a sequential population of two trapping levels, but more work must be done on these systems to fully elucidate the mechanism.

5.2 IMAGE PROCESSING

Photorefractive polymers can also be considered for the large number of image processing applications that have been proposed for inorganics such as optical limiting, phase conjugation, pattern recognition, novelty filtering, optical amplification, optical interconnection, etc. Here again it is the ease of manufacturing and low cost as well as possibly improved performance that make photorefractive polymers worthy of consideration. Some parameters of importance are, depending upon the application, large beam coupling gain, broad wavelength sensitivity, high speed, fast erasure, and good optical quality (low scattering from imperfections). The spontaneous grating erasure that occurs for some photorefractive polymers with shallow traps is not a severe problem for these applications, although the sublinear intensity behavior that results[55],[69] can affect scaling to the high intensity regime.

One application that might particularly benefit from photorefractive polymers is the imaging of defects in artificial periodic structures such as VLSI memory chips[100],[101]. The technique utilizes the intrinsic nonlinear response of the photorefractive material to enhance the signal from weak defects at the expense of the strong, bright spikes from the periodic structure. Usually, fast optics are required to utilize the presently available material sensitivities. Maximum sensitivity to small defects and a reduction in the need for fast optics are achieved with a large-area photorefractive material, and polymers may be useful in this regard if the recording sensitivity is high enough.

While the new photorefractive polymers have equaled some inorganic materials in gain coefficient and speed (see the next section), the properties are still inadequate in some respects. First, the operational wavelength range of photorefractive polymers is to the red of the strong optical absorption of the nonlinear optical chromophore, requiring wavelengths longer than about 600 nm to be used. Thus at this time, applications in the red and infrared would be more likely to succeed than broadband image processing applications in the visible. Another issue is the need for ΓL values to be greater than about 5-10 for applications depending upon beam coupling such as optical limiting. Much larger effective thicknesses or much higher gain coefficients are required in order to produce good optical limiting performance. The multilayer or rod-like geometries mentioned above may provide a solution. In particular, a major milestone not yet achieved is the direct observation of beam fanning in photorefractive polymer materials. Finally, the speed of the current materials needs to be several orders of magnitude larger in order to produce sufficient gain for fast pulse applications.

6. COMPARISON TO INORGANIC CRYSTALS AND FUTURE PROSPECTS

Despite the fact that photorefractive polymers were first identified less than four years ago, their performance at present is comparable to that of some of the well-known and much more extensively studied inorganic photorefractive materials, such as $LiNbO_3$, $Bi_{12}SiO_{20}$ (BSO), $KNbO_3$, and $BaTiO_3$[5]-[11]. These different materials can be compared using several measures. A direct comparison of two of the most relevant photorefractive properties, steady-state grating diffraction efficiency η_{ss} and growth rate $1/\tau$, is shown in Fig. 13, where η_{ss} is plotted as a function of $1/\tau$ on a log-log scale. The η_{ss} values are for 1 mm thick samples, measured directly for the inorganic crystals and extrapolated for the organic systems (assuming $\eta \approx d^2$), and the growth rates are for write intensities of 1 W/cm^2. With the exception of PVK:FDEANST:TNF and one or two other very new materials (not shown), essentially all of the polymeric photorefractive systems lie in the lower left of Fig. 13.

Consideration of some of the limitations of the current photorefractive polymers provides a guide to issues which must be addressed in future research efforts. The problems may be divided roughly into two areas: fabrication-related issues and

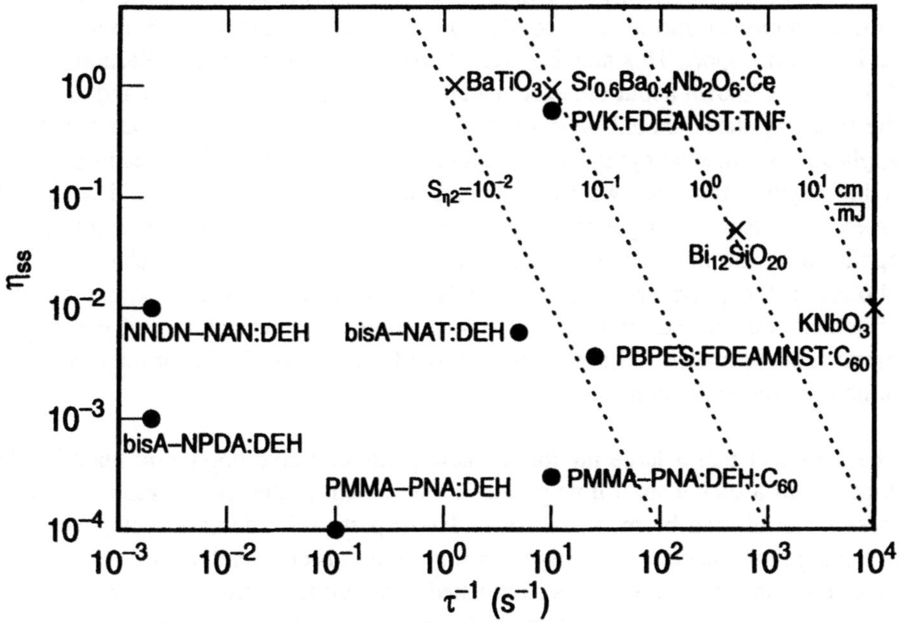

Fig. 13. Comparison of several inorganic photorefractive materials and a selection of photorefractive polymers. Saturation diffraction efficiency η_{ss} *versus* growth rate τ^{-1}, scaling to 1 mm thickness and 1 W/cm^2 writing intensity. The lines are contours of constant sensitivity $S_{\eta 2}$. Reprinted from Ref. [56].

mechanistic issues. On the fabrication side, in order to achieve larger diffraction efficiencies, useful beam-coupling gain, and improved angular selectivity, methods must be devised for making thicker samples while not sacrificing the poling-induced nonlinearity. Samples with multiple layers forming a stratified volume holographic optical element[91] can be used to achieve this, as discussed in Section 5. Other configurations, such as a rod-like geometry which has provided advantages for strontium barium niobate crystals[93], should be easy to achieve with photorefractive polymers and should be considered. Waveguide geometries provide special difficulties because as described above, the poling direction must have a component along the space-charge field. This means that poling normal to the waveguide plane while sending both optical beams through the waveguide should yield only small interactions. One solution to this problem would be to pole the waveguide normal to the plane and send one beam through the optical waveguide as usual, but then to send in the second beam as a bulk mode from outside of the waveguide. Since thin films are easy to fabricate from photorefractive polymers, these concepts merit further investigation.

On the mechanistic side, the sensitivity, or photorefractive response per unit time, must be improved further by a combination of enhancements in both optical nonlinearity and transport. This area has been discussed in some detail in Section 4.1. and 4.3. Little is known about the exact nature of the traps, and the depth and density of the trapping states should be optimized to provide long dark lifetimes for storage applications. Since holograms are usually partially erased during the reading process, fixing mechanisms and two-color writing mechanisms[95]-[97] should be investigated. The issue of fatigue in a large number of read-write cycles is important for applications and must be addressed. Finally, the structure-property relationships for photorefractive polymers are only partially established, so mechanistic and modeling efforts should continue in order to provide improved understanding of the ultimate limits of performance and eventually to provide predictive ability for the formulation of photorefractive polymers.

The very fact that a large number of new photorefractive polymeric materials has appeared in a such a short time period is a testament to the relative ease with which these materials can be made, which is due in part to the flexibility of polymers in accepting multiple dopants in large concentration. It is clear that the ultimate limits of these materials have not yet been reached, and further multidisciplinary efforts to devise new photorefractive polymers should continue in order to take advantage of the low dielectric constant, low cost, and flexibility of this new class of advanced materials.

ACKNOWLEDGEMENTS

We wish to acknowledge the Advanced Research Projects Agency, Defense Sciences Office, contract No. DAAB07-91-C-K767 for partial support of this work, and we thank G. C. Bjorklund, R. D. Miller, J. C. Scott,J. J. Stankus, R. J. Twieg, and C. A. Walsh for stimulating discussions and support throughout this work.

REFERENCES

[1] F. S. Chen, J. Appl. Phys. **38**, 3418 (1967).

[2] See for example *Photorefractive Materials and Their Applications I & II*, edited by P. Günter and J.-P. Huignard (Springer Verlag, Berlin, Germany,

1988).

[3] S. Ducharme, J. C. Scott, R. J. Twieg, and W. E. Moerner, Postdeadline Paper, Optical Society of America Annual Meeting, (Boston, MA, November 5-9, 1990).

[4] S. Ducharme, J. C. Scott, R. J. Twieg, and W. E. Moerner, Phys. Rev. Lett. **66**, 1846 (1991).

[5] P. Günter, Phys. Rep. **93**, 199 (1982).

[6] J. Feinberg, in *Optical Phase Conjugation*, edited by R. A. Fisher (Academic Press, New York, 1983), pp. 417-443.

[7] T. J. Hall, R. Jaura, L. M. Connors, and P. D. Foote, Prog. Quant. Electr. **10**, 77 (1985).

[8] G. C. Valley, M. B. Klein, R. A. Mullen, D. Rytz, and B. Wechsler, Ann. Rev. Mater. Sci. **18**, 165 (1988).

[9] J. Feinberg, Phys. Today **41**, 46 (1988).

[10] M. P. Petrov, S. I. Stepanov, and A. V. Khomenko, *Photorefractive Crystals in Coherent Optical Systems* (Springer-Verlag, Berlin, Germany, 1991).

[11] P. Yeh, *Introduction to Photorefractive Nonlinear Optics* (John Wiley, New York, 1993).

[12] H. J. Eichler, P. Günter, and D. W. Pohl, *Laser-Induced Dynamic Gratings*, Springer Series in Optical Sciences, Vol. 50 (Springer, Berlin, Germany, 1986).

[13] See for examples *Photochromism*, edited by G. H. Brown, Techniques of Chemistry, Vols. I-III (Wiley-Interscience, New York, 1971).

[14] G. C. Valley, J. Opt. Soc. Am. B **1**, 868 (1984).

[15] See *Nonlinear Optical Properties of Organic Molecules and Crystals*, Vols. 1 and 2, edited by D. S. Chemla and J. Zyss (Academic Press, Orlando, 1987).

[16] S. M. Silence, F. Hache, M. C. J. M. Donckers, C. A. Walsh, D. M. Burland, G. C. Bjorklund, R. J. Twieg, and W. E. Moerner, Proc. Soc. Photo-Opt.

Instrum. Engr. **1852**, 253 (1993).

[17] K. Sutter, J. Hullinger, and P. Günter, Sol. St. Commun. **74**, 867 (1990).

[18] K. Sutter and P. Günter, J. Opt. Soc. Am. B **7**, 2274 (1990).

[19] K. Sutter, J. Hulliger, R. Schlesser, and P. Günter, Opt. Lett. **18**, 778 (1993).

[20] M. Eich, B. Reck, D. Y. Yoon, C. G. Willson, and G. C. Bjorklund, J. Appl. Phys. **66**, 3241 (1989).

[21] P. J. Melz, J. Chem. Phys. **57**, 1694 (1972).

[22] L. Onsager, Phys. Rev. **54**, 554 (1958).

[23] B. Ries, G. Schönherr, H. Bässler, and M. Silver, Phil. Mag. B **49**, 259 (1984); B. Ries, G. Schönherr, H. Bässler, and M. Silver, Phil. Mag. B **49**, 597 (1984).

[24] H. Scher and S. Rackovsky, J. Chem. Phys. **81**, 1994 (1984); S. Racovsky and H. Scher, J. Chem. Phys. **89**, 7242 (1988).

[25] E. A. Silinsh and H. Inokuchi, Chem. Phys. **149**, 373 (1991).

[26] R. R. Chance and C. L. Braun, J. Chem. Phys. **64**, 3573 (1976).

[27] C. L. Braun, J. Chem. Phys. **80**, 4157 (1984).

[28] M. Stolka, in *Encyclopedia of Polymer Science and Engineering,* Vol. 11 (John Wiley, New York, 1988), pp. 154-175.

[29] J. X. Mack, L. B. Schein, and A. Peled, Phys. Rev. B **39**, 7500 (1989).

[30] L. B. Schein, Phil. Mag. B **65**, 795 (1992).

[31] M. A. Abkowitz, Phil. Mag. B **65**, 817 (1992).

[32] P. M. Borsenberger, Phys. Stat. Sol. (b) **173**, 671 (1992).

[33] H. Bässler, Phys. Stat. Sol. (b) **175**, 15 (1993).

[34] J. C. Scott, L. Th. Pautmeier, and W. E. Moerner, J. Opt. Soc. Am. B **9**, 2059 (1992).

[35] J. C. Scott, L. Th. Pautmeier, and W. E. Moerner, Synth. Met. **54**, 9 (1992).

[36] J. C. Scott, L. Th. Pautmeier, W. E. Moerner, C. A. Walsh, S. M. Silence, T. J.

Matray, and R. J. Twieg, Proc. Mat. Res. Soc. **277**, 135 (1992).

[37] M. Stolka and M. A. Abkowitz, Proc. Mat. Res. Soc. **277**, 15 (1992).

[38] S. M. Silence, J. C. Scott, F. Hache, E. J. Ginsburg, P. K. Jenkner, R. D. Miller, R. J. Twieg, and W. E. Moerner, J. Opt. Soc. Am. B **10**, 2306 (1993).

[39] See M. B. Klein, Chapter 7 in *Photorefractive Materials and Their Applications I*, edited by P. Günter and J.-P. Huignard (Springer Verlag, Berlin, 1988), pp. 195-236.

[40] G. C. Valley and M. B. Klein, Opt. Engr. **22**, 704 (1983).

[41] A. D. Buckingham, Adv. Chem. Phys. **12**, 107 (1967).

[42] J. Oudar and D. Chemla, J. Chem. Phys. **66**, 2664 (1977).

[43] A. Willetts, J. E. Rice, D. M. Burland, and D. P. Shelton, J. Chem. Phys. **97**, 7590 (1992).

[44] D. M. Burland, J. Downing and J. Michl, Mol. Cryst. Liq. Cryst. Sci. Technol.-Sec. B **3**, 195 (1992).

[45] D. J. Williams, in *Nonlinear Optical Properties of Organic Molecules and Crystals*, Vols. 1 and 2, edited by D. S. Chemla and J. Zyss (Academic Press, Orlando, 1987), Chapter II-7.

[46] For a recent review, see D. M. Burland, R. D. Miller, and C. A. Walsh, Chem. Revs. **94**, 31 (1994).

[47] W. E. Moerner, S. M. Silence, F. Hache, and G. C. Bjorklund, J. Opt. Soc. Am. B **11**, 320 (1994).

[48] A. Partovi, A. Kost, E. M. Garmire, G. C. Valley, and M. B. Klein, Appl. Phys. Lett. **56**, 1089 (1990).

[49] D. D. Nolte, D. H. Olson, G. E. Doran, W. H. Knox, and A. M. Glass, J. Opt. Soc. Am. B **7**, 2217 (1990).

[50] J. E. Millerd, S. D. Koehler, E. M. Garmire, A. Partovi, A. M. Glass, and M. B. Klein, Appl Phys. Lett. **57**, 2776 (1990).

[51] A. Agranat and Y. Yacoby, J. Opt. Soc. Am. B **5**, 1792 (1988).

[52] A. Agranat, V. Leyva, and A. Yariv, Opt. Lett. **14**, 1017 (1989).

[53] L. B. Schein, *Electrophotography and Development Physics* (Springer, Berlin,

Heidelberg, 1988).

[54] D. M. Sturmer and D. W. Heseltine, in *The Theory of the Photographic Process,* 4th ed., edited by T. H. James (Macmillan, New York, 1977), pp. 194-234.

[55] S. M. Silence, C. A. Walsh, J. C. Scott, and W. E. Moerner, Appl. Phys. Lett. **61**, 2967 (1992).

[56] W. E. Moerner and S. M. Silence, Chem. Revs. **94**, 127 (1994).

[57] S. M. Silence, M. C. J. M. Donckers, C. A. Walsh, D. M. Burland, W. E. Moerner, and R. J. Twieg, Appl. Phys. Lett. **64**, 712 (1994).

[58] C. A. Walsh and W. E. Moerner, J. Opt. Soc. Am. B **9**, 1642 (1992); C. A. Walsh and W. E. Moerner, J. Opt. Soc. Am. B **10**, 753 (1993).

[59] Sandalphon, B. Kippelen, N. Peyghambarian, S. R. Lyon, A. B. Padias, and H. K. Hall, Jr., Opt. Lett. **19**, 68 (1994).

[60] J. P. Huignard and A. Marrakchi, Opt. Commun. **38**, 249 (1981).

[61] S. M. Silence, M. C. J. M. Donckers, C. A. Walsh, D. M. Burland, R. J. Twieg, and W. E. Moerner, to appear in Appl. Opt. **33**, 2218(1994).

[62] L. B. Schein, A. Rosenberg, and S. L. Rice, J. Appl. Phys. **60**, 4287 (1986).

[63] M. Liphardt, A. Goonesekera, B. E. Jones, S. Ducharme, J. M. Takacs, and L. Zhang, Science **263**, 367 (1994).

[64] J. Obrzut, M. J. Obrzut, and F. E. Karasz, Synth. Met. **29**, E103 (1989).

[65] O. M. Gelsen, D. D. C. Bradley, H. Murata, T. Tsutsui, S. Saito, J. Rühe, and G. Wegner, Synth. Met. **41**, 875 (1991).

[66] Y. Zhang, Y. Cui, and P. N. Prasad, Phys. Rev. B **46**, 9900 (1992).

[67] M. C. J. M. Donckers, S. M. Silence, C. A. Walsh, F. Hache, D. M. Burland, W. E. Moerner, and R. J. Twieg, Opt. Lett. **18**, 1044 (1993).

[68] M. Pope and C. E. Swenberg, *Electronic Processes in Organic Crystals* (Clarendon, Oxford, 1982), pg. 748.

[69] S. M. Silence, C. A. Walsh, J. C. Scott, T. J. Matray, R. J. Twieg, F. Hache, G. C. Bjorklund, and W. E. Moerner, Opt. Lett. **17**, 1107 (1992).

[70] W. E. Moerner, C. A. Walsh, J. C. Scott, S. Ducharme, D. M. Burland, G. C. Bjorklund, and R. J. Twieg, Proc. Soc. Photo-Opt. Instrum. Engr. **1560**, 278

(1991).

[71] S. Ducharme, B. Jones, J. M. Takacs, and L. Zhang, Opt. Lett. **18**, 152 (1993).

[72] B. Kippelen, Sandalphon, N. Peyghambarian, S. R. Lyon, A. B. Padias, and H. K. Hall, Jr., Electron. Lett. **29**, 1873 (1993).

[73] K. Tamura, A. B. Padias, H. K. Hall, Jr., and N. Peyghambarian, Appl. Phys. Lett. **60**, 1803 (1992).

[74] N. Peyghambarian, K. Tamura, B. Kippelen, Y. Kawabe, F. Jarka, S. Mazumdar, D. D. Guo, H. K. Hall, Jr., and A. B. Padias, Proc. Mat. Res. Soc. **277**, 145 (1992).

[75] L. Yu, W. Chan, Z. Bao, and S. X. F. Cao, Macromol. **26**, 2216 (1993).

[76] L. Yu, W. Chan, Z. Bao, and S. X. F. Cao, Polym. Preprints **33**, 398 (1992).

[77] D. von der Linde and A. M. Glass, Appl. Phys. **8**, 85 (1975).

[78] A. M. Glass, Opt. Engr. **17**, 470 (1978).

[79] P. Yeh, Appl. Opt. **26**, 602 (1987).

[80] A. M. Glass, M. B. Klein, and G. C. Valley, Appl. Opt. **26**, 3189 (1987).

[81] N. V. Kukhtarev, V. B. Markov, M. Soskin, and V. L. Vinetskii, Ferroelectrics **22**, 949 (1979); N. V. Kukhtarev, V. B. Markov, M. Soskin, and V. L. Vinetskii, Ferroelectrics **22**, 961 (1979).

[82] G. C. Valley and J. F. Lam, in *Photorefractive Materials and Their Applications I*, edited by P. Günter and J.-P. Huignard (Springer Verlag, Berlin, 1988), pp. 75-98.

[83] A. Twarowski, J. Appl. Phys. **65**, 2833 (1989).

[84] A. Mozumder, J. Chem. Phys. **60**, 4300 (1974); A. Mozumder, J. Chem. Phys. **60**, 4305 (1974).

[85] L. Pautmeier, R. Richert, and H. Bässler, Phil. Mag. B **63**, 587 (1991).

[86] W. Helfrich, in *Physics and Chemistry of the Organic Solid State III*, edited by D. Fox, M. M. Labes, and A. Weissberger (Wiley-Interscience, New York,

1967), pp. 22-25.

[87] J. S. Schildkraut and A. V. Buettner, J. Appl. Phys. **72**, 1888 (1992).

[88] J. S. Schildkraut and Y. Cui, J. Appl. Phys. **72**, 5055 (1992).

[89] D. Mahgerefteh and J. Feinberg, Phys. Rev. Lett. **64**, 2195 (1990).

[90] Y. Wang, Nature **356**, 585 (1992).

[91] R. V. Johnson and A. R. Tanguay, Jr., Opt. Lett. **13**, 189 (1988).

[92] J. J. Stankus, S. M. Silence, W. E. Moerner, and G. C. Bjorklund, Opt. Lett. **19**, 1480 (1994).

[93] L. Hesselink and S. Redfield, Opt. Lett. **13**, 877 (1988).

[94] L. Hesselink and M. C. Bashaw, Opt. and Quant. Electron. **25**, S611 (1993).

[95] G. C. Bjorklund, C. Bräuchle, D. M. Burland, and D. C., Alvarez, Opt. Lett. **6**, 159 (1981).

[96] T. P. Carter, C. Bräuchle, V. Y. Lee, M. Manavi, and W. E. Moerner, Opt. Lett. **12**, 370 (1987).

[97] D. von der Linde, A. M. Glass, and K. F. Rodgers, J. Appl. Phys. **147**, 217 (1976).

[98] W. E. Moerner, W. Lenth, and G. C. Bjorklund, Chapter 7 of *Persistent Spectral Hole-Burning: Science and Applications,* edited by W. E. Moerner, Topics in Current Physics Vol. 44 (Springer, Berlin, Heidelberg, 1988).

[99] S. M. Silence, R. J. Twieg, G. C. Bjorklund, and W. E. Moerner, Phys. Rev. Lett. **73**, 2047.

[100] C. Uhrich and L. Hesselink, Opt. Lett. **17**, 1087 (1992).

[101] L. Hesselink, private communication.

Additional references of interest added in proof:

L. Yu, Y. Chen, W. K. Chan and Z. Peng, Appl. Phys. L. **9**, 2489 (1994)

Y. Zhang, C. A. Spencer, S. Ghosal, M. K. Casstevens and R. Burzynski, J. Appl. Phys. **76**, 671 (1994)

M. E. Orcyzk, B. Swedek, J. Zieba and P. N. Prasad, SPIE Proc. **2285**, 166 (1994)

W. E. Moerner, Nature **371**, 475 (1994)

K. Meerholz, B. L. Volodin, Sandalphon, B. Kippelen and N. Peygambarian, Nature **371**, 497 (1994)

6.

NEAR-RESONANT PHOTOREFRACTIVE EFFECTS IN BULK SEMICONDUCTORS

James E. Millerd, Elsa M Garmire*, Afshin Partovi**
MetroLaser, 18006 Skypark Circle, Suite 108, Irvine, CA, 92714
* University of Southern California, Center for Laser Studies, University Park, DRB17, Los Angeles, CA 90028
** AT&T Bell Laboratories, 600 Mountain Ave., Murray Hill, NJ 07974

1. INTRODUCTION TO PHOTOREFRACTIVE SEMICONDUCTORS

1.1 Background

In an effort to utilize photorefractives in practical applications, it is necessary to choose a material on the basis of cost, performance, durability and operating parameters. Semiconductor materials offer many advantages over their ferroelectric counterparts in certain applications and, as such, represent a very important class of photorefractive materials. They are characterized by 1) fast response times, typically micro. to millisecond buildup times for modest intensities, 2) sensitivity in the near-infrared, 0.5 - 1.5 μm, 3) comparatively low cost and 4) high optical quality. In addition, photorefractive sensitivities of these materials, defined as index change per incident energy density, is two to three orders of magnitude larger than photorefractive ferroelectrics. Fast response times are essential to applications such as image processing, optical information processing, real-time holographic interferometry, phase conjugation and beam combining. Near infrared sensitivity translates to compatibility with diode lasers, a rugged, compact and inexpensive light source. Compatibility with such a source is crucial to the development of practical commercial systems. Semiconductor growth technology is a well-developed commercial field, and the effects of various dopants on semiconductors is well characterized. Large bulk crystals can be grown currently at fractions of the cost of ferroelectric crystals. This is important for development as well as for mass production potential. Semiconductors can be grown with excellent optical quality, because the growth process does not introduce strain-related striations. They remain stable over a wide range of operating temperatures, although performance may be affected. They are mechanically hard and durable, and compatible with existing opto-electronic technology. In addition, they do not suffer from problems of optical activity and require no poling.

With all these advantages why are photorefractive semiconductors not finding widespread use? The answer is that their maximum refractive index change, produced in response to holographic exposure and carrier diffusion, is relatively small. Even though they have good sensitivity (index change per absorbed photon) the small saturated refractive index change results in poor diffraction efficiency and low wave-mixing gains. Diffraction efficiency can be increased with longer interaction lengths. However, as will be discussed later, the necessary condition for self-pumped phase conjugation is that the photorefractive gain coefficient, Γ, (linearly proportional to change in refractive index) must be larger than the absorption coefficient, α, both of which are independent of interaction length. (In fact, for a ring self-pumped phase conjugate mirror $\Gamma > 2\alpha$.) It is also worth noting that the advantage of fast response time makes semiconductors poor materials for long term optical data storage because the hologram decays quickly because of thermal processes.

Over the past decade, there has been a great deal of effort to increase the photorefractive gain coefficient in semiconductors to overcome this limitation[1-6]. This chapter will present a summary of the different techniques used to increase wave mixing gains in CW or long pulse recordings and discuss the relative merits of each. In particular, near-resonant effects such as working near the band-edge[7] and the temperature-intensity resonance[8,9] (occurring in materials having carriers of opposite sign responsible for thermal and photoconductivity) will be covered in detail. The issue of photorefractive response at large modulation index (or large signal effects) in materials with applied electric fields will be introduced and analyzed. Large signal effects are significant in all the gain enhancement techniques presented here and have been adequately modeled only recently. Finally, a theoretical and experimental investigation of self-pumped phase conjugation, an important application for photorefractive semiconductors, will be presented.

An excellent review of photorefractive semiconductors was given by Glass and Strait[10] and rather than duplicate their work, this section is intended to be a supplement. We will provide a brief summary and then go on to describe the advances made since this first text was written. To the applications engineer or scientist, this section will hopefully present a useful summary of the current semiconductor photorefractive materials available and their characteristics, provide an explanation of the gain enhancement techniques discovered so far and their limitations, and provide a benchmark to measure future developments in photorefractive semiconductors.

1.2 Photorefractive Beam Coupling

A large fraction of the work in photorefractive research involves two wave mixing or beam coupling. In this process, energy is exchanged unidirectionally from one beam to another. Both two- and four-wave mixing involve the same dynamics in the writing of index gratings, so the study of one reveals information about the other. Because two wave mixing is a relatively straightforward experimental procedure, as shown in Figure 1, it is used extensively to characterize materials.

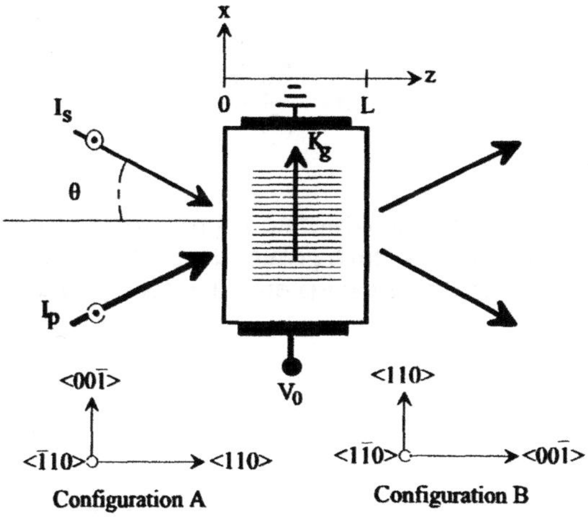

Figure 1. Basic arrangement and designations for two beam coupling in semiconductors. Typically the pump wave intensity, Ip, is much stronger than the signal wave, Is. Crystal orientation shown for electron dominated sample. Crystal configuration B suppresses Pockels linear electro-optic beam coupling.

As described in the first chapter of this volume, beam coupling results from the periodic intensity pattern created inside a crystal caused by the coherent interference of two beams. Charged free carriers are generated in the bright regions and they drift or diffuse into the darker regions. The free carriers leave behind fixed charges of opposite sign. The resulting charge separation produces an electric field inside the crystal with the same period as the incident intensity pattern but spatially shifted in phase. Through the Pockels electro-optic effect, a periodic variation in refractive index is created in response to the electric field. When the spatial phase shift of the index grating is $\pm\pi/2$ with respect to the intensity fringe pattern, the grating gives rise to unidirectional beam coupling.

Summarizing the derivation of the first chapter, the incident intensity pattern caused by the two beams interfering inside the volume of the crystal can be written as,

$$I(x) = I_0 \, [\, 1 + m \cos(k_g x)] \tag{1}$$

where I_o is the combined intensity of the pump (or reference) and signal beams, $I_o = I_p + I_s$, $k_g = 2\pi/\Lambda$ is the grating vector, $\Lambda = \lambda/(2\sin\theta)$ is period of the grating, λ is the free-space wavelength of the light, θ is the half angle between beams and

$$m = \frac{2\sqrt{I_p I_s}}{(I_p + I_s)} \tag{2}$$

is the modulation index. In the steady-state approximation it is typically assumed that the optical generation rate is much greater than the thermal generation rate. The space-charge field generated inside the material in response to the steady state optical intensity pattern is

$$E = E_{sc}[\exp(ik_g x + \phi) + \text{C.C.}] \tag{3}$$

where C.C. denotes complex conjugate and only the first order Fourier harmonics of each spatially varying quantity are considered. It will be seen later that it is necessary to include higher order harmonics when using large applied fields at high modulation index. The well known solutions are:[10]

$$E_{sc} = m\xi\,(h\nu)E_D \frac{1 + \dfrac{E_D}{E_q} + \dfrac{E_{dc}^2}{E_D E_q}}{\left(1 + \dfrac{E_D}{E_q}\right)^2 + \left(\dfrac{E_{dc}}{E_q}\right)^2} \,, \tag{4}$$

$$\tan \phi_{sc} = \frac{E_D}{E_{dc}}\left(1 + \frac{E_D}{E_q} + \frac{E_{dc}^2}{E_D E_q}\right) \tag{5}$$

where

$$E_D = \frac{2\pi k_b T}{\Lambda e}, \qquad E_q = \frac{N_t \Lambda e}{2\pi\varepsilon} \tag{6}$$

and E_{dc} is the externally applied electric field, e is the electron charge, k_b the Boltzman constant, T is temperature, ε is the static dielectric constant, N_t is the effective trap density in the material and $\xi(h\upsilon)$ is the so-called electron hole competition factor. E_D is known as the diffusion-limited field and represents the maximum space-charge field that may build up due to diffusion of carriers from the bright spots. E_q is the trap-limited field and represents the maximum field that may be produced if all the available charge inside each grating period is separated.

By using this solution along with the electro-optic effect and the wave equation, the energy transfer between the two beams may be obtained. The expression for the increase in intensity of the signal beam is

$$I_s = I_{so}\frac{(\beta+1)e^{\Gamma L}}{\beta + e^{\Gamma L}}, \tag{7}$$

where L is the interaction length, β is the ratio of pump to signal beam and

$$\Gamma = \frac{2\pi n_0^3 r_{eff}}{\lambda \cos\theta}\frac{\operatorname{Im} E_{sc}}{m}, \tag{8}$$

where Im E_{sc} represents the imaginary (or π/2 shifted) component of the space-charge field, θ' is the internal angle and r_{eff} is the effective electro-optic coefficient that depends on crystal orientation and beam polarization. With the proper choice of crystal geometry, cross polarization coupling is possible in semiconductors [11]. This type of coupling, where the diffracted light is polarized orthogonal with respect to the polarization of the incident or zero order beams, allows for background suppression and high signal to noise.

Without an applied field, it can be seen from eq.(5), that the phase of the space-charge field is at exactly $\pi/2$. In this case, the dominant mechanism for carrier movement is diffusion and, because of the symmetrical nature of the process, the charge distribution remains in phase with the intensity pattern. The space-charge field is then 90° out of phase, because it is proportional to the integral of charge along the x direction (Poisson's equation).

These formulas are based on a single-defect model which, in general, is not valid. In many cases, and particularly at low temperature, the role of multiple defects must be considered. This requires a numerical solution of the full rate equations as outlined in chapter 1. However, under a surprising number of conditions, usually at or near room temperature, the single defect model can provide good agreement with measured photorefractive behavior. The effects of temperature-dependent trap occupation can be accounted for heuristically by inclusion of the so-called electron-hole competition factor. By including generation of both carrier types, it is found that each carrier type tries to build up a grating of opposite sign. This effect can be summarized using the electron-hole competition factor [12,8]. Under the assumption of a large grating period compared to a carrier diffusion length, $\xi(h\upsilon)$ can be written as

$$\xi(h\upsilon) = (n_T\sigma_n - p_T\sigma_p)I_o/(e_n n_T + e_p p_T), \tag{9}$$

where I_o is the total intensity, n_T and p_T are the density of available electrons and holes at their respective trap sites, e_n and e_p are the total emission rate and σ_n and σ_p are the optical cross sections for the electron and hole, respectively. The total emission rate is the sum of optical and thermal processes so $e_n = \sigma_n I_o + e_n^{th}$, where e_n^{th} is the thermal emission rate of the electrons. This factor is often hidden inside r_{eff}. If one type of carrier is dominant then $\xi = \pm 1$, depending on the sign of the carrier. For grating spacings comparable to the carrier diffusion lengths, ξ has a dependence on Λ, caused by the different diffusion lengths for each carrier[12]. In many materials, including InP:Fe and GaAs, the role of multiple defect levels has been found to give a strong temperature dependence of the electron-hole competition factor that can cause the dominant photocarrier to change with temperature and thus change the direction of the wave-mixing gain [13]. There may also be a strong wavelength dependence of the electron-hole competition factor. For example, in GaAs at room temperature, the photorefractive response for $\lambda < 1.06$ μm is electron dominated and for $\lambda > 1.15$ μm is hole dominated [14].

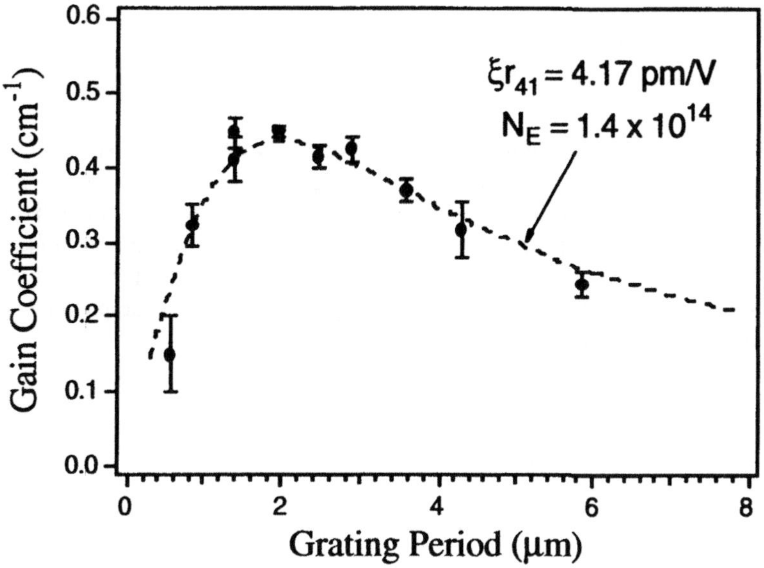

Figure 2. Beam coupling gain coefficient as a function of grating period for ZnTe at 633 nm wavelength. The dashed line is a theoretical fit to a single carrier, single trap level model. (Ref. [15]).

The maximum space-charge field for a given material occurs at an optimum grating spacing which can be found by differentiating eq.(4) with respect to Λ. For a material with a single carrier type, the optimum value is $\Lambda_{opt} = 2/e(k_bT/N_T)^{1/2}$. Figure 2 shows the dependence of the gain coefficient, which is linearly related to the space-charge field, as measured in a ZnTe sample [15]. From this measurement the effective trap density and effective electro-optic coefficient can be determined. Comparison with values of r_{41} measured using other means gives an indication of the electron-hole competition in the sample. In this case, ξ is approximately 0.93, which indicates very little electron hole competition. Combining this information with independent electro-optic modulation experiments to determine the crystollographic orientation of the sample, one can determine whether the dominate photocarriers are electrons or holes [16]. The trap density is lower in this material than typical photorefractive semiconductors, which results in a lower maximum gain coefficient.

To investigate the optical power necessary to saturate the photorefractive effect, beam coupling response can be measured as a function of incident intensity. In a material where a single carrier model is adequate, photoconductivity must overcome thermal or "dark" conductivity to achieve the maximum photorefractive effect. In the single carrier approximation, the intensity dependence of the photorefractive effect can be described by

$$\Gamma = \frac{\Gamma_0 I}{I + I_{sat}} \tag{10}$$

where Γ_0 is the saturated gain coefficient measured at large intensity and I_{sat} is the saturation intensity (i.e., the irradiance at which photoconductivity equals dark conductivity). Figure 3 shows the intensity dependence of the normalized gain coefficient (Γ/Γ_0) at a grating spacing of 1.5 μm for a CdTe sample [17]. The fit uses eq.(10) and a dark conductivity of 110 $\mu W/cm^2$. This small value, typical of semiconductors, results in a very high efficiency, because very little power is needed to overcome thermal conductivity.

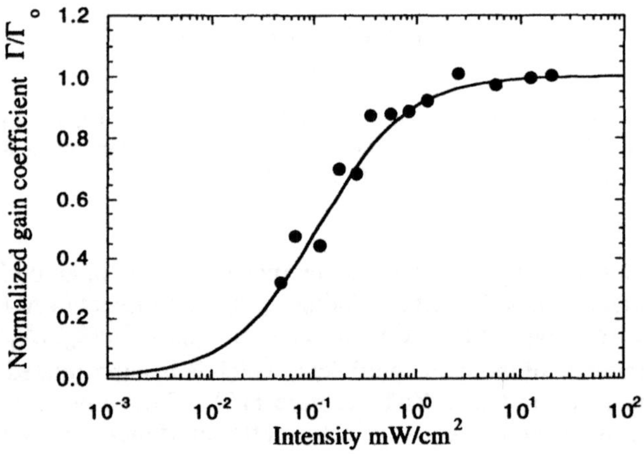

Figure 3. Normalized beam coupling gain coefficient (Γ/Γ_0) as a function of incident intensity in CdTe.

1.3 Materials

The requirements that have traditionally been associated with photorefractive materials have been (i) presence of sufficient density of deep levels to create a space-charge field, and (ii) lack of an inversion center of symmetry permitting a Pockels electro-optic effect. A high density of deep levels can be accomplished through doping in semiconductors with impurities that cause mid-gap defects. A great deal of knowledge already exists on the energy levels occupied by various impurities in many different semiconductor compounds[18].

The majority of photorefractive semiconductor investigations have been with the III-V materials GaAs and InP, mainly due to their ready availability in bulk semi-insulating form. CdTe and ZnTe appear to offer superior photorefractive properties and operate at more desirable wavelengths. However, they have only recently become available and little experimental work has been done to grow high quality II-VI materials. GaP has been explored as a visible (red) photorefractive material and has been shown to have a photorefractive response similar to the other III-V materials. Advances in ternary and quaternary semiconductor growth techniques, both bulk and thin film, offer the possibility for "band-gap engineered" materials that can be grown for maximum performance at most any desired wavelength. CdMnTe is an example of a recently explored ternary material that has demonstrated photorefractive response in the desirable 600 nm to 1 μm range. Growth of ultra thin (100 Å) quantum well structures introduces an additional mechanism (layer thickness) to engineer the bandgap of these materials (see chapter 7 of this book).

A convenient figure of merit for photorefractive materials is sensitivity $S = L\cos(\theta')/[4\pi I\tau(1-e^{-\alpha L})]$, where τ is the grating formation time, α is the absorption coefficient of the material and L is the interaction length. By assuming $L = 1$ cm, S would correspond to index change per absorbed energy per unit volume. Strictly speaking, this number is only valid for measurements made with unity modulation index. Without externally applied fields, the gain coefficient is independent of modulation index. However, for situations where applied fields were used, the gain coefficient measured at low modulation index (high beam ratios) will be larger than that measured at unity, thus overestimating the sensitivity figure of merit.

Two other useful figures of merit are the normalized gain coefficient, Γ/α, and the normalized electro-optic coefficient, $n_0^3 r_{eff}/\varepsilon$. Γ/α is a particularly important figure of merit for applications such as self-pumped phase conjugation where gain must exceed loss by a factor of two. The normalized electro-optic coefficient corresponds to the index change per separated charge and is a rough measure of efficiency.

1.3.1 GaAs

Gallium arsenide can be made semi-insulating and photorefractive through use of the stoichiometry-related EL2 center [19] or by doping with chromium (GaAs:Cr) [20]. The growth technology for GaAs is arguably the most advanced of all the compound semiconductors. However, photorefractive quality can vary considerably between samples obtained from even the same boule. Photorefractive response has been measured from close to its band-edge at 915 nm[7] to 1.3 μm [21]. Based on absorbance measurements, GaAs:Cr should have a photorefractive response as far into the infrared as 1.8 μm [20]. Gain coefficients as large as 0.4 cm^{-1} have been demonstrated in diffusion-type recordings and as high as 16 cm^{-1} with the application of external electric fields and near-resonant effects [7].

Application of electric fields to GaAs results in spatial non-uniformities. It has been found experimentally that for electric field values up to 2 kV/cm high field domains exist across the sample that travel toward the positive electrode. These domains are attributed to negative differential resistivity arising from field-enhanced capture of carriers by deep level traps [22]. At applied fields larger than 2 kV/cm, the fields are essentially uniform. However, above 3-5 kV/cm the mobility (and hence drift length) of the electrons is severely limited, due to inter-valley scattering [23]. This limits the amount of space charge increase that can be gained by applying electric fields. In practice, the highest applied fields that can be used in GaAs without causing field nonuniformities is about 4 kV/cm.

1.3.2 InP

Indium phosphide has been demonstrated to be photorefractive by doping with iron (InP:Fe)[20] and titanium (InP:Ti) [24]. Photorefractivity has been demonstrated from near the band-edge at 970 nm[9] out to 1.32 μm [25]. Titanium-doped material is believed to have a photorefractive response past 2 μm, based on photoconductivity data. In addition, the bi-polar conductivity found in some InP:Fe samples is not seen in InP:Ti material, mitigating electron-hole competition and optimizing photorefractive response.

Net gain ($\Gamma > \alpha$) has been demonstrated in InP:Fe at a wavelength of 1.32 μm without using applied fields [25]. The gain coefficient of 0.27 cm^{-1} was almost twice as large as the absorption coefficient, $\alpha = 0.145\ cm^{-1}$. Using applied fields, gain coefficients as large as 30 cm^{-1} were measured at 980 nm where, the absorption coefficient was 5 cm^{-1}.

The role of multiple defects has been found to be important when modeling the photorefractive behavior of certain InP samples [13]. A two-defect model has been shown to more accurately model the temperature-dependent photorefractive effect in InP:Fe rather than a single Fe defect model. The temperature-dependent concentration of these multiple trap levels leads to dramatic changes in electron-hole competition, which can result in sign reversal or quenching of the coupling coefficient. The concentration of this second defect appears to be enhanced near the seed end of the boule. However, in many cases, particularly for samples where photoconductivity is strongly dominated by holes, the single defect model can give adequate results. Indeed, the single defect, two band model is adequate to describe the temperature-intensity balance that can take place in InP:Fe, resulting in large wave-mixing gains in the presence of dc applied fields.

Field domains, similar to those measured in GaAs, have also been measured in InP. However, they have only been observed at low temperature ($T < 77$ K) [26]. The domains are a result of negative differential resistivity caused by a field-enhanced carrier capture into a deep level (not associated with Fe). Thermal ionization at room temperature depopulates this trap level and field domains caused by this mechanism are not present. A second mechanism for generating field domains has been reported at room temperature in InP, when used in a double pumped phase conjugate arrangement. In this case, field domains arise from negative differential resistivity caused by the sudden increase of intensity fringe contrast inside the crystal as the DPCM achieves threshold [27]. Effects of these domains can be mitigated using a two-zone DPCM, which is discussed in a subsequent section.

1.3.3 GaP

Growth of semi-insulating gallium phosphide (GaP) has been challenging. However, Kuroda et. al. [28] have reported photorefractivity in a nominally undoped sample. GaP has the highest bandgap of the III-V semiconductors (2.26 eV ~550 nm), permitting operation into the visible. Beam coupling coefficients of 0.33 cm^{-1} were reported at a wavelength of 633 nm. With applied electric fields and moving gratings, gain coefficients as high as 2.5 cm^{-1} have been reported [29]. Absorption measurements indicate that photorefractive response should extend out to 1 μm, but no measurements have been reported.

1.3.4 CdTe

Photorefractivity has been demonstrated in cadmium telluride by doping with vanadium (CdTe:V)[17] and titanium (CdTe:Ti) [30]. CdTe has by far the largest experimentally measured sensitivity figure of merit, *S* (index change per absorbed photon). Photorefractivity has been measured at wavelengths from 1.3 μm to as long as 1.5 μm [17], which is the longest wavelength demonstrated in compound semiconductors to date.

In the absence of applied fields, gain coefficients as large as 0.7 cm^{-1} have been measured in CdTe at a wavelength of 1.32 μm. Using an AC field for enhancement, gain coefficients as large as 10 cm^{-1}, with a field of 10 kV/cm and a beam ratio $\beta = 10^4$, have been demonstrated[31]. The growth of high quality semi-insulating CdTe crystals has, so far, been difficult to reproduce. With improved growth techniques CdTe could become an important photorefractive material.

1.3.5 ZnTe

Vanadium-doped zinc-telluride (ZnTe:V) has been shown to be photorefractive in the wavelength region from 0.63 - 1.3 μm. [15] This material has the highest measured value of, $n_0^3 r_{eff}/\varepsilon$ (corresponding to the electro-optic index change per separated charge), of all the compound semiconductors. This suggests that if high trap densities can be introduced, gain coefficients comparable to CdTe could be obtained. It is sensitive in a spectral region compatible with a wide range of laser diodes. Beam coupling has been demonstrated using a GaAs-GaAlAs diode laser at 830 nm.

The maximum gain coefficient measured at 633 nm was 0.45 cm^{-1} and fell to 0.12 cm^{-1} at 830 nm. This sharp fall-off in gain could be caused by wavelength dependent electron-hole competition, possibly caused by multiple defect levels. Photoluminescence data also gives evidence of multiple defect levels in this material. To date, however, no investigation of the temperature dependence of the photorefractive effect has been made to confirm this.

1.3.6 CdS

Photorefractivity has been reported in semi-insulating cadmium sulfide (CdS) at a wavelength of 633 nm [32]. The bandgap of this material is 2.47 eV and should allow operation into the green (~520 nm). CdS belongs to the 6 mm symmetry group and as such is one of the few semiconductor materials not in the $\bar{4}3m$ group. It has three non-zero electro-optic coefficients, $r_{13} = r_{23}$, r_{33} and $r_{42} = r_{51}$ and is uniaxial, $n_o = 2.46$, $n_e = 2.48$. Gain coefficients as high as 0.3 cm^{-1} were measured at 633 nm where the background absorption was 0.5 cm^{-1}. The grating formation time was measured to have a sub-linear dependence on writing intensity due to the interaction of shallow traps in the material. The good sensitivity and normalized gain coefficient measured with diffusion fields make this a promising material.

1.3.7 Alloys and other materials

One of the more interesting prospects for new photorefractive materials is in the use of alloyed semiconductors. Ternary and quaternary materials can be grown with "engineered" band-gaps by selecting the proper stoichiometry. III-V and II-VI alloys have been grown epitaxially, with thin layers or quantum wells deposited on substrates and made semi-insulating through ion implantation (see chapter 7 of this book).

Sefler et. al.[33] have reported on photorefractivity in the bulk ternary material $Cd_{0.9}Mn_{0.1}Te$ grown nominally undoped and recently, $Cd_{0.45}Mn_{.0.55}Te$ has been grown semi-insulating by doping with vanadium [34]. Photorefractivity has been demonstrated at wavelengths between 0.7 μm and 1.0 μm, which are compatible with laser diode technology. Because this is a dilute magnetic semiconductor, magneto-optic interactions such as the Faraday effect are present. Sefler et. al.[33] demonstrated that at low temperatures, (to increase the magnetic susceptibility), a magnetic field can control the direction and magnitude of photorefractive energy transfer by rotating the polarization of the beams inside the crystal.

The linear electro-optic coefficient of MnTe is less than that of CdTe while its bandgap is larger. This suggests that the material should be grown to have a bandgap as small as possible for a given application. For example, operation with a 780 nm laser diode source would be optimized with a material having its bandgap near 750 nm because of increased linear and quadratic electro-optic and magneto-optic effects. The combination of all these effects along with the ability to tailor bandgaps make this material system one of the most interesting for future investigation.

Many other compound II-VI semiconductors remain to be investigated, including ZnS, ZnSe, and CdSe. If these compounds can be grown semi-insulating and photoconductive, the possibility exists for photorefractivity. Of the II-VI materials, however, CdTe has the highest sensitivity figure of merit, suggesting that this material should continue to be explored. The wide possibilities of substitutional dopants, native growth-related defects and alloying within the III-V and II-VI material systems leaves a vast potential for future research in photorefractive semiconductor materials growth and characterization.

A comparison of semiconductor photorefractive materials is provided in Table 1. The parameter values were taken from actual measurements, as noted by the references. Missing parameters are so noted. Three figures of merit were calculated based on the published values; the normalized gain coefficient, Γ/α, the normalized electro-optic coefficient, $n_0^3 r_{eff}/\varepsilon$, and sensitivity, *S*. The value for Γ_{meas} is the maximum measured in the specified reference. In some cases, the grating formation time τ_{meas} may have been measured with a lower gain which results in a smaller sensitivity. Table 1 includes performance figures for both diffusion type recordings and drift type recordings where externally applied electric fields have been used to enhance the gain coefficient. Notice that the use of these techniques increases Γ/α dramatically while the sensitivity figure of merit drops. Indeed, the increase in gain coefficient is offset by an increase in grating buildup time.

Table 1 Performance comparison of photorefractive semiconductors and enhancement techniques used to date. Measured or calculated parameters were taken from references as noted. Dashed lines indicate parameters not reported.

Material	λ	$\xi n_0^3 r_{eff}$	α	Γ_{meas}	τ_{meas}	I_p	Γ/α	$S(\times 10^3)$
(enhancement*)[Ref.]	μm	pm/V (/ ε)	cm^{-1}	cm^{-1}	ms	mW/cm^2		$n\ cm^3/J$
GaAs:Cr[19]	1.06	43 (3.3)	1.2	.4	.02	4000	0.33	60
InP:Fe[35]	1.06	27 (2.23)	1.0	.15	.1	1070	0.15	19
InP:Fe[25]	1.32	36.3 (3.0)	.145	.273	—	2000	1.88	—
GaP[28]	.633	44 (3.7)	2.0	.33	3	30	0.165	21
CdTe:V[17]	1.5	91 (8.9)	1.8	0.6	2.8	2.6	0.33	510
$Cd_{.45}Mn_{.55}Te$[34]	.74	—	1.5	.3	2	100	0.2	11
ZnTe[15]	.633	110 (11)	5.2	.45	.015	4700	0.086	32
CdS[32]	.633	40 (4)	0.5	0.3	1.0	80	0.6	48
GaAs (MG)[4]	1.06	—	1.65	7	40	50	4.2	37
CdTe:V (AC)[31]	1.32	91 (8.9)	2	10	10	75	5.0	162
InP:Fe (T-I)[8]	1.06	—	2	6	—	10	3.0	—
GaAs:Cr (ER, MG)[7]	.939	—	3.5	16.3	—	140	4.65	—
InP:Fe (ER T-I)[36]	.97	—	5.1	32	100	120	6.3	21
BSO[37]	.514	101 (1.8)	1.6	.33	20	50	0.20	1.7
$BaTiO_3$[38]	.633	3780 (6.3)	1.0	20	2500	500	20	.13
BSO (MG)[39]	.568	84 (1.5)	.13	10	100	150	77.0	25

ER: ERPR, MG: moving grating, AC: ac field, T-I: temperature intensity resonance.

2. BEAM COUPLING GAIN ENHANCEMENT TECHNIQUES

To compensate for the small electro-optic coefficients and to achieve larger gain coefficients, researchers have investigated many methods of increasing the magnitude of the photorefractive space-charge field. Various techniques such as application of an external electric field [1,2], the moving grating technique [3,4], application of an ac electric field [5,6], and use of a temperature-dependent resonance [8] have so far been used to this end. In GaAs, far from the band-edge, a maximum two-beam coupling gain coefficient of 7.0 cm^{-1} has been obtained by using the moving grating technique [4]. In CdTe a maximum gain coefficient of 10 cm^{-1} has been obtained using AC fields [31]. In InP:Fe, by taking advantage of the ambipolar conductivity of the material, a temperature-dependent resonance in the space-charge field has been used to obtain gain coefficients far from the band-edge as high as 11.3 cm^{-1} [8]. Through an additional enhancement mechanism, utilizing band-edge resonance, gains as high as 16.3 and 31 cm^{-1} have been achieved in GaAs and InP respectively. These results represent a significant improvement over the diffusion case (gain coefficient $\Gamma\sim0.6$ cm^{-1} [17]) and are comparable to the gain coefficients obtained in materials such as $BaTiO_3$ ($\Gamma\sim 20$ cm^{-1} [38]). The high gains are desirable for achieving compact, high resolution, and efficient photorefractive devices for self-pumped phase conjugation, image processing, and optical switching applications.

2.1 DC Field

The application of a simple DC field will increase the magnitude of the space-charge field, as can be seen from eq.(4). It will, however, also reduce the phase shift from $\pi/2$, as seen from eq.(5). The reason for the enhancement in magnitude is that the carrier movement is now assisted by drift, so charge may be effectively moved at larger grating spacings. Unfortunately, the symmetry is now broken, so that the carriers move preferentially in one direction, causing the index grating to shift in phase.

Figure 4 illustrates the dependence of space-charge field magnitude and phase on an applied dc field. For this calculation it was assumed that $E_D = 0.1E_q$; that is, the grating spacing was large enough so that diffusion played a small role, except for very small applied fields. Notice that even though the phase is shifting from $\pi/2$, the imaginary component continues to increase for all values of electric field. Several methods have been successfully employed to effectively shift the grating back to the optimum $\pi/2$ phase shift. These methods will be discussed in the following section.

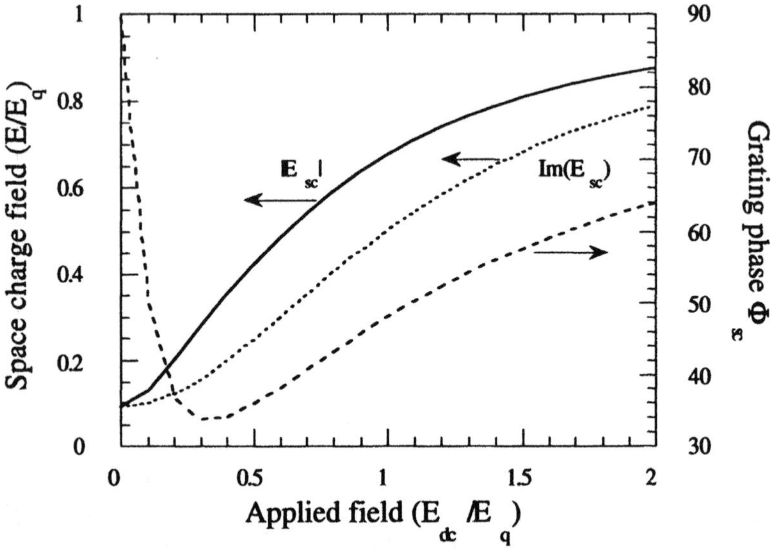

Figure 4. Dependence of space-charge field magnitude and phase on externally applied field strength. Plots are calculated from eq.(4) and eq.(5) where E_D is taken to be $0.1E_q$. Field quantities are normalized by E_q, the trap limited field. Im(E_{sc}) is the imaginary component of the space charge.

2.2 Moving Gratings

By introducing a small frequency shift in one of the two beams, the interference pattern will no longer be stationary in time, but rather will move in one direction or the other, depending on the sign of the frequency shift. If the speed and direction of the interference pattern are matched carefully it can compensate for the change in grating phase and restore the $\pi/2$ phase shift. In order to describe the situation mathematically, it is necessary to consider both the time and spatial dependence of the space-charge field [40]. By introducing a frequency shift, $\delta\omega$, in one of the beams, the grating will have a velocity,$v = \delta\omega/k_g$. Allowing the quantities N_T, n, J and E_{sc} to be time-dependent in the material equations, (see chapter 1), permits finding an optimum grating velocity for the grating velocity [39]

$$v_{opt} = -\frac{eN_t\sigma I_o}{\varepsilon k_g^2 E_{dc}}. \tag{11}$$

To arrive at this solution it is necessary to assume that 1) the electron drift length is larger than the grating spacing, $\mu_n \tau E_{dc} > \Lambda$; 2) the applied field is much less than the trap limited field, $E_{dc} << E_q$, and 3) the diffusion field is negligible, $E_D << E_{dc}$ These requirements are typically satisfied, because conditions outside these limits represent regions of small enhancement. The reason that the grating velocity is inversely proportional to the applied field can be understood by considering a model in which carriers are not affected by diffusion. In the absence of an applied field, no charge distribution builds up because carriers recombine at the same place they are generated. In the other limit of very large applied fields, that is, when the drift length is very large compared to the grating spacing, carriers will move across many grating periods before falling into a trap. Because the average drift distance is so large, the width of the distribution is also large. This leads to a nearly uniform distribution of carriers which, by symmetry, leads to the desired $\pi/2$ phase shift, in much the same way as in the case of diffusion-only. As the electric field is lowered and the drift length approaches the grating spacing, an asymmetry in carrier distribution develops. To maintain the desired phase shift of the fundamental component, the grating must move and hence the inverse proportionality to electric field arises in this regime. The dependence on intensity arises simply because carriers are generated at a faster rate with increasing intensity, and therefore the process of rewriting the grating takes a shorter amount of time; the grating must move faster to follow it.

It is interesting to note that in the limit of very small applied fields, $E_{dc} << E_D$, the optimum velocity becomes

$$v_{opt} = -\frac{e\mu_n n\, E_{dc}}{2\varepsilon_o \varepsilon_s K\, E_D}. \tag{12}$$

The fact that the velocity is linear in applied field illustrates that, for small applied fields, the net drift of charged carriers may be compensated by moving the grating by the appropriate amount.

At the optimum fringe velocity, the space-charge field is given by [4]:

$$E_{sc}(v = v_{opt}) = m\frac{E_{dc}}{2}\frac{[1+(\frac{E_{dc}}{E_m})^2]^{1/2} + \frac{E_{dc}}{E_m}}{1 + \frac{E_{dc}^2}{E_m E_q}} \tag{13}$$

where $E_m = 1/(k_{g_n}\tau_r)$. The carrier lifetime $\tau_r = 1/(\gamma_r N_T)$, where γ_r is the carrier recombination constant and μ_n is the carrier mobility (electrons in GaAs[41]). At the optimum fringe velocity and the optimum grating period,

$$\Lambda_{opt} = \frac{2\pi E_{dc}}{N_T}\sqrt{\frac{\varepsilon \varepsilon_s \mu}{e \gamma_r}}$$

the space-charge field is given by [39]:

$$E_{sc}(\Lambda=\Lambda_{opt}, v=v_{opt}) = m\frac{E_{dc}}{2}\left(\frac{e\mu_n}{\varepsilon\gamma_r}\right)^{1/2} . \quad (14)$$

The grating is moved at the rate of approximately one grating period per grating build-up time. In this way, the grating always has a $\pi/2$ phase shift with respect to the intensity pattern and can efficiently transfer energy between the beams. To achieve a moving grating, one of the beams in the beam-coupling experiment must be shifted in frequency. This is obtained by phase modulating the beam. Many techniques are available for this [42,43]. The technique most commonly used is Doppler shifting of the pump beam by reflecting it off a piezoelectrically driven mirror [4,39]. Gain coefficients as large as 8 cm^{-1} [4] have been measured in GaAs using moving gratings.

For practical applications the moving grating technique is inconvenient because of the need to introduce a frequency shift that should change with applied field, intensity and grating spacing as shown in eq.(11). In addition, the frequency shifts introduced by piezo-driven mirrors or electro-optic phase modulators have the problem of "glitching" when the mirror reaches the end of its travel (or the modulator its maximum voltage) and is reset. Finally, some important applications of photorefractive materials such a self-pumped phase conjugation, do not allow frequency shifts of individual beams because all the pump beams are derived from the single input beam.

2.3 Temperature-Intensity Enhancements

Another mechanism similar to moving gratings, for obtaining the desired $\pi/2$ phase shift, is the so-called temperature intensity resonance [8]. The resonance requires that thermal conductivity and photoconductivity be dominated by different carrier types. Because of this rather unusual requirement, the effect has only been reported in InP:Fe. However, it should occur in certain wavelength regions of GaAs and may be possible in other materials with appropriate deep levels. To describe the physics of the resonance, it is necessary to include both electron and hole optical and thermal generation and use a one-defect, two-band model (Chapter 1) The full expression for the space-charge field has been derived by Picoli et. al. [8] and is given by,

$$E_{sc} = \frac{imI_o\left(\sigma_p p_T - \sigma_n n_T\right)}{e_n n_T\{[1/E_q]+[1/(E_D - iE_{dc})]\} + e_p p_T\{[1/E_q]+[1/(E_D + iE_{dc})]\}},\quad(15)$$

where

$$E_q = (e\Lambda/2\pi\varepsilon)[n_T p_T/(n_T + p_T)].\quad(16)$$

The imaginary component of the space-charge field is found to be a maximum when,

$$e_n n_T = e_p p_T,\quad(17)$$

where e_n is the total emission rate for electrons, e_p is the total emission rate for holes, n_T is the concentration of traps containing electrons (Fe^{2+} in the case of InP) and p_T is the concentration of traps with holes available (Fe^{3+}). It is found that when the resonance condition is fulfilled, the space-charge field is purely imaginary ($\pi/2$ shifted). Notice that if the thermal emission rates are not taken into account (i.e. $e_n^{th} = e_p^{th} = 0$), the fulfillment of the resonance condition eq.(17) also means that the electron-hole competition factor ξ eq.(9) is zero, resulting in a zero gain coefficient. If, as in the case of InP:Fe, thermal and optical processes are dominated by different carrier types, then ξ will be non-zero. For typical InP:Fe samples the thermal emission rate of holes and the photoionization rate of electrons near room temperature can be neglected, so the resonance condition becomes

$$e_n^{th} n_T = \sigma_p p_t I_0,\quad(18)$$

where e_n^{th} is the thermal emission rate of electrons, σ_p is the optical cross section for holes and I_o is the total intensity. The left side of eq.(18) is a function of temperature alone, and the right side is a function of intensity, so that it is possible to fulfill this condition by adjusting temperature and intensity accordingly. Under these conditions the ξ factor becomes equal to 1/2.

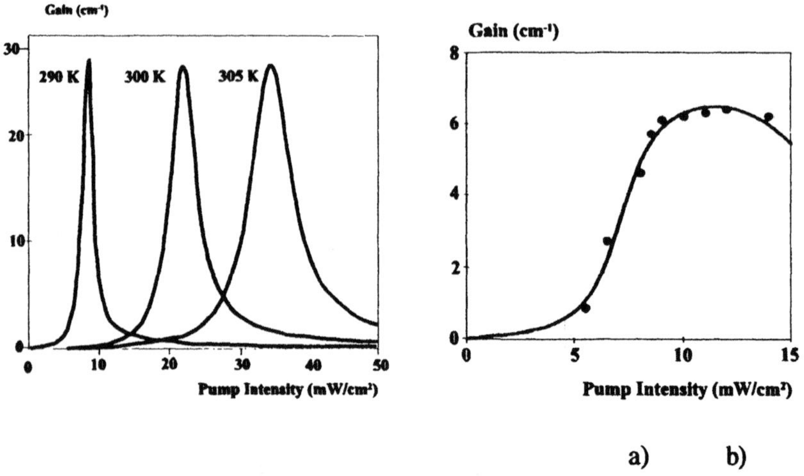

Figure 5. a) Theoretical curves for the gain coefficient vs. pump intensity at 1.06 μm, for three values of sample temperature. Eq.(15) was used with $\Lambda = 10$ μm, $E_{dc} = 10$ kV/cm. b) Fit of experimentally measured gain coefficient vs. pump intensity in an InP:Fe sample (T = 290 K, $\Lambda = 6$ μm, $E_{dc} = 8$ kV/cm) fit to a theoretical curve, calculated by integrating along the thickness of the sample. (Ref. [8])

Combining eq.(8) and eq.(15) the gain coefficient can be calculated as a function of temperature and intensity. Figure 5a plots the gain coefficient as a function of pump intensity for three different sample temperatures with trap densities, thermal and optical cross sections typical of InP:Fe. In practice, the resonance peaks are not as sharply peaked, caused by absorption along the interaction length of the material. Figure 5b shows a typical plot of the gain measured as a function of intensity.

By substituting an expression for the thermal emission rate of the electron in eq.(18) it is possible to write the optimum intensity as an explicit function of temperature,

$$I_{opt} = (n_T/p_T)\,(hc/\lambda)\,T^2\,D\,\exp(-E_{na}/k_bT) \qquad (19)$$

where $D = 3.6\times10^8$ mW/cm^{2}K^2, $E_{na} = 0.62$ eV is the thermal activation energy as measured from DLTS[44], h is Planck's constant, and c is the speed of light. The only free parameter in this expression is the factor n_T/p_T, or the ratio of Fe^{2+} to Fe^{3+}. Figure 6 shows the optimum measured pump beam intensity for the sample used in Ref.[9], as the sample temperature was changed. The line is a fit to the data using eq.(19), where the ratio n_T/p_T is used as a free parameter. From the fit, the ratio is found to be 0.2, which is close to the ratio of 0.3 reported by Ozkul et.

al.[45] for a different sample. Because this function is exponential in temperature, it is impressive that the magnitudes are in agreement.

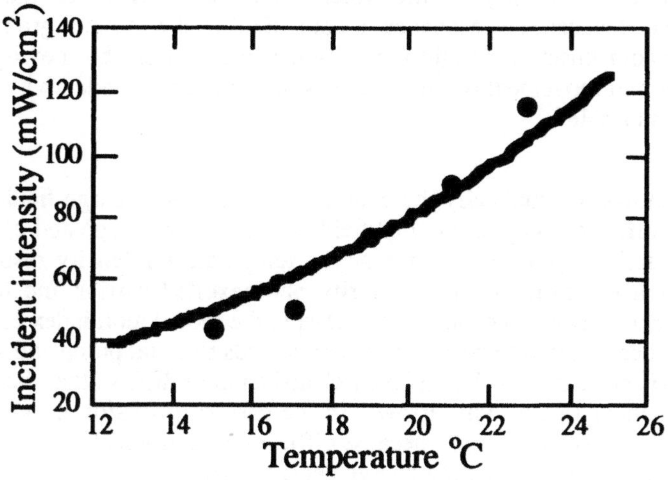

Figure 6. Optimum intensity vs. sample temperature for 2.2 mm InP:Fe sample.

The measured gain coefficient is really an average along the length of the crystal. Because of the moderate absorption near the band-edge and the resonance of gain with intensity, the peak gain coefficient occurs somewhere inside the crystal, and may be higher than the measured gain coefficient. Indeed, it has been found that even larger gain coefficients can be measured using a thinner sample (see Figure 18).

Further insight can be gained into the nature of the resonance by approaching the problem in a manner similar to the moving grating technique. By allowing one of the beams to have a frequency shift, we solve for the optimum grating velocity in a similar fashion to eq.(11):

$$v_{opt} = -\frac{e}{\varepsilon k_g^2 E_{dc}}\left(e_n^{th} n_T - \sigma_p p_T I_o\right), \tag{20}$$

which, in the absence of thermal emission, is identical to eq.(11). If the resonance condition eq.(18) is met, then the optimum velocity of the grating becomes zero. The thermal emission of electrons counteract the drift of the photoexcited holes, producing a grating with no net drift. The reason the grating does not wash out entirely is because one carrier type is produced in response to the light intensity and therefore contains the modulation information, while the other carrier type is produced by uniform thermal excitation. Because the second carrier is produced uniformly and contains no information about the modulated light intensity, it offsets only the average density of the other carrier and preserves the modulation information. Because the resonance is equivalent to the moving grating case, it is possible to compensate for slightly off-resonance conditions by moving the grating. The direction of movement is governed by whether the net conductivity is dominated by electrons or holes.

Figure 7 shows schematically the evolution of the space-charge field for the three cases of a) diffusion only, b) applied field with one carrier type and c) applied field with two carrier types generated near the temperature-intensity resonance. It is assumed that electrons are the majority photoexcited carrier and holes are the dominant thermally excited carrier. For clarity, the modulation index of the intensity pattern has been chosen to be 1, which corresponds to equal pump and signal beams. The free carrier density is shown "before" diffusion or drift so that their motion can be shown schematically with arrows. In the case of diffusion only, 7a, electrons move equally in both directions, because diffusion is isotropic. The electrons uncover positive charge and the resulting charge density distribution $\rho_{fixed}(x)$ is periodic and in phase with the intensity grating. Through Poisson's equation, the space-charge electric field E_{sc} is then shifted by $\pi/2$ with respect to the charge distribution, giving rise to the optimum beam coupling condition.

If we now ignore diffusion and apply a small applied field (equivalent to having a grating spacing much larger than the diffusion length), the electrons preferentially move in one direction, as shown in Figure 7b. This results in a distorted charge distribution shifted by an amount ϕ. The fundamental Fourier component of the space-charge field is then no longer at exactly $\pi/2$ but at $\pi/2 + \phi$. The final case, Figure 7c, shows the temperature-intensity resonance where electrons are the photoexcited carrier and holes are the thermally excited carrier. Notice that the free carrier distribution "before" drift and diffusion is made up from a periodic distribution of optically excited electrons, n, and a uniform distribution of thermally excited holes, p. The net free carrier distribution alternates in sign each half period. This change in free carrier sign is the key to maintaining the charge density in phase with the intensity pattern because carrier movement is symmetric.

The temperature-intensity resonance is equivalent to the moving grating technique, but has the advantages of not needing to introduce frequency shifts onto the beams and being insensitive to applied field and grating spacing. It is, however, sensitive to the optical intensity, which is an added restriction from the diffusion case.

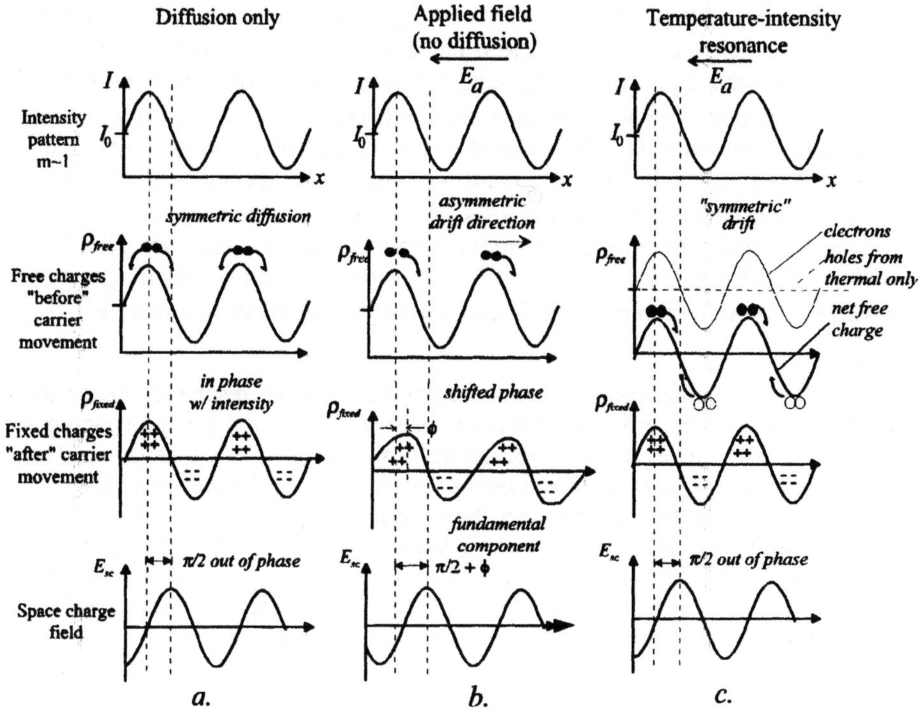

Figure 7. Evolution of the space-charge field for three different cases. a) diffusion alone, b) applied electric field and one carrier type (electrons), c) temperature-intensity resonance case (thermally excited holes). Graphs in vertical direction are 1) incident intensity pattern, 2) free carrier density "before diffusion or drift", 3) charge density after diffusion, drift and trapping, 4) resulting space-charge field.

2.4 AC Field

The asymmetry caused by the DC field and subsequent shift in grating phase can be avoided by periodically alternating the direction of the electric field. If the field is cycled much faster than the grating buildup time, τ_g, but less than the carrier lifetime, τ_r, then each carrier sees a constant field, but no substantial drift of the entire grating is caused. This is equivalent to the case of Figure 8a. If the period of the square wave field is T_{ac} then[46]

$$\tau_r < T_{ac} < \tau_g \tag{21}$$

for the AC field technique to be effective. As T_{ac} approaches τ_g the grating begins to move with the field and part of the electrical signal will be transferred to the optical beams. At the other limit, that is $T_{ac} < \tau_r$, a single carrier experiences no net drift, because it sees fields in both directions over its lifetime. Violation of this condition will lead to a reduction in enhancement but no signature of the ac signal will be imposed on the beams. Because E_{sc} is proportional to the applied field squared (see eq.(4)), the coupling direction is not affected by the sign of the applied field. It is, however, sensitive to the magnitude, so it is important to use square wave fields with fast switching times (rise time $\tau_{rise} << T_{ac}$) in order to avoid having a signature of the AC field transferred to the optical beams during the finite switching time.

The AC field technique has the decided advantage that no resonance needs to be maintained because the enhancement is insensitive to intensity, grating spacing and applied field magnitude. Grating formation time is related to intensity, so that as intensity is increased, care must be taken to ensure that the frequency of the square wave is still sufficiently high to meet the criteria of eq.(21). The most significant problem using the ac field enhancement is the electronic switching of large voltages at high speeds. The complexity of the switching circuit and the unwanted electrical interference that accompanies high frequency alternating currents could be a prohibitive factor when using this technique in practical applications.

The theory for square wave enhancement is described by Stepanov et al.[46]. The space-charge field is given by

$$E_{sc} = \frac{-imE_D}{1+k_g^2 L_s^2}\left(\frac{1+[k_g L_{Ea} E_a]/[E_D(1+k_g^2 L_D^2)]}{1+[k_g^2 L_{Ea} L_e]/[(1+k_g^2 L_D^2)(1+k_g^2 L_s^2)]} \right), \qquad (22)$$

where $k_g = 2\pi/\Lambda$, $L_{Ea} = \mu\tau E_a$ is the carrier drift length, $L_D = (D\tau)^{1/2}$ is the carrier diffusion length, $L_s = (\varepsilon\varepsilon_0 k_b T/e^2 N_t)^{1/2}$ is the screening length, $L_e = \varepsilon\varepsilon_0 E_a/eN_t$, μ is the carrier mobility, τ is the carrier lifetime, E_a is the applied field magnitude and D is the carrier diffusion coefficient.

Walsh et. al.[47] have shown analytically that the measured gain coefficient using a slew-rate limited ac waveform can be significantly less than that predicted for a perfect square wave. In semiconductor materials, where the grating formation time is small, even at modest intensities, it is necessary to produce high frequency square waves in order to fulfill the condition given by eq.(21). As the driving frequency increases, it becomes difficult to maintain sharp square waves, because capacitive losses limit the slew rate. Ziari et. al. [31] have shown the dramatic effects of slew rate limiting in CdTe and have also confirmed that another phenomenon, field shielding[48], would also impose reductions on the measured gain.

Field shielding arises when the ac period is comparable to or longer than the dielectric relaxation time and non-uniformities exist in the illumination across the width of the crystal. Charged carriers can build up in the darker regions and produce a field that opposes the applied one. This can be overcome by insuring uniform illumination (which may not always be possible) or by reducing the ac period so that $T_{ac} << \tau_{die}$, where τ_{die} is the dielectric relaxation time ($\tau_{die} = \varepsilon\varepsilon_0/\sigma$, where σ is the total conductivity caused by thermal and optical excitation). In semiconductors, dielectric relaxation times of less than 1 microsecond are typical for intensities on the order of 100 mW/cm^2, requiring MHz frequencies to overcome shielding. Field shielding will also be a problem in DC field enhancements where illumination is non-uniform, particularly because of such effects as shadowing by electrodes. Figure 8 shows measured and predicted performance of a CdTe:V sample, where both field shielding and slew-rate limiting are taken into account. Though both effects were important, slew-rate limiting was clearly the most severe limitation under these experimental conditions.

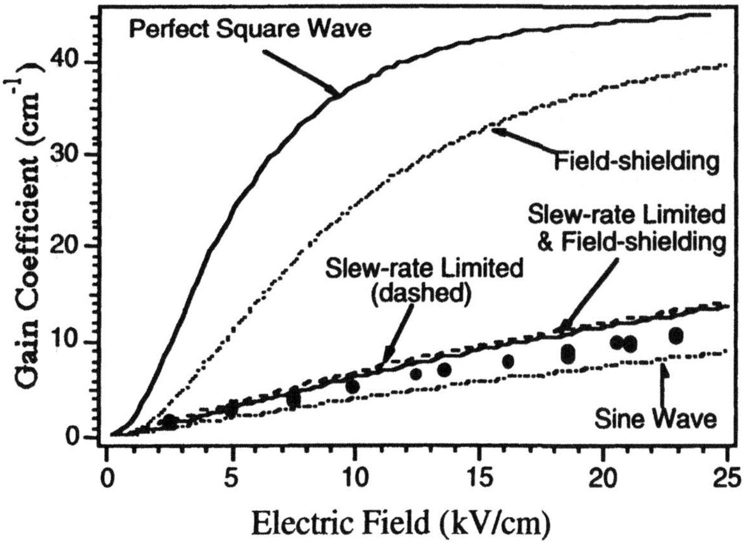

Figure 8. Gain coefficient vs. applied field for CdTe:V using AC field enhancement, f = 230kHz, Λ = 7.5 m I_p=75 mW/cm^2. Slew rate rise time was 0.7 sec. The simulation curves for perfect square wave, slew rate limited, field shielding are compared with experimental data at a beam ratio of 10^4. (Ref. [31])

2.5 Band-edge Enhancement

2.5.1 Introduction

The application of external electric fields serves to increase the overall magnitude of the space-charge field, while enhancement techniques such as moving grating, temperature-intensity resonance and ac fields optimize the $\pi/2$ spatial component of the space-charge field. Most of the work concerned with increasing the photorefractive effect in semiconductors, as well as oxide materials, has used these space charge enhancement techniques along with the conventional Pockels electro-optic effect. Band-edge enhancement increases the magnitude of the index gratings, and hence the gain coefficient, through the use of nonlinearities (in addition to the Pockels effect) that are present in semiconductors.

The application of an electric field to a material will change its absorption coefficient and refractive index near its band-edge[49]. This effect, known as the Franz-Keldysh effect, occurs because an electric field E tilts the conduction and valence bands of a material, thereby increasing the probability of photon-assisted tunneling across the band-gap. The change in absorption coefficient (electroabsorption) and refractive index (electrorefraction) have been measured in InP and GaAs [50], Si [51], and quantum well structures [52]. Electroabsorption and electrorefraction are related through the Kramers-Kronig relationship. Although electroabsorption (EA) is strong near the band-edge, it decreases rapidly with increasing wavelength for space-charge fields typical of the photorefractive effect (<100 kV/cm). The peak of the electrorefraction (ER) spectrum, however, occurs at longer wavelengths and decreases more slowly with increasing wavelength, so that (ER) is significant at wavelengths where the background absorption is relatively small. It is therefore possible to form strong electrorefractive gratings in spectral regimes with small background absorption. The magnitude of the index change caused by ER can be many times that of the linear or quadratic Pockels effect, for moderate electric fields (~ 20 kV/cm).

Using the space-charge field generated in photorefractive semiconductors through the usual drift and diffusion processes, it is possible to write gratings using the large, near-band-edge electrorefraction nonlinearities. This method has the additional advantage that in the proper geometry, the conventional electro-optic grating can also add to the electrorefractive grating to result in even larger nonlinearities.

The process of photorefractivity at wavelengths near the band-edge can be divided into a discussion of the new band-edge electrorefraction photorefractivity (ERPR) and its combination with the conventional Pockels electro-optic photorefractivity (EOPR). These two processes are discussed separately.

2.5.2 Electrorefractive photorefractivity

Below the band-gap, the Franz-Keldysh electrorefractive change in the refractive index of a semiconductor due to application of a field can be approximated by a quadratic equation $\Delta n_{ER} \propto E^2$ [53]. This is different from the Pockels electro-optic effect where $\Delta n_{EO} \propto E$. The sign of the refractive index change caused by band-edge electrorefraction (ER) does not depend on the direction of the electric field. Under diffusion-only conditions (no applied field), for a sinusoidal photorefractive space-charge field, the induced refractive index pattern will, therefore, have twice the spatial frequency of the space-charge field. Because such a grating is not Bragg matched in the direction of the writing beams, it is not possible to transfer energy in two-beam-coupling with such a field. This problem can be overcome by application of an external field.

The intersection of two mutually coherent optical beams in a photorefractive crystal generates a sinusoidal intensity pattern that results in a space-charge field (E_{sc}) through the usual drift and diffusion processes (see illustration in Figure 7). Figure 9 illustrates the formation of the ERPR grating. With the application of a dc electric field, E_{dc}, the total electric field is $E_{tot} = E_{dc} + E_{sc}$. In general, application of an external field increases the magnitude of the space-charge field and changes the phase shift between the intensity and space-charge field patterns. Figure 9 shows Δn_{ER} for two directions of applied field and for no applied field. Without applied field, the induced ERPR grating has twice the spatial frequency of the space-charge field pattern and cannot transfer energy. It can, however, be read via a third beam at the appropriate Bragg angle in a four-wave-mixing geometry. Application of a bias electric field lifts the degeneracy and the refractive index grating has the same periodicity as the space-charge field.

The two polarizations of the applied electric field change the phase of the ERPR refractive index grating 180 degrees with respect to the space-charge field. If the phase of the space-charge field is shifted with respect to the intensity pattern, through a combination of diffusion, drift and external techniques such as moving grating, the ERPR grating can couple energy. Because of the phase dependence of the refractive index on applied electric field, the direction of energy transfer is dependent on the field direction. This is in contrast to the conventional photorefractive effect, where application of a field can only increase the magnitude of the energy transfer without affecting its direction. The direction of EOPR energy transfer is only dictated by the crystal orientation and the dominant photocarrier species [16].

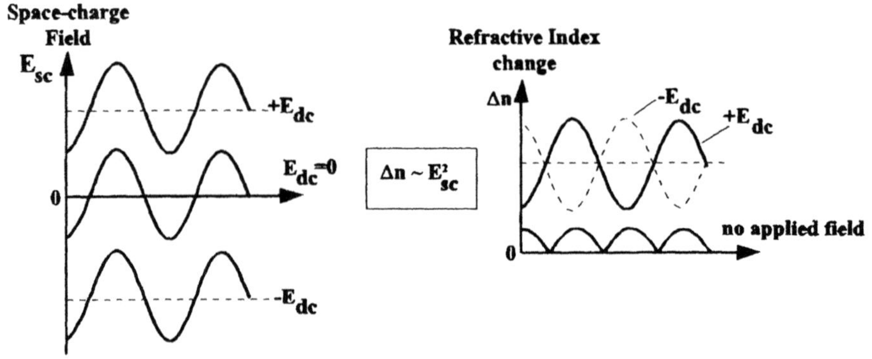

Figure 9. ERPR grating formation. Without applied field, the grating period is doubled. The polarity of applied field controls coupling direction through the phase of index grating.

2.5.3 Band-edge electroabsorption and electrorefraction

Figure 10 shows the change in absorption coefficient caused by applied electric fields from 1 to 12 kV/cm in an indium alloyed GaAs sample. The background (no applied field) absorption spectrum for this sample is also shown. It can be seen that large electroabsorption $\Delta\alpha$ occur at a spectral region where the absorption coefficient is rapidly increasing with decreasing wavelength. The change in absorption coefficient above the band-gap was inferred by fitting the data to a broadened electroabsorption theory [54] so that the index change in the presence of a field could be calculated.

The change in refractive index caused by a change in the absorption coefficient can be determined using the Kramers-Kronig relation:

$$\Delta n\,(h\nu) = \frac{ch}{\pi} P\int_0^\infty \frac{\Delta\alpha\,(h\nu')}{(h\nu')^2 - (h\nu)^2} d\,(h\nu'), \tag{23}$$

where ν is the optical frequency at which Δn is calculated, h is Planck's constant, $\Delta\alpha$ is the change in the absorption coefficient measured at frequency ν', and P is the Cauchy principal value of the integral:

$$P\int_0^\infty = \lim_{\delta \to 0}\left(\int_0^{h\nu-\delta} + \int_{h\nu+\delta}^\infty\right). \tag{24}$$

A set of discrete $\Delta\alpha_i$ measurements can be made continuous through use of linear interpolation between each pair of $h\nu_i$, $h\nu_{i+1}$ points. The integration can then be carried out as described in Ref. [55].

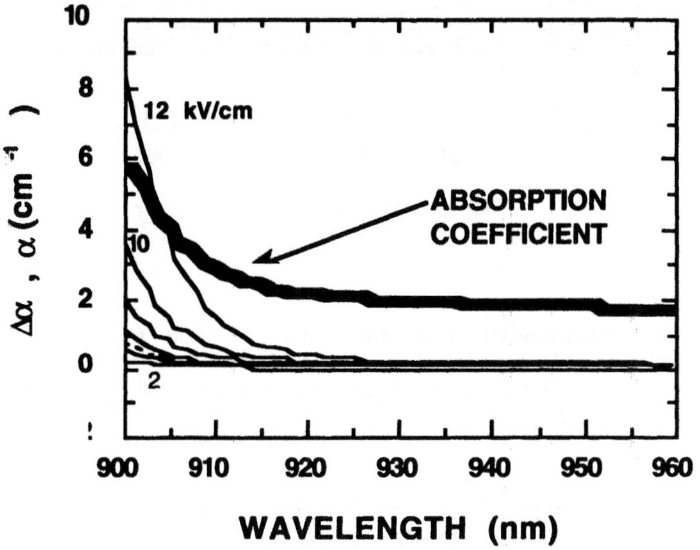

Figure 10. Franz-Keldysh electroabsorptive change in the absorption coefficient as a function of applied field and wavelength. The zero-voltage absorption curve is also shown. The electroabsorption curves are for fields of E = 2 to 12 kV/cm in 2 kV/cm increments, with larger values corresponding to larger fields.

The result of applying the Kramers-Kronig relation to the electroabsorption spectrum is shown in Figure 11. The absorption spectrum of the sample is also shown. For photorefractive applications, large Δn and small α values are desirable. We can see from Figure 11 that at $\lambda > 905$ nm, these requirements are simultaneously satisfied. However, if the change in the absorption coefficient caused by electroabsorption is included, then the optimum wavelength region appears to be $\lambda > \sim 910$ nm.

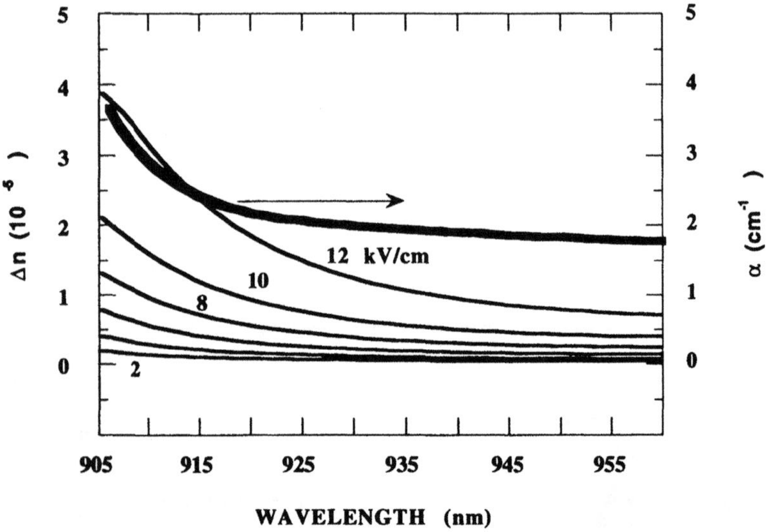

Figure 11. Franz-Keldysh electrorefraction as a function of wavelength and applied electric field in undoped GaAs calculated by applying the Kramers-Kronig relationship to data of Figure 10. The zero-voltage absorption spectrum is also shown.

In the absence of a simple expression for the dependence of band-edge electrorefractive Δn on electric field (E) and wavelength (λ), Δn can be written as:

$$\Delta n(\lambda,E) = n_1(\lambda)|E| + n_2(\lambda)\, E^2 + \cdots \quad , \tag{25}$$

where $n_1(\lambda)$ and $n_2(\lambda)$ are respectively the linear (non-Pockels) and quadratic components of $\Delta n(\lambda,E)$. The absolute value sign on the electric field ensures that the same Δn is obtained regardless of sign of the field.

A quadratic fitting routine was used at every λ in Figure 11 to obtain values for $n_1(\lambda)$ and $n_2(\lambda)$. Figure 12 shows the dispersion of $n_1(\lambda)$ and $n_2(\lambda)$ in the range studied. These curves provide a convenient way of obtaining $\Delta n_{ER}(\lambda)$ for any applied field and can be used in the same way the Pockels electro-optic coefficient r_{41} is used in determining $\Delta n_{EO} = n_o^3\, r_{41} E/2$.

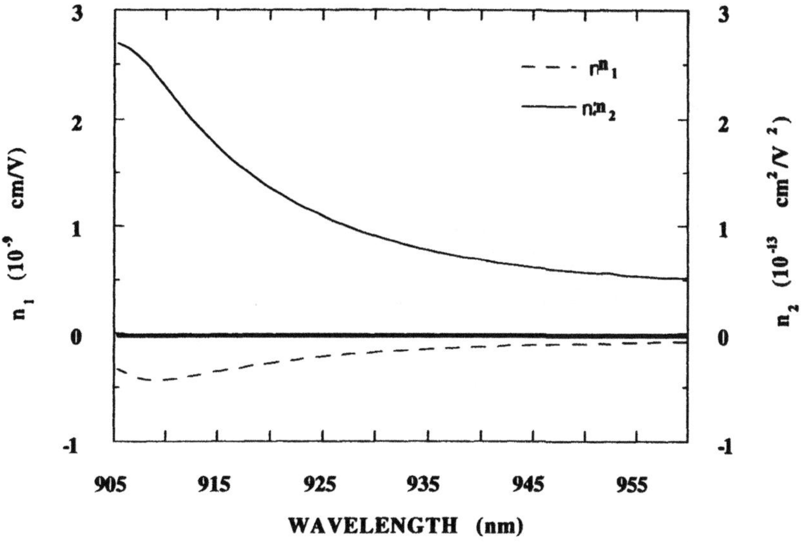

Figure 12. Dispersion of n_1 and n_2 for a quadratic equation approximation to electrorefraction in undoped GaAs as described in eq. (25).

Figure 13 is a comparison of Δn_{ER} at 905 nm calculated from data shown in Figure 11 and values calculated by computing the first two terms in eq.(25), with n_1 and n_2 obtained from Figure 12. The good agreement shows that the first two terms in eq.(25) are sufficient for the field range studied.

Figure 14 is a comparison of the electrorefractive index change Δn_{ER}(= 905 nm) and the Pockels electro-optic index change Δn_{EO}, using $n_o = 3.55$ and $r_{41} = 1.5 \times 10^{-12}$ V/cm. It can be seen that at fields of $E > \sim 13$ kV/cm, the ER effect becomes larger than the EO effect because of its quadratic electric field dependence. It is therefore at large space-charge electric field values (such as achieved with moving grating or temperature stabilization) that using the electrorefractive effect provides dramatic improvements in the beam-coupling gain coefficient.

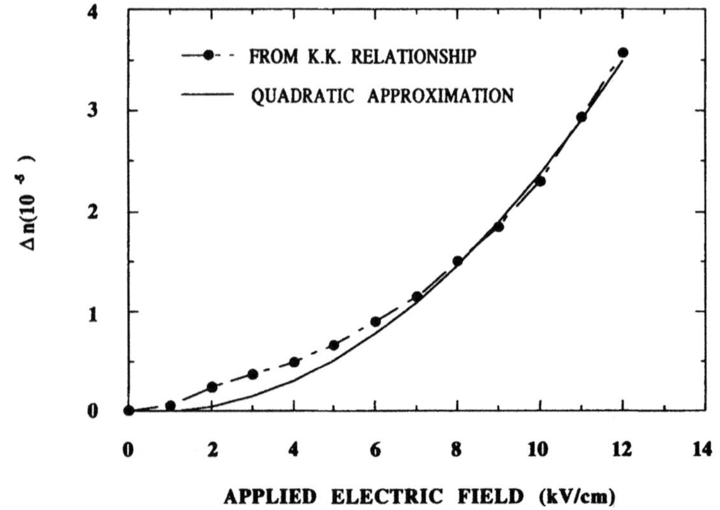

Figure 13. Electrorefraction as a function of applied field at 905 nm wavelength in undoped GaAs. The line is a quadratic fit eq.(25) with the values for n_1 and n_2 shown in Figure 12.

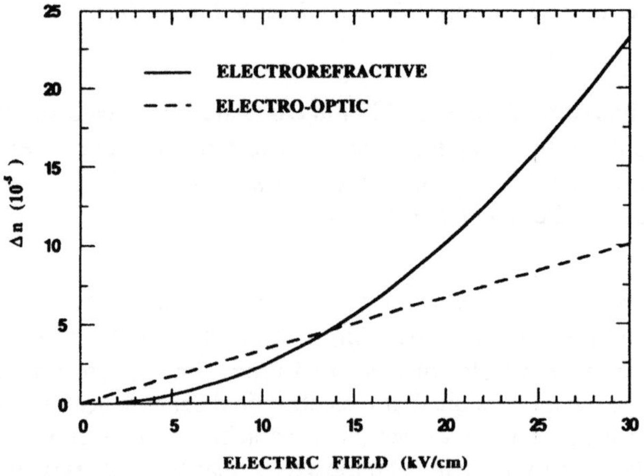

Figure 14. Comparison of electrorefractive and electro-optic change in the refractive index as a function of applied electric field at λ = 905 nm wavelength.

Electroabsorption can also be fit to a quadratic equation:

$$\Delta\alpha(\lambda,E) = \alpha_1(\lambda)|E| + \alpha_2(\lambda)\, E^2 \ . \tag{26}$$

The values of $\alpha_1(\lambda)$ and $\alpha_2(\lambda)$ are shown in Figure 15. Although electroabsorption can be used to write an absorptive grating, such a grating would exist in a spectral region with high background absorption and is not desirable for most photorefractive applications. We will therefore not discuss these gratings here. However, the data in Figure 15 can be used to estimate the magnitude of such gratings and may also be used to obtain values for electroabsorption in the spectral range shown.

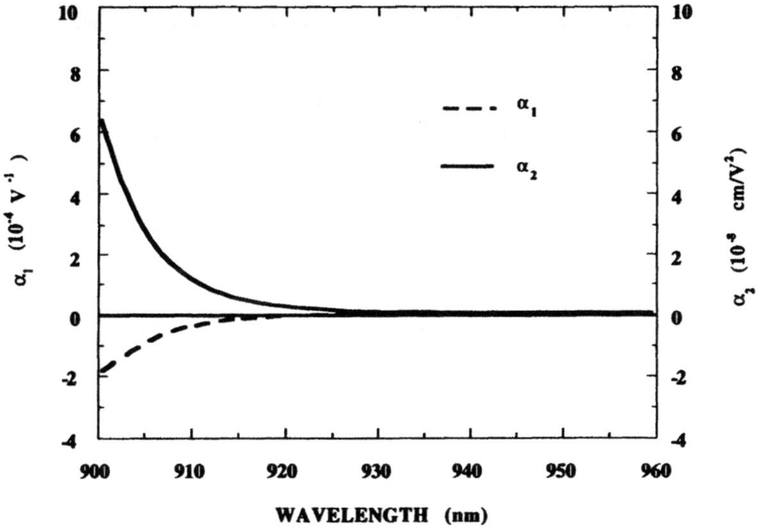

Figure 15. Dispersion of α_1 and α_2 for the quadratic equation approximation to electroabsorption in undoped GaAs as described in eq.(26).

2.5.4 Beam coupling using band-edge photorefractivity.

In the presence of an applied electric field, the component of the photorefractive space-charge that is spatially phase-shifted by $\pi/2$ from the intensity pattern, and therefore takes part in beam-coupling, is given by eq.(4) and eq.(5). Note that this expression is completely independent of the sign of E_{dc}. The electro-optic beam-coupling gain coefficient, $\Gamma_{EO} \propto E_{sc}$, is unaffected by the direction of the applied field. The total electric field can be written as:

$$E_{tot} = E_{dc} + E_{sc} \cos (k_g x) \, . \tag{27}$$

The corresponding electrorefractive change in the refraction index is given by

$$\Delta n_{ER}(E_{tot}, \lambda) = n_1(\lambda)[E_{dc} + E_{sc} \cos(k_g x)] + n_2(\lambda) \, [E_{dc} + E_{sc} \cos(k_g x)]^2 \, , \tag{28}$$

where $n_1(\lambda)$ and $n_2(\lambda)$ are given in Figure 12. For $E_{dc} > E_{sc}$, we can write:

$$\Delta n_{ER}(E_{tot},\lambda) = \{n'_1(\lambda)E_{dc} + n_2(\lambda)(E_{dc}^2 + \frac{E_{sc}^2}{2})\} \quad \text{:independent of x,}$$

$$+ \{[n'_1(\lambda) + 2n_2(\lambda)E_{dc}]E_{sc} \cos (k_g x)\} \quad \text{: with } k_g \text{ spatial frequency,}$$

$$+ \{\frac{n_2(\lambda)}{2} E_{sc}^2 \cos(2k_g x)\} \quad \text{: with } 2k_g \text{ spatial frequency,}$$

where

$$n'(\lambda) = \begin{cases} n_1(\lambda) \; if \; E_{dc} > 0 \\ -n_1(\lambda) \; if \; E_{dc} < 0 \end{cases} . \tag{29}$$

Only the second bracket in eq.(29) will contribute to beam-coupling energy transfer. Thus the ERPR gain coefficient is

$$\Gamma_{ER}(E_{dc},\lambda) = 2(\frac{2\pi}{\lambda \cos\theta}) \, [n'_1(\lambda) + 2n_2(\lambda)E_{dc}]E'_{sc} . \tag{30}$$

where $E'_{sc} = \mathrm{Im}(E_{sc}/m)$ is the $\pi/2$-shifted component of the space-charge field. In order to isolate ERPR, a crystal and/or beam polarization that rules out conventional electro-optic gratings must be used. Many different crystal geometries can be found that will not produce EO gratings. Most of the photorefractive work in cubic crystals is done in the so-called "holographic cut", with faces of $\{110\},\{\bar{1}10\},\{001\}$. Using a standard cut crystal in the geometry shown as Configuration B in Figure 1, an S polarized beam will experience a zero Pockels electro-optic effect. The gain measured in this orientation is therefore due only to the electrorefractive effect.

The direction of EOPR gain is determined only by the majority photocarrier species and the crystal orientation. The dominant photocarrier species in photorefractive media can thus be found by carrying out a single beam electro-optic phase retardation experiment to determine the direction of the electric field for a positive index change and thus fixing the crystal orientation, and then combining this result with the direction of energy transfer in a beam-coupling experiment [16]. Using this technique, other researchers have determined the majority photocarrier species at $\lambda =$ 1064 nm to be electrons in undoped GaAs [16] and holes in InP [35]. Note that unlike the EOPR case, near the band-edge only one experiment (i.e. ERPR beam-coupling) is necessary to determine the majority photocarriers. This technique is therefore simpler to perform and may prove useful for such determinations.

Figure 16 shows the dependence of the ERPR gain coefficient Γ_{ER} on the wavelength for different applied electric fields in a GaAs sample. Although there is considerable scatter in the data, the increase in gain coefficient near the band-edge is observed at all applied voltages. The data were taken by fixing the wavelength and increasing the field. Below 910 nm, large gain coefficients were observed but the increased photoconductivity resulted in large currents and crystal heating, so that accurate measurements could not be made. The maximum gain coefficient observed was 2.8 cm^{-1} at $\lambda = 922$ nm with $E_{dc} = +14$ kV/cm. This gain coefficient is comparable in magnitude to the best previous dc electric field results so far ($\Gamma \sim 2.6\ cm^{-1}$) using EOPR at similar grating spacing and applied fields in GaAs [2].

Figure 16 also shows the evaluation of eq.(28) for the E_{sc} values calculated from eq.(4) for $2\theta = 15.3°$, $N_t = 3.0 \times 10^{15}\ cm^{-3}$ and electric field values applied in the experiments. Considering the uncertainty in the Δn_{ER} data, in the value of N_t, and in the exact value of the electric field inside the sample and its spatial distribution, the agreement between theory and experiment can be considered good. Figure 16 also shows the absorption spectrum for this sample on the same scale.

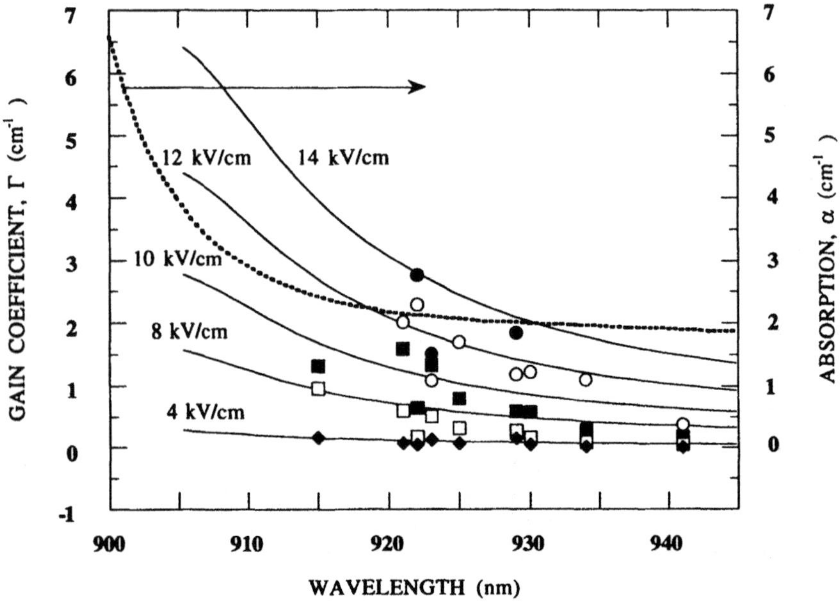

Figure 16. Measured and calculated Beam-coupling gain coefficient for ERPR in GaAs as a function of wavelength and applied electric field of 14 (l), 12 (m), 10 (n), 8(o), and 4 kV/cm (u).The grating period was 3.49 μm at $\lambda = 923$ nm wavelength. The incident pump beam, $I_{po} = 17$ mW/cm^2, and the signal beam, $I_{so} = 1.2$ mW/cm^2, were kept constant for different wavelengths.

2.5.5 Combined ERPR and EOPR effect

The Franz-Keldysh effect exists for any crystal orientation and optical beam polarization. It is therefore possible to choose the crystal geometry, beam polarization, and direction of applied field so that the conventional EO grating adds to the ER grating for even larger beam-coupling gains. Furthermore, techniques such as moving grating and temperature stabilization that can be used to increase the $\pi/2$ component of the space-charge field can similarly be used to enhance the ERPR or the combined ERPR and EOPR gain coefficients.

The combined ERPR and EOPR gain coefficient can be obtained by summing the two components:

$$\Gamma_{BE}(E_{dc},\lambda) = 2(\frac{2\pi}{\lambda\cos\theta})\,[n'_1(\lambda) + 2n_2(\lambda)E_{dc} + \frac{1}{2}\,n_0^3 r_{eff}]\,E'_{sc}. \quad (31)$$

Figure 17 shows the combined ERPR and EOPR as well as ERPR-only gain coefficients for $N_t = 3.0 \times 10^{15}$ cm^{-3} , 2 = 15.3° , E_{dc} = 14 kV/cm, n_o = 3.55, r_{eff} = 1.5 pm/V. It can be seen that the ERPR component can be more than twice as large as the EOPR component near the band-edge.

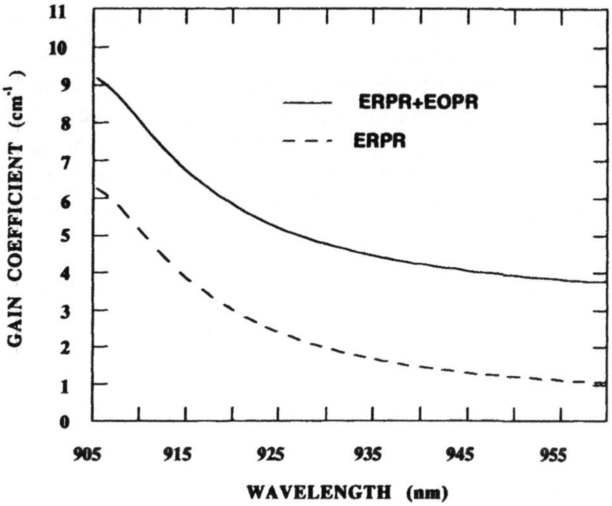

Figure 17. Wavelength dependence of combined ERPR and EOPR and EOPR alone for an applied field of E_{dc} = 14 kV/cm.

2.5.6 Combined ERPR Beam-coupling with moving grating

We can see from Figure 14 that the ER change in the refractive index exceeds the Pockels electro-optic contribution for fields of $E > 13$ kV/cm. In order to take full advantage of the ERPR effect, large space-charge fields are desirable. The moving grating technique is one means of achieving large space-charge fields.

Using a moving mirror in GaAs to obtain moving gratings, Partovi et. al.[7] obtained a large gain coefficient of Γ = 16.3 cm^{-1} (I_{po} = 140 mW/cm^2, I_{so} = 29 µW/cm^2, Λ = 7.0 µm, λ = 939 nm, E_{dc} = +10 kV/cm) by optimizing the piezo mirror velocity. The 4 mm interaction length resulted in $\gamma \equiv I_s$ (with pump) / I_s

(without pump) = 600. The net gain achieved was $\Gamma - \alpha = 13.3\ cm^{-1}$, resulting in more than 200 times net amplification of the signal beam.

The optimum grating velocity depends on the grating build-up time, which in turn depends on the optical beam intensity. Near the band-edge where the absorption coefficient is moderate ($\alpha \sim 3.0\ cm^{-1}$), considerable variation in the beam intensity as it travels through the crystal occurs. Only 30% of the incident beam exits a 4.0 mm long crystal with $\alpha = 3.0\ cm^{-1}$ absorption coefficient (neglecting Fresnel reflections that could be overcome by anti-reflection coatings). Such a variation in intensity means that the grating build-up time and therefore the optimum grating velocity varies considerably through the length of the crystal. The overall optimized velocity is obtained by averaging the optimum at every region in the length of the crystal. Using shorter crystals would result in a more effective use of the moving grating technique and produce larger gain coefficients. This has been experimentally proven in BSO, where the gain coefficient was increased from $\Gamma = 7.0\ cm^{-1}$ to $\Gamma = 12.0\ cm^{-1}$ by shortening the length of the crystal from 10 mm to 1.27 mm [39].

2.5.7 Combined ERPR and the temperature-intensity enhancement

Band-edge enhancement can be combined with the temperature-intensity resonance to obtain very large gains in InP:Fe. The techniques are well suited for each other because they both require DC fields to operate. The combination offers a convenient method for improving gains, because it requires the use of only a thermoelectric device and a dc field.

Millerd et al.[9] combined the temperature-intensity resonance with ERPR in InP:Fe. Beam coupling experiments were performed using an Argon-pumped Ti-Sapphire laser tuned near 970 nm. Two silver paste electrodes were painted on the sample and contacted to copper electrodes, which in turn were each mounted on thermoelectric coolers. The temperature was stabilized within 1°C using a feedback loop that monitored thermocouples mounted near the edge of the electrodes. The direction of the crystal and applied DC field were oriented so that the contributions of Pockels and Franz-Keldysh effects were additive. The samples were anti-reflection coated, $R < 0.5\%$, to eliminate spurious gratings from Fresnel reflections. The electrode spacing was 2 mm and the beam diameter was expanded to 5 mm to avoid field screening effects caused by non-uniform illumination in the transverse direction.

Figure 18 shows the gain coefficient as a function of wavelength and applied field for a fixed grating spacing of 5 μm and a temperature of 19°C. The pump intensity was adjusted to achieve maximum gain at each wavelength. Both absorption and gain increase with photon energy (decreasing wavelength), producing a maximum net gain in the region 970-985 nm.

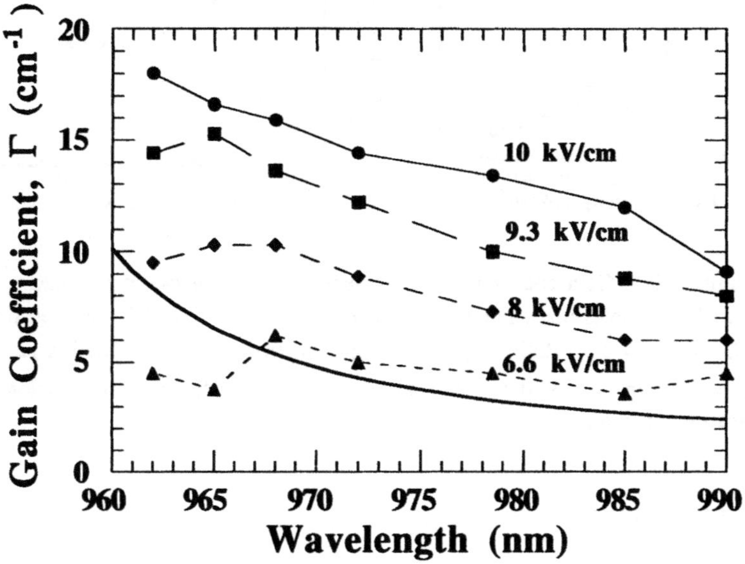

Figure 18 Two beam coupling gain coefficient measured as a function of wavelength for InP sample. Intensity was adjusted at each point to achieve maximum gain T = 19°C, β = 1000. Room temperature absorption is shown as the heavy line.

The gain coefficient measured as a function of incident intensity, for two different sample lengths at a fixed temperature, is shown in Figure 19a. Large beam intensity ratios were used ($\beta = 1.6 \times 10^6$, the ratio of pump-to-signal intensity) to avoid pump depletion and large-signal effects. The two samples were cut from neighboring positions in the same boule. The smaller gain coefficient measured in the longer sample arises primarily from the inability to maintain the optimum intensity over the entire length of the absorbing sample. Thus, the measured gain coefficient is a spatial average.

The effect of grating spacing on gain coefficient at a fixed intensity and temperature is shown in Figure 19b. The gain coefficient increases rapidly with grating spacing and shows a maximum near $\Lambda = 9\ \mu m$. This particular optimum should change with the applied field.

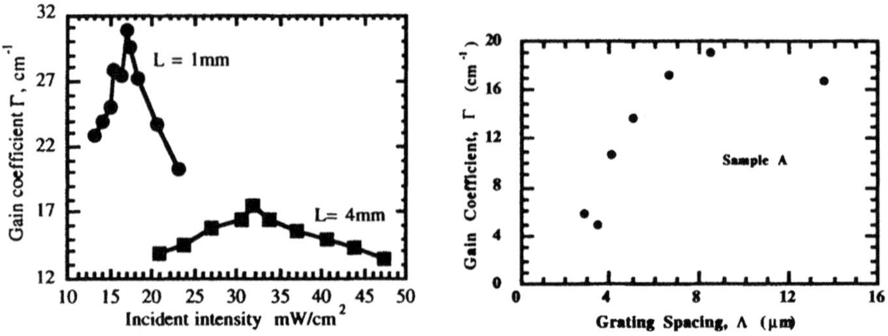

Figure 19.a) Gain coefficient vs. incident intensity for two wave mixing in two InP samples maintained at constant temperature: $T = 20^{o}C$, Applied field $E_{dc} = 10$ kV/cm, Ratio of pump to signal intensity $\beta = 10^{6}$, Grating period $\Lambda = 7.7$ mm, $\lambda = 970$ nm, $\alpha = 5$ cm^{-1} b) Gain coefficient as a function of grating spacing.

2.6 Comparisons

Each of the enhancement techniques has relative advantages and disadvantages. Table 2 summarizes the different space-charge enhancement techniques and their relative merits. The moving grating technique is the least attractive from the standpoint that the optimum velocity is a function of intensity, applied field strength and the grating spacing. Its strengths are that it can be used with any sample, regardless of carrier type, and can be combined with the band-edge enhancement.

The temperature-intensity resonance is not sensitive to applied field or grating spacing but does require the thermal and photo-conductivity of the sample to be dominated by different carrier types; a condition that cannot be fulfilled in all materials at any given wavelength. The AC field technique is most attractive from the standpoint that no resonance must be maintained, making it insensitive to intensity, applied field and grating spacing. However, it cannot be used with the band-edge enhancement because of the directional dependence of the gain coefficient on applied field.

The band-edge enhancement technique of ERPR has characteristics (e.g. direction of beam-coupling gain that depends on external field direction) that may prove useful in new devices. With the recent introduction of photorefractive semiconductor alloys, it is possible to band-gap engineer materials to take advantage of ERPR gains. One unique feature of CdTe is the fact that it may be alloyed with mercury to reduce its bandgap or with manganese to increase its bandgap. It should be possible to use ERPR over a wide range of wavelengths by tailoring the bandgap of the material to the desired energy.

For the application of semiconductors to holographic interferometry, where beam coupling is not necessary or perhaps even desired [34], considerable enhancements in diffraction efficiency can be made by simply using dc fields and working near the band-edge.

Table 2. Comparison of enhancement techniques

Enhancement technique	Controlling parameter	Insensitive to intensity	Insensitive to applied field	Insensitive to grating spacing	Can use ERPR
Moving grating	v - grating velocity	no	no	no	yes
Temp-intensity	temperature of sample	no	yes	yes	yes
AC field	switching of high voltage	yes	yes	yes	no

3. PHOTOREFRACTIVE RESPONSE AT HIGH MODULATION DEPTHS

The application of external electric fields combined with the enhancement techniques have produced large two-wave mixing gains in semiconductor and sillenite materials. However, the largest gains are produced only for very weak signal beams [4,39,56]. Experimentally, it has been observed that as the signal beam intensity becomes comparable to the pump beam, a sharp decrease is measured in the effective gain coefficient, which cannot be solely explained by pump depletion. This effect is significant for any application where a substantial fraction of the pump wave is coupled into the signal, such as self pumped phase conjugation, in which the build-up of the phase conjugate wave will reduce the effective gain and self-limit its performance. In fact, large signal effects must be taken into account when predicting performance of most any application where external fields are applied.

Standard linear photorefractive theory, valid for small modulation index, shows that the magnitude of the fundamental component of the space-charge field is linear with modulation index m. The gain coefficient, Γ, is therefore insensitive to modulation index, because $\Gamma \sim E_{sc}/m$, where E_{sc} is the magnitude of the fundamental $\pi/2$ spatially shifted space-charge field. Evidence that the gain coefficient is a function of modulation index implies that the space-charge field must vary nonlinearly with m. In the simplest physical description, this nonlinearity is caused by the clamping of the space-charge field magnitude to the applied field [57-60].

Several different approaches have been taken to model this nonlinear behavior. The finite difference model developed by Brost[61] accurately describes the performance of $Bi_{12}TiO_{20}$ (BTO) in the presence of applied AC fields, at all modulation depths and grating spacings. Because the model is not limited to steady state and calculates the full space-charge field (within the finite resolution of the element size), it allows the calculation of the buildup of all harmonics. The model has led to a simple empirical formula that allows direct calculation of the large signal effects for materials using AC field enhancement, in terms of the effective trap density, electro-optic coefficient and carrier drift length. It has also demonstrated experimentally that, under certain conditions, the analytical model of Swinburne et al.[57] can provide a reasonable estimate of the nonlinear behavior.

3.1 Analytic Models

The analytic model of Swinburne et al.[57] provides an expression for the local space-charge field, $E_{sc}(z)$, as a function of modulation index $m(z)$, applied field, and grating spacing. The model, which was derived by ignoring the effects of diffusion and trap limitation (valid for grating spacings much larger than the drift length), can be summarized as:

$$E_{sc}(z) = E_a \left(\frac{(1 + 2[\,p\,m(z)]^2)^{1/2} - 1}{p\,m(z)} \right) \tag{32}$$

where z is the location along the interaction length and $p = L_{Ea}\,k_g$ is the so-called enhancement factor (see eq.(22)). The enhancement factor represents the enhancement in space-charge field gained by using AC fields, relative to DC fields of the same magnitude. To obtain a fit to experimental gains measured in a crystal of finite thickness, it is necessary to account for pump depletion. This can be done by performing a numerical integration of the coupled wave equations

$$dA_s/dz = -(1/4)\,\Gamma(m)\,A_p$$

$$dA_p/dz = (1/4)\,\Gamma(m)\,A_s \qquad (33)$$

where A_s and A_p are the amplitude coefficients of the signal and pump waves respectively. The gain coefficient, $\Gamma(m)$, is given by replacing the space-charge field, E_{SC}, in eq.(8) with the new expression given by eq.(32). Integration is necessary to take into account the changing value of m along the crystal length. In this way, the energy in the beams at the end of the crystal may be calculated as a function of input beam ratio. If losses are important, as in the case of self-pumped phase conjugate mirrors, the absorption coefficient, $-\alpha$, must also be included in eq.(33).

3.2 Numerical Models

To solve the problem more exactly, it is necessary to use a numerical model that calculates the local space-charge field by solving the basic material generation and transport equations. From this calculation the local space charge dependence on beam ratio can be obtained.

Brost et. al. [61] generated a numerical model for the case of ac fields that assumes a single set of recombination centers and one species of free carrier. Grating formation was calculated spatially by dividing one grating period into N cells and applying periodic boundary conditions. The magnitude and phase of the fundamental component of the space-charge field was obtained by Fourier decomposition of the calculated space-charge field.

The resulting calculation of the local space-charge field, $E_{sc}(m)$, could then be used in eq.(8) and integrated in eq.(33) to provide a comparison with measured data. Upon further investigation of the model's results it was found that: 1) in the limit of small m the results agreed with eq.(22); 2) the functional form of the large signal effects could be described by an empirical equation

$$E_{sc}(m) = E_{ac}\,\frac{\exp(m)}{a_f\;m}\,[1- \exp(-a_f m)] \qquad (34)$$

where E_{ac} was given by eq.(22) and a_f is a fitting parameter; and finally 3) there was a simple relationship to calculate a_f,

$$a_f = 2.19 \frac{E_{ac}}{(E_a + E_D)m} \quad . \tag{35}$$

Eq.(35) is a slightly modified form of the empirical function first suggested by Refregier et al.[39] Although determined empirically, this function allows direct calculation of the space-charge field at any grating spacing and modulation depth if N_t, r_{eff}, L_D and E_a are known. It should be noted that the model is not valid for all values of the applied field. Studies indicate that the expression is valid if the applied field is greater than twice the diffusion limited field and less then the trap limited field (i.e. valid for $2E_D < E_a < E_q$). This does not impose a serious restriction, however, because the regions outside the boundaries represent regimes where enhancement from external fields is small and are therefore not of much practical interest. Typically, the conditions are satisfied for grating spacings larger than 1 μm and applied fields larger than 2 kV/cm.

3.3 Experimental Comparison and Discussion

Large signal effects are caused by the same mechanism in both semiconductors and sellenites; therefore, the study of these effects in the sellenite BTO is directly applicable to the semiconductor materials. To evaluate the models it is necessary to measure intensity gain, γ, as a function of input beam ratio β, defined as the ratio of pump to signal beam intensities. Figure 20 shows measurements in a BTO sample at 5 μm grating spacing, for several applied fields. The thin line marked "pump depletion only" is the falloff in gain expected from standard pump depletion theory. The increased rate of falloff is caused by the nonlinear relationship between the space-charge field and the modulation index. Note that β and m are related by $m = (2\beta^{1/2})/(\beta+1)$. The thick lines in Figure 3 are the fit using the empirical model, eq.(34) and eq.(35), which has no free parameters. The two top curves in Figure 22 (discussed below) were also fit with this model. The model agrees well with measured data, suggesting that the assumptions of a single trap were valid. Further validation of the numerical model could be accomplished by introducing a third beam at the appropriate Bragg angles and monitoring the diffraction efficiency of the higher order harmonics.

The analytic model of eq.(32) was unable to fit the data in Figure 20a because the grating period was smaller than the drift length. The fit at grating spacings much larger than the drift length using eq. (32) is shown in Figure 20b. It is interesting to note that the enhancement factor, p, gives the saturated value of the gain coefficient (at large beam ratios) and also allows the calculation of all large signal effects. This illustrates the fact that the large signal effects are directly tied to the amount of space charge enhancement. Indeed, for grating spacings comparable to and smaller than the drift length, where this model is invalid, p may be chosen such that it gives the correct saturated gain; in doing so, it also provides a reasonable estimate of the falloff with beam ratio. This fact is quite useful because it indicates that the performance at all modulation depths may be modeled by measuring only the saturated gain coefficient at each grating spacing of interest.

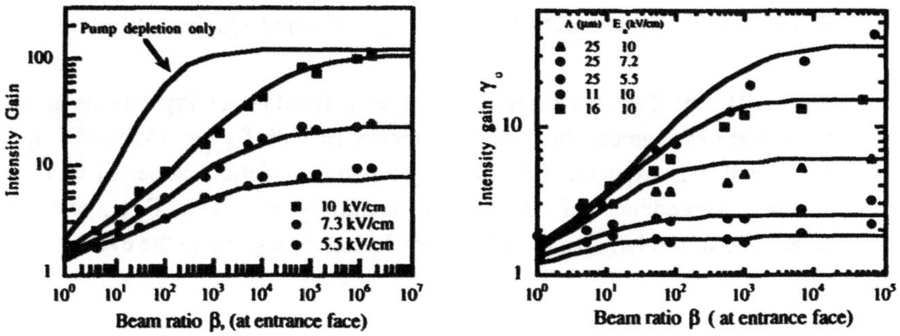

Figure 20. a) Two wave mixing gain as a function of input pump to signal beam ratio β. L = 1.37mm Λ = 5.5 μ. Heavy lines are results of numerical model, thin line is standard pump depletion theory for 10 kV/cm case. b) Fit of analytic theory of eq.(32).

Large signal effects in InP have been examined by measuring the intensity amplification, γ_0, as a function of beam ratio (See Figure 21 where total intensity = 32 mW/cm^2, T=20°C, E_{dc} = 10 kV/cm, λ = 970 nm, L = 4 mm). The combined intensity of the two beams was maintained constant to ensure the temperature-intensity resonance. The thin line in Figure 21 shows the fall-off in amplification expected from pump depletion under the linear approximation. The data was fit using the nonlinear model of eq.(34). To account for both pump depletion and large signal effects, the coupled wave equations were integrated numerically as in eq.(35). For a grating spacing of 7.7 μm, it was found that a_f is approximately 3.7, when the saturated gain, Γ_{sat} (measured in the limit of small m), is 18.5 cm^{-1} (see heavy line in Figure 21a). This fitting procedure was repeated for different grating spacings and the values obtained for a_f and Γ_{sat} are plotted in Figure 21b. Note that a_f and Γ_{sat} follow the same trend, being large at comparable grating spacings. These measurements show that the values of Γ_{sat} and a_f are coupled to each other.

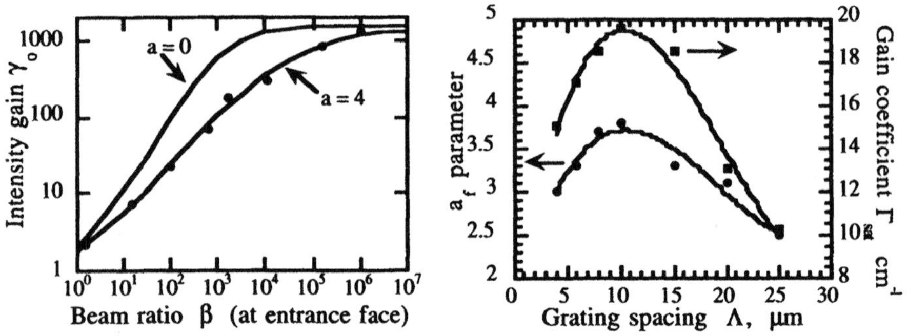

Figure 21. a) Two wave mixing gain as a function of input beam ratio: Circles are experiment, heavy line is theory fit with Γ_{sat} = 18.5 cm^{-1}, a_f = 4, thin line is expected fall off due to pump depletion alone. b) Grating spacing dependence of the saturated gain coefficient, Γ_{sat}, and the magnitude of the large signal effects, a_f. Lines are a guide to the eye.

The difference between the two models can be further appreciated by explicitly factoring out the space-charge field dependence on modulation index. This can be done by rewriting equations (32) and (34) as $E_{sc} = E_a f(m)$ where f(m) contains the entire m dependence, with appropriate scale factors so that E_a is the maximum space-charge field obtained in the limit of small m. In the case of the standard theory given by equation 4, f(m) is simply equal to m itself and thus f(m) vs. m is a straight line. From our numerical model, the expression for f(m) obtained from eq. (34) is f(m) = $[1/a_f]\exp(m)[1-\exp(-a_f m)]$. A closely related expression for the DC field moving grating case was determined by Refregier[39]. Figure 22 shows f(m) for both of the models as well as the exact numerical solution. The large signal effects are evident as the rolling over or clamping of f(m).

The time response measurements of two-beam coupling show that small beam ratios have smaller time constants than larger ones. The shorter effective time constant for small beam ratio is due to the fact that the space-charge field reaches a clamped condition early in its buildup phase. Larger beam ratios require a smaller space-charge field to obtain the same gain, so the field grows for a longer time before clamping. This shows good qualitative agreement with the numerical calculations of the time evolution of the microscopic space charge as a function of beam ratio. However, a quantitative comparison is difficult because of the time-dependent change in the modulation index caused by beam coupling.

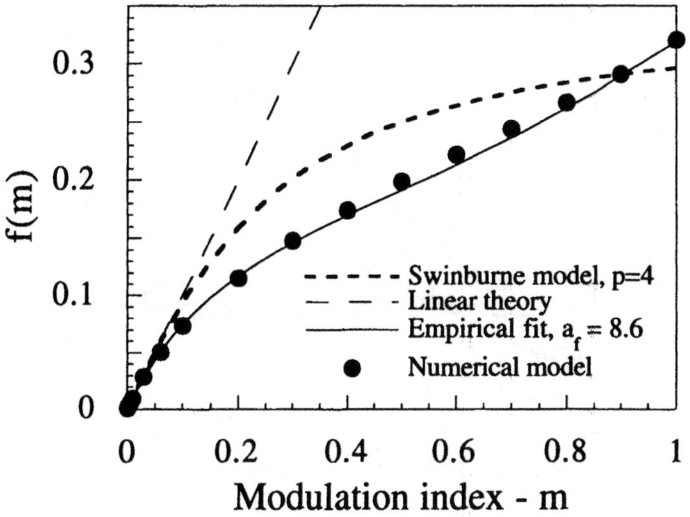

Figure 22. Comparison of functional forms of f(m) for the different models. Values for a_f and p are best fit for = 5.5 μm data.

The effectiveness of using an applied sinusoidal waveform rather than a square waveform was also investigated experimentally. Figure 26 shows the data for a 12.5 kV/cm zero-to-peak sine wave (squares) as well as for two square-wave fields of smaller magnitude. Comparison with the numerical calculations for the square wave fields was done by using eq.(34) in the coupled wave equations, where a_f was calculated using eq.(35). To obtain a_f for the sinusoidal case, the functional form of eq.(34) was fit to the numerical results using a_f as a free parameter. For the sine wave case a_f was found to be 1.8 which is considerably lower than the calculated values of 8.8 and 8.1 for the 10 kV/cm and 5.5 kV/cm square wave fields. The low value for a_f actually results in slightly super-linear behavior of the space-charge field at high modulation index.

Although the magnitude of the enhancement is significantly smaller than that from square waves, the sine wave produces constant amplification over 5 orders of magnitude change in beam ratio, apparently immune to large signal effects. In addition, it can be more easily generated using resonantly tuned circuits. Notice that even though the 5.5 kV/cm square wave has a higher saturated gain than the sine wave, at large modulation (small β) the square wave gain drops below that of the sine wave. Because the gains actually cross, it is apparent that the sine wave produces more uniform amplification. These advantages could prove useful in situations where linear amplification of high contrast images is necessary.

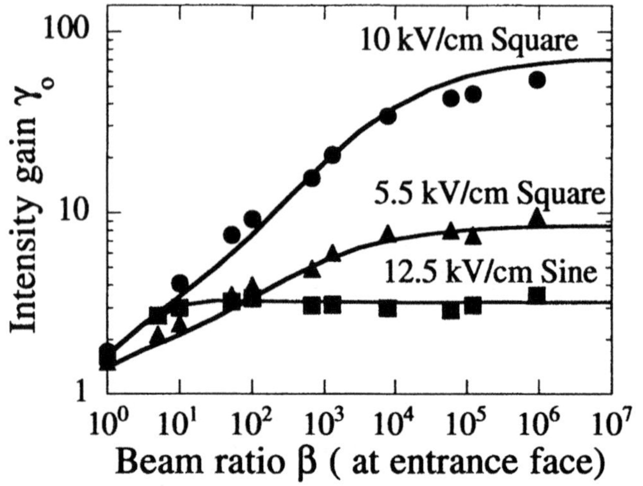

Figure 23. Comparison of large signal beam coupling performance for the case of sinusoidal and square waves. $\Lambda = 3.5\ \mu m$

The development of a numerical model to calculate large signal performance of photorefractive materials using AC square wave enhancement at all grating spacings and modulation depths has lead to a simple empirical function. The function predicts the performance of photorefractive materials at all modulation depths and agrees well with measurements. Also, the model of eq.(34) was shown to give reasonable agreement with measurements at grating spacings much larger than the carrier drift length. These models can be useful tools in calculating performance of other photorefractive materials, and in particular semiconductors, using large applied fields.

4. SEMICONDUCTORS AS SELF-PUMPED PHASE CONJUGATE MIRRORS

4.1 Background

The photorefractive self pumped phase conjugate mirror (SPCM) was first demonstrated by White et al.[62] in 1982. A photorefractive crystal was placed inside a linear resonator and a phase conjugate of the input beam was generated. The process uses a four wave mixing geometry, where the pump beams are derived from amplified scattering of the incoming signal beam. Feinberg,[63] Cronin-Golomb[64],

Chang[65] and others have since introduced alternative geometries. Under ideal conditions, phase conjugate reflectivities can approach 100% [66]. Using $BaTiO_3$, SPCM reflectivities have been measured on the order of 80%.[67] SPCMs have been demonstrated using semiconductor and sillenite materials. However, the measured reflectivities for these materials were surprisingly low (no larger than 11%)[68], even though gain coefficients are comparable to those in $BaTiO_3$ (as large as 35 cm^{-1}) [36].

The relatively poor performance of the semiconductors and sillenite is due to the combination of linear absorption (typically higher than in $BaTiO_3$), as well as the nonlinear photorefractive response described in the preceding section. The effect of absorption and nonlinear photorefractive response on the double pumped phase conjugate mirror (DPCM) and the self pumped phase conjugator have been examined by Wolffer et al.[69] and Millerd et. al.[70] and follow similar approaches.

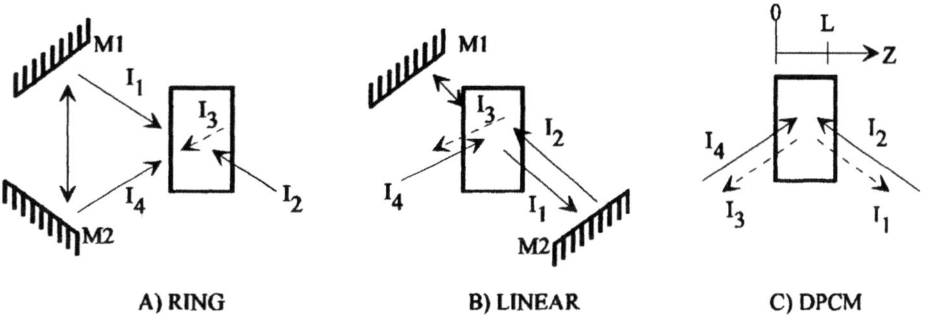

Figure 24. Geometry and beam designations for the ring and linear SPCM and the DPCM.

The theory for self-pumped phase conjugation using photorefractive four-wave mixing has been developed by Cronin-Golomb et. al.[66] The geometry and designation of the beams for the ring and linear SPCM and the DPCM are shown in Figure 27. These two resonators were considered because they offer the lowest thresholds for oscillation and because external electric fields are difficult to apply in other geometries.

4.2 Solution for Lossless Material

The reflectivity of the SPCM can be derived from a coupled wave analysis where pump depletion is considered but absorption is neglected [66]. Calculation of reflectivity for the mirror may be found by including the proper boundary conditions and requires solution of an equation that may have multiple roots [71]. Reflectivity of the ring mirror is plotted as a function of coupling strength for various values of external mirror reflectivity in Figure 25. The threshold condition of the ring mirror is given by,

$$\Gamma L = \frac{2(M+1)}{(M-1)} \ln\left(\frac{M+1}{2M}\right) \tag{36}$$

where $M = M_1 M_2$ is the product of the intensity reflectivities of the two external mirrors. In the limit of perfect external mirrors, M = 1, the threshold coupling strength is 2. For the case of the linear mirror the threshold condition is

$$\Gamma L = \ln M, \tag{37}$$

which has a threshold of zero for perfect external mirrors.

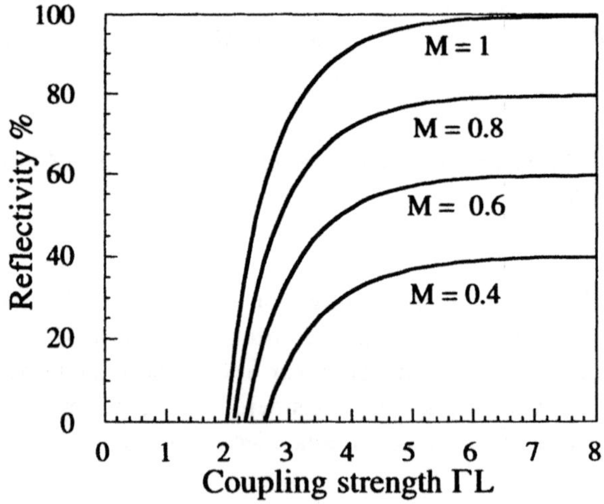

Figure 25. Theoretical reflectivity of the lossless ring SPCM as a function of coupling strength for several values of external mirror reflectivity M.

4.3 Effect of Absorption

The assumption that the loss coefficient is zero is clearly inappropriate for the case of the band-edge enhancement, because its existence requires moderate loss. In a typical InP crystal operating at 970 nm, the one-way transmission loss over a 2 mm crystal length is on the order of 50%. To evaluate the effects of loss on the SPCM it is necessary, in general, to solve the coupled equations numerically. It is convenient to introduce a material figure of merit, η, defined by

$$\eta = \alpha/\Gamma\,, \tag{38}$$

where α and Γ are the loss and gain coefficients respectively. Doing this allows SPCM performance to be calculated as a function of sample length for a given material parameter. Numerical solutions to the coupled equations can be obtained in a straightforward manner in the case of the ring mirror, because the boundary conditions at $z = L$ are well-defined.

Results for the ring mirror are shown in Figure 26 for different values of η. It can be seen that for a given value of η there is an optimum length of sample such that any further increase will reduce performance due to the overall increase in loss.

It was found that under certain conditions, the effects of absorption may be approximated by lumping the absorption losses into the external mirror reflectivity.

$$M = M_1 M_2 \exp(-2\eta\Gamma L) \tag{39}$$

This approximation is valid when the coupling strength is well above threshold and when the photorefractive gain is not sensitive to intensity. For large coupling strengths the reflectivity is limited by the loss and becomes equal to the effective transmission, calculated using eq.(39). Comparisons with the numerical solutions show this approximation is good at all coupling strengths for $\eta < 0.1$ (see data points in Figure 26). For larger η values, this approximation begins to underestimate reflectivity at coupling strengths near threshold.

Investigation of the coupling strength necessary for threshold in the ring mirror reveals that for $\eta > 0.48$ there is no length of sample for which oscillation may be achieved. This agrees with the results of Wolffer et. al.[69] for the double pumped phase conjugator that oscillation requires $\Gamma > 2\alpha$. The similarities between the two configurations has been previously identified by Cronin-Golomb [72]. Unlike laser resonators, simply having net gain, $\Gamma-\alpha > 0$, is not enough to guarantee oscillation.

Because of this requirement, the ratio α/Γ (or Γ/α) provides a good Figure-of-merit to determine a material's ability to achieve threshold in a ring SPCM.

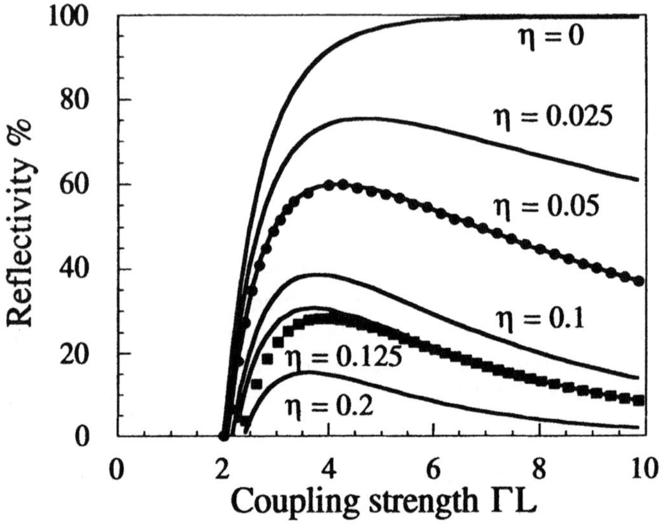

Figure 26. Effects of absorption on the ring SPCM reflectivity. The parameter $\eta = \alpha/\Gamma$, the relative amount of absorption. M = 1. Points are the results using the approximation of eq.(11).

The effect of loss on the linear mirror can be modeled using an approximation similar to eq.(39). Here the effective reflectivity for each mirror is given by the product of its reflectivity and the loss accrued in one pass through the sample. The ring mirror has a higher maximum reflectivity. However, the linear mirror still has a lower threshold. The reason that loss has a more severe effect on the linear mirror may be because the effective pump beams must make multiple passes in the linear resonator while only two passes are made in the case of the ring.

4.4 Large Signal Effects

Large signal effects (the reduction in apparent gain coefficient with increasing modulation index) also have significant impact on the operation of the self-pumped mirror. The function f(m) may be substituted for m in the equations for the SPCM to account for this effect. Again, a numerical solution is necessary. Results for the lossless ring SPCM are shown in Figure 27a for several values of a_f. The main effect of a_f is to decrease the rate at which reflectivity rises with coupling strength; the reflectivity still can grow large with a long sample.

With the addition of absorption, however, the reflectivity shows a maximum at an optimum coupling strength (sample length). Figure 27b shows the effect of both loss and large signal effects on an SPCM for an a_f value typical of InP. Although both physical phenomena reduce performance, loss appears to be the greatest problem because it limits the maximum reflectivity regardless of sample length.

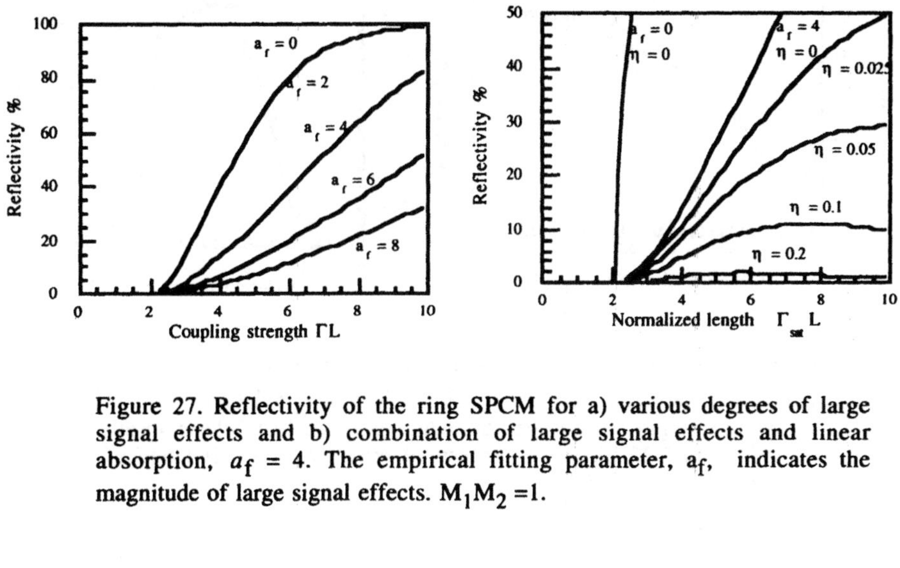

Figure 27. Reflectivity of the ring SPCM for a) various degrees of large signal effects and b) combination of large signal effects and linear absorption, a_f = 4. The empirical fitting parameter, a_f, indicates the magnitude of large signal effects. M_1M_2 =1.

4.5 InP:Fe Self Pumped Phase Conjugate Mirror

A comparison of the theory and measured performance of a ring SPCM using InP:Fe is shown in Figure 28a. The absorption coefficient at 970 nm is 5 cm^{-1}, which gives $\eta = 0.27$ (at the 7.7 μm grating spacing). Clearly, based on absorption alone, we can see from Figure 27a that our reflectivity is reduced to less than 15% (L = 4 mm). The addition of large signal effects, demonstrated in Figure 27b, further reduces the reflectivity to less than 1%. The measured reflectivity for the ring SPCM using the 4 mm sample is shown in Figure 28a as the data, and the calculated reflectivity is shown as the line. The coupling strength was varied by changing the applied field. Figure 28b shows the reflectivity as the intensity is tuned through the temperature-intensity resonance. The relationship between applied voltage and coupling strength was separately measured in a two-beam coupling experiment and used to obtain the scaling of coupling strength, shown in Figure 27. Although a_f is a function of applied field, (a_f is related to Γ_{sat} as discussed above), it is assumed that a_f is a constant ($a_f = 3.7$ as measured at 10 kV/cm) in order to simplify calculations. From measurements shown in Figure 21b, we would expect a_f to vary between 3 and 3.7 over the experimental range, which results in a 20% overestimate of a_f at low field values. Using this approximation, the agreement is good, especially at predicting the relative magnitude of the reflectivity. Without including absorption and large signal effects into the calculations, one would predict reflectivities on the order of 80%. Clearly these are important corrections to the theory.

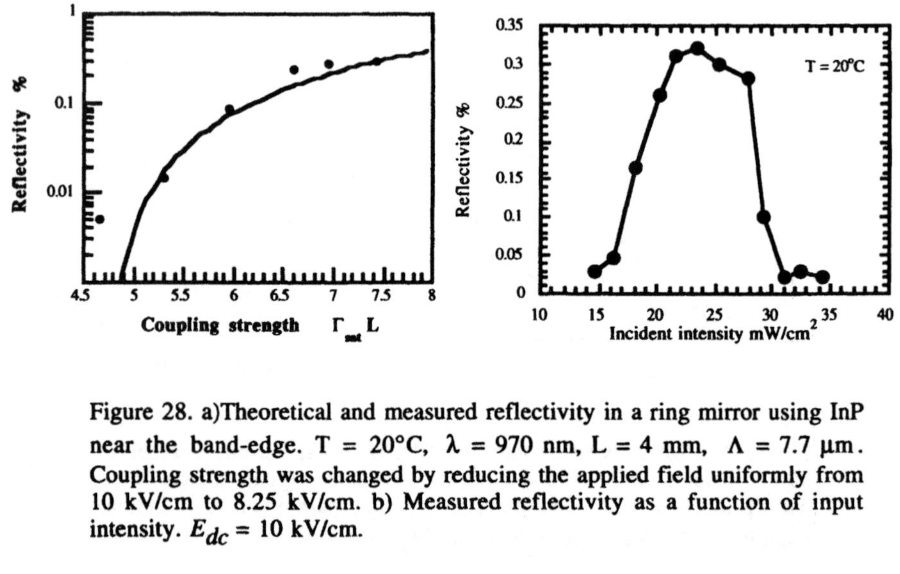

Figure 28. a)Theoretical and measured reflectivity in a ring mirror using InP near the band-edge. T = 20°C, λ = 970 nm, L = 4 mm, Λ = 7.7 µm. Coupling strength was changed by reducing the applied field uniformly from 10 kV/cm to 8.25 kV/cm. b) Measured reflectivity as a function of input intensity. E_{dc} = 10 kV/cm.

Bylsma et al.[68] demonstrated a ring SPCM using InP and AC fields at a wavelength of 1.3 µm with a maximum reflectivity of 11%. The gain coefficient was reported to be 2.5 cm^{-1} for their 17 mm crystal. However, the absorption was only 0.15 cm^{-1}, so that η = 0.06. If we assume that their large signal effects are roughly the same as those measured in Figure 21, the calculated reflectivity for this system is about 10% (see Figure 27), in agreement with their results.

4.6 Double Phase Conjugate Mirrors

The double phase conjugate mirror (DPCM) is another important application for photorefractive semiconductors. The DPCM can couple two mutually incoherent beams into the phase conjugate of each other. The beams need not be the same wavelength, although chromatic aberrations will be introduced if they are not. This has many possible practical applications, including the coupling of single mode fibers[27] and phase locking of laser diode arrays[73][74].

A DPCM was demonstrated using InP:Fe and the band-edge resonance combined with the temperature-intensity resonance. The input conversion efficiency (defined here as the ratio of the diffracted phase conjugate beam to the incident beam) is plotted vs. input intensity in Figure 29. The input beams on both sides of the crystal were of equal intensity. The measured output conversion efficiency (defined here as the ratio of diffracted beam intensity to the total transmitted intensity) has a peak value of 10%. The maximum conversion efficiency of the DPCM was 1.3% while that of the ring was only 0.35%. This difference is believed to be mainly due to the improved uniformity in illumination along the length of the sample inherent in the DPCM design; allowing a larger average coupling strength. Similar performance enhancement may be possible in the ring mirror by introducing a temperature gradient along the length of the crystal [8].

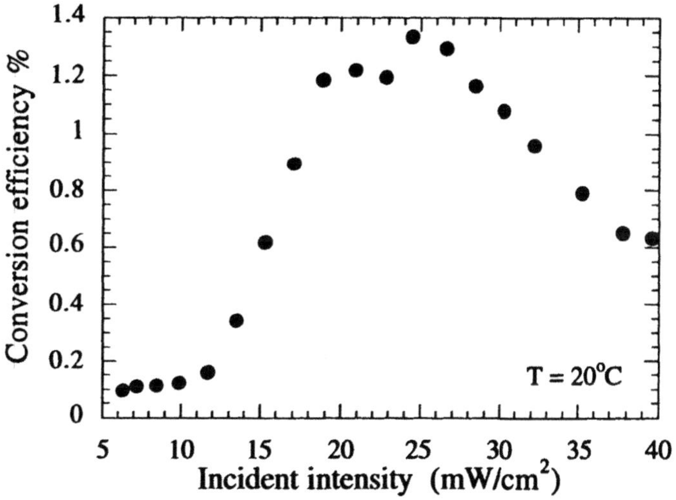

Figure 29. Measured input conversion efficiency from a double pumped phase conjugate mirror in InP. T = 20°C, E_a = 10 kV/cm, λ = 970 nm, L = 4 mm, Λ = 7.7 µm. Coupling strength was changed by reducing the applied voltage.

Wolffer et. al.[75] have demonstrated a double color pumped oscillator (DCPO) which is essentially a DPCM except that the two input wavelengths (I_2 and I_4) are different. The DCPO is particularly interesting for the application of optical interconnects. The output conversion efficiency of the device (using InP and dc fields with temperature-intensity resonance) was measured as a function of wavelength for each input beam. The data were also compared with a numerical model where both absorption and large signal effects were taken into account. Figure 30 shows the conversion efficiency with one beam fixed at 1.047 μm and the other scanned from the band-edge to longer wavelengths. These results clearly illustrate that, even though the gain coefficient is enhanced near the band-edge, it does not compensate for the increase in absorption.

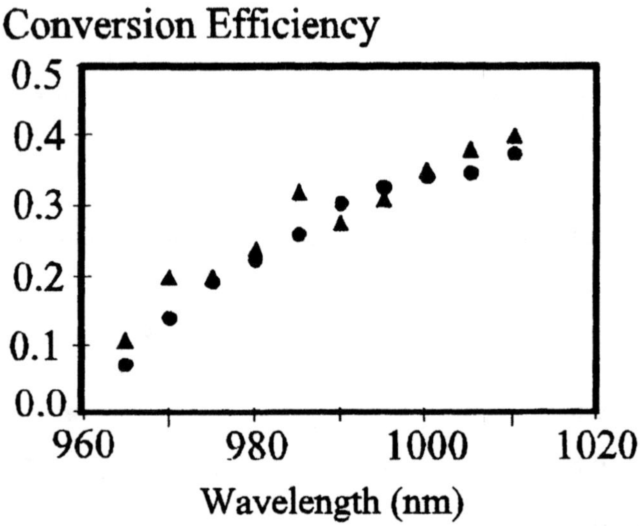

Figure 30. Measured output conversion efficiency for DCPO as a function of second beams wavelength. exp.- Triangles, calc.- circles. λ_1 was fixed at 1.047 m. Applied dc field 10 kV/cm; = 7 m.(Ref. [75])

Wolffer et. al.[27] have demonstrated a two-zone DPCM that exhibits high quality, high reflectivity phase conjugation. The arrangement for the conjugator is shown in Figure 31a. It consists of two back-to-back DPCMs coupled with a lens. This arrangement eliminates two problems found in DPCMs: conical diffraction and instability domains, both of which degrade the reflectivity and conjugate fidelity.

The experiment used alternating electric fields for enhancement and worked at a wavelength well away from the band-edge where the absorption was low. This allowed a relatively high throughput (54%). Figure 31b shows the diffraction efficiency as a function of time. A steady state value of 60% is obtained, which corresponds to a conversion efficiency of 30%. The maximum global coupling efficiency, as measured through the fiber at detector 2, is 12%, which corresponds to roughly 40% efficiency in coupling the diffracted light into the fiber. The 40% coupling is a measure of the phase conjugate fidelity.

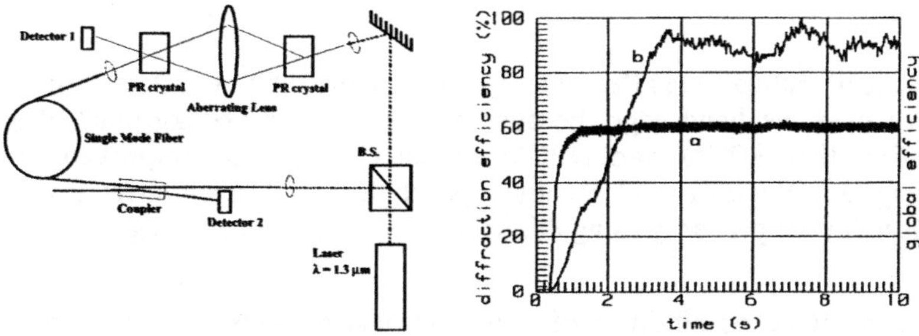

Figure 31. a) Experimental setup for two-zone DPCM for coupling of light into single mode fiber. b) Measured diffraction efficiency as a function of time during grating formation. (Ref. [27])

5. SUMMARY

Semiconductor materials offer many advantages over their ferroelectric counterparts, such as fast response times, sensitivity in the near-infrared, comparatively low cost and high optical quality. The application of electric fields and use of near resonant effects can dramatically increase two-wave mixing gains, making them comparable to ferroelectrics.

The application of simple dc fields can be combined with the temperature-intensity resonance or moving gratings to increase wave-mixing gains. These techniques are complicated by the need to maintain resonance over the entire length of an absorbing sample. AC fields, though more difficult to produce, have the advantage that the increased gain coefficient can be maintained over any length of crystal.

The increased gain coefficients, resulting from applied fields at large grating spacings, are accompanied by a saturation effect at large modulation index, caused by a clamping of the space-charge field to the applied field. These so-called large signal effects must be accounted for when modeling device performance. Several models have been shown to adequately predict these effects.

The band-edge electro-refractive photorefractive effect makes it possible to increase the gain coefficient, Γ_{sat}, without increasing large signal effects, a_f. This is because an additional index change is created without increasing the magnitude of the space-charge field. Although operating near the band-edge is accompanied by an increase in absorption, the net gain, Γ−α, increases as the wavelength approaches the band-edge. High net gains have been measured in semiconductors that are competitive with ferroelectric crystals for applications such as optical amplification and information processing.

For practical applied fields (less than 15 kV/cm), the ratio of Γ/α has been found to decrease near the band-edge, because of the increased background absorption, suggesting that SPCM and DPCM reflectivity will be improved at longer wavelengths. Experimental results support this, and researchers investigating DPCMs today are primarily using AC fields away from the band-edge.

Finally, the wide possibilities of substitutional dopants, native growth related defects and alloying within the III-V and II-VI material systems leaves a vast potential for future research in photorefractive semiconductor materials growth and characterization.

Acknowledgements

We would like to thank the following individuals for their valuable contributions to the research reviewed here: G. Brost, S. Trivedi, M. Ziari, M. B. Klein, A. Kost, P. Gravey and N. Wolffer.

REFERENCES

1. G. Albanese, J. Kumar, and W.H. Steier, "Investigation of the photorefractive behavior of chrome-doped GaAs by using two-beam coupling," Opt. Lett., **11**, 650 (1986).
2. D.T.H. Liu, L.-J. Cheng, M.-F. Rau, and F.-C. Wang, "Photorefractive gain in GaAs under a dc electric field," Appl. Phys. Lett., **53**, 1369 (1988).
3. J. Kumar, G. Albanese, and W.H. Steier, Opt. Commun., **63**, 191 (1987).
4. B. Imbert, H. Rajbenbach, S. Mallick, J.P. Herriau and J.P. Huignard, "High photorefractive gain in two beam coupling with moving fringes in GaAs:Cr crystals." Optics Lett., **13**, 327 (1988).
5. J. Kumar, G. Albanese, W.H. Steier, and M. Ziari, "Enhanced two-beam mixing gain in photorefractive GaAs using alternating electric fields," Opt. Lett., **12**, 120 (1987).
6. M.Klein, S.W. McCahon, T.F. Boggess and G.C. Valley, "High-accuracy, high reflectivity phase conjugation at 1.06 μm by four-wave mixing in photorefractive gallium arsenide," J. Opt. Soc. Am. B **5**, 2467 (1988).
7. A. Partovi, A. Kost, E Garmire, G. C. Valley and M. B. Klein, "Band-edge photorefractive effect in semiconductors." Appl. Phys. Lett. **56**, 1089 (1990).

8. G. Picoli, P. Gravey, C. Ozkul and V. Vieux, "Theory of two wave mixing gain enhancement in photorefractive InP:Fe: A new mechanism of resonance." J. Appl. Phys. **66**, 3789 (1989).
9. J. E. Millerd, S. D. Koehler, E. M. Garmire, A. Partovi, A. M. Glass and M. B. Klein, "Photorefractive gain enhancement in InP:Fe using band edge resonance and temperature stabilization", Appl. Phys. Lett. **57**, 2776 (1990).
10. *Photorefractive Materials and their applications I,* Edited by P. Günter, and J.-P. Huignard, Topics in Applied Physics, Vol. **61**, ch8, (Springer-Verlag, New York 1987).
11. P. Yeh, "Photorefractive two-beam coupling in cubic crystals," J. Opt. Soc. Am. B., **4**, 1382 (1987).
12. F. P. Strohkendl, J.M.C Jonathan and R.W. Hellwarth, "Electron hole competition in photorefractive gratings." Opt. Lett, **11**, 312 (1986).
13. R.S. Rana, D.D. Nolte, R.Steldt and E.M. Monberg, "Temperature dependence of the photorefractive effect in InP:Fe: role of multiple defects," J. Opt. Soc. Am. B, **9**, 1614 (1992).
14. A. Partovi, E. Garmire, G. Valley and M.Klein, "Photorefractive caharacterization of deep level compensation in semi-insulating GaAs," Appl. Phys. Lett., **55**, 2701 (1989).
15. M. Ziari, W. H. Steier, P. Ranon, S.Trivedi and M. B. Klein, "Photorefractivity in vanadium-doped ZnTe," Appl. Phys. Lett., **60**, 1052 (1992).
16. A.M. Glass, M.B. Klein, and G.C. Valley, Electron. Lett., **21**, 220 (1985).
17. A. Partovi, J. Millerd, E. Garmire, M. Ziari, W. H. Steier, S. Trivedi and M. B. Klein, "Photorefractivity at 1.5 μm in CdTe:V." Appl. Phys. Lett. **57**, 846 (1990).
18. J. I. Pankove, *Optical Processes in Semiconductors*, (Dover Publications, New York 1971).
19. M.B. Klein, "Beam coupling in undoped GaAs at 1.06 μm using the photorefractive effect," Opt. Lett., **9**, 350 (1984).
20. A.M. Glass A.M. Johnson, D.H. Olsen, W. Simpson and A.A. Ballman, " Four-wave mixing in semi-insulating InP and GaAs using the Photorefractive effect," Appl. Phys. Lett., **44**, 948 (1984).
21. L.J. Cheng and D.T.H.Liu, "Optical computing and image processing using photorefractive gallium arsenide," Jap. J. Appl. Phys., **29**, 1297 (1990).
22. H. Rajbenbach, J.M. Verdiell, and J.-P. Huignard, "Visuallization of electrical domains in semi-insulating GaAs:Cr and potential use for variable grating mode operation," Appl. Phys. Lett., **53**, 541 (1988).
23. K. Seeger, Semiconductor Physics, Springer Verlag, Berlin 1989
24. D.D. Nolte, D.H. Olsen, E.M. Monberg, P.M. Bridenbaugh and A.M.Glass, "Optical and photorefractive properties of InP:Ti: a new photorefractive semiconductor," Opt. Lett., **14**, 1278 (1989).
25. J.Strait, J.D. Reed, A.Saunders, G.C. Valley and M. B. Klein, "Net gain in photorefractive InP:Fe at $\lambda = 1.32$ μm without and applied field," Appl. Phys. Lett., **57**, 951 (1990).

26. D.D. Nolte, D.H. Olson and A.M. Glass, "Spontaneous current oscillations in optically pumped semi-insulating InP," J. Appl. Phys., **68**, 4111 (1990).
27. N.Wolffer, P. Gravey, "High quality phase conjugation in a double phase conjugate mirror using InP:Fe at 1.3 μm", Opt. Comm., **107**, 115 (1994).
28. K.Kuroda, Y. Okazaki, T. Shimura, H. Okamura, M. Chihara, M. Itoh and I. Ogura, "Photorefractive effect in GaP," Opt. Lett., **15**, 1197 (1990).
29. J. Ma, Y. Taketomi, Y. Fainman, J. Ford, S. Lee and K. Chino, "Moving grating and dc external field in photorefractive GaP at 633 nm," Opt. Lett., **16**, 1081 (1991).
30. R.B. Byslma, P.M. Bridenbaugh, D.H. Olsen and A.M. Glass, "Photorefractive properties of doped cadmium telluride," Appl. Phys. Lett., **51**, 889 (1987).
31. M. Ziari, W. Steier, P. Ranon, M. Klein and S. Trevedi" Enhancement of the photorefractive gain at 1.3-1.5 μm in CdTe using alternating electric fields," J. Opt Soc. Am. B, **9**, 1461 (1992).
32. P. Taybati, J. Kumar and S. Scott, "Photorefractive effect at 633 nm in semi-insulating cadmium sufide," Appl. Phys. Lett., **59**, 3366 (1991).
33. G.Sefler, E. Oh, R. Rana, I. Miotkowski, A. Ramdas and D. Nolte, "The Faraday photorefractive effect in a diluted magnetic semiconductor," Opt. Lett., **17**, 1420 (1992).
34. M.Ziari, W.H.Steier, J.Millerd, J. Swienton, G. Brost K. Madge and S.Trivedi "Observation of the photorefractive effect and demonstration of real-time interferometry in CdMnTe:V" IEEE conf. proc. on Nonlinear Optics: Materials, Fundamentals and Applications Hawaii, PD4 (1994).
35. G.C. Valley, S.W. McCahon and M.B. Klein, "Photorefractive measurement of photoionization and recombination crossections in InP:Fe", J. Appl. Phys., **64**, 6684 (1988).
36. J. Millerd, E. Garmire, M. B. Klein, "Self pumped phase conjugation in InP:Fe using bandedge resonance and temperature stabilization: theory and experiments." Opt. Lett., **17**, 100 (1992).
37. A. Marrakchi and J. P. Huignard, "Diffraction efficiency and energy transfer in two-wave mixing experiments with $Bi_{12}SiO_{20}$ crystals," Appl. Phys., **24**, 131 (1981).
38. D. Rak, I. Ledoux and J.P. Huignard, "Two wave mixing and energy transfer in BaTiO3 Application to laser beamsteering", Opt. Comm., **49**, 302 (1984).
39. P. Refregier, L.Solymar, H. Rajbenbach, and J.P. Huignard, "Two beam coupling in photorefractive $Bi_{12}SiO_{20}$ crystals with moving grating: Theory and experiments." J. Appl. Phys., **58**, 45 (1985).
40. G. C. Valley, "Two wave mixing with an applied field and a moving grating," J. Opt. Soc. Am. B, **1**, 868 (1984).
41. G. C. Valley, H. Rajbenbach amd H.J. Von Bardeleben, "Mobility-lifetime product of photoexcited electrons in GaAs," Appl. Phys. Lett. **56**, 364 (1990).
42. H.Z. Hu, Appl Opt. **22**, 2052 (1983).
43. M. P.Kothiyal and C. Delisle, "Optical frequency shifter for heterodyne interferometry using counterrotatingwave plates," Opt. Lett., **9**, 319 (1984).

44. G. Bremond, N.Nouaihat, G.Guillot and B. Cockayne, "Fe deep level spectroscopy in InP", Solid State Comm., **41**, 477 (1982).
45. C. Ozkul, G. Picoli, P. Gravey and N. Wolffer, " High gain coherent amplification in thermally stabilized InP:Fe crystals under dc fields", Appl. Opt., **29**, 2711 (1990).
46. S. Stepanov and M. Petrov "Efficient unstationary holographic recording in photorefractive crystals under an external alternating electric field." Optics Comm., **53**, 292 (1985).
47. K. Walsh, A.K. Powell, C. Stace and T. J. Hall, "Techniques for the enchancement of space-charge fields in photorefractive materials," J. Opt. Soc. Am. B, **7**, 288 (1990).
48. W.H. Steier, J. Kumar, M. Ziari, "Infrared power limiting and self switching in CdTe," Appl. Phys. Lett., 53, 840 (1988).
49. V.W. Franz, Z. Naturforch, and L.V. Keldysh, Sov. Phys. JETP **7**, 788 (1958).
50. T.E. Van Eck, L.M. Walpita, W.S.C. Chang, and H.H. Wieder, Appl. Phys. Lett. **48**, 451 (1986).
51. R.A. Soref, and B.R. Bennett, IEEE JQE, QE-**23**, 123 (1987).
52. D.A.B. Miller, D.S. Chemla, T.C. Damen, A.C. Gossard, W. Wiegmann, T.H. Wood, and C.A. Burrus, Phys. Rev. Lett., **53**, 2173 (1984).
53. A. Alping, and L.A. Coldren, J. Appl. Phys., **61**, 2430 (1987).
54. D.E. Aspnes, Phys. Rev., **153**, 972 (1967).
55. A. Kost, M. Kawase, E. Garmire, H.C. Lee, A. Danner, A. Hariz, and P.D. Dapkus, Proc. SPIE, **881**, 122 (1988).
56. S.I. Stepanov and S.L. Sochava, " Effective energy transfer in a two-wave interaction in $Bi_{12}TiO_{20}$," Sov. Phys. Tech. Phys., **32**, 1054 (1987).
57. G.A. Swinburne, T. J. Hall and A. K. Powell, "Large modulation effects in photorefractive crystals," IEE Proc. Int. Conf. on Holographic Systems, Components and Applications, Bath, U.K. Sept. 11-13, **175** (1989).
58. E. Ochoa, F. Vachss, L. Hesselink, "Higher order analysis of the photorefractive effect for large modulation depths." J. Opt. Soc. Am. A , **3**, 181, (1986).
59. F. Vachss, L. Hesselink, "Nonlinear photorefractive response at high modulation depths." J. Opt. Soc. Am. A, **5**, 690 (1988).
60. L. B. Au, L. Solymar, "Higher harmonic gratings in photorefractive materials at large modulation with moving fringes." J. Opt. Soc. Am. A, **7**, 1554 (1990).
61. G. Brost "Photorefractive grating formation at large modulation with alternating electric fields" J. Opt. Soc. Am. B., **9**, 1454 (1992).
62. J. O. White, M. Cronin-Golomb, B. Fisher and A. Yariv, "Coherent oscillation by self-induced gratings in the photorefractive crystal $BaTiO_3$.", Appl. Phys. Lett., **40**,, 450 (1982).
63. J. Feinberg, "Self pumped, continuous wave phase conjugator using internal reflection." Optics Lett., **7** 486 (1982).
64. M. Cronin-Golomb, B. Fisher, J. O. White and A. Yariv, "Passive phase conjugate mirror based on self-induced oscillation in an optical ring cavity," Appl. Phys. Lett., **42**, 919 (1983).

65. T. Y. Chang and R. W. Hellwarth, "Optical phase conjugation by backscattering in barium titanate." Optics Lett. **10**, 408 (1985).
66. M. Cronin-Golomb, B. Fisher, J. O. White and A. Yariv, "Theory and applications of four-wave mixing in photorefractive media", IEEE J. Quant. Elect., QE-**20**, 12 (1984).
67. B. Wechsler, M. Klein, C. Nelson and R. Schwartz, "Spectroscopic and photorefractive properties of infrared sensitive rhodium-doped barrium titanate," Opt. Lett., **19**, 536 (1994).
68. R. B. Bylsma, A. M. Glass, D. H. Olson and M. Cronin-Golomb, "Self-pumped phase conjugation in InP:Fe".", Appl. Phys. Lett., **54**, 1968 (1989).
69. N. Wolffer, P. Gravey, J.Y. Moisan, C.Laulan and J.C. Launay, "Analysis of double pumped phase conjugate mirror interaction in absorbing photorefractive crystals: applications to BGO:Cu". Opt. Comm., **73**, 351 (1989).
70. J. Millerd, E. Garmire and M. Klein, "Investigation of photorefractive self-pumped phase conjugate mirrors in the presence of loss and high modulation depth," J. Opt. Soc. Am. B, **9**, 1499 (1992).
71. M. Cronin-Golomb, B. Fischer, J. White and A. Yariv, "Passive (self-pumped) phase conjugate mirror: Theory and experimental investigation," Appl. Phys. Lett., **41**, 689 (1982).
72. M. Cronin-Golomb, "Almost all transmission grating self-pumped phase conjugate mirrors are equivalent." Opt. Lett., **15**, 897 (1990).
73. S. MacCormack and J. Feinberg, "High-Brightness output from a laser-diode array coupled to a phse-conjugating mirror" Opt. Lett., **18**, 211 (1993).
74. M. Segev, S. Weiss and B. Fischer, "Coupling of diode laser arrays with photorefractive passive phase conjugate mirrors", Appl. Phys. Lett., **50**, 1397 (1987).
75. N. Wolffer, P. Gravey, G. Picoli and V. Vieux, " Double phase conjugated mirror and double colour pumped oscillator using band-edge photorefractivity in InP:Fe," Opt. Comm., **89**, 17 (1992).

7

PHOTOREFRACTIVE QUANTUM WELLS AND THIN FILMS

D. D. Nolte and M. R. Melloch*
Dept. of Physics and
*School of Electrical Engineering
Purdue University
West Lafayette, IN 47907

1. INTRODUCTION

The photorefractive effect consists of three roughly separate physical processes: photoexcitation of charge carriers; electronic transport and trapping; and an electro-optic effect. Of these three processes, the first and the last explicitly involve the interaction of light with matter. One way to improve photorefractive performance is to enhance the optical interactions, i.e., by operating near an optical resonance of the material. This enhances both the photogeneration of charge carriers and the electro-optic effect. The electro-optic effect is one limit of an optical nonlinearity, and nonlinearities are strongest near a resonance.

In a semiconductor, the fundamental optical resonance is the photo-excitation of an electron-hole pair into a Coulomb bound state, called an exciton. At low temperatures, the excitonic optical transition appears as a peak or series of peaks in the absorption spectrum at the edge of the fundamental absorption gap. Absorption at the bandedge is large, and is highly sensitive to electric fields. These advantageous optical properties are diminished at room temperature by interaction with phonons, but can be regained by artificially structuring the semiconductor into quantum well structures.

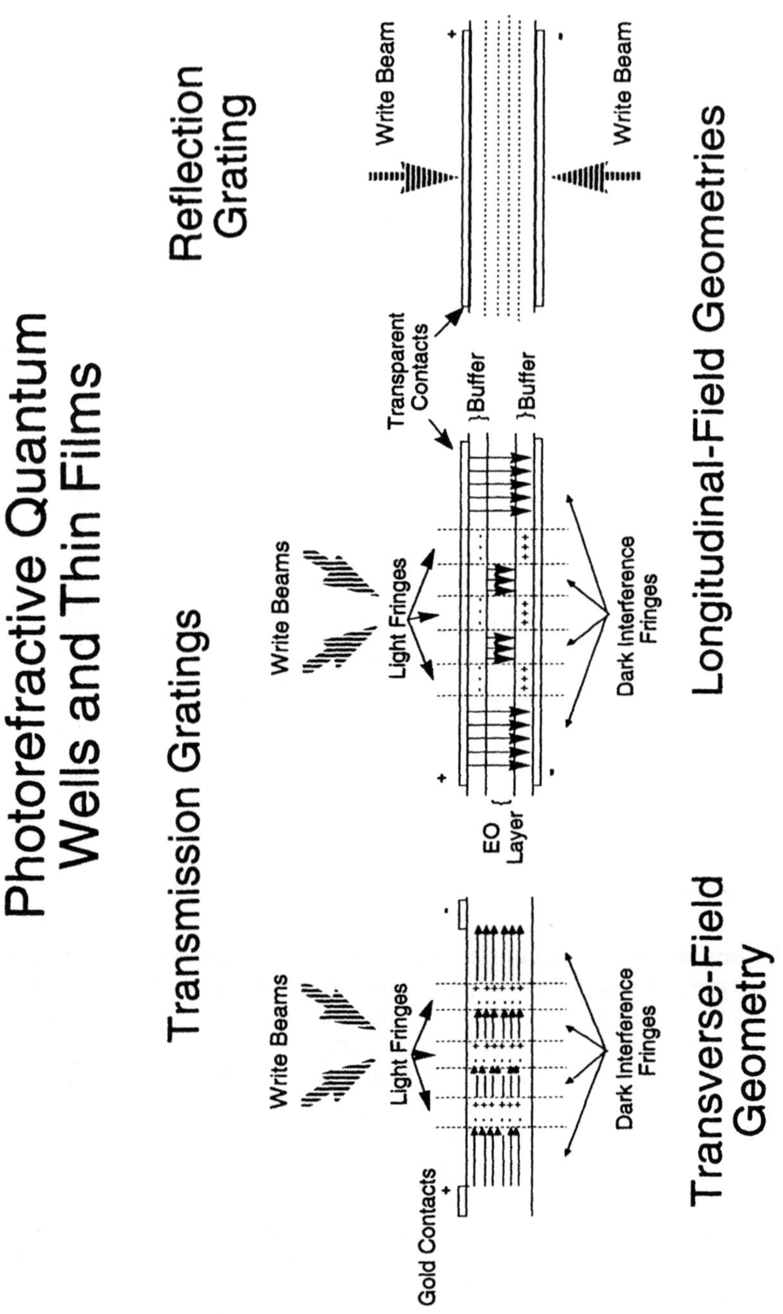

Fig. 1 Photorefractive quantum-well field and grating geometries.

In the previous chapter of this volume, the absorption and electro-optic effects of near-resonant excitation of semiconductors were discussed. The resonant enhancements of the bandedge extend to photon energies tuned below the bandedge, in the range of the exponential absorption tail.

In this chapter, we describe the physics of photorefractive semiconductor thin films and quantum well structures operating with photon energies tuned directly at or near the excitonic transitions. The large absorption at the excitonic transition energy produces short attenuation depths that constrain these structures to operate as thin films with thicknesses of only several microns. These thin films also require large applied electric fields because the excitonic electro-optic effects are quadratic in electric field, and are relatively weak at low field strengths, but are extremely large at high field strengths. Diffusion fields, that play such a crucial role in two-wave mixing in nonresonant photorefractive materials, play only a secondary role in the photorefractive thin films, influencing spatial resolution, but not contributing significantly to photorefractive gratings.

These two physical constraints, thin interaction lengths and high applied fields, define a set of photorefractive quantum well geometries. Electric fields can be applied in the plane of the thin film, or perpendicular to the film. Likewise, the propagating laser beams can enter the same side of the thin film, producing transmission gratings, or can enter from opposite sides, forming reflection gratings. These combinations are shown in Fig. 1. The field direction is specified by transverse or longitudinal geometries. The propagation directions are specified by transmission or reflection geometries. In this chapter, we discuss the physics, operation and applications of these photorefractive quantum well geometries.

2. DIFFRACTION FROM THIN FILMS

Diffraction from holographic thin films is not constrained to match Bragg angles, as is necessary for diffraction from bulk holograms. Photorefractive thin films perform as planar diffraction gratings because the thickness of the film is typically smaller than the hologram fringe spacing. This aspect of photorefractive quantum wells make them easier to operate than bulk photorefractive crystals, especially during nondegenerate four-wave mixing, because no alignment of the probe laser beam is necessary to observe a diffraction order. The principles of diffraction from thin films are described by the Raman-Nath diffraction theory that was originally developed to describe diffraction of light by sound waves [1] (as in acousto-optic modulators [2]) and undulating surfaces [3, 4, 5]. Raman-Nath diffraction in photorefractive quantum wells causes two-wave mixing in addition to diffraction into multiple higher orders.

2.1 Raman-Nath Diffraction

Raman-Nath diffraction is treated using Fraunhofer diffraction theory with wavefront modulation. A thin film with a dielectric grating modulates the phase and amplitude of an incident plane wave that passes through the film, as shown in Fig. 2. If the thickness of the film L is much smaller than the fringe spacing Λ, then effects of the finite thickness can be ignored, and the entire influence of the film can be expressed as a one-dimensional modulation of the wavefront. The diffracted amplitude at the far-field observation point P is calculated by integrating the complex amplitude of the modulated wavefront over one fringe period.

The diffraction angles θ_d for the multiple diffraction orders are given by the optical path difference

$$k_{vac}\,\Lambda\left(\sin\theta_d + \sin\theta_{in}\right) = M\,2\pi \tag{1}$$

where k_{vac} is the vacuum propagation constant, M is the diffraction order and θ_{in} is the incident angle of the probe wave. This can also be expressed as a ratio of the free-space propagation constant and the grating spatial frequency

$$\sin\theta_d = -\sin\theta_{in} + M\frac{K}{k_{vac}} \tag{2}$$

where the grating constant is $K = 2\pi/\Lambda$.

The dielectric grating in the thin film can consist of an index grating, an absorption grating, or both. In the case of semiconductor thin films and quantum wells, the electroabsorption observed for photon energies near the excitonic transition is accompanied by a change in the refractive index, required by the Kramers-Kronig relation. Therefore, photorefractive quantum wells have absorption and index gratings that contribute approximately equally to the diffracted signal, although with different wavelength dependences. In this analysis, we assume a sinusoidal modulation of the complex refractive index

$$\tilde{n}(x) = \tilde{n}_0 + \tilde{n}_1\cos\left(Kx + \phi\right) \tag{3}$$

The phase shift ϕ is the photorefractive phase shift that represents the spatial shift of the index grating relative to the interference fringe pattern writing the grating. Finite phase shifts lead to finite energy transfer during photorefractive two-wave mixing. The complex first-order Fourier coefficient is expressed in terms of the Fourier coefficient of the real index of refraction and the absorption as

$$\tilde{n}_1 = n_1 + i\kappa_1 = n_1 + i\frac{\lambda}{4\pi}\alpha_1 \qquad (4)$$

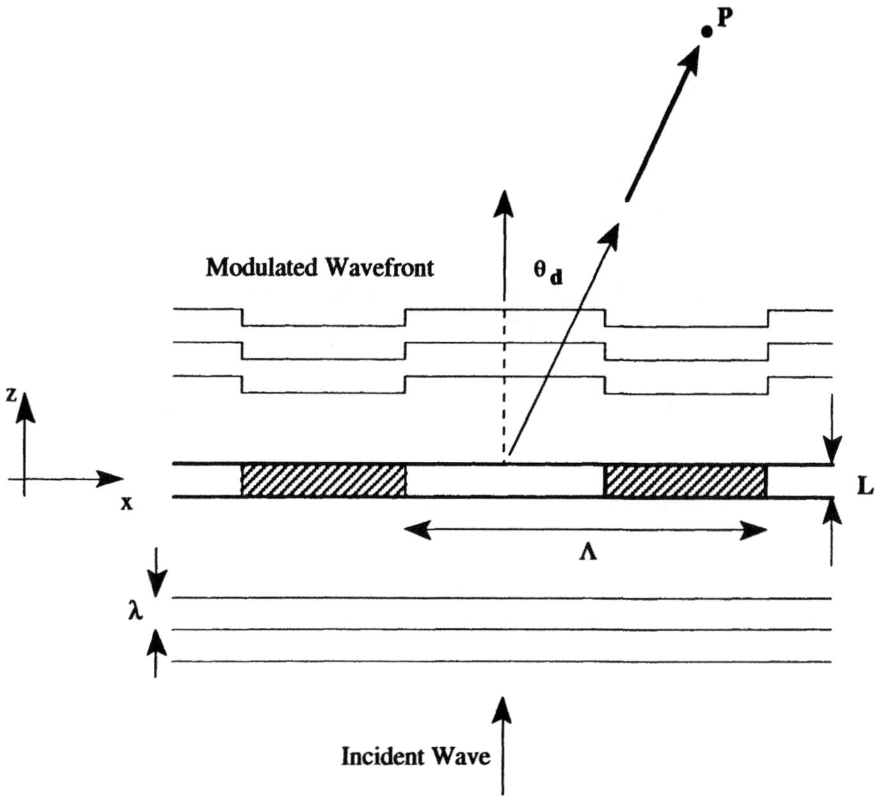

Fig. 2 Wavefront modulation by a dielectric grating in a thin film. The wave diffracted into the angle θ_d is detected at the far-field observation point P, and is calculated by a Fraunhofer integral over one fringe period Λ.

where κ_1 and α_1 are the first-Fourier extinction and absorption coefficients. A plane wave traversing the thin film at an incident angle θ_{in} experiences a complex amplitude modulation given by

$$E_t(x) = E_i\, e^{i(\delta_0 + \delta_1 \cos(Kx + \phi))} \qquad (5)$$

where the complex phase factors are

$$\delta_0 = \frac{2\pi\, n_0\, L}{\lambda \cos\theta'} + i\, \frac{\alpha_0\, L}{2\cos\theta'} \tag{6}$$

and

$$\delta_1 = \left(\frac{2\pi\, n_1\, L}{\lambda \cos\theta'} + i\, \frac{\alpha_1\, L}{2\cos\theta'} \right) \tag{7}$$

where $\sin\theta' = \frac{1}{n_0}\sin\theta_{in}$ is the internal angle, and L is the thickness of the film. Using the Bessel function identity

$$e^{i\delta_1 \cos(Kx+\phi)} = \sum_{N=-\infty}^{\infty} J_N(\delta_1)\, e^{iN(Kx+\phi+\pi/2)} \tag{8}$$

the transmitted field can be written as

$$E_t(x) = E_i\, e^{i\delta_0} \sum_{N=-\infty}^{\infty} J_N(\delta_1)\, e^{iN(Kx+\phi+\pi/2)}\,. \tag{9}$$

The Fraunhofer integral to calculate the diffracted amplitude in the Mth-order at the far-field point P in Fig. 2 is

$$E_M = \frac{1}{\Lambda} \int_{-\frac{\Lambda}{2}}^{\frac{\Lambda}{2}} E_t(x)\, e^{-ikx\left(\sin\theta_d + \sin\theta_{in}\right)}\, dx\,. \tag{10}$$

Inserting eq.(9) into eq.(10) and using eq.(2) gives

$$E_M = E_i\, e^{i\delta_0} \sum_{N=-\infty}^{\infty} J_N(\delta_1)\, e^{iN\left(\phi+\pi/2\right)} \frac{1}{\Lambda} \int_{-\frac{\Lambda}{2}}^{\frac{\Lambda}{2}} e^{iNKx} e^{-iMKx} dx \tag{11}$$

which is only nonzero when N = M. The diffracted amplitude is therefore

$$E_M = E_i\, e^{i\left(\delta_0 + M\phi + M\pi/2\right)}\, J_M(\delta_1)\,. \tag{12}$$

The diffraction efficiency for Mth-order diffraction is defined as the ratio of the diffracted intensity relative to the incident intensity

$$\eta_M = e^{-\alpha_0 L} J_M(\delta_1) J_M(\delta_1^*) . \tag{13}$$

When the spatial modulation of the refractive index is small, the above expression can be simplified using

$$J_M(x) \approx \frac{x^M}{2^M M!} \tag{14}$$

for small x, where M is any nonnegative integer. In photorefractive quantum wells the diffraction efficiency is typically smaller than 5%, making the expansion in eq.(14) valid. Therefore, to good approximation for photorefractive quantum wells, the diffraction efficiency is

$$\eta_M = \frac{1}{2^{2M} M!^2} \left[\left(\frac{2\pi n_1 L}{\lambda \cos\theta'} \right)^2 + \left(\frac{\alpha_1 L}{2 \cos\theta'} \right)^2 \right]^M \exp\left(\frac{-\alpha_0 L}{\cos\theta'} \right) \tag{15}$$

from the combined index and absorption gratings. The first-order signal is the largest and therefore the most interesting for applications. The first-order diffraction efficiency is

$$\eta_{\pm 1} = \left[\left(\frac{\pi n_1 L}{\lambda \cos\theta'} \right)^2 + \left(\frac{\alpha_1 L}{4 \cos\theta'} \right)^2 \right] \exp\left(\frac{-\alpha_0 L}{\cos\theta'} \right) . \tag{16}$$

The two common mixing conditions for photorefractive thin films are degenerate and non-degenerate mixing, shown in Fig. 3. Thin films have a special advantage for non-degenerate four-wave mixing, because the thickness of the film is typically comparable to or smaller than the absorption depth. Therefore any photon energy larger than the fundamental absorption edge in the semiconductor can be used to write gratings. This makes it possible to have wide-band operation for hologram writing. The read beams, on the other hand, are constrained to have photon energies near the excitonic absorption edge, which are compatible with inexpensive semiconductor diode lasers.

In degenerate mixing the write beams perform as the read beams, reading the same hologram that they write. In degenerate mixing the diffraction orders from one beam also overlap with orders from the other beam, which are all mutually coherent but with relative phase shifts [6]. The overlap and coherence of the multiple orders has its most important consequences for the zero-order beams. The coherent interference between the first-order and zero-order beams produces two-wave mixing [7]. In the case of a photorefractive phase shift $\phi = \pi/2$, there is constructive interference for one, and destructive interference for the other, resulting in the nonreciprocal transfer of energy from one beam to the other. This phenomena constitutes photorefractive gain, which is the basis of many photorefractive applications [8].

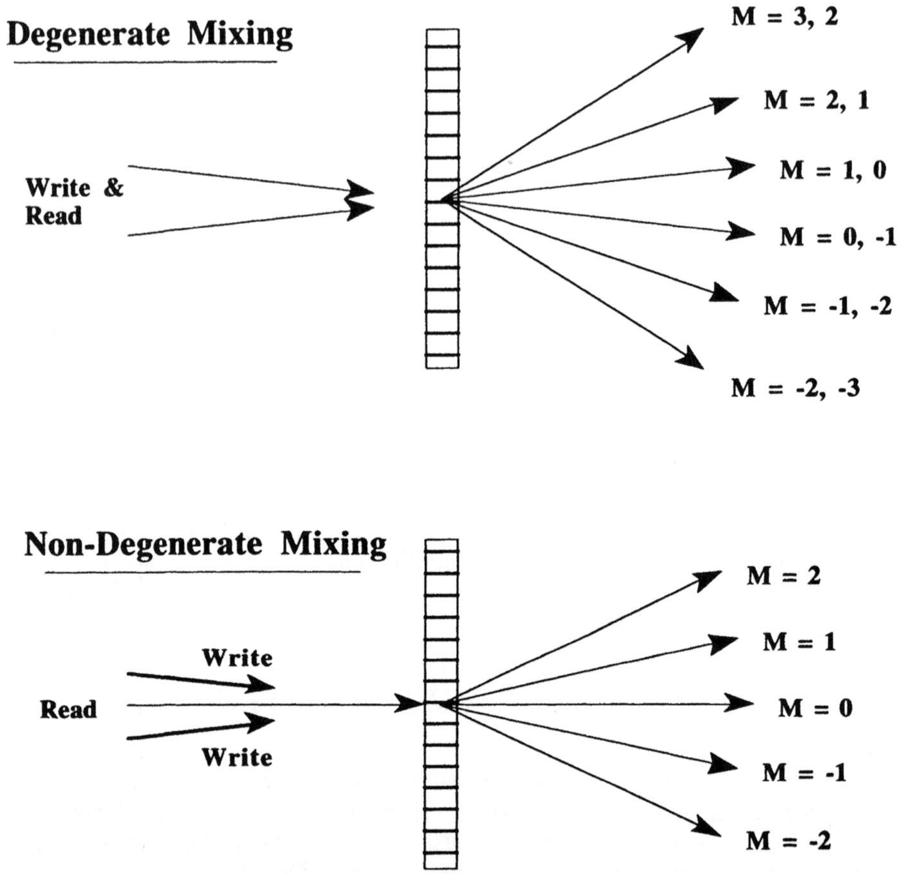

Fig. 3 Degenerate and non-degenerate mixing in thin holographic films. In degenerate mixing, several orders overlap. The overlap with the zero-order beams produces two-wave mixing.

2.2 Two-Wave Mixing

During degenerate mixing, the first-order diffraction from one pump beam propagates in the same direction as the other pump beam. The diffracted and pump beams interfere coherently and transfer energy nonreciprocally between them. Two-wave mixing in photorefractive quantum wells has several unique features that set it apart from two-wave mixing in bulk photorefractive materials. The difference arises from the combined absorption and refractive index gratings, and from the quadratic dependence of the electro-optic effects on electric field.

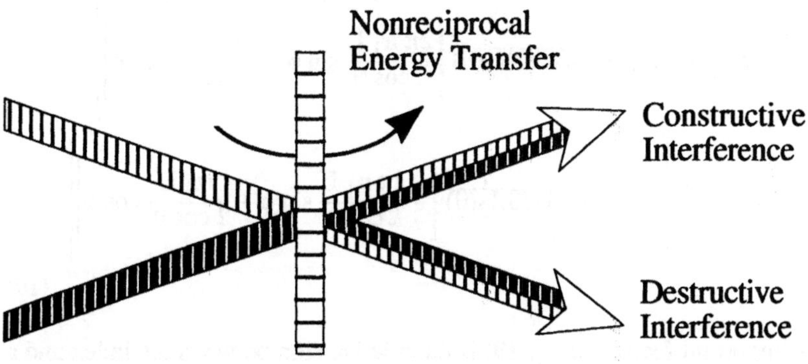

Fig. 4 Field amplitudes during two-wave mixing. The coherent interference between the zero-order and first-order beams leads to cross-modulation and energy transfer.

The pump and signal beams during two wave mixing are shown in Fig. 4. The sign of the photorefractive phase shift determines the direction of energy transfer. Using eq.(12) the amplitude of the first beam after passing through the dielectric film is

$$E_1(L) = J_0(\delta_1)\, E_1(0)\, e^{i\delta_0} + J_1(\delta_1)\, E_2(0)\, e^{i(\delta_0 + \phi + \pi/2)} \quad (16)$$

The corresponding intensity for I_1 is

$$I_1(L) = \eta_0\, I_1(0)\, e^{-\alpha L} + \eta_1\, I_2(0)\, e^{-\alpha L} + 2\sqrt{I_1(0)\, I_2(0)}\, e^{-\alpha L} \left\{ \mathrm{Re}\left[J_1(\delta_1)\, J_0(\delta_1^*)\right] \sin\phi - \mathrm{Im}\left[J_1(\delta_1^*)\, J_0(\delta_1)\right] \cos\phi \right\} \quad (17)$$

A similar expression holds for intensity I_2

$$I_2(L) = \eta_0\, I_2(0)\, e^{-\alpha L} + \eta_1\, I_1(0)\, e^{-\alpha L} + 2\sqrt{I_1(0)\, I_2(0)}\, e^{-\alpha L} \left\{ \mathrm{Re}\left[J_{-1}(\delta_1)\, J_0(\delta_1^*)\right] \sin\phi + \mathrm{Im}\left[J_{-1}(\delta_1^*)\, J_0(\delta_1)\right] \cos\phi \right\} \quad (18)$$

When the dielectric grating is weak, the expansion in eq.(14) can be used to express the intensities as

$$I_1(L) = e^{-\alpha_0 L}\left[I_1(0) + \sqrt{I_1(0)I_2(0)}\left[\frac{2\pi\, n_1\, L}{\lambda \cos\theta'}\sin\phi + \frac{\alpha_1\, L}{2\cos\theta'}\cos\phi\right]\right]$$

$$I_2(L) = e^{-\alpha_0 L}\left[I_2(0) + \sqrt{I_1(0)I_2(0)}\left[-\frac{2\pi\, n_1\, L}{\lambda \cos\theta'}\sin\phi + \frac{\alpha_1\, L}{2\cos\theta'}\cos\phi\right]\right]$$

(19)

The important feature of eq.(19) is the relative sign between the index and absorption grating terms for the two beams. The contribution to the transmitted intensity from the absorption grating is symmetric with respect to the photorefractive phase shift ϕ, while the contribution from the index grating is asymmetric. This feature makes it possible to measure the contributions from the absorption and index gratings separately.

Photorefractive gain in a photorefractive thin film can be expressed in terms of the ratio of the mixed intensity to the unmixed intensity, analogous to the case for bulk photorefractive materials. The gain ratio is

$$\gamma = \frac{I_1(L)}{I_1(0)\, e^{-\alpha_0 L}} = 1 + \sqrt{\frac{I_2(0)}{I_1(0)}}\left[\frac{2\pi\, n_1\, L}{\lambda\cos\theta'}\sin\phi + \frac{\alpha_1\, L}{2\cos\theta'}\cos\phi\right].$$

(20)

This can also be expressed in terms of the beam intensity ratio $\beta = I_2(0)/I_1(0)$

$$\gamma = \frac{I_1(L)}{I_1(0)\, e^{-\alpha_0 L}} = 1 + \frac{\beta}{1+\beta}\left[\frac{4\pi\, n_m\, L}{\lambda\cos\theta'}\sin\phi + \frac{\alpha_m\, L}{\cos\theta'}\cos\phi\right]$$

(21)

where the maximum grating amplitudes are $n_m = n_1/m$ and $\alpha_m = \alpha_1/m$ for the modulation ratio m of the interference fringes writing the dielectric grating.

3. SEMICONDUCTOR QUANTUM WELLS AND THIN FILMS

The invention of molecular beam epitaxy [9-11] opened a tremendous range of possible structures and devices for semiconductor thin films and multilayers. The parameter space for these artificially constructed materials is the full range of bandgap energies, combined with the nearly unlimited permutations of multilayer sequences, ranging from monolayers and quantum wells to homogeneous thin films. The parameter space of epitaxial III-V semiconductor growth is shown in Fig. 5. The bandgaps and lattice constants change with alloy concentration. The most heavily studied multilayer systems are those in which the lattice constants match closely, such as the AlGaAs/GaAs system. However, advances in strained-layer growth have made almost any combination possible. These semiconductor epilayers have excellent crystalline quality, with the ability to control dopant impurities at will. Such control of material properties is unprecedented in modern solid state and materials research. By using this parameter space, material properties can be engineered to fit specific applications, such as photorefractive quantum wells and thin films.

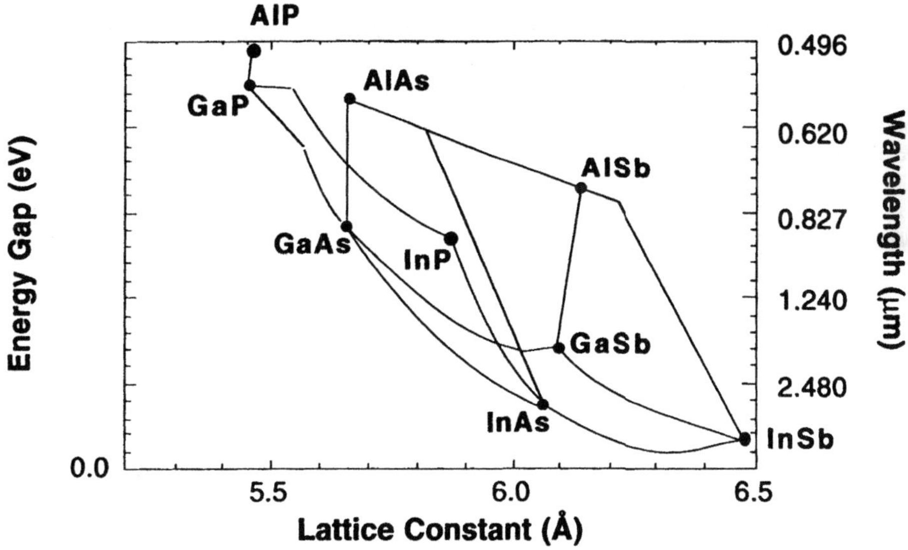

Fig. 5 Bandgaps of the III-V semiconductors as functions of lattice parameter and alloy composition.

Semiconductor thin films and quantum wells offer several advantages for photorefractive applications. One general advantage of photorefractive semiconductors is carrier mobilities significantly larger than in the photorefractive ferroelectric oxides and sillenites [12, 13]. This makes photorefractive semiconductors significantly faster for real-time image applications. Epitaxy can control mobilities to varying degrees. For instance, carrier mobilities parallel to the growth axis can be engineered over several orders of magnitude, including strong nonlinear dependences of carrier velocity on electric field. A second advantage of semiconductor thin films is the strong electroabsorption at the fundamental bandgap. The excitonic component of the bandedge absorption is sensitive to electric fields and can be tailored by bandgap engineering. Absorption changes of many thousand inverse centimeters under fields of $10^4 - 10^5$ V/cm produce refractive index changes of several percent. These advantages, and the ability to tailor them to the needs of photorefractive applications, make photorefractive quantum wells and thin films a versatile new field of photorefractive material research.

3.1 Excitons in Semiconductors

The fundamental optical response of a semiconductor is the photo-excitation of an electron-hole pair into a bound quantum state called an exciton. The electron and hole are bound by their mutual Coulomb attraction into a hydrogenic-like state. The exciton is the principle electromagnetic many-body excitation of a semiconductor, and is the starting point for understanding coulomb correlations in optical spectra.

At temperatures near absolute zero the optical spectrum of a semiconductor for photon energies near the bandgap is dominated by a set of discrete absorption lines corresponding to the hydrogenic series

$$E_n = E_g - \frac{e^2}{2\,\varepsilon\,\varepsilon_0\, n\, a_n} \tag{22}$$

where the Bohr radius is

$$a_n = n\,\frac{\hbar^2\,\varepsilon\,\varepsilon_0}{e^2\, m_r^*} \tag{23}$$

n is the principle quantum number, and the reduced mass is $1/m_r^* = 1/m_e^* + 1/m_h^*$, where m^*_e and m^*_h are the electron and hole effective masses. The exciton Rydberg constant is typically a thousand times smaller than the hydrogen Rydberg because of the large dielectric constants in semiconductors and the small effective masses,

$$Ry_{ex} = \frac{m^*}{\varepsilon^2}\, Ry_H \approx 10^{-3}\, Ry_H\,. \tag{24}$$

The exciton Rydberg is shown in Fig. 6 as a function of bandgap for several semiconductors. Semiconductors with larger bandgaps have heavier effective masses and smaller dielectric constants, both of which act to increase the exciton binding energy. Also shown in the figure is the thermal energy k_BT at room temperature. At finite temperatures, the discrete excitonic transition lines are broadened by interaction with phonons. More strongly bound excitons are less sensitive to thermal broadening, but are consequently less sensitive to electric fields.

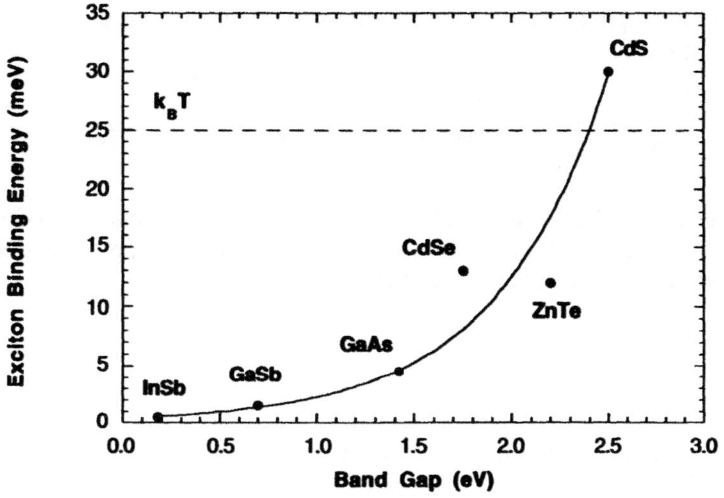

Fig. 6 Excitonic binding energy as a function of bandgap for several direct gap semiconductors.

The role of the excitonic transition in the optical response of a semiconductor can be treated semi-classically. The semiclassical dielectric function is

$$\varepsilon' = 1 + f\,\frac{4\pi\,Ne^2}{m}\,\frac{\omega_0^2 - \omega^2}{\left(\omega_0^2 - \omega^2\right)^2 + \omega^2\,\Gamma^2}$$

$$\varepsilon'' = f\,\frac{4\pi\,Ne^2}{m}\,\frac{\omega\,\Gamma}{\left(\omega_0^2 - \omega^2\right)^2 + \omega^2\,\Gamma^2} \qquad (25)$$

where Γ is the Lorentzian linewidth and f is the oscillator strength of the transition. Both Γ and f play important roles in the strength of the excitonic transition and in the response of the exciton to electric fields. The linewidth and oscillator strength can vary significantly, and are affected by sample quality, sample structure, and temperature, all of which can be controlled.

The oscillator strength of a Wannier exciton, such as in GaAs, is expressed as a product of electric dipole matrix elements between ground and excited states [14]

$$\begin{aligned} f_{n0} &= \frac{2}{m\,\hbar\omega_{0n}} \left(e^{ikr}\,\vec{p}\right)_{0n} \left(e^{ikr}\,\vec{p}\right)_{n0} \\ &\approx \frac{2m}{\hbar^2}\left(E_n - E_0\right)\left|U_{nk}(0)\right|^2 \left|z_{nm}\right|^2 \end{aligned} \tag{26}$$

where $|U_{nk}(0)|$ is the amplitude of the effective-mass envelope function at the origin. The dipole element z_{nm} is

$$z_{nm} = \int a_{n0}^{*}(\vec{r})\, z\, a_{m0}(\vec{r})\, d\vec{r} \,, \tag{27}$$

evaluated between two Wannier wavefunctions. Accurate calculations of the oscillator strength are difficult, but it is possible to make simple approximations. For instance the envelope function can be evaluated as

$$\left|U_{nk}(0)\right|^2 \approx \frac{\Omega_0}{\pi\, a_n^3} \tag{28}$$

where Ω_0 is the volume of the unit cell. Furthermore, the dipole element z_{nm} is approximately 1 Å between Wannier functions (approximately equal to an atomic radius). In GaAs the oscillator strength of the n = 1 exciton transition is therefore approximately

$$f_{GaAs} \approx \frac{\Omega_0}{\pi\, a^3} \approx 10^{-5} \ldots 10^{-4} \,. \tag{29}$$

The oscillator strength decreases as the cube of the exciton Bohr radius. Therefore smaller excitons that are more tightly bound (i.e. in semiconductors with larger bandgaps) have larger oscillator strengths.

Using the estimate for the oscillator strength in GaAs, the contribution to the refractive index and absorption can be approximated by

$$\begin{aligned} n &\approx \frac{\omega_p^2\, f}{4\,\omega_0\,\Gamma} \approx 0.2 \\ \alpha &\approx \frac{\omega_p^2\, f}{n\, c\, \Gamma} \approx 10^4\ \mathrm{cm}^{-1} \end{aligned} \tag{30}$$

where the ion plasma frequency is $\omega_p = 15$ eV and a typical room-temperature value of 6 meV is chosen for Γ. These values are the upper limit of changes that can be expected when the excitonic transition is strongly perturbed by an electric field.

The coulombic attraction between a photoexcited electron and hole is present even when the electron-hole pair is excited into an unbound continuum state. In fact, Coulomb correlations produce large enhancements in the continuum absorption over the simple one-electron absorption spectrum. This many-body enhancement was first derived analytically by Elliot [15]. The absorption at the bandedge can be expressed as [16]

$$\alpha(\omega) = \alpha_{bound} + \alpha_{cont} \tag{31}$$

where

$$\begin{aligned} \alpha_{bound} &= \alpha_{ex} \frac{\hbar\omega}{Ry_{ex}} \sum_{n=1}^{\infty} \frac{4\pi}{n^3} \delta\left(\Delta + 1/n^2\right) \\ \alpha_{cont} &= \alpha_{free}\, C(\omega)\, \Theta(\Delta) \end{aligned} \tag{32}$$

are the contributions from the excitons and the continuum respectively, α_{free} is the free-carrier absorption in the absence of Coulomb attraction, $\Theta(\Delta)$ is the Heaviside step function, and

$$\Delta = \frac{\left(\hbar\omega - E_g\right)}{Ry_{ex}} \tag{33}$$

is the scaled energy. The coefficient α_{ex} is a measure of the exciton oscillator strength, in which the n^{-3} dependence on the exciton oscillator strength has been shown explicitly. The Sommerfeld or Coulomb enhancement factor is given by

$$C(\omega) = \frac{x\, e^x}{\sinh(x)} \tag{34}$$

where $x = \pi/\sqrt{\Delta}$. The Coulomb enhancement is significant, leading to a strong deviation from $\sqrt{\Delta}$ expected for single-particle absorption. At room temperature, the exciton linewidth is comparable to the exciton binding energy, and the continuum edge is broadened. The absorption edge in GaAs at room temperature is shown schematically in Fig. 7. The discrete transition is unresolvable from the continuum, producing only a washed-out absorption edge.

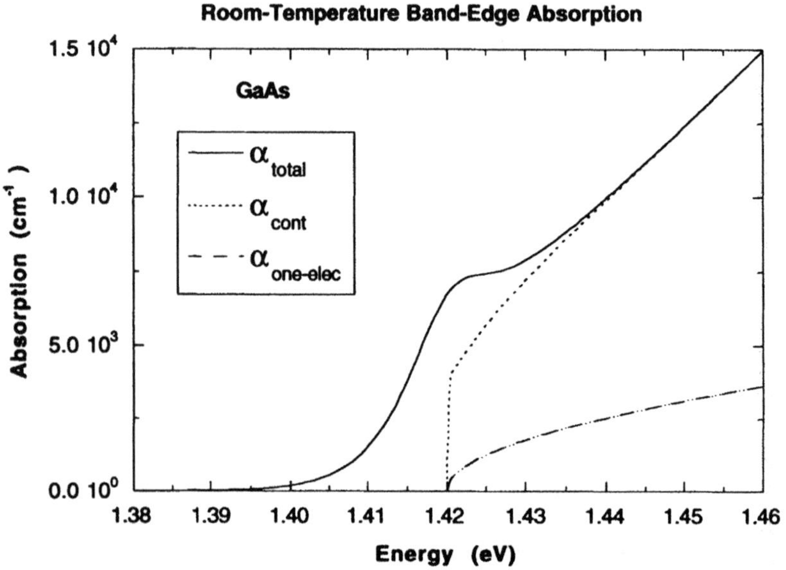

Fig. 7 Contributions to GaAs absorption as a function of photon energy at room-temperature, showing the one-electron absorption compared with the many-electron continuum and excitonic contributions. The excitonic contribution is significantly broadened.

3.2 Semiconductor Quantum Wells

A resonance with less damping will exhibit larger nonlinearities for the same stimulus. In the case of excitonic damping, the phonon broadening at room temperature in bulk semiconductors reduces the response of the excitonic absorption to applied electric fields. Therefore, although the electroabsorption from room-temperature excitons represents one of the largest electro-optic effects per interaction length in any optical material, its advantages are dissipated at room temperature. Cooling the semiconductor sample can reduce the thermal broadening, which may be reasonable for some low-temperature applications. However, for room temperature operation, other means of reducing the broadening is needed. One method is quantum confinement in reduced dimensionality semiconductor heterostructures. The most studied heterostructures, and the easiest to fabricate, are semiconductor quantum wells.

3.2.1 Confined energy levels

A semiconductor quantum well structure consists of multiple layers of semiconductors with different alloy content. For instance, layers of GaAs form low-bandgap quantum wells between higher-bandgap AlGaAs quantum barriers. The bandgap discontinuities at the interfaces form the confinement potentials. The AlGaAs/GaAs quantum well system is one of the most studied quantum well systems because AlAs and GaAs are nearly lattice-matched, making it possible to grow unstrained quantum wells. The direct and indirect bandgaps as functions of aluminum fraction are given by [17, 18]

$$
\begin{aligned}
E_g^{\Gamma}(x) &= 1.425 + 1.155x + 0.37x^2 \\
E_g^{X}(x) &= 1.911 + 0.005x + 0.245x^2 \\
E_g^{L}(x) &= 1.734 + 0.574x + 0.055x^2
\end{aligned}
\tag{35}
$$

and are shown in Fig. 8. The bandedge discontinuity $Q_{v,c}$ is the fraction of the bandgap difference that falls across the valence band and across the conduction band. In AlGaAs/GaAs, there was considerable debate concerning the bandgap discontinuities, but most work puts the value close to 65% for the direct bandgap discontinuity in the conduction band, with 35% of the discontinuity in the valence band, remaining relatively constant as a function of aluminum fraction x. The bandedge discontinuities, which form the confinement potentials of the quantum wells, are shown in Fig. 9 for $0 < x < 0.45$.

The real-space bandgap diagram of an $Al_{0.3}Ga_{0.7}As$/GaAs quantum well is shown to scale in Fig. 10. The energy eigenstates in a semiconductor quantum well can be calculated to lowest order by considering the quantum well to be a textbook example of a single symmetric quantum well with a finite barrier. The additional complication (at this simplest level) is the different effective masses in the well and the barrier. This alters the usual condition for the continuous first derivative of the wavefunction to

$$\frac{1}{m_w^*}\frac{\partial \psi_w}{\partial z} = \frac{1}{m_b^*}\frac{\partial \psi_b}{\partial z} \tag{36}$$

which ensures the continuity in probability current density across the interface [19, 20]. The energy eigenvalues of the quantum wells are given by the solution to the equations

$$\cos\varphi - \frac{m_b^*}{m_w^*}\frac{k_w}{k_b}\sin\varphi = 0 \quad \text{for even states}$$
$$\cos\varphi + \frac{m_w^*}{m_b^*}\frac{k_b}{k_w}\sin\varphi = 0 \quad \text{for odd states} \qquad (37)$$

where $\varphi = \frac{1}{2}k_w L_w$. The ground state solutions to eq.(37) were used to calculate the transition energy difference between the confined electron and hole ground state levels, shown in Fig. 11 from Ref. [21]. Confinement shifts are typically on the order of 50 meV in the AlGaAs/GaAs multiple quantum well systems. Also included in the figure are the experimentally measured transition energies for several quantum well structures with 75 Å quantum wells and varying barrier height.

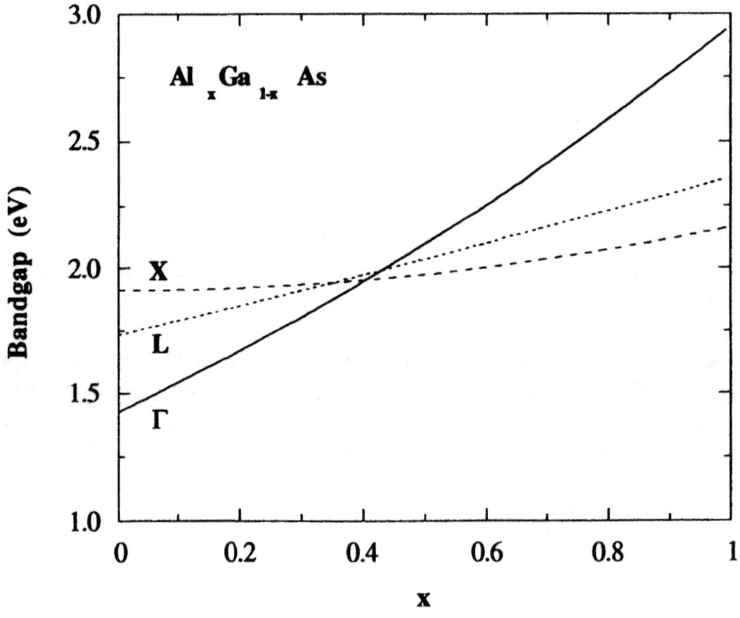

Fig. 8 Direct and indirect bandgaps as functions of aluminum fraction for $Al_xGa_{1-x}As$.

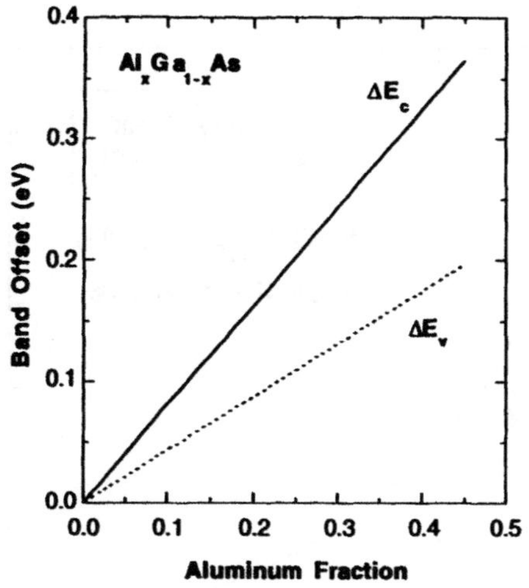

Fig. 9 Confinement bandedge discontinuities for the conduction and valence bands as functions of aluminum fraction.

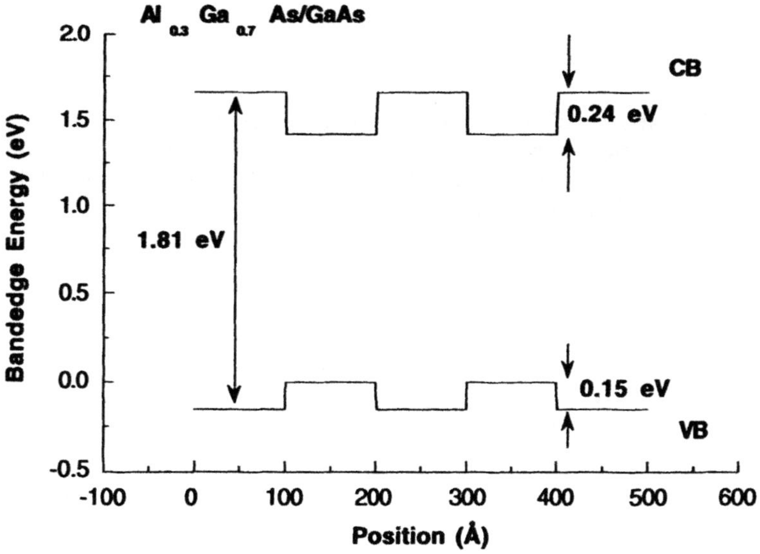

Fig. 10 Bandgap diagram of a $Al_xGa_{1-x}As/GaAs$ multiple quantum well structure drawn to scale for x = 30%.

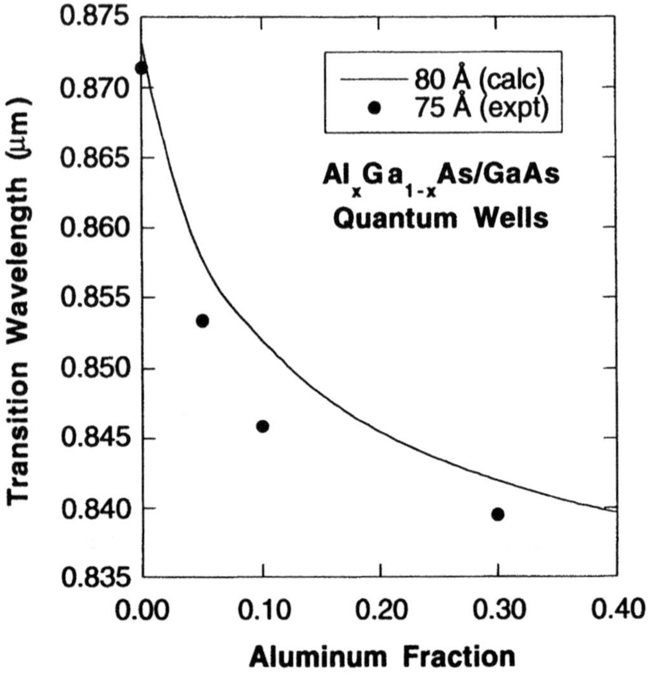

Fig. 11 Heavy-hole to electron transition wavelength as a function of aluminum fraction in the barriers. The calculations are for 80 Å wells and wide barriers, ignoring exciton effects [21]. The experimental data are from 75 Å wells and 100 Å barriers.

3.2.2 Quantum-confined excitons

Excitons confined in quantum wells have stronger binding energies and therefore are less easily thermally ionized at room temperature. The increased binding energy also increases the transition oscillator strength. Both these effects produce relatively sharp resonant transitions at room temperature, and sharper resonances create stronger electro-optic effects.

In the limit of a 2-dimensional Schrödinger equation, the 2-dimensional hydrogenic Rydberg is four times larger than the 3-dimensional Rydberg. This is because the confined wavefunctions of the electron and hole have a larger Coulomb overlap integral. In practice, the excitons never become 2-dimensional. That would require infinitely high and infinitely narrow quantum wells, which cannot be realized with bandgap engineering in semiconductors. For AlGaAs/GaAs, the maximum increase in the exciton binding due to quantum confinement is approximately 2.5 times. The maximum enhancement for an aluminum fraction of 30% occurs for a well width of approximately 50Å [22]. Below that width, the finite barrier height allows the confined wavefunctions to extend into the barrier, reducing the Coulomb overlap integral between the electron and hole.

Quantum confinement is also a sensitive function of the aluminum composition of the barriers. The absorbance of quantum-confined excitons in GaAs wells of constant thickness, but with different aluminum fractions in the barriers are shown in Fig. 12. As the aluminum fraction is decreased, the exciton transition energy approaches the bandgap energy of GaAs. Sharp resonant excitonic absorption can be seen even for an aluminum fraction in the barrier of only 5%.

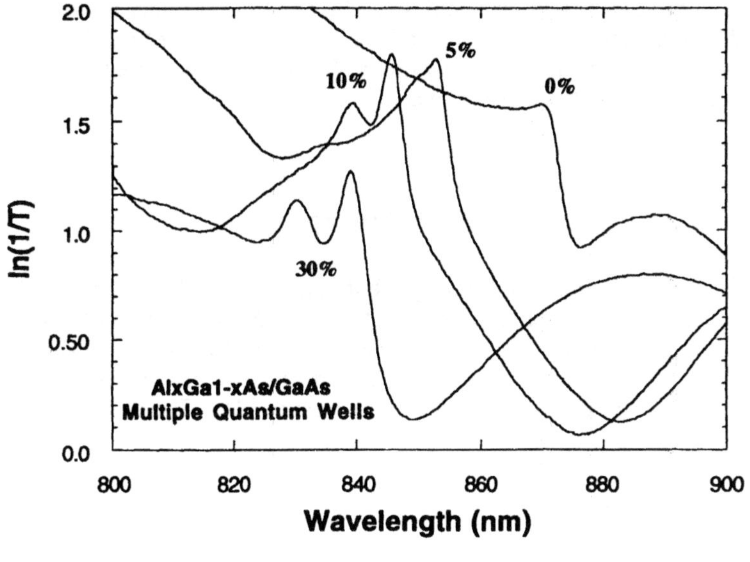

Fig. 12 Excitonic absorbance of quantum confined excitons in 75 Å wells with 100 Å barriers at room temperature. Note the strong Fabry-Perot structures below the bandgap that are created by the thin film structure [23].

4. EXCITON ELECTRO-OPTICS

The primary advantage of resonant interband photorefractive effects is the large electro-optic properties of the excitons. The principle electro-optic effect for excitons is electroabsorption: electric fields modify the absorption lineshape and energy of the excitonic transition. In quantum wells, there are two electroabsorption mechanisms: 1) field broadening, and 2) the Stark shift. Field broadening increases the linewidth of the transition lineshape, and the Stark shift moves the transition to lower energies. Both effects alter the absorption, leading to a wavelength-dependent $\Delta\alpha(\lambda)$ and an associated wavelength-dependent refractive index change $\Delta n(\lambda)$ by the Kramers-Kronig relation. Both effects are quadratic in the electric field, and require large electric field magnitudes to have appreciable effects on the absorption.

Because the growth direction of a quantum-well structure defines a symmetry axis, there are two different electric field geometries for quantum-confined exciton electroabsorption. The electric field can be applied in the plane of the quantum well (perpendicular to the growth direction), or the electric field can be applied parallel to the growth direction (perpendicular to the quantum well planes). The field in the plane of the wells produces field-broadening of the exciton, but only a weak Stark shift. The field perpendicular to the wells produces the quantum-confined Stark shift, for which field-broadening plays a secondary role. Each of the field orientations have relative merits for photorefractive performance.

4.1 Electric-Field-Induced Broadening

Electric fields broaden the transition lineshape, but conserve oscillator strength. This leads to a characteristic electroabsorption spectrum $\Delta\alpha(\lambda)$ that is approximately a second derivative of a gaussian transition lineshape. Experimentally, Gaussian lineshapes are often found to give a better description of broadened excitonic lineshapes at room temperature than Lorentzian lineshapes [24]. The absorption for a Gaussian transition is

$$\begin{aligned} \alpha(h\nu) &= \frac{\omega}{cn}\,\varepsilon'' \\ &\approx \sqrt{\pi\ln(2)}\,\frac{\hbar\,\omega_p^2}{cn}\,\frac{f}{\Gamma}\exp\left(-4\ln(2)\left(h\nu - h\nu_0\right)^2/\Gamma^2\right) \end{aligned} \tag{38}$$

where Γ is the full width at half maximum, ω_p is the ionic plasma frequency, f is the oscillator strength and ν_0 is the center frequency. The normalization constant satisfies the f-sum rule. In this expression, the extinction coefficient is assumed to be small relative to the refractive index, which is true for bandedge transitions in III-V semiconductors.

When an electric field is applied in the plane of the quantum wells, in the transverse-field geometry, the linewidth Γ is increased by $\Delta\Gamma$. The electroabsorption spectra for the AlGaAs/GaAs materials shown in Fig. 12 are shown in Fig. 13. Refractive index changes are associated with the absorption changes, related through the Kramers-Kronig relation

$$\Delta n(\lambda) = \frac{1}{2\pi^2} P \int_0^{\infty} \frac{\Delta\alpha(\lambda')\, d\lambda'}{\lambda^2 - \lambda'^2} \tag{39}$$

The refractive index changes that accompany Fig. 13 are shown in Fig. 14.

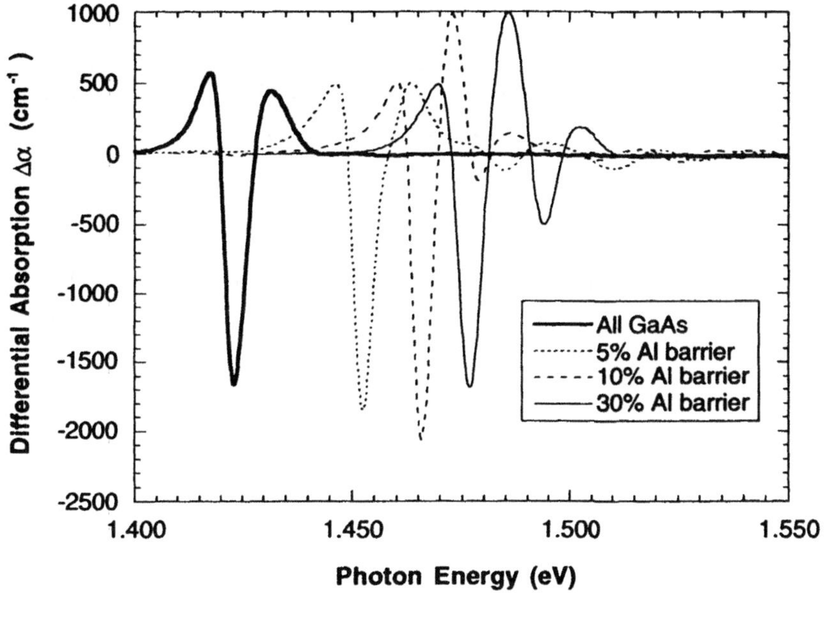

Fig. 13 Electroabsorption for transverse electric fields for 75 Å quantum wells with barriers of varying aluminum fraction. The applied electric field is 10 kV/cm [23].

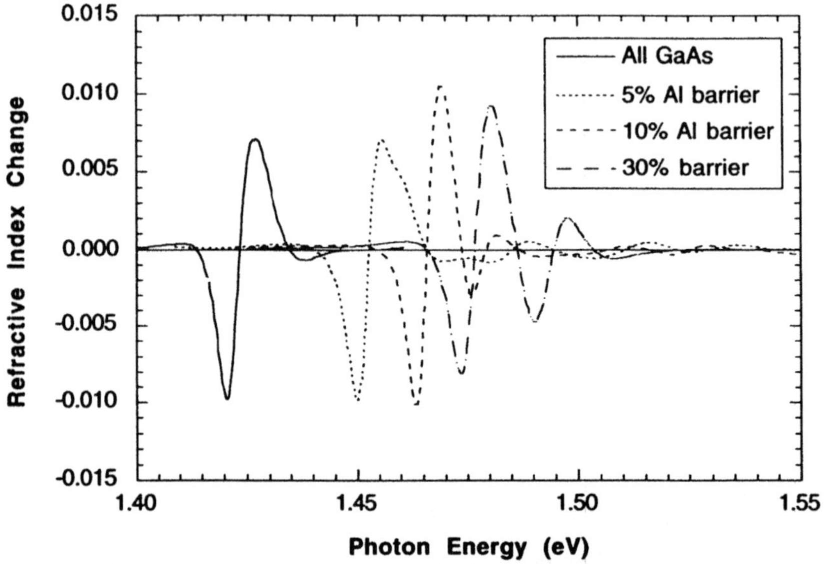

Fig. 14 Electrorefraction for the transverse field data in Fig. 13 [23].

For a Gaussian lineshape, the magnitude of the electroabsorption is

$$\Delta\alpha_p \approx \sqrt{\pi \ln(2)}\,\frac{\hbar\,\omega_p^2}{cn}\,f\,\frac{\Delta\Gamma}{\Gamma^2} \tag{40}$$

in the limit of weak fields. The most important feature in this expression is the inverse dependence on the square of the linewidth Γ. Relatively small increases in the thermal width can lead to a large decrease of the maximum electroabsorption and electrorefraction. Quantum confinement therefore is important because it reduces the effects of thermal ionization. Quantum confinement also increases the oscillator strength f. On the other hand, it reduces the sensitivity to applied field, given by $\Delta\Gamma$. Consequently, the strong change in the excitonic absorption through quantum confinement, shown in Fig. 12, leads to only a minimal increase for the peak electroabsorption in Fig. 13 in the transverse field geometry. Therefore, thin-film GaAs, without quantum barriers, is also a viable candidate for interband photorefractive effects.

4.2 Quantum-Confined Stark Effect

The strongest electro-optic effect is obtained when the electric field is applied perpendicular to the quantum well planes. In this geometry, the electric field introduces a linear potential to the quantum-well potential, shown in Fig. 15. The linear potential forms triangular potential wells, reducing both the electron and hole confinement energies and leading to a decrease in the transition energy. A shift of the transition energy by as much as 20 meV occurs for a field strength of 10^5 V/cm [25].

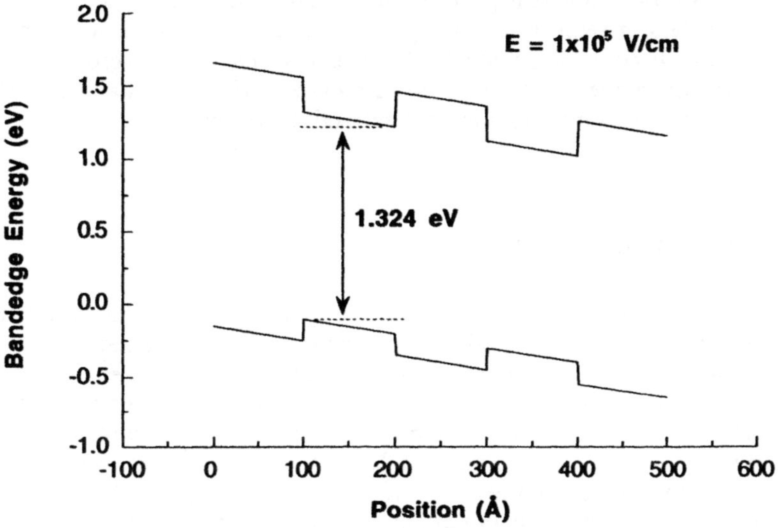

Fig. 15 Quantum-well potentials drawn to scale in a field of 10^5 V/cm. Transition wavelengths are red-shifted under an applied field.

For the Gaussian lineshape in eq.(38) the magnitude of the electroabsorption from the quantum-confined Stark effect is

$$\Delta\alpha_p \approx 4\ln(2)\sqrt{\frac{\pi}{2e}}\,\frac{\hbar\omega_p^2}{cn}\,f\,\frac{\Delta h\nu_0}{\Gamma^2} \tag{41}$$

in the limit of weak fields, where $\Delta h\nu_0$ is the Stark shift. The electroabsorption depends inversely with the square of the exciton zero-field linewidth, as was the case with field-broadening, favoring narrower linewidths. The optimum performance of photorefractive quantum wells therefore relies on the narrowest possible linewidths of quantum-confined excitons.

5. DEFECT ENGINEERING AND SEMI-INSULATING MATERIAL

As pointed out in Chapter 1, the photorefractive effect is a defect process that requires high defect densities and semi-insulating materials. Until recently, epitaxially-grown films were not semi-insulating as-grown. One technique for rendering the epilayers semi-insulating was radiation damage to compensate residual shallow dopants. More recently, a new class of semi-insulating epitaxial materials, grown at low substrate temperatures, has become available [26]. The low-temperature-growth (LTG) materials have advantages, such as ultra-fast carrier lifetimes, that make them particularly attractive as components in photorefractive structures.

5.1 Proton Implantation

The key to the development of photorefractive quantum wells, that made them different from quantum-well electroabsorption modulators, was the recognition that proton implantation could be used to make the quantum wells semi-insulating without adversely affecting the lineshape of the exciton or its electroabsorption properties [27]. The protons themselves play no role in making the material semi-insulating, but the radiation damage induced during the implant compensates shallow dopants and pins the Fermi level near the middle of the bandgap. This is the only significant difference between a photorefractive quantum well and the quantum well modulators with all their varieties, including self-electro-optic-effect devices (SEEDs) [28]. By making the quantum wells semi-insulating, it is no longer necessary to physically reticulate the structure into a planar array of modulators with individual electrical addresses defined with photolithography: in the photorefractive quantum wells the holography *is* the photolithography, but it occurs in real-time and is dynamic!

Proton implantation is only one type of ion implantation that can make the semiconductor semi-insulating, but it has several advantages. First, protons are a natural by-product of ion implantation of other species and therefore require no special configuration for ion implantors. Second, it is the lightest ion and therefore has the longest penetration depth, making it possible to implant relatively thick quantum-well structures. Third, the effects of proton implants have been well studied because of their use for electrical isolation of GaAs integrated circuits [29].

The implant depth and deviation for proton implantation are shown in Fig. 16. For a typical implant energy of 150 keV, the implant depth is 1.14 microns with a deviation of 0.2 microns. The implant depth in the figure represents where the proton stops, on average. Some implant damage extends from the surface to the stopping range as the proton loses energy through scattering. However, the proton dumps most of its energy within one deviation of the average range.

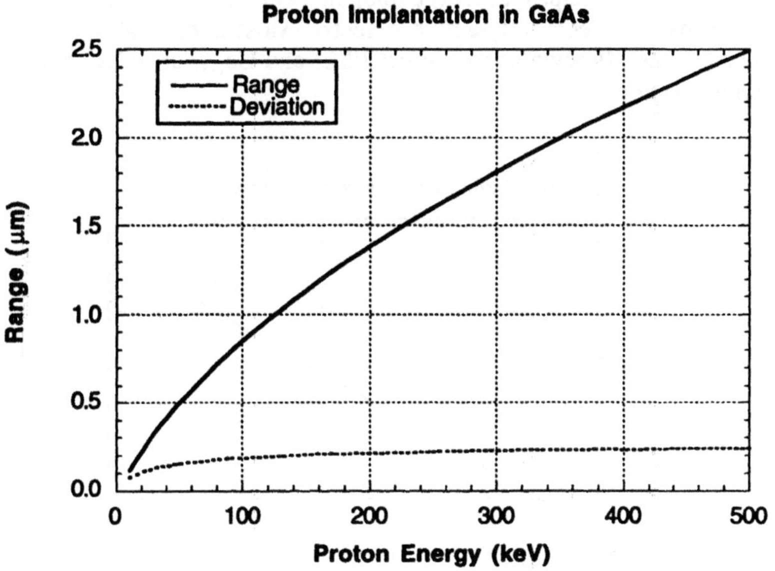

Fig. 16 Range and deviation of proton implants in GaAs as a function of implant energy.

The principle quantities of interest for photorefractive effects in proton-implanted thin films are the number densities and identities of the radiation defects created by the proton implantation. It is important to know how many radiation defects are created, and what fraction are acceptors or donors. This information provides the effective number of traps that can participate in storing space-charge during photorefractive mixing.

A simple dynamical calculation of the number of knock-out atoms per implanted proton can be made [30-33]. The total number of knock-out atoms is

$$N_{ko} \approx \frac{10^{-3}(E_0 - E_c) + 0.5E_c}{E_d} \tag{42}$$

where E_0 is the energy of the projectile, E_c = 1 keV is the energy at which the nuclear stopping power equals the electronic stopping power, and E_d = 15 eV is the displacement energy for a Ga or As ion. The number of knock-out atoms as a function of implant energy is shown in Fig. 17. At an energy of 150 keV there are approximately 43 knock-out atoms per proton. Not all of these displaced atoms in GaAs remain after the implantation. Many of the point defects formed by the proton implant are vacancy-interstitial pairs that recombine during or shortly after the implant. Many other of the displaced atoms form electrically neutral defect complexes that cannot participate in charge trapping. Only a fraction of the displaced atoms form permanent electrically-active defects. Most likely stable defects are antisites, such as As_{Ga} (known as the defect EL2) and Ga_{As}, or defect complexes with residual impurities.

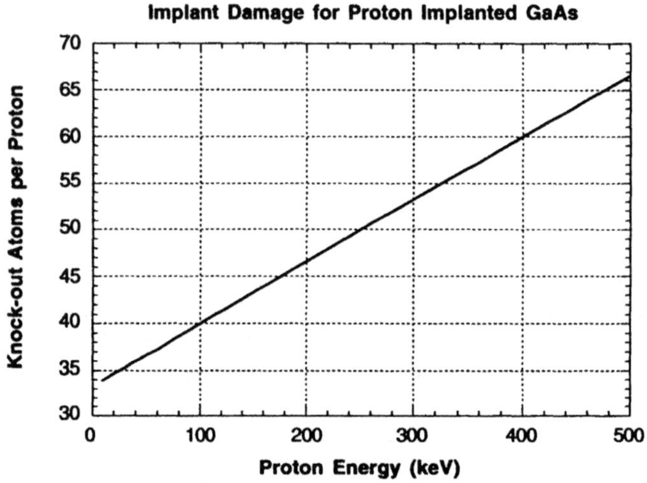

Fig. 17 Number of knock-out atoms in GaAs as a function of proton implant energy (after Refs.[30, 31]).

Proton implantation produces an auto-compensation mechanism that compensates both n-type and p-type material and pins the Fermi level near midgap. The implant damage consists of both deep acceptors and deep donors. An interesting feature of radiation damage in semiconductors is the universal behavior of the Fermi level pinning. Both the deep acceptors and the deep donors pin the Fermi level at an energy between 0.6 and 0.8 eV above the top of the valence band [34], producing relatively uniform properties for the semi-insulating material after implant, regardless of the starting dopants. However, the compensation efficiency as a function of proton dose is different for n-type and p-type material.

Accurate limits of the compensation capability of proton implantation have been obtained by studying the proton dose needed to compensate known shallow dopant concentrations. Proton implantation studies in n-type doped GaAs have shown that protons with an energy in the range of 150 keV remove approximately 3 conduction electrons per proton [30, 35]. Implantation into p-type GaAs shows a smaller removal rate. Protons with 300 keV energy remove approximately one valence band hole per implanted proton [36]. These studies set lower bounds on the number of radiation-induced donors and acceptors from proton implantation. The upper bounds are limited by the number of knock-out atoms. Therefore the defect numbers produced by proton implantation are

$$3\,/\,\text{proton} < N_D < 43\,/\,\text{proton}$$
$$1\,/\,\text{proton} < N_A < 43\,/\,\text{proton}$$

For an implant of 150 keV, with an average implant depth of 1.2 μm and a dose of $[H^+] = 10^{12}$ cm^{-2}, the defect densities fall in the range

$$2.5 \times 10^{16}\,\text{cm}^{-3} < N_D < 3.6 \times 10^{17}\,\text{cm}^{-3}$$
$$8.3 \times 10^{15}\,\text{cm}^{-3} < N_A < 3.6 \times 10^{17}\,\text{cm}^{-3}$$

These densities are at least an order of magnitude larger than the residual shallow dopant densities in undoped MBE materials. Therefore, proton-implanted devices are self-compensating, and shallow defects play a negligible role in determining the compensation ratio and the effective number of traps. The compensation ratio in proton implanted devices is therefore an intrinsic property, and will be uniform from growth to growth or system to system.

The lifetime of carriers in photorefractive quantum wells plays an important role in the spatial resolution of the thin films, as will be discussed in Section 6.0 for the transverse geometry and Section 7.0 for the longitudinal-field geometry. Studies of carrier lifetimes in proton implanted GaAs have been motivated by a search for ultra-fast photoconductors with sub-picosecond response times [37-39]. Carrier lifetimes in GaAs are shown in Fig. 18 as functions of proton implant dose. The values vary from 500 psec to 500 fsec for doses ranging from 10^{10} cm^{-2} to 10^{16} cm^{-2}. The values were measured using transient optical spectroscopy of the recovery time after excitation by an intense pulse.

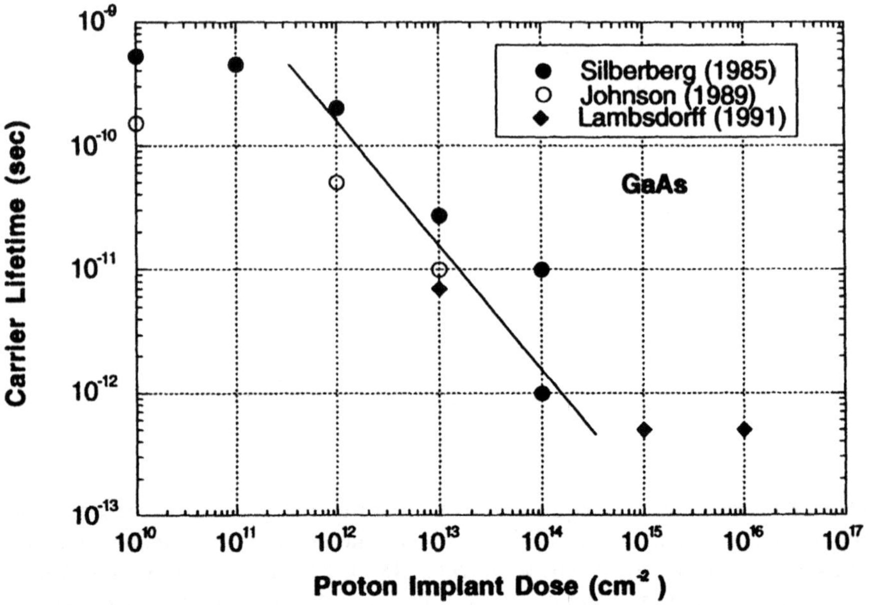

Fig. 18 Carrier lifetime as a function of proton implant dose in GaAs. From Refs. [37-39].

Although high spatial resolution in the photorefractive quantum wells requires as many defects as possible to reduce the free-carrier lifetimes and minimize their diffusion lengths, when the implant dose is too large the implant damage adversely broadens the excitonic transition peak and reduces the electro-optic effect. Therefore, there is a trade-off between spatial resolution and mixing efficiency. The maximum defect density that can still allow reasonable excitonic electro-optic properties is defined by the Mott criterion

$$N_M \frac{4}{3} \pi a_B^3 \approx 1 \tag{43}$$

where $a_B = 117$ Å is the Bohr radius of the exciton (in GaAs) and N_M is the number density of defects. This criterion states that the excitons are disrupted when there is on average one defect per exciton volume. Using the Mott criterion in GaAs of $N_M \approx 1 \times 10^{17}$ cm^{-3} sets an upper limit of

$$\left[H^+\right]_{max} \approx 2 \times 10^{12} \text{ cm}^{-2} \tag{44}$$

for the maximum allowable proton implant dose that still permits favorable electro-optic properties of the excitonic bandedge transition.

5.2 Epitaxy Growth at Low Substrate Temperatures

As stated previously, one key parameter that defines the performance of photorefractive quantum wells is the carrier lifetime. The lifetime determines the transport lengths, which influence the photorefractive sensitivity and the spatial resolution. Proton implantation at the maximum dose of $[H^+] = 2\times10^{12}\ cm^{-2}$ produces a carrier lifetime of about 50 psec.

In the transverse-field geometry, the diffusion length determines the device resolution. We show in the next section that the cut-off fringe spacing in this geometry is $\Lambda_c = 2\pi L_D$, where L_D is the diffusion length

$$L_D = \sqrt{\frac{k_B T}{e} \mu \tau} \ . \qquad (45)$$

A lifetime of 50 psec yields a diffusion length of 1 μm and a device resolution of approximately 6 μm.

In the longitudinal transmission geometry there are different concerns about carrier lifetimes. The longitudinal geometry relies on carrier trapping in the buffer layers that sandwich the electro-optic quantum well layers. If the transit time is smaller than the trapping time, then the quantum efficiency of the device will be low. The transit time of a carrier across a 1 μm buffer at high field is approximately 5 psec. Therefore it is important to find a material that has a trapping time in the range of 1 psec that can be incorporated into the photorefractive quantum well structure.

A newly developed semiconductor material that satisfies this ultrafast trapping lifetime is a material called low-temperature-growth (LTG) GaAs. This material is otherwise ordinary GaAs, but grown by MBE at reduced substrate temperatures (200 °C). The reduced growth temperature (followed by higher temperature post-growth anneals) results in semi-insulating material (a requirement for photorefractive effects) with an excess of arsenic that can be manipulated to control carrier lifetimes into the subpicosecond regime.

5.2.1 Arsenic Precipitate Engineering

Low-temperature-growth GaAs was introduced to reduce parasitic side- and back-gating in GaAs integrated circuits [26, 40, 41]. During the growth, excess arsenic (up to around 1.5%) is incorporated into the crystal [42]. The as-grown material is conducting because of defect-band transport [43, 44], but becomes semi-insulating after mild post-growth anneals [26, 44]. Stronger post-growth anneals, at temperatures in the range of 700 °C to 900 °C for 30 seconds, allows the excess arsenic to form precipitates [45] with well-controlled size and spacing.

Low-temperature-growth GaAs has generated a lively controversy and considerable confusion in the literature. The confusion arises from the many varieties of LTG material. There is no single LTG material with fixed properties. All the electrical, optical and structural properties depend on the growth temperature, the post-growth anneal temperature, and the anneal time. Weakly annealed materials can have very different properties than strongly annealed materials, and conclusions made about one type cannot be categorically applied to the others.

Apart from the problem of comparing like materials, there is a controversy over the mechanism that makes LTG material semi-insulating. Two camps have evolved with different mechanisms to explain why the Fermi level is pinned in the middle of the bandgap. These two mechanisms are: 1) the compensating point defect model; and 2) the buried Schottky model. The compensating point defect model assumes that As-related point defects in the As-rich material compensate shallow dopants [43], pinning the Fermi level close to their ionization levels, which are midgap. The buried Schottky model assumes that the metallic As precipitates act as internal Schottky barriers with a Schottky pinning level midgap [46, 47]. The Schottky contacts deplete carriers within a depletion length around each precipitate. If the depletion regions overlap, then the material is fully depleted and is semi-insulating.

There is a growing consensus that the precipitates deplete n-type material, but that point defects compensate p-type material. In nominally undoped crystals one or the other mechanism may play a stronger role in the compensation, depending on growth and annealing temperatures. However, the presence of large point defect concentrations as well as the presence of the arsenic precipitates must be considered together when attempting to describe and understand any of the electrical and optical properties of LTG material. In no cases can point defects or precipitates be ignored, although one or the other may dominate different processes.

The detailed control of the precipitate densities and sizes, and their introduction into different layers of heterostructures, provides unprecedented control over the properties of these composite materials. This detailed control is called precipitate engineering [48-51]. Average precipitate spacings are shown in Fig. 19 as functions of anneal temperature for two different samples grown at different substrate temperatures of 250 °C and 320 °C with excess arsenic concentrations of 0.9% and 0.3%, respectively. The precipitate spacing controls the carrier lifetime.

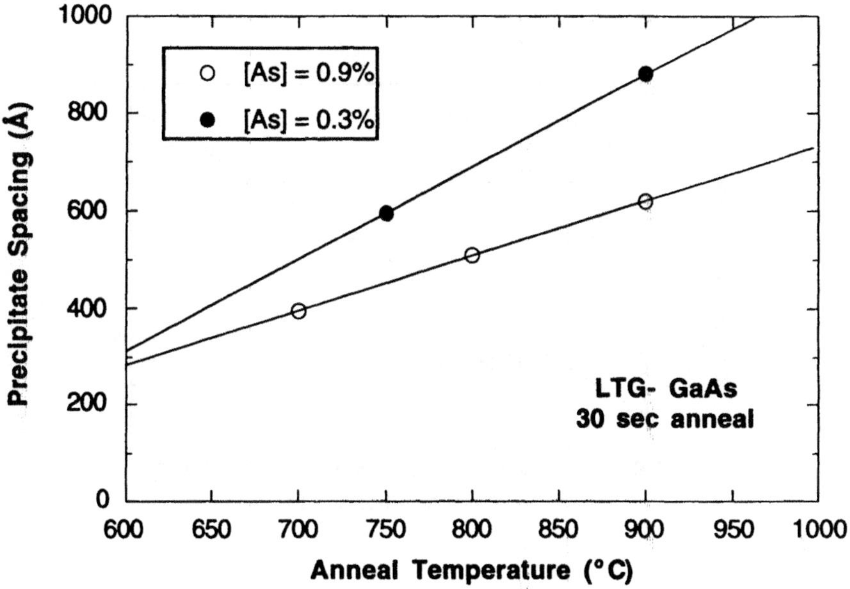

Fig. 19 Average precipitate spacing as a function of anneal temperature for two different As-fractions in LTG GaAs.

5.2.2 Carrier Lifetimes

It was recognized early that low-temperature-growth materials would have ultra-fast carrier lifetimes because of the high defect densities. The ultra-short lifetimes made this material ideal for ultrafast applications, such as the generation of ultra-fast electric pulses and freely-propagating Terrahertz radiation. Photoconductive transmission lines fabricated on LTG material launched electrical pulses with widths of around 1 psec [52, 53]. Freely propagating electromagnetic pulses were generated and detected using LTG material (with precipitates) with 0.7 psec pulse widths [54].

The carrier lifetimes in annealed LTG GaAs are shown in Fig. 20 as a function of precipitate spacing [55]. The lifetime varies from 2 psec to 10 psec for precipitate spacings between 400 Å and 1000 Å. The lifetime depends on the square of the precipitate spacing, which is consistent with diffusion to the precipitates before recombination. The electron diffusion length for an assumed electron mobility of 4000 cm^2/Vs is shown in Fig. 21. Diffusion lengths down to 1000 Å are possible in LTG GaAs. The LTG materials are therefore ideal for photorefractive thin films with high-spatial resolution [56].

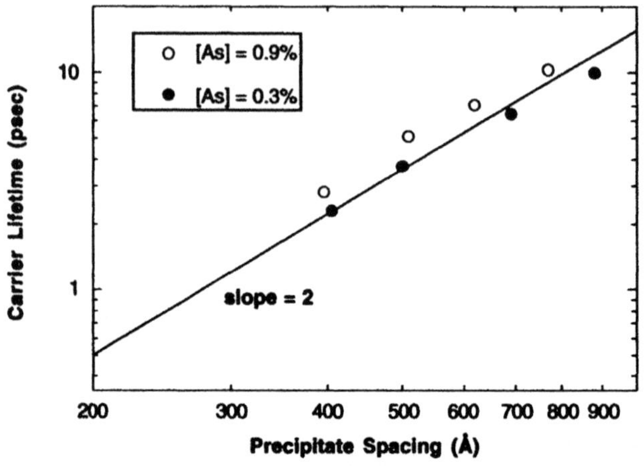

Fig. 20 Carrier lifetime as a function of precipitate spacing in LTG GaAs for two excess arsenic fraction [55].

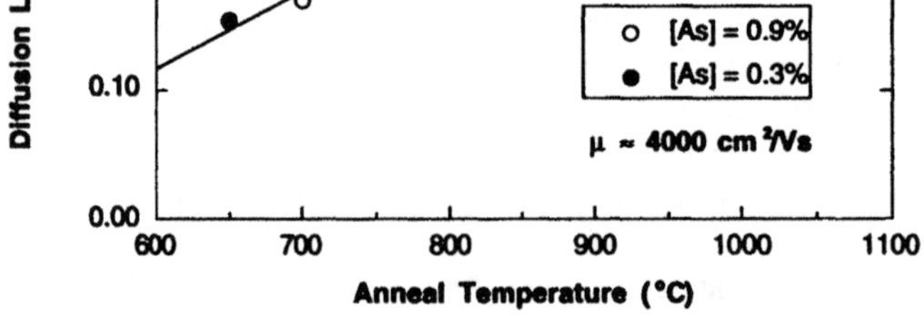

Fig. 21 Estimated diffusion length for electrons as a function of anneal temperature in LTG GaAs. An electron mobility of 4000 cm^2/Vs is assumed.

6. TRANSVERSE-FIELD GEOMETRY

The interband photorefractive process in photorefractive quantum wells and thin films has aspects different than the traditional photorefractive process in bulk photorefractive materials. The quadratic electro-optic effect, which has been described in Chapter 6 and Section 4.0 of this chapter, is the first important difference. The second important difference is the interband generation of charge carriers in the quantum well structures. In bulk photorefractive process, the charge carriers that separate into space-charge fields are generated by photoionization of defects. A balance is therefore established between the trapping of charge at defect sites and the re-emission of that charge by photoionization. In photorefractive quantum well structures, on the other hand, the photoionization of defects contributes negligibly to the charge generation. The carriers are generated by interband transitions that form electron-hole pairs. The interband photorefractive effect leads to qualitatively different behavior, which is described in this section.

6.1 Photorefractive Transport

6.1.1 Interband Photorefractive Effect

The transition processes of the interband photorefractive effect in a quantum well structure are shown in Fig. 22. The light that generates the free carriers can be resonant with quantum-confined states in the quantum wells, or can have any photon energy larger than that value. The thin films typically have thicknesses that are comparable to the absorption depth of the light. This makes the photorefractive thin films broad-band devices in which the photorefractive gratings can be written with any photon energy larger than the bandgap. This above-bandgap holography makes the devices extremely flexible for diverse applications that can use almost any visible laser as the source of photorefractive gratings. Once the carriers are generated, they quickly thermalize to a distribution that is divided between the quantum well and the quantum barrier. This distribution will in part depend on the applied electric field that accelerates and heats the carriers.

In the transverse-field geometry the carrier transport is in the plane of the quantum wells and barriers. Electrons heated by the field can gain sufficient energy to transfer to the indirect bandedges (not shown in the figure), which produces the well-known velocity saturation and transport nonlinearity that lead to Gunn oscillations. The free carriers trap at defects, which can capture both electrons and holes, although with different capture probabilities. Photoionization of the defects plays a negligible role.

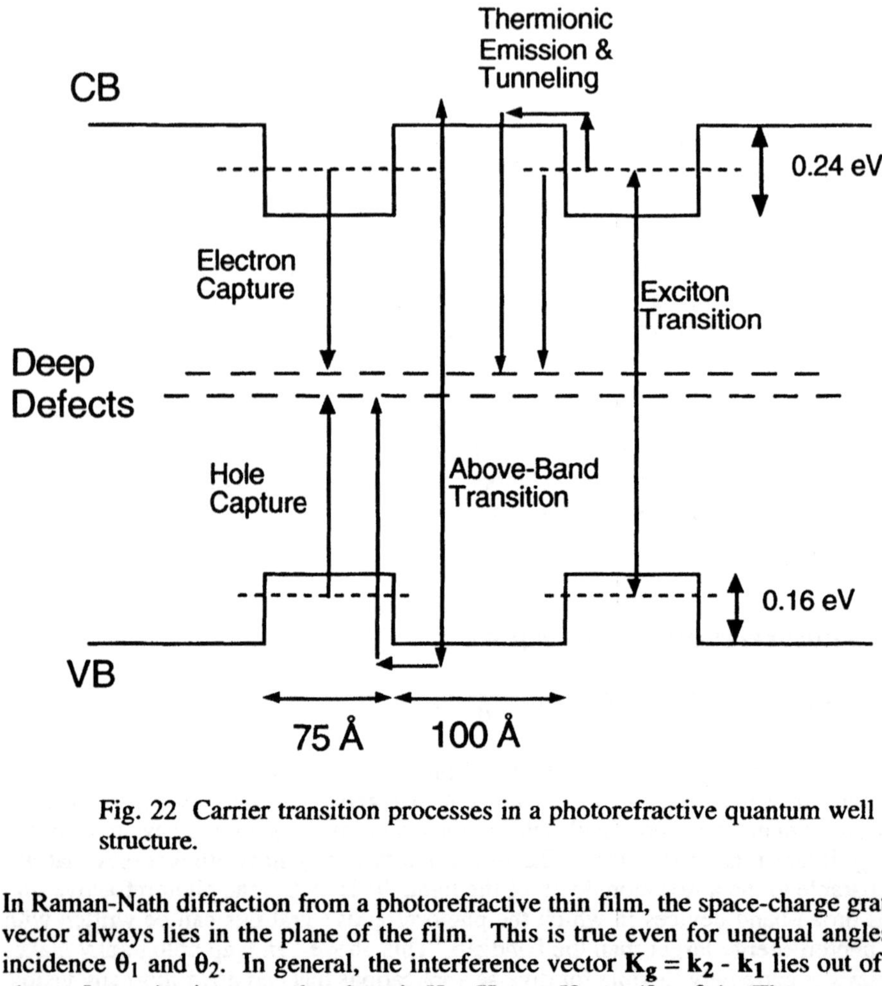

Fig. 22 Carrier transition processes in a photorefractive quantum well structure.

In Raman-Nath diffraction from a photorefractive thin film, the space-charge grating vector always lies in the plane of the film. This is true even for unequal angles of incidence θ_1 and θ_2. In general, the interference vector $\mathbf{K_g} = \mathbf{k_2} - \mathbf{k_1}$ lies out of the plane. Its projection onto the plane is $K = \mathbf{K_g}\,\mathbf{x} = K_g \cos(\theta_1 - \theta_2)$. The component of the grating vector in the plane is $K = 2\pi/\Lambda$, where the fringe spacing Λ is

$$\Lambda = \frac{\lambda}{\left|\sin\theta_1 - \sin\theta_2\right|} \tag{46}$$

This fringe spacing ensures perfect phase matching in Raman-Nath diffraction for two-wave mixing. Because the transport (carrier drift in the applied field) is parallel to the grating vector K, the photorefractive transport properties of the transverse-field geometry can be fully described in terms of one-dimensional transport equations.

The transport equations, continuity equation and Gauss' law for the transverse-field geometry are

$$\frac{\partial n}{\partial t} - \nabla \frac{j_e}{e} = I\alpha + \beta_e N_D^0 - \sigma_e n v_e N_D^+ - \gamma_{eh} np \tag{47}$$

$$\frac{\partial p}{\partial t} + \nabla \frac{j_h}{e} = I\alpha + \beta_h N_D^+ - \sigma_h p v_h N_D^0 - \gamma_{eh} np \tag{48}$$

$$j_e = e \mu_e n E + k_B T \mu_e \nabla n \tag{49}$$

$$j_h = e \mu_h p E + k_B T \mu_h \nabla p \tag{50}$$

$$\frac{\partial}{\partial t}\left(n + N_A^- - p - N_D^+\right) = \frac{\nabla}{e}\left(j_e + j_h\right) \tag{51}$$

$$\nabla E = -\left(\frac{e}{\varepsilon \varepsilon_0}\right)\left(n + N_A^- - p - N_D^+\right) \tag{52}$$

where v_e and v_h are the electron and hole thermal velocities. The first two equations are the generation-recombination equations. The third and fourth equations define the current densities. The fifth equation is the expression of charge conservation, and the sixth equation is Gauss's law. In these equations, e and h refer to electron properties and hole properties, respectively, where α is the absorption coefficient for electron-hole pair production, I is the laser intensity, j is the current density, β is the carrier thermal emission rate, σ is the defect capture cross-section, and γ_{eh} is the electron-hole recombination cross-section. The free carrier concentrations are n and p. The photocarrier generation rate $I\alpha$ is common to both the electron and hole populations, as is the electron-hole recombination rate $\gamma_{eh}np$. The recombination cross-section γ_{eh} includes radiative recombination as well as non-radiative recombination through recombination defects. In the one-dimensional approximation, eqs.(47) to (52) are equivalent to the photorefractive transport equations for resonant excitation of bulk materials [57] and Chapter 6 of this volume.

For the sake of illustration, we assume that deep donors N_D compensate shallow acceptors N_{sa}, and the number of donors is

$$N_D = N_D{}^0 + N_D{}^+ \tag{53}$$

This choice is opposite to the choice of dominant defect in Ref. [58], but does not lead to significantly different behavior. However, the photorefractive behavior of photorefractive quantum wells does depend on the dominant photocarrier. Although electrons and holes are generated in equal numbers, they have different capture rates. Therefore the steady-state concentration of one carrier is likely to dominate over the other. Although the identity of the defects are important for determining the capture rates, deep acceptors or deep donors can each have large capture cross sections for either electrons or holes. Therefore, holes can dominate the photoconductivity whether the dominant deep defect is an acceptor or a donor. In this sense, the identity of the deep defect is not relevant to the photorefractive properties of the quantum well. In our simulations here, we will assume capture cross sections for the deep acceptor that cause holes to be the dominant photocarrier.

The zero-order solutions for n_0 and p_0 in eqs.(47) to (52) are solved by assuming homogeneous illumination I_0. These solutions are

$$n_0 = \frac{I_0 \alpha + (I_0 s_e + \beta_e)(N_{sd} - n_0 + p_0)}{\gamma_e (N_A - N_{sd} + n_0 - p_0) + \gamma_{eh} p_0} \tag{54}$$

$$p_0 = \frac{I_0 \alpha + (I_0 s_h + \beta_h)(N_A - N_{sd} + n_0 - p_0)}{\gamma_h (N_{sd} - n_0 + p_0) + \gamma_{eh} n_0} \tag{55}$$

where α is the absorption coefficient for generation of electron-hole pairs. The values for α are in the range of 10^4 cm^{-1}, compared to typical absorption coefficients of ~1 cm^{-1} for deep defects. The electron-hole pair production rate is therefore much larger than photoionization rates from the deep-level defects, and to lowest-order the carrier concentrations are

$$n_0 = \frac{I_0 \alpha}{\gamma_e (N_A - N_{sd})} = I_0 \alpha \tau_e$$

$$p_0 = \frac{I_0 \alpha}{\gamma_h N_{sd}} = I_0 \alpha \tau_h \tag{56}$$

where $\tau_{e,h}$ are the electron and hole lifetimes. A laser intensity of 1 mW/cm^2 will typically generate free carrier concentrations in the range of 10^{10} to 10^{12} cm^{-3}, which are typically much smaller than N_A or N_{sd}.

The first-order solutions of eqs.(47) to (52) are obtained in the small modulation amplitude approximation. The parameters can be expressed as $n = n_0 + n_1$, $p = p_0 + p_1$ and $E = E_0 + E_1$, where E_0 is the applied electric field. The first-order terms are assumed to vary as exp(iKx), under steady-state excitation. The solution is expressed in terms of the hole transition rates in Table 6.1.

Table 6.1 Hole Transition Rates

Transition rate:	$\Gamma_{Ih} = s_h I_0 + \beta_h + \gamma_h p_0$
Ion Recombination rate:	$\Gamma_{Rh} = \gamma_h(N_A + n_0 - p_0)$
Drift rate:	$\Gamma_{Eh} = K \mu_h E_0$
Diffusion rate:	$\Gamma_{Dh} = K^2 k_B T \mu_h/e$
Direct Recombination rate:	$\Gamma_{ehh} = \gamma_{eh} n_0$
Dielectric relation rate:	$\Gamma_{dih} = e \mu_h p_0/\varepsilon \varepsilon_0$

The corresponding transition rates for electrons need special attention. Electrons in GaAs experience significant deviations from linear transport. Under the high fields of 10 kV/cm typical of the transverse-field geometry, the electrons are accelerated to sufficient energies that they are transferred from the high-mobility direct gap to the low-mobility indirect gap. This leads to velocity saturation for fields larger than 3 kV/cm. Electron transport nonlinearity occurs in several, but not all semiconductors. Therefore to be most general, it is important to include notation that can be used for all cases of linear and nonlinear transport. A field-dependent electron mobility can be defined as the ratio of the drift velocity to the electric field

$$\mu(E) = \frac{v(E)}{E} \tag{57}$$

For this definition, velocity saturation v(E) -> const produces a finite mobility. On the other hand, the differential mobility dv/dE vanishes for velocity saturation. These features can be included in the transition rates for the electrons:

Table 6.2 Electron Transition Rates

Transition rate:	$\Gamma_{Ie} = s_e I_0 + \beta_e + \gamma_e n_0$
Ion Recombination rate:	$\Gamma_{Re} = \gamma_e(N_A - N_D^+ + n_0 - p_0)$
Drift rate:	$\Gamma_{Ee} = K v(E)$
Diffusion rate:	$\Gamma_{De} = K^2 k_B T \mu_e/e$
Direct Recombination rate:	$\Gamma_{eeh} = \gamma_{eh} p_0$
Dielectric relation rate:	$\Gamma_{die} = F(E) e n_0/\varepsilon \varepsilon_0$

where F(E) includes all the effects of the transport nonlinearity, which will be described in section 6.1.3.

The steady-state photorefractive rate equations in the steady-state small-signal analysis become

$$\begin{bmatrix} -i\Gamma_{Ee}+\Gamma_{De}+\Gamma_{Ie} \\ \quad +\Gamma_{Re}+\Gamma_{eeh} & -\Gamma_{Ie}+\Gamma_{ehh} & -\Gamma_{die}+\Gamma_{Ie} \\ -\Gamma_{Ih}+\Gamma_{eeh} & \begin{matrix} i\Gamma_{Eh}+\Gamma_{Dh}+\Gamma_{Ih} \\ +\Gamma_{Rh}+\Gamma_{ehh} \end{matrix} & \Gamma_{dih}-\Gamma_{Ih} \\ \Gamma_{Ie}+\Gamma_{Re}+\Gamma_{Ih} & -\Gamma_{Ie}-\Gamma_{Ih}-\Gamma_{Rh} & \Gamma_{Ie}+\Gamma_{Ih} \end{bmatrix} \begin{bmatrix} n_1 \\ p_1 \\ N_1 \end{bmatrix}$$

$$= I_1 \begin{bmatrix} \alpha + s_e\left(N_D^+ + p_0 - n_0\right) \\ \alpha + s_h\left(N_A - N_D^+ - p_0 + n_0\right) \\ s_e\left(N_D^+ + p_0 - n_0\right) - s_h\left(N_A - N_D^+ - p_0 + n_0\right) \end{bmatrix} \qquad (58)$$

where $N_1 = \varepsilon_0\varepsilon iKE_1/e$ has units of number density and is the first Fourier coefficient of the space-charge density. In our simulations we use the parameters given in Table 6.3

Table 6.3 Simulation Parameters for PRQW

$s_e = 1\text{x}10^{-17}\ \text{cm}^2$
$s_h = 1\text{x}10^{-16}\ \text{cm}^2$
$\sigma_e = 1\text{x}10^{-14}\ \text{cm}^2$
$\sigma_h = 5\text{x}10^{-15}\ \text{cm}^2$
$\gamma_{eh} = 1\text{x}10^{-10}\ \text{s}^{-1}\ \text{cm}^{-3}$
$\mu_e = 6000\ \text{cm}^2/\text{Vs}$
$\mu_h = 400\ \text{cm}^2/\text{Vs}$

6.1.2 Linear Transport

The solutions of E_1 from eq.(58) are shown in Fig. 23 as functions of fringe spacing for two different defect concentrations. The defect densities bracket the range of defect densities caused by a proton implant dose of $[H^+] = 1x10^{12}$ cm^{-2}. These are N_A = $2x10^{17}$ cm^{-3} and $N_A = 2x10^{16}$ cm^{-3} with a compensation ratio $r = N_D/N_A = 0.5$ and other parameters given in Table. 6.3. These parameters are chosen to be representative, but are not intended to be the exact values of the photorefractive quantum wells. A characteristic fringe spacing is observed in the figure, which is strongly dependent on defect concentration. Defect densities have a direct impact on transport lengths because of the dependence of carrier lifetime on defect density. Lower defect densities lead to larger characteristic lengths Λ_c because the carrier lifetime increases. This characteristic length defines the resolution of a photorefractive device, and is therefore an important feature of photorefractive quantum wells.

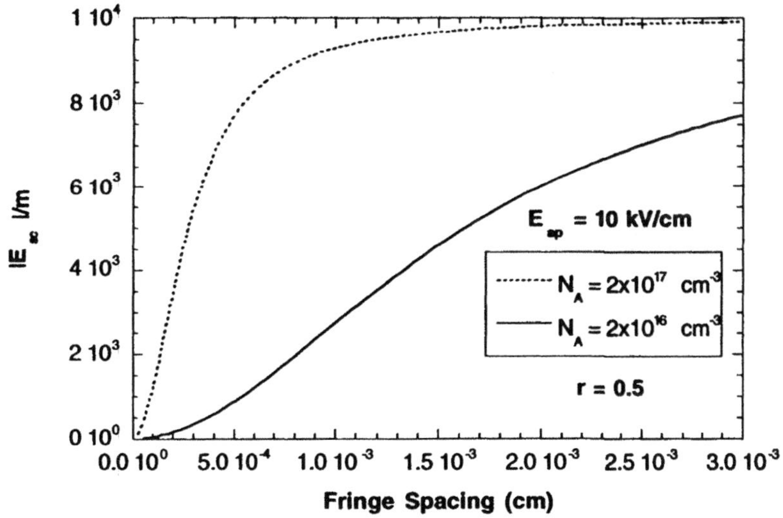

Fig. 23 Magnitude of the space-charge field for two defect densities that represent the range of defect densities for a proton dose of 10^{12} cm^{-2}.

The origin of the characteristic length can be understood by considering the transition rates in eq.(58). Several of the rates are negligibly small, such as the transition rates Γ_{Ie}, Γ_{Ih}, the dielectric relaxation rates Γ_{die}, Γ_{dih} and the direct recombination rates Γ_{eeh}, Γ_{ehh}. These transition rates are small because of the large absorption across the bandgap.

By keeping only the largest terms in the solution to eq.(58), the solution for the space-charge density is

$$N_1 = -I_0 \alpha \left[\frac{-\Gamma_{Rh}\left(-i\Gamma_{Ee} + \Gamma_{De}\right) + \Gamma_{Re}\left(i\Gamma_{Eh} + \Gamma_{Dh}\right)}{\Gamma_{dih}\Gamma_{Rh}\left(-i\Gamma_{Ee} + \Gamma_{De} + \Gamma_{Re}\right) + \Gamma_{die}\Gamma_{Re}\left(i\Gamma_{Eh} + \Gamma_{Dh} + \Gamma_{Rh}\right)} \right] \tag{59}$$

This is expressed in terms of the space-charge field as

$$E_1 = -E_0 \left[\frac{K L_{Ee} + iK^2 L_{De}^2 + K L_{Eh} + iK^2 L_{Dh}^2}{L_{Eh}\left(K^2 L_{De}^2 + 1\right) + L_{Ee}\left(K^2 L_{Dh}^2 + 1\right)} \right] \tag{60}$$

where the drift and diffusion lengths are given by

$$L_E = \mu \tau E$$

$$L^2{}_D = \mu \tau k_B T/e \tag{61}$$

The expression for the space-charge field can be written directly in terms of the applied field E_0 and the diffusion field E_D as

$$E_1 = -\frac{1}{\mu_h \tau_h \mu_e \tau_e K} \left[\frac{\mu_e \tau_e \left(E_0 + iE_D\right) + \mu_h \tau_h \left(E_0 - iE_D\right)}{2E_D + \frac{1}{\mu_e \tau_e K} + \frac{1}{\mu_h \tau_h K}} \right] \tag{62}$$

where $E_D = k_B TK/e$. In eq. (62), the role of the $\mu\tau$ product for electrons and holes can be seen explicitly. The most important feature of eq.(62) is the diffusion field E_D in the denominator. In spite of the fact that $E_0 >> E_D$, diffusion plays a crucial role in limiting space-charge fields under resonant excitation.

The characteristic fringe spacing observed in Fig. 23 is given by

$$\Lambda_C = 2\pi \left\langle L_D \right\rangle \tag{63}$$

where the bipolar diffusion length is

$$\langle L_D \rangle^2 = \frac{k_B T}{e} \langle \mu\tau \rangle \tag{64}$$

with the bipolar $\mu\tau$ product defined by

$$\langle \mu\tau \rangle = \frac{1}{\frac{1}{2}\left(\frac{1}{\mu_e \tau_e} + \frac{1}{\mu_h \tau_h}\right)} \tag{65}$$

The resolution of the photorefractive quantum wells operating in the transverse Franz-Keldysh geometry is therefore limited by the bipolar diffusion length. This limitation is easily addressed by using higher proton implant doses to reduce the diffusion length and to achieve better device resolution, or to use low-temperature-growth GaAs with ultra-fast carrier lifetimes.

6.1.3 Nonlinear Transport

In spite of the almost universal use of linear velocity-field relations in photorefractive simulations, in practice almost all materials have nonlinear transport under high fields. Nonlinear transport can have a profound effect on photorefractive effects, especially on the photorefractive phase shift.

In semiconductors, such as GaAs and InP, the electrons have a strong nonlinear velocity vs. field behavior. The nonlinear transport is caused by electron heating by large electric fields. Under high fields, the electron temperature T_e can be significantly higher than the lattice temperature T_L because the electrons gain kinetic energy through acceleration in the electric field. The electron temperature and the drift velocity v are related to the energy relaxation time τ_r by balancing the Joule heating of the electron gas with energy relaxation

$$J \cdot E = eE\, v(E) = \frac{3}{2} k_B \left(\frac{T_e - T_L}{\tau_r}\right) \tag{66}$$

The electron temperature is therefore

$$T_e = T_L + \frac{2 e \tau_r v(E)}{3 k_B} E \tag{67}$$

where v(E) is the electric-field-dependent drift velocity. Under electric fields of 10 kV/cm, the electron temperature can exceed the lattice temperature by several hundred degrees centigrade.

All semiconductors have this property of hot electrons at high fields, however a few also have upper conduction bands that are relatively close in energy to the direct conduction band minimum. Therefore electrons heated by the electric field can be promoted into the indirect conduction band minima. Electrons in these minima have significantly lower mobilities, leading to a strongly nonlinear velocity vs. field dependence. The velocity is therefore a weighted average over the electron density in the lower valley n_1 with high mobility μ_1 and the upper valleys n_2 with low mobilities μ_2

$$v(E) = \left(\frac{\mu_1 n_1 + \mu_2 n_2}{n_1 + n_2} \right) E \approx \frac{\mu_1 E}{1 + n_1 / n_2} \tag{68}$$

where the approximation is made because $\mu_1 >> \mu_2$. The ratio of the electron populations is given by a Boltzman factor

$$\frac{n_1}{n_2} = R \exp\left(-\frac{\Delta U}{k_B T_e} \right)$$

where ΔU is the energy difference between the lower and upper band minima, R is the ratio of band density of states, and the electron temperature T_e is included explicitly.

The electron temperature therefore becomes

$$T_e = T_L + \frac{2 e \tau_r \mu_1}{3 k_B} \frac{E^2}{\left[1 + R \exp\left(-\Delta U / k_B T_e \right) \right]} \tag{69}$$

with a corresponding nonlinear velocity

$$v(E) = \frac{\mu_1 E}{\left[1 + R \exp\left(-\Delta U / k_B T_e \right) \right]} \tag{70}$$

For GaAs, $R \approx 94$ and $\Delta U \approx 0.31$ eV. Using these parameters, the electron temperature is shown in Fig. 24 as a function of electric field, and the electron velocity is shown in Fig. 25.

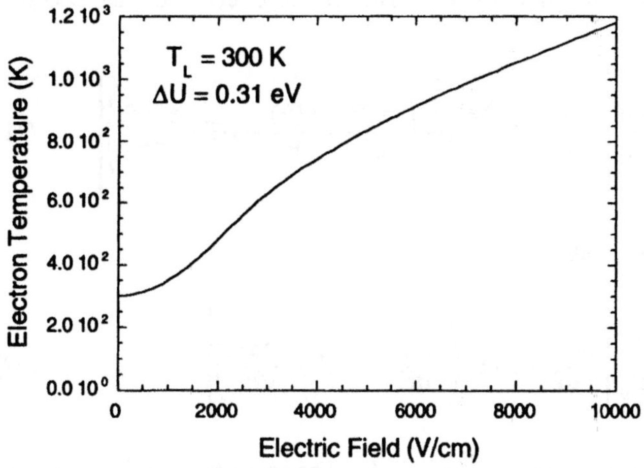

Fig. 24 Electron temperature simulated as a function of applied electric field.

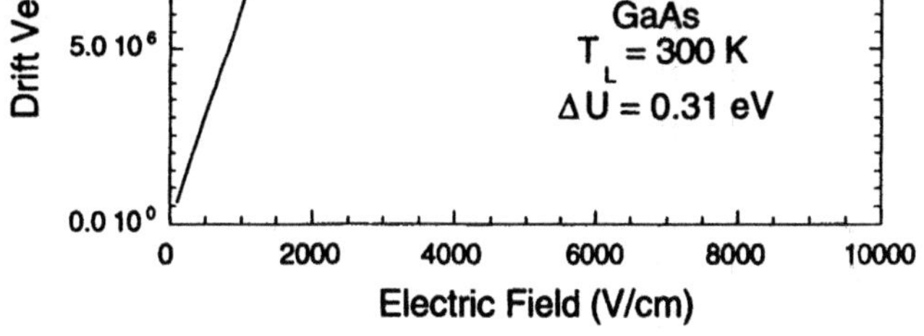

Fig. 25 Electron velocity simulated as a function of applied field showing velocity saturation near 4 kV/cm.

In the interband photorefractive effect, the transport nonlinearity is incorporated directly into the transport rates, without further modifications of eq.(58). In all the rates in Table 6.2, except for the dielectric relaxation rate, the low-field linear electron mobility is replaced by $\mu(E) = \frac{v(E)}{E}$. The dielectric relaxation rate becomes

$$\Gamma_{die} = \frac{en_0}{\varepsilon\varepsilon_0}\left[\frac{dv(E)}{dE} + \frac{iKk_BT_e}{eE}\left(\frac{dv(E)}{dE} - \frac{v(E)}{E} + v(E)\frac{d\ln T_e}{dE}\right)\right] \tag{71}$$

which can also be written as

$$\Gamma_{die} = \frac{en_0}{\varepsilon\varepsilon_0}\left[\frac{dv(E)}{dE} + i\,E_D\left(\frac{d\mu(E)}{dE} + \mu(E)\frac{d\ln T_e}{dE}\right)\right] \tag{72}$$

where the diffusion field is $E_D = Kk_BT_e/e$. The important nonlinear contribution for high fields in either eq. (71) or (72) is the differential mobility dv/dE. When the velocity saturates, the differential mobility vanishes. If the diffusion term in eq.(72) is small, then the electron dielectric relaxation nearly vanishes: Velocity saturation disables dielectric relaxation while still permitting large drift rates! This feature of transport nonlinearity has important implications for the photorefractive effect. If the charge cannot relax, then the space-charge is driven to saturation, with a corresponding large phase shift.

The solution to the photorefractive transport equations that replaces eq.(60) is

$$E_1 = -E_0\left[\frac{L_{Ee} + L_{Eh} + iK(L_{De}^2 - L_{Dh}^2)}{L_{Eh}(K^2L_{De}^2 + 1) + L_{Ee}^{nonlin}(K^2L_{Dh}^2 + 1) + iL_{Eh}K\left(L_{Ee}^{nonlin} - L_{Ee}\right)}\right] \tag{73}$$

where

$$L_{Ee} = v(E)\,\tau_e \tag{74}$$

is the electron drift length with the field-dependent drift velocity, and the nonlinear contribution presents an additional nonlinear transport length

$$L_{Ee}^{nonlin} = \left(\frac{dv(E)}{dE} + i\frac{d\mu(E)}{dE}E_D\right)E\,\tau_e \tag{75}$$

The nonlinear drift length of the electrons is a complex quantity, and may not have an immediate physical interpretation.

A new mechanism for the photorefractive phase shift is immediately apparent in eq.(73). If the transport is linear, then the imaginary term in the denominator vanishes and there is no phase shift for fields below the saturation field E_{max}. However, when transport nonlinearity is strong, then the imaginary term can become large. If it becomes comparable to or larger than the real terms in the denominator, then a large photorefractive phase shift results. The phase shift as a function of field is shown in Fig.26 using the parameters in Table 6.3.

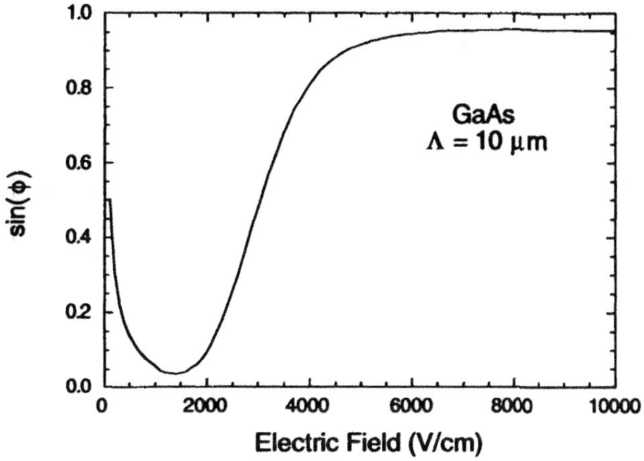

Fig. 26 Phase shift simulated as a function of applied field including the hot electron transport nonlinearity in GaAs.

The response time of the photorefractive quantum well is also improved by the nonlinear transport. The response time as a function of applied electric field is shown in Fig. 27 for the parameters in Table 6.3, comparing the linear transport results with the nonlinear transport. Increasing electric field slows down the device response time because larger fields take longer times to build up, and because the carriers overshoot the dark fringes. In the case of hot electron transport, the transport nonlinearity limits the fringe overshoot, leading to a faster response. The zero-field limit is equal to the dielectric relaxation time of 370 nsec for an illumination of 1 mW/cm^2. The device response time slows to several tens of microseconds under the large operating fields.

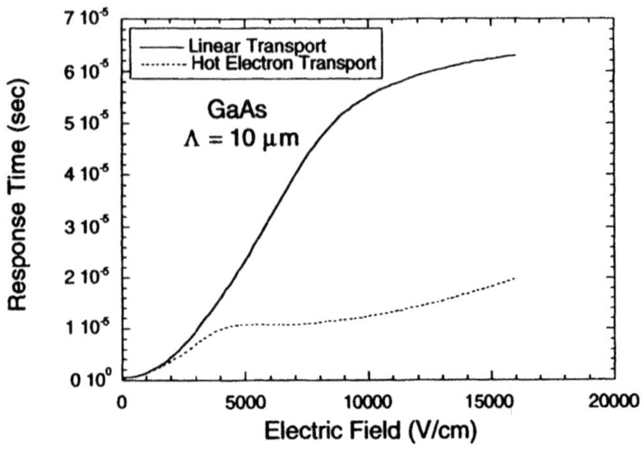

Fig. 27 Photorefractive response time comparing linear and nonlinear transport, simulated as a function of applied electric field for an intensity of 1 mW/cm^2. The zero-field limit is 370 nsec.

6.2 Transverse-Field Sample Growth and Fabrication

The transverse-field geometry is the simplest photorefractive quantum well geometry to fabricate. It requires no buffer layers or doped contacts, and electrical contacts can be applied after the sample has been fully processed.

A typical layer design is shown in Fig. 28. The growth begins with a removal layer. In the AlGaAs/GaAs quantum well system, the excitonic transition energies are larger than the bandgap of the GaAs substrate. Therefore the substrate is opaque and must be removed for transmission studies. Two techniques are used to remove the substrate: a substrate etch with a stop-etch layer or a lift-layer. Devices can be grown with both options, as in Fig. 28. The stop etch layer is typically 4000 Å of $Al_{0.5}Ga_{0.5}As$. The lift-layer is typically 200 Å of AlAs. Following the removal layer, the multiple quantum well layers are grown. In the figure, there are 75 Å quantum wells with 100 Å quantum barriers. The aluminum fraction in the barriers is 10 %. The multiple quantum well layer is capped with a thin 50 Å layer of GaAs to protect the surface after growth.

After growth, the wafer is implanted with protons at a dose of 10^{12} cm^{-2} and an implant energy between 140 – 200 keV. The wafer is then cleaved. When the substrate is removed by etching, the sample is first epoxied to a glass slide with the epi-layer down, and the GaAs substrate is lapped to approximately 50 microns. An etch of dilute NH_4OH is used to etch the remaining substrate to the AlGaAs stop-etch layer. The stop-etch can be removed by a dilute hydrofluoric flash etch. When the epi-layer is removed by lift-off, the sample is coated with black wax on the epi side and placed in a dilute hydrofluoric etch. After the epi-layer has lifted away from the substrate, it is Van der Waals bonded to a glass slide. For more details of the lift process see Refs.[59, 60].

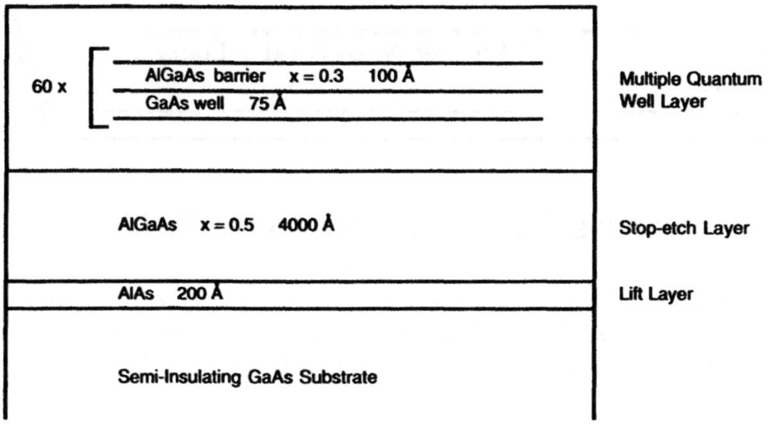

Fig. 28 Epitaxy structure for a transverse-field photorefractive quantum well device.

The final transverse-field device is shown in Fig. 29. The contacts can be gold, gold-germanium, gold-zinc or indium deposited by evaporation. Gold contacts are the simplest to fabricate, but diffused indium contacts give the best ohmic contacts to semi-insulating semiconductors. The electrical contacts define a window, typically with a 1 mm electrode spacing, through which the laser beams are steered during mixing experiments. Voltages up to 1500 Volts are applied across the window, producing electric field strengths up to 15 kV/cm.

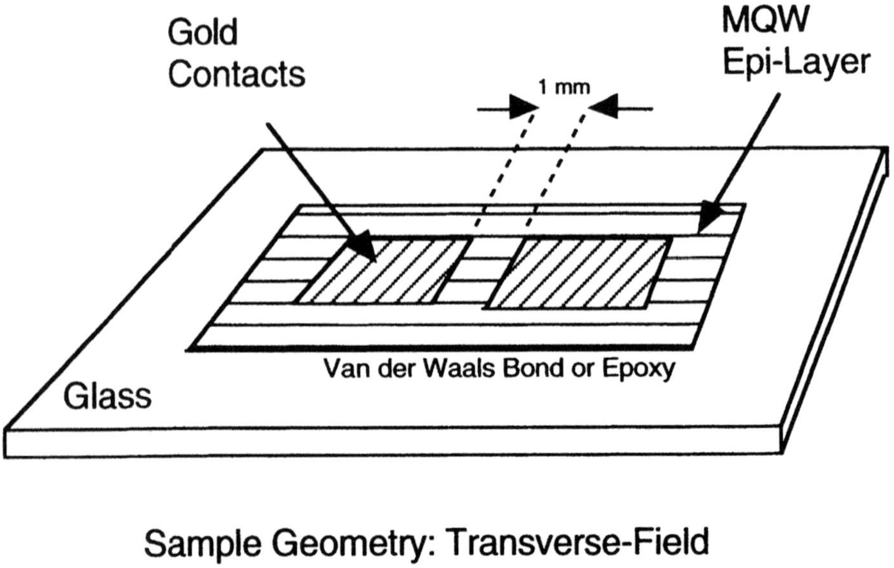

Fig. 29 Transverse-field transmission geometry photorefractive quantum well device. The epilayer has been removed from the growth substrate and has been fastened to a transparent glass slide.

6.3 Transverse-Field Two-Wave Mixing

6.3.1 Photorefractive Gain

The two-wave mixing experimental setup is shown in Fig. 30. The semi-insulating multiple quantum well film (SIMQW) is biased by a DC power supply that can reverse its polarity. The pump beam I_2 is chopped by a mechanical chopper and the signal beam is detected by a silicon photodiode and lock-in amplifier. The measured two-wave mixing gain, from eq.(21), is

$$\gamma = \frac{I_1(L)}{I_1(0)\, e^{-\alpha(E)L}} = 1 + \frac{\beta}{1+\beta}\left[\frac{4\pi\, n_m\, L}{\lambda \cos\theta'} \sin\phi(E) + \frac{\alpha_m\, L}{\cos\theta'} \cos\phi(E)\right] \tag{76}$$

where the electric-field dependence is included explicitly. Two features of the transverse-field geometry that make the interpretation of two-wave mixing particularly straightforward are the quadratic electro-optic effect and the change in sign of the photorefractive phase shift when the direction of the applied electric field is changed. When the electric field changes sign, the phase shift changes sign, but the refractive index grating keeps its sign because of the quadratic electro-optic effect. Therefore, simply by changing the sign of the electric field, it is possible to separate the effect of the absorption grating amplitude α_m from the refractive index amplitude n_m.

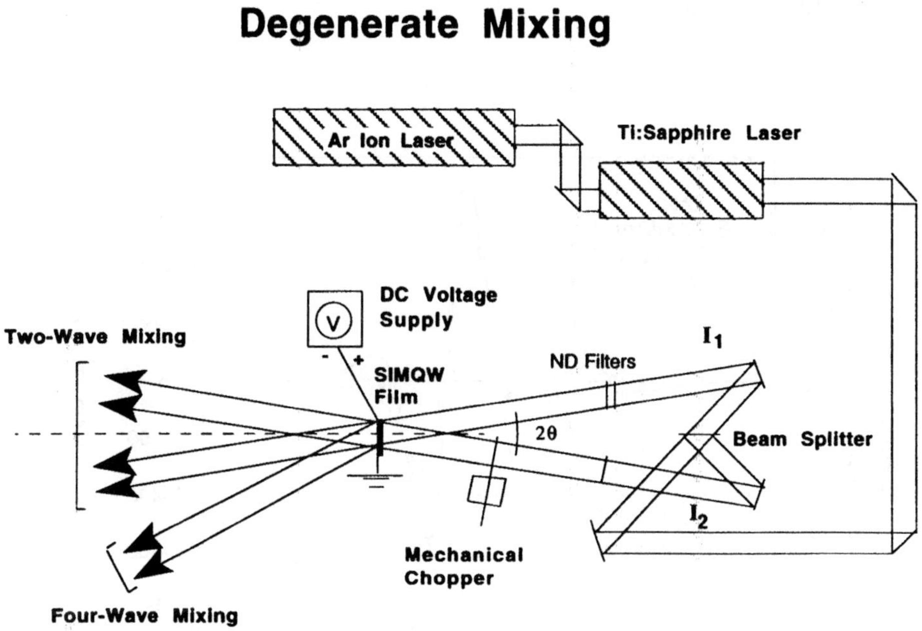

Fig. 30 Two-wave and four-wave mixing experimental geometry. Only one of the diffracted beams is shown. In thin films, the two-wave and four-wave mixing efficiencies can be measured simultaneously.

The photorefractive gain and refractive index are measured experimentally by taking the difference between the positive and negative field directions

$$\begin{aligned}\gamma_n &= \frac{\gamma(+E) - \gamma(-E)}{2} \\ &= \frac{\beta}{1+\beta} \frac{4\pi\, n_m L}{\lambda \cos\theta'} \sin\phi \\ &= \frac{\beta}{1+\beta} \Gamma L\end{aligned} \tag{77}$$

where

$$\Gamma = \frac{4\pi\, n_m}{\lambda \cos\theta'} \sin\phi \tag{78}$$

is the photorefractive gain. The photorefractive gain for the transverse geometry is shown in Fig. 31 compared with the refractive index derived from the electroabsorption from the same sample. The agreement is good to within a multiplicative constant.

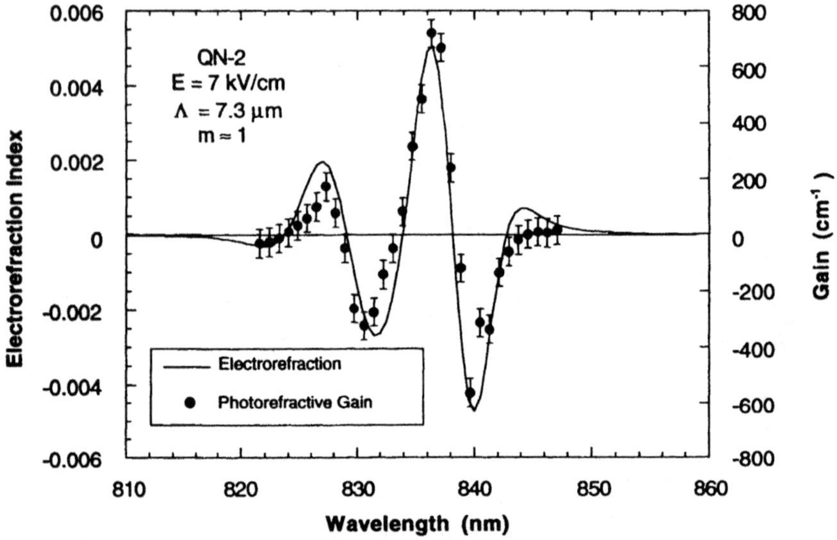

Fig. 31 Experimental photorefractive gain for the transverse-field geometry compared with the electrorefraction derived by Kramers-Kronig analysis of the experimental electroabsorption.

The absorption grating, measured by the same technique, has a subtle complication. In the definition of eq.(21) the signal intensity with mixing is compared with the signal in the absence of mixing. However, the absorption $\alpha(E)$ that appears in the exponential term in the denominator is a function of the modulation technique. For instance, this average absorption changes, even for constant applied electric field, depending on whether the grating is present or not. Therefore in the lock-in detection experiment with a chopper or a modulated DC electric field, part of the modulation appears in this average absorption and is detected as cross-beam modulation. Fortunately, this contribution is symmetric with electric field, so only contributes to the effective electroabsorption and does not influence eq.(77) for the refractive index grating. Details of the effective electroabsorption grating can be found in Refs.[7, 58].

6.3.2 *High-Sensitivity Optical Nonlinearity*

The photorefractive quantum-well structures operate with ultra-low light intensities. For example, the nonlinear optical response of free-carrier gratings in quantum wells require kW/cm^2 to saturate the absorption at room temperature[61]. In contrast, a light intensity in Fig. 32 of 10 $\mu W/cm^2$ (comparable to that found in a dimly-lit room) generates a large refractive index change in the photorefractive thin films. Such low power requirements for these thin films brings them fully into the realm of practicality for marketable optical processing devices. The ultra-low light intensity does reduce the grating development speed. However, response rates greater than 10 kHz are typical for the photorefractive quantum wells, and are acceptable for image processing applications.

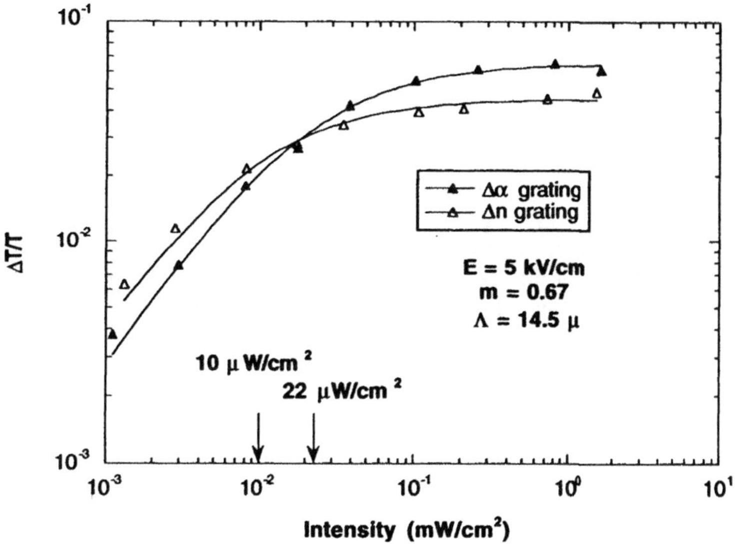

Fig. 32 Intensity dependence of the two-wave mixing signal in the transverse-field geometry. The ultra-low saturation intensity of 10 $\mu W/cm^2$ is the result of large absorption and long lifetimes, both of which can be engineered in the photorefractive quantum wells.

The saturation intensity I_{sat} is the intensity for which the photoconductivity exceeds the dark conductivity, i.e. the gratings are fully saturated. The saturation intensity for the quantum wells is given by

$$I_{sat} = n_d/\alpha\tau \tag{79}$$

where n_d is the dark carrier density, α is the absorption coefficient and τ is the carrier lifetime. The ideal of ultra-low-intensity nonlinear optics is achieved through small n_d, large absorption α and long carrier lifetimes τ. The saturation intensity can be controlled by engineering each of the parameters in eq. (79). The radiation-damage from proton-implants or Cr substitutional impurities introduced during growth can be adjusted to control the carrier lifetimes. The defect densities control the position of the Fermi level, which affects the value of n_d. The magnitude of the bandgap also determines n_d, and can be controlled by adjusting the composition of the wells and barriers. The absorption and electroabsorption similarly can be maximized by adjusting the width of the quantum wells and barriers. By engineering the bandgap and defects, the saturation intensity can be varied by several orders of magnitude.

The nonlinear sensitivity S is expressed as the index change per incident photon per area, given by

$$S = \frac{\Delta n}{I_{sat}} = \frac{\Delta n\,\alpha\tau}{n_d} \tag{80}$$

using eq. (79). Light intensities of approximately 10 to 100 $\mu W/cm^2$ fully develop (saturate) the photorefractive gratings with refractive index changes in the range $\Delta n = 10^{-2}$ to 10^{-1}. This leads to a sensitivity of $S = 10^3\ cm^2/W$, which is comparable with the sensitivity of SEED devices [62]. This is a consequence of the common underlying physical origin that photorefractive quantum wells and SEED devices share. However, the photorefractive devices require no artificial reticulation of the surface into discrete pixels: The holography is the photolithography!

The speed of the photorefractive quantum wells depends on several features of electronic transport. Each of the three geometries has different response times that are functions of the geometry of the structures and of the fringe spacing Λ. However, the dielectric relaxation time is the common limiting response time for each structure. The grating response time is given by

$$\tau_R = \frac{\varepsilon\varepsilon_0}{(I\alpha\tau + n_d)e\mu} f(K, E) \tag{81}$$

where I is the photon flux density, ε is the dielectric constant, E is the applied electric field, K is the grating spatial frequency and μ is the bipolar mobility that is highly anisotropic in the quantum well structures. The response time at the saturation intensity for weak fields is approximately 1 kHz. The device response slows down at high fields because the carriers overdrift the dark fringes, which decreases the writing efficiency. The field dependence of the response time in eq.(81) is [63]

$$f(K_g, E) = 1 + \left(K_g L_E\right)^2 \tag{82}$$

where $L_E = \mu\tau E$ is the bipolar drift length, which can also be tailored by controlling defect densities. In the limit $\mu\tau E << d$, where d represents a fringe spacing or a device thickness in the respective geometries, the function $f(K, E) = 1$.

The response time decreases inversely with the carrier mobility. The vertical mobility $\mu_\perp$ can be controlled in the longitudinal geometries by engineering the vertical transport through the quantum wells [64, 65]. In the transverse geometry, the transverse mobility $\mu_\parallel$ is primarily controlled by ionized impurity scattering. The response time for large fringe spacing Λ at the saturation intensity I_{sat} is $\tau_R = 1$ msec. For a larger intensity of 10 mW/cm^2, the response time is $\tau_R = 1$ µsec.

An alternate sensitivity S' is defined by the index change per writing energy, where the writing energy is given by $E_{wr} = I\tau_R$. This sensitivity is a combined figure-of-merit that multiplies the sensitivity S by the response speed of the device. The sensitivity is

$$S' = \frac{\Delta n}{I\,\tau_R} = \frac{\Delta n\,\alpha\,e\,\mu\tau}{\varepsilon\,\varepsilon_0} \tag{83}$$

where eq.(82) has been used in the appropriate limit where $f(K_g, E) = 1$. A typical writing energy of $E_{wr} = 10$ nJ/cm^2 gives $S' \approx 10^6$ cm^2/J, which is again comparable to values for SEED devices.

6.4 Transverse-Field Four-Wave Mixing

6.4.1 Degenerate Mixing

One of the benefits of Raman-Nath diffraction in thin films is simultaneous two-wave and four-wave mixing, without requiring Bragg matching of pump or probe beams. In degenerate four-wave mixing, the grating writing beams diffract into multiple overlapping orders, shown in Fig. 3. Details of the degenerate four-wave mixing experimental geometry and the multiple diffraction orders are given in Refs. [6, 58].

A key figure-of-merit for holographic devices is the spatial resolution, i.e. the minimum fringe spacing that the material can support. The spatial resolution is important for applications, not simply for more information density, but because many holographic applications operate with the holographic medium in a Fourier plane. For instance, the Fourier transform of 1 cm^2 contains spatial frequency information down to one or two microns. In the photorefractive quantum wells, the spatial resolution is determined by transport lengths, and is not typically limited by trap concentration. In the transverse-field geometry, for example, the spatial resolution is limited by eq.(63) for linear transport, and by eq.(75) for nonlinear transport.

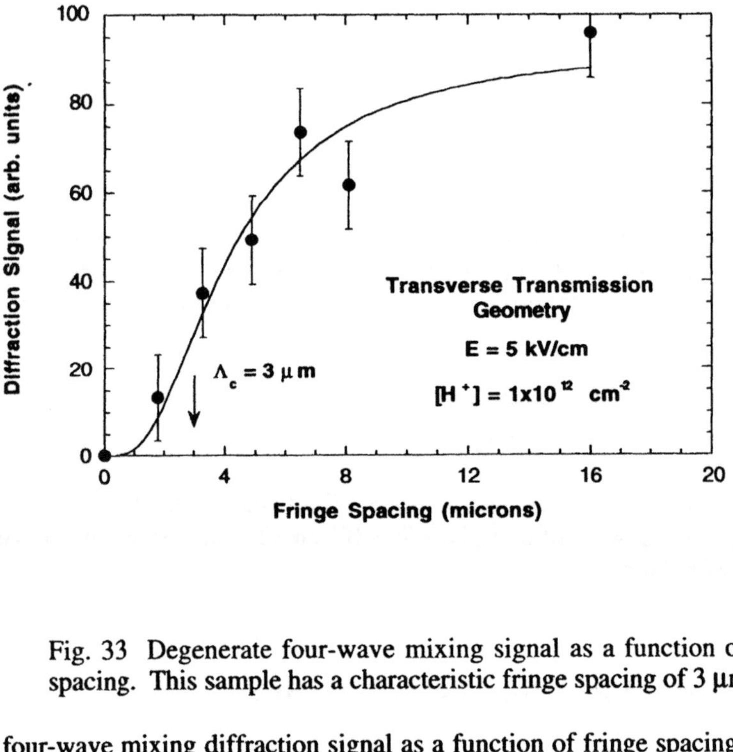

Fig. 33 Degenerate four-wave mixing signal as a function of fringe spacing. This sample has a characteristic fringe spacing of 3 μm.

The four-wave mixing diffraction signal as a function of fringe spacing is shown in Fig. 33 for a sample implanted with a proton dose of $[H^+] = 1\text{x}10^{12}$ cm^{-2} at an energy of 140 keV. The functional dependence is approximately

$$\eta(K) \propto \left[\frac{1}{1+\left(\Lambda_c/\Lambda\right)^2}\right]^2 \tag{84}$$

where the term in the brackets is proportional to the space-charge field modulation, and the diffraction efficiency depends on the square of the space-charge field. The characteristic fringe spacing fit to the data is $\Lambda_c \approx 3\ \mu m$. However, this value underestimates the spatial resolution required for Fourier-plane applications, which require flat spatial spectral response. The data in the figure has a useful spectral range only for fringe spacings greater than 8 μm. Much smaller spatial resolutions are possible using low-temperature-growth GaAs [56]. A free-carrier lifetime of 500 fsec translates to $\Lambda_c \approx 0.6\ \mu m$ for Fourier-plane applications.

6.4.2 *Photorefractive Phase Shift*

Two-wave mixing and degenerate four-wave mixing can be measured simultaneously, as shown in Fig. 30, and used to extract the photorefractive phase shift as a function of applied electric field. The diffraction efficiency for a wavelength that coincides with a zero in the electroabsorption will be caused entirely by the refractive index grating. The diffraction efficiency furthermore depends only on the magnitude of the index grating, and not its phase. Two-wave mixing, on the other hand, depends on the same parameters, but is sensitive to the phase shift. For instance γ_n is given by eq.(77) and the diffraction efficiency can be expressed as

$$\sqrt{\eta} = \frac{\pi n_1 L}{\lambda \cos \theta'} . \tag{85}$$

Therefore combining the diffraction measurement with the two-wave mixing allows us to write

$$\sin \phi = \frac{1}{2\sqrt{\beta}} \frac{\gamma_n}{\sqrt{\eta}} . \tag{86}$$

It is therefore possible to measure the photorefractive phase shift directly, and track its evolution as a function of applied electric field.

The trigonometric sine of the photorefractive phase shift is shown in Fig. 34 as a function of electric field for two samples, an AlGaAs/GaAs multiple quantum well, and a 1 μm thin film of GaAs. The behavior of the GaAs thin film is striking because the phase shift approaches π/2 for fields above 3 kV/cm. This phase shift is the shift that produces the largest photorefractive gains. It is also illustrative that the characteristic field for the phase shift is approximately equal to the field in GaAs that leads to negative differential conductance. The theory presented in Refs. [66, 67] and section 6.1.3 predicts this behavior based on hot-electron effects. The multiple quantum well sample has a smaller saturated phase shift than the bulk GaAs film, which may be an indication that the transport nonlinearity is weaker in the heterostructures. This nonlinear transport contribution to the photorefractive properties of photorefractive quantum wells is currently being studied Ref.[23].

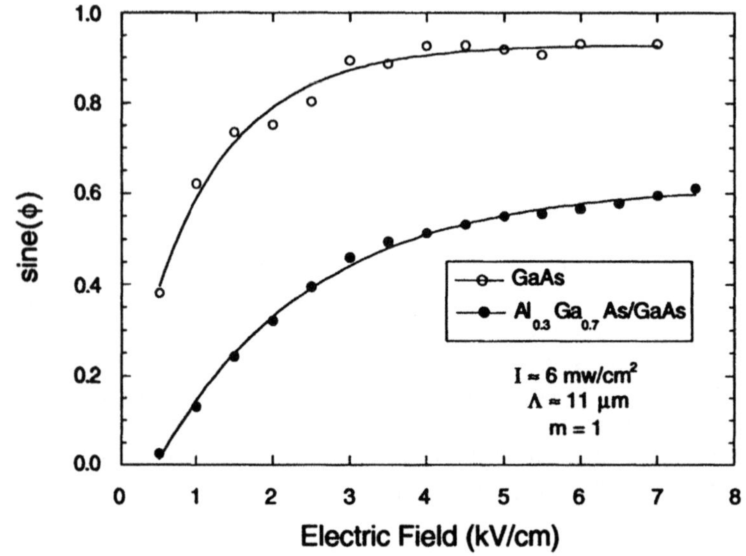

Fig. 34 Sine of the photorefractive phase shift as functions of electric field for thin-film GaAs and a multiple quantum well structure.

6.4.3 Nondegenerate Mixing

A benefit of photorefractive thin films is an interaction length that is comparable to the absorption depth, even for photon energies significantly larger than the bandgap [68]. Photorefractive quantum wells therefore are broad-band devices that can use any near-IR and visible laser source to write the holographic gratings, as long as the probe laser is tuned near the excitonic transition energy. The nondegenerate mixing geometry is shown in Fig. 35 using a HeNe laser as the write laser.

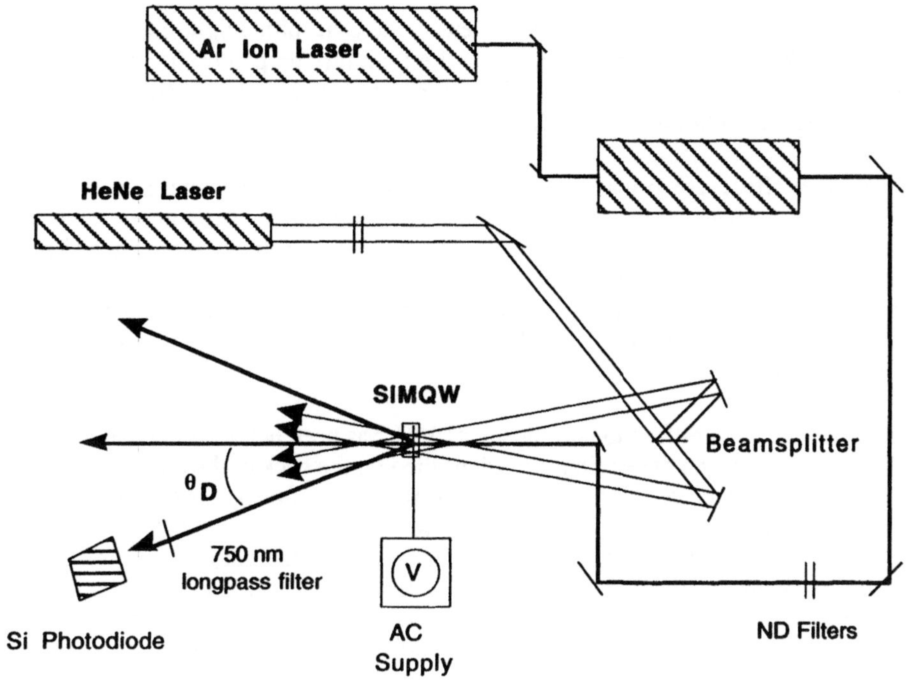

Fig. 35 Nondegenerate four-wave mixing experimental arrangement. The above-gap laser writes the fringe pattern, which is read by the infrared probe. Nondegenerate diffraction efficiencies are larger than degenerate diffraction efficiencies.

Nondegenerate four-wave mixing experiments have produced the largest diffraction efficiencies for the transverse transmission geometry. The nondegenerate diffraction efficiencies are larger than degenerate diffraction efficiencies under identical conditions. The nondegenerate gratings are also relatively more difficult to erase with the probe beam, making it possible for a weaker write beam to control a stronger probe beam. The more robust nondegenerate gratings may be related to the efficiency of space-charge trapping in the barriers for above-bandgap excitation Ref. [68]. However, the details of this process involve a full three-dimensional transport calculation, including charge tunneling into and emitted over the barriers. This problem has not yet been addressed theoretically.

The largest diffraction efficiencies in photorefractive quantum wells have been observed in the transverse-field geometry in nondegenerate mixing when multiple beam interference effects produce Fabry-Perot resonances that occur near the excitonic transition energy. Quantum-well Fabry-Perot modulators use the sensitive interferometric balance of light amplitudes reflected from the front and back surfaces of the quantum-well modulator to produce large contrast ratios [69-74]. This same idea can be used for diffractive devices, producing large output diffraction efficiencies, defined as the ratio of diffracted light intensity to transmitted light intensity (which has been taken as a standard definition in the literature on photorefractive quantum wells). The key feature in diffractive asymmetric Fabry-Perot structures is the interferometric control of phase [75].

Fabry-Perot effects are virtually unavoidable in photorefractive quantum-well devices because the devices are composed of thin planar films of semiconductor materials. The high refractive index of the semiconductor produces relatively pronounced Fabry-Perot fringes in transmission as well as reflection. We have been able to demonstrate that these naturally occurring fringes can be used to advantage during four-wave mixing to produce relatively large diffraction efficiencies in both transmission and reflection [76].

A maximum output diffraction efficiency of 15% has been obtained in the transverse-field geometry using a dielectric mirror deposited on the back of the device to form an asymmetric Fabry Perot. The experimental diffraction efficiency is shown in Fig. 36 as a function of wavelength. The transition wavelengths of the excitonic transitions are shown in the figure. The most striking feature in the data is the pronounced diffraction peak on the long wavelength side of the excitonic transitions. The wavelength of the maximum diffraction efficiency corresponds with the location of the Fabry-Perot fringe. On this scale, the diffraction from wavelengths near the exciton transitions is negligible.

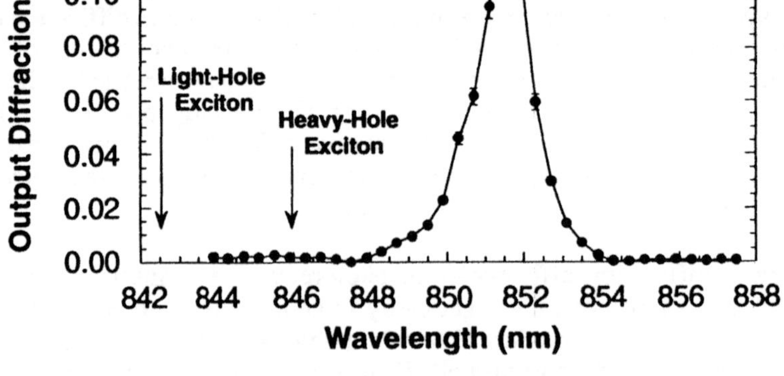

Fig. 36 Output diffraction efficiency from a transverse-field photorefractive quantum-well asymmetric Fabry-Perot [77].

In future work, output diffraction efficiencies in the reflection asymmetric Fabry-Perot structures can become arbitrarily large, with corresponding maximum input diffraction efficiencies approaching 6%. (Input diffraction efficiencies are defined as the ratio of the diffracted light intensity to the incident light intensity, thereby including insertion loss and device absorption.)

7. LONGITUDINAL-FIELD TRANSMISSION GEOMETRY

The photorefractive longitudinal-field transmission geometry has the potential for the largest quantum-well diffraction efficiencies because it operates at high field strengths using the quantum-confined Stark effect. Absorption changes as large as 8000 cm^{-1} are possible in this geometry in AlGaAs/GaAs multiple quantum wells, with corresponding refractive index changes of several percent.

7.1 Longitudinal-Field Transport

The transport properties of the longitudinal-field photorefractive quantum wells have little in common with the transverse-field geometry discussed in the previous section. The fundamental difference between the transverse and longitudinal transmission geometries is the direction of transport relative to the orientation of the grating vector. In the transverse geometry the transport is parallel to the grating vector, similar to traditional photorefractive transport. In the longitudinal transmission geometry, the transport is perpendicular to the grating vector. Therefore in the longitudinal geometry it is necessary to incorporate insulating buffer layers for the trapping of space-charge to screen applied electric fields.

The basic geometry for the longitudinal-field devices is shown in Fig. 37. An electric field is applied perpendicular to the plane of the thin film. Charges generated by photons in the illuminated region drift parallel to the field and are trapped at the interface between the electro-optic layer and the insulating buffer layer. These trapped charges screen the applied field in the illuminated regions. The net electric field is therefore spatially modulated in the plane of the film, parallel to the grating vector. The optical properties are modulated through the electro-optic effect in the electro-optic layer.

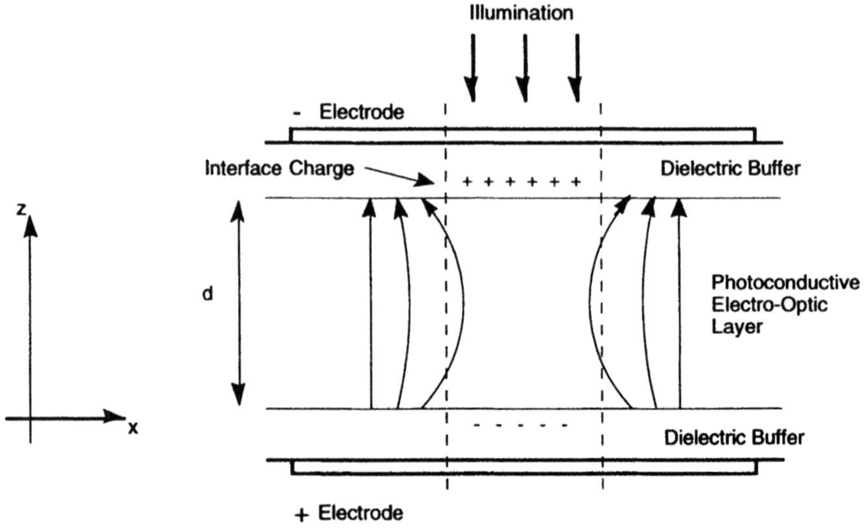

Fig. 37 Field screening by trapped space-charge in the longitudinal-field geometry.

The insulating buffer layers trap space-charge, but also perform as insulating stand-offs from the equipotential electrode. Not shown in the figure are image charges induced in the electrodes by the trapped space-charge. The complex interplay among the trapped charges and induced charges raises important issues of temporal response [78] and spatial resolution [79, 80] for the performance of the longitudinal geometry.

7.1.1 Equivalent Circuit and Frequency Dependence

The longitudinal-field geometry operates through transients. In a steady-state field under steady (but spatially modulated) illumination, the field will drop predominantly across the insulating buffer layers, preventing field penetration into the electro-optic layer. To introduce a field into the electro-optic layer, there must be transient changes in field or illumination.

The usual mode of operation uses an alternating applied electric field. The effective circuit of the photorefractive quantum well structure is shown in Fig. 38.

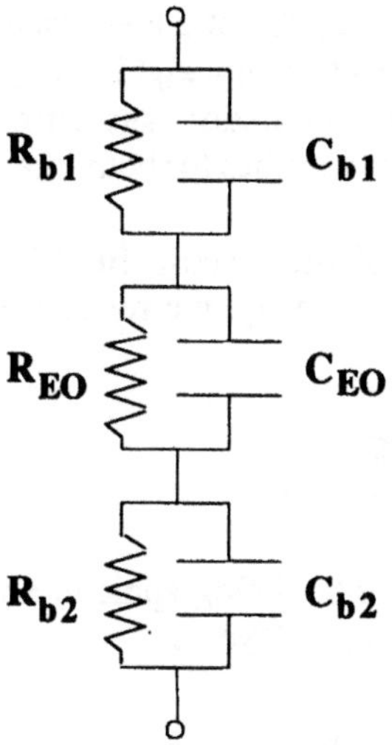

Fig. 38 Equivalent circuit of the longitudinal-field photorefractive quantum well structure.

The impedance of each layer is

$$\begin{aligned} Z &= \frac{R}{1 + i\omega RC} \\ &= \frac{R}{1 + i\omega\tau_{die}} \end{aligned} \tag{87}$$

where τ_{die} is the dielectric relaxation time of the layer. The dielectric relaxation time of the electro-optic layer is intensity dependent, and therefore is spatially modulated by the intensity interference fringes. The electric field amplitude in the electro-optic layer is

$$E_{EO} = \frac{V_{ap}/L}{1 + \frac{2R_b}{R_{EO}} \frac{1 + i\omega\tau_{die}^{EO}}{1 + i\omega\tau_{die}^{b}}} \tag{88}$$

which is implicitly a function of position through the intensity-dependent dielectric relaxation time. When the buffer layers are insulators, this expression reduces to

$$E_{EO} = \frac{i\omega \frac{C_b R_{EO}}{2}}{1 + i\omega\left(\tau_{die}^{EO} + \frac{C_b R_{EO}}{2}\right)} \frac{V_{ap}}{L} \tag{89}$$

for which the electric field in the electro-optic layer vanishes for ω -> DC. At high frequency, the electric field reduces to $E_{E0} = E_0/(1+2C_{EO}/C_b)$, which is independent of intensity. Therefore, the spatial modulation (and diffraction efficiency) caused by the holographic interference fringes vanishes for both low and high frequencies.

To find the spatial modulation of the electric field in the electro-optic layer in response to the spatially varying intensity, the resistance of the electro-optic layer can be described by

$$R_{EO}(x) = \frac{R_{EO}^{av}}{1 + m\cos Kx} \tag{90}$$

where m is the modulation index of the holographic interference fringes. When m is small, then the field in the electro-optic layer is

$$E_{EO}(x) = E_0 + E_1 \cos Kx \tag{91}$$

where

$$E_1 = -i\, m \frac{V_{ap}}{L} \frac{\omega \frac{C_b}{2} R_{EO}^{av}}{\left[1 + i\omega R_{EO}^{av}\left(\frac{C_b}{2} + C_{EO}\right)\right]^2} \tag{92}$$

is the coefficient of the spatial modulation of the electric field. The frequency that produces the maximum response is

$$\omega = \frac{1}{R_{EO}^{av}\left(\frac{C_b}{2} + C_{EO}\right)} \tag{93}$$

The amplitude of the electric field at this frequency is

$$E_1 = -\frac{1}{2} m \frac{V_{ap}}{L} \frac{1}{1 + \frac{2\,C_{EO}}{C_b}} \tag{94}$$

which is 180° out of phase with the applied voltage. Larger values of C_b yield larger values of spatially modulated field. However, larger C_b also slows down the device. A reasonable trade-off for best speed vs. diffraction efficiency is $C_b = 2C_{EO}$. In this case the optimum modulation frequency is approximately equal to the inverse of the dielectric relaxation time of the electro-optic layer. The optimum frequency for the best diffraction efficiency is also linearly dependent on intensity. The field modulation is shown in Fig. 39 as a function of frequency for several intensities.

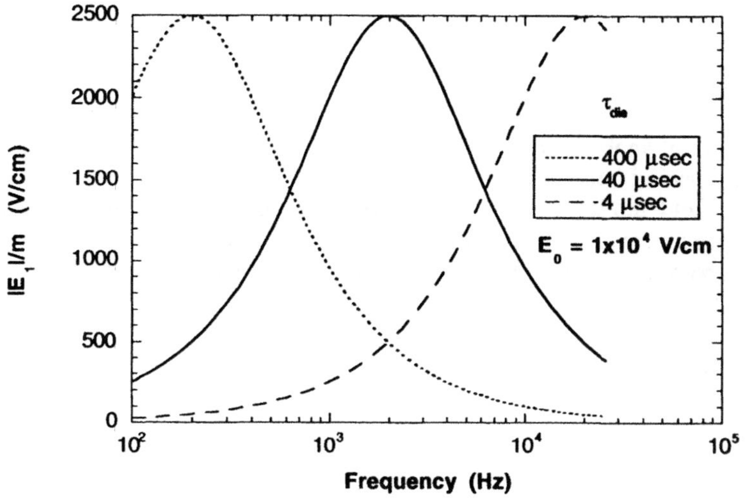

Fig. 39 Modulated field vs. frequency for three different intensities (corresponding to different dielectric relaxation times in the electro-optic layer) in a longitudinal-field photorefractive quantum well structure for buffers with $C_b = 2C_{EO}$.

Even under ideal conditions and for the optimum frequency, the spatially modulated field magnitude in the electro-optic layer is a factor of four smaller than the applied electric field. Because the diffraction efficiency depends on the fourth-power of the field strength, the diffraction efficiencies for sinusoidal modulation are smaller by a factor of $4^4 = 250$ than for transient diffraction efficiencies (in response to a square pulse). Therefore, quasi-cw operation of the longitudinal-field geometry yields diffraction efficiencies that are comparable to the DC diffraction efficiencies of the transverse-field geometry.

This time-response highlights an important practical aspect about the transverse- and longitudinal-field geometries. The differences in the transport and trapping between the transverse-field and the longitudinal-field geometries lead to substantially different operating conditions. The transverse-field geometry operates with frequencies from DC to kHz with full modulation of the applied electric field. The longitudinal geometry operates optimally only near a specific frequency defined by the light intensity on the device and with only a fractional modulation of the applied field. Therefore, despite the larger electro-optic effects of the quantum-confined Stark effect, the transverse-field and the longitudinal-field geometries lead to comparable quasi-cw diffraction efficiencies. It is only during transient operation with relatively low duty cycles that the full strength of the Stark effect can be used when operating the longitudinal geometry.

7.1.2 Spatial Resolution

The equivalent circuit analysis is only valid in the limit of very long fringe spacings. When the fringe spacing becomes comparable to the thickness of the device ($\Lambda \approx L$) then the spatial field modulation E_1 is reduced by fringing fields and by lateral transport of charge carriers. In addition, a geometric limit fixed by the device thickness is caused by fringing fields. The modulation transfer function, limited by the geometric fringing fields, is

$$m = \exp\left(-KL/2\right) \tag{95}$$

yielding a minimum fringe spacing $\Lambda = \pi L$. A 2-micron layer thickness therefore has a 6 micron minimum fringe spacing as its geometric limit.

Another limit to spatial resolution, in addition to the geometric limit, is the role of lateral transport. The trapped sheet charge at the interface between the electro-optic layer and the buffer is

$$\sigma_{sc} = \sigma_0 + \sigma_1 \cos Kx \tag{96}$$

For simplicity, we assume that the space-charge modulation at the lower interface is the same as the upper interface, although this condition can be relaxed to include more general cases. This space-charge produces space-charge field components [81]

$$E_z^{sc}(x,z) = \frac{-1}{\varepsilon\varepsilon_0}\left[\sigma_0 + \sigma_1 e^{-K\frac{L}{2}} \cosh(Kz)\cos(Kx)\right] \tag{97}$$

$$E_x^{sc}(x,z) = \frac{1}{\varepsilon\varepsilon_0}\left[\sigma_1 \sinh(Kz)\cosh\left(K\frac{L}{2}\right)\sin(Kx)\right] \tag{98}$$

The lateral electric field E_x^{sc} is the source of lateral drift. The lateral transport of charge carriers allows charge to trap in the dark fringes, screening the applied field everywhere and erasing the spatially modulated field that produces photorefractive diffraction. This theoretical problem has been addressed by Ref. [80].

Once again, the differences in the transport geometries between the transverse-field and longitudinal-field geometries have important consequences in device operation. In the transverse-field photorefractive quantum-well geometry the spatial resolution is limited only by the defect densities. In principle, fringe spacings down to 0.1 microns should be possible using low-temperature-growth materials [56]. On the other hand, the longitudinal-field geometry is fundamentally limited by the geometry factor in eq.(95), and may be further limited by lateral diffusion. In practical circumstances, the device resolution for the longitudinal geometry will be limited to values greater than 3 microns.

7.2 Longitudinal-Field Device Designs

The longitudinal-field transmission geometry has considerable design flexibility and has lead to several variant device designs. The central design consideration for the longitudinal geometry is charge trapping in the buffer or at the buffer interfaces. The variety of device designs stems from the possible choices of buffer layers and interfaces.

7.2.1 Oxide and other Insulating Buffer Layers

The first working buffer designs used thin insulating oxide layers with semi-insulating multiple quantum wells and a semi-insulating semiconductor buffer layer sandwiched between them [78, 82]. Carriers arriving at the semiconductor-oxide interface are trapped at interface defects and at defects within the semiconductor adjacent to the oxide. The quantum wells in these structures were $Cd_{0.34}Zn_{0.66}Te/ZnTe$ with 90 periods of 51 Å CdZnTe wells and 122 Å ZnTe barriers. The total MQW thickness was 1.55 microns. Diffraction efficiencies in these structures approached output diffraction efficiencies of $\eta \approx 1.5\times10^{-2}$. The input diffraction efficiency, which includes absorption and insertion loss, was approximately $\eta \approx 5\times10^{-4}$.

The next family of buffers retained the thin oxide insulators, but included semi-insulating semiconductor buffers on both sides of the MQW layer [83]. One of these semiconductor buffers was grown at low substrate temperatures to provide large trap concentrations and short carrier lifetimes, described in section 5.0. The multiple quantum wells in this structure were $Al_{0.29}Ga_{0.71}As/GaAs$ with 150 periods of 100 Å GaAs wells and 35 Å AlGaAs barriers. The total MQW thickness was 2.03 microns. Diffraction efficiencies in these structures approached output diffraction efficiencies of $\eta \approx 3\times10^{-2}$. The input diffraction efficiency, which includes absorption and insertion loss, was approximately $\eta \approx 1\times10^{-2}$. The quantum well structure, including the well and barrier widths, was chosen to optimize carrier escape from the quantum wells, improving the spatial resolution and maximizing the diffraction efficiency [65]. The output diffraction efficiency for this device structure is shown in Fig. 40 as a function of fringe spacing. The functional fit to the data varies according to eq. (95). The cut-off fringe spacing is approximately $\Lambda_C \approx 30$ microns, compared with 3 microns for the transverse-field geometry in Fig. 33.

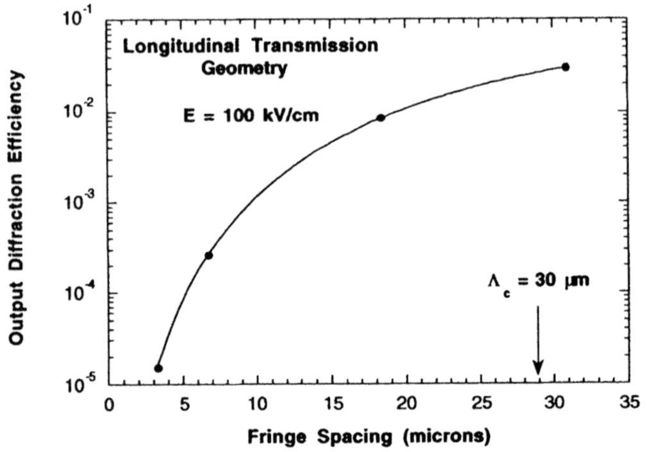

Fig. 40 Output diffraction efficiency as a function of fringe spacing for a field of 100 kV/cm for AlGaAs/GaAs multiple quantum wells. From Ref. [65]

7.2.2 Photorefractive p-i-n and p-i-p Quantum-Well Diodes

There is some motivation to develop a monolithic photorefractive semiconductor structure that does not include high resistivity non-semiconductor buffer layer designs. Such a device would consist only of semiconductor layers that are grown by molecular beam epitaxy. This structure would not require the additional processing and deposition steps of a non-semiconductor material. The key to monolithic performance is finding a semiconductor material that can replace the high-resistivity oxide layers. The advantage of the oxide layer was a nearly 100% quantum efficiency for trapping carriers, because the carriers cannot escape through the high bandgap dielectric to the contacts.

The first all-semiconductor longitudinal geometries used low-temperature-growth (LTG) layers to sandwich the multiple quantum well layer [84], with metal contacts placed in direct contact with the LTG layer. The LTG layers, discussed in section 5.0, have high trap concentrations and ultrafast trapping lifetimes [55]. The carrier lifetimes in the LTG buffer can therefore be shorter than the transit time across the buffer, providing efficient trapping. The output diffraction efficiencies of these structures [84] were lower than for the structures that included a non-semiconductor dielectric buffer. The output diffraction efficiencies approached $\eta \approx 3x10^{-4}$. The input diffraction efficiency, which includes absorption and insertion loss, was approximately $\eta \approx 8x10^{-5}$.

p+ AlGaAs [30%] 1000 Å	Doped upper contact
i- AlGaAs [30%] 1000 Å	Spacer to prevent depletion of doped contact
LTG AlAs 1000 Å	Blocking Layer to prevent carrier sweepout to the contact
LTG AlGaAs [30%] 7000 Å	Trapping Layer using ultrafast lifetimes of LTG material
150 periods 100 Å GaAs Wells 35 Å AlGaAs Barriers [30%]	2 μ EO layer consisting of multiple quantum wells
LTG AlGaAs [30%] 7000 Å	The LTG bandgaps can be graded to build-in a retarding effective field to reduce carrier transit time
LTG AlAs 1000 Å	
i- AlGaAs [30%] 1000 Å	
n+ AlGaAs [30%] 2000 Å	Doped lower contact

Fig. 41 All-semiconductor longitudinal-field p-i-n diode geometry using doped contacts and LTG layers to trap space-charge.

Recent all-semiconductor structures with improved diffraction efficiencies used doped layers in contact with the LTG buffer layers. In addition, the LTG buffers had stepped bandgaps that increase the transit time through the buffer, reduce the dark conductivity, and improve the quantum efficiency by providing more time to trap the carriers. The device structure is shown in Fig. 41.

Three different structures of this type are possible: p-i-p, n-i-n and p-i-n structures. The output diffraction efficiency of a p-i-p structure is shown in Fig. 42 as a function of photon wavelength for a square-wave field of 130 kV/cm. A peak output diffraction efficiency of $\eta \approx 0.7\%$ is attained, with an input diffraction efficiency (that includes absorption and insertion loss) of $\eta \approx 7\text{x}10^{-4}$. As discussed above, the transient response to a low-duty-cycle square pulse would produce transient diffraction efficiencies higher than these. The p-i-p design has suffered from large leakage currents that produce Joule-heating of the device during operation.

The frequency dependence of a photorefractive p-i-n quantum-well diode is shown in Fig. 42 for a 35 V reverse-biased square wave applied across the junction. The buffer thicknesses are equal in this structure and are equal to half the thickness of the electro-optic multiple quantum-well layer. These conditions are the same used for Fig. 39 (simulation). The maximum response was observed at a frequency of 2 kHz, corresponding to a dielectric relaxation time of 40 μsec in the multiple quantum well.

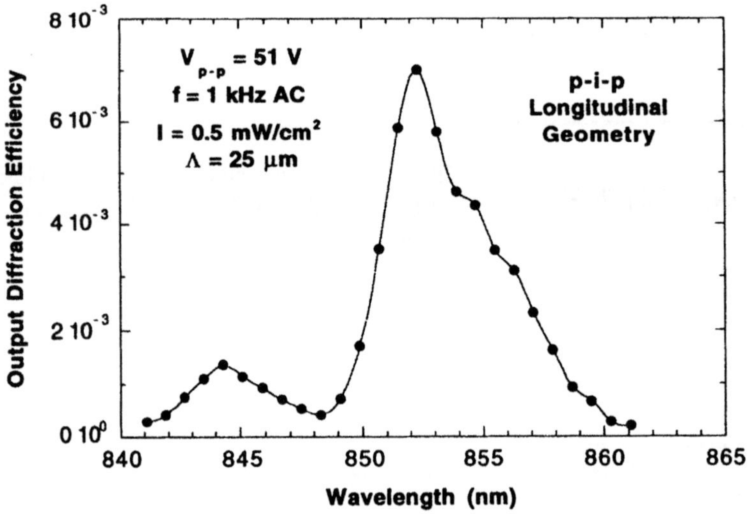

Fig. 42 Output diffraction efficiency of a p-i-p AlGaAs/GaAs all-semiconductor structure at a 1 kHz square-wave field of 130 kV/cm.

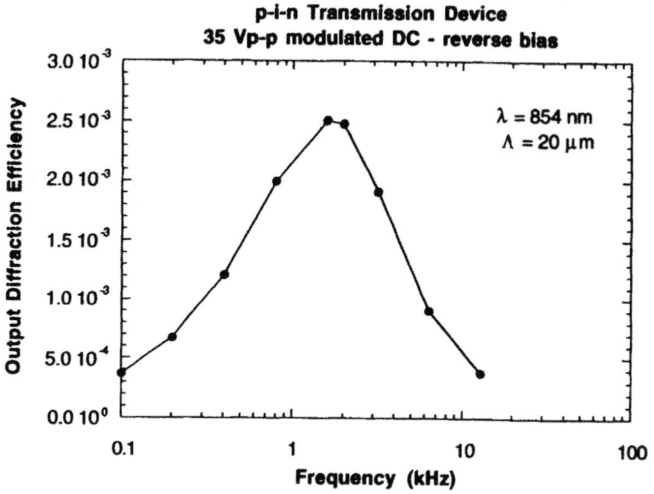

Fig. 43 Frequency-dependence of four-wave mixing in a p-i-n photorefractive diode.

8. APPLICATIONS

Photorefractive quantum wells are enjoying increased interest with respect to applications. They have several features that make them attractive as dynamic holographic films. For instance, they operate with low light powers at high speeds, they have relatively large diffraction efficiencies, they are electrically switched and optically addressed, they are easy to fabricate using standard semiconductor processing techniques, and are easy to operate. These advantages have lead to several demonstrations in optical applications including high-speed image correlators [85], as a femtosecond auto-correlator [86], and for time-to-space mapping of femtosecond pulses [87]. This last application is an example of femtosecond spectral holography [88] that promises to provide new methods for high-speed data transmultiplexing for fiber-optic telecommunications applications. Spectral holography requires a versatile dynamic holographic medium compatible with laser diode sources, for which photorefractive quantum well devices appear ideally suited.

These examples of applications represent only a beginning of the usefulness of photorefractive quantum-well devices. Many of the optical architectures currently being explored for all-optical processing require holographic elements. The photorefractive quantum-well devices, because they can tap from the already established semiconductor technology, are naturally integrated with laser diode sources and drive electronics. They also have tremendous engineering flexibility by atomic-layer growth, and are poised to become choice devices for advanced optical systems.

Acknowledgments: This work on photorefractive quantum wells could not have been accomplished without the substantial effort and insight of Q. N. Wang, R. M. Brubaker, and K. M. Kwolek. Support for this work has been supplied by NSF, AFOSR and Rome Laboratories.

References:

[1] C. V. Raman and N. S. N. Nath, "The diffraction of light by high frequency sound waves", Proc. Indian Acad. Sci., vol. 2, p. 406 (1935)

[2] E. I. Gordon, "A review of acousto-optical deflection and modulation devices", Proc. IEEE, vol. 54, p. 1391 (1967)

[3] D. S. Subbaramaiya, "Diffraction of light by ripples on liquid surfaces", Proc. Ind. Acad. Sci., vol. 6, p. 333 (1938)

[4] E. P. Ippen, "Diffraction of light by surface acoustic waves", Proc. IEEE, vol. 55, p. 248 (1967)

[5] E. Salzmann and D. Weismann, "Optical detection of Rayleigh waves", J. Appl. Phys., vol. 40, p. 3408 (1969)

[6] Q. N. Wang, D. D. Nolte and M. R. Melloch, "Spatial-Harmonic gratings at high modulation depths in photorefractive quantum wells", Opt. Lett., vol. 16, p. 1944 (1991)

[7] Q. N. Wang, D. D. Nolte and M. R. Melloch, "Two-wave mixing in photorefractive AlGaAs/GaAs quantum wells", Appl. Phys. Lett., vol. 59, p. 256 (1991)

[8] P. Günter and J.-P. Huignard, Photorefractive Materials and Their Applications I and II, in Topics in Applied Physics **61**, (Springer-Verlag, Berlin, 1988)

[9] A. Y. Cho, Thin Solid Films, vol. 100, p. 291 (1983)

[10] A. Y. Cho and J. R. Arthur, Prog. Solid State Chem., vol. 10, p. 157 (1975)

[11] L. Esaki, "A bird's-eye view on the evolution of semiconductor superlattices and quantum wells", IEEE J. Quant. Electron., vol. QE-22, p. 1611 (1986)

[12] A. M. Glass, A. M. Johnson, D. H. Olson, W. Simpson and A. A. Ballman, "Four-wave mixing in semi-insulating InP and GaAs using the photorefractive effect", Appl. Phys. Lett., vol. 44, p. 948-950 (1984)

[13] A. M. Glass and J. Strait, "The photorefractive effect in semiconductors", in Photorefractive materials and their applications I **61**, p. 237-262 (Springer-Verlag, Berlin, 1988)

[14] R. S. Knox, *Theory of Excitons* (New York, Academic Press, 1963)

[15] R. J. Elliot, Phys. Rev., vol. 108, p. 1384 (1957)

[16] H. Haug and S. W. Koch, *Quantum Theory of the Optical and Electronic Properties of Semiconductors* (Singapore, World Scientific, 1990)

[17] H. C. Casey and M. B. Panish, in Heterostructure Lasers p. (Academic, New York, 1978)

[18] H. J. Lee, L. Y. Juravel and J. C. Woolley, Phys. Rev. B, vol. 21, p. 659 (1980)

[19] G. Bastard and J. A. Brum, "Electronic states in semiconductor heterostructures", IEEE J. Quant. Electron., vol. QE-22, p. 1625 (1986)

[20] G. Bastard, "Theoretical investigations of superlattice band structure in the envelope-function approximation", Phys. Rev. B, vol. 25, p. 7584 (1982)

[21] G. Bastard, *Wave Mechanics Applied to Semiconductor Heterostructures* (New York, Halsted, 1988)

[22] R. L. Greene, K. K. Bajaj and D. E. Phelps, Phys., Rev. B, vol. 29, p. 1807 (1984)

[23] R. M. Brubaker, Q. N. Wang, D. D. Nolte and M. R. Melloch, unpublished, (1995)

[24] D. S. Chemla, D. A. B. Miller, P. W. Smith, A. C. Gossard and W. Wiegmann, "Room temperature excitonic nonlinear absorption and refraction in GaAs/AlGaAs multiple quantum well structures", IEEE J. Quant. Electron., vol. QE-20, p. 265 (1984)

[25] D. A. B. Miller, D. S. Chemla, T. C. Damen, A. C. Gossard, W. Wiegmann, T. H. Wood and C. A. Burrus, "Electric field dependence of optical absorption near the band gap of quantum well structures", Phys. Rev. B, vol. 32, p. 1043 (1985)

[26] F. W. Smith, A. R. Calawa, C. Chen, M. J. Mantra and L. J. Mahoney, IEEE Electron. Dev. Lett., vol. EDL-9, p. 77 (1988)

[27] A. M. Glass, D. D. Nolte, D. H. Olson, G. E. Doran, D. S. Chemla and W. H. Knox, "Resonant photodiffractive four-wave mixing in semi-insulating GaAs/AlGaAs quantum wells", Opt. Lett., vol. 15, p. 264 (1990)

[28] D. A. B. Miller, "Quantum-well Self-Electro-Optic Effect Devices", Opt. and Quant. Electron., vol. 22, p. S61 (1990)

[29] D. C. D'Avanzo, "Proton isolation for GaAs integrated circuits", IEEE Trans. Electron. Dev., vol. ED-29, p. 1051 (1982)

[30] B. Schwartz, L. A. Koszi, P. J. Anthony and R. L. Hartman, "Thermal annealing of proton-bombarded GaAs and AlGaAs", J. Electrochem. Soc., vol. 131, p. 1703 (1984)

[31] J. F. Gibbons, Proc. IEEE, vol. 60, p. 1062 (1972)

[32] P. Sigmund, Appl. Phys. Lett., vol. 14, p. 114 (1969)

[33] G. H. Kinchin and R. S. Pease, "The displacement of atoms in solids by radiation", Re. Prog. Phys., vol. 18, p. 2 (1955)

[34] W. Walukiewicz, "Mechanism of Fermi-level stabilization in semiconductors", Phys. Rev. B, vol. 37, p. 4760 (1988)

[35] B. R. Pruniaux, J. C. North and G. L. Miller, "Compensation of n-type GaAs by Proton Bombardment", in Proceedings of Ion Implantation in Semiconductors, p. (Springer-Verlag, 1971)

[36] J. C. Dymet, J. C. North and L. A. D'Asaro, "Optical and electrical properties of proton-bombarded p-type GaAs", J. Appl. Phys., vol. 44, p. 207 (1973)

[37] Y. Silverberg, P. W. Smith, D. A. B. Miller, B. Tell, A. C. Gossard and W. Wiegmann, "Fast nonlinear optical response from proton-bombarded multiple quantum well structures", Appl. Phys. Lett., vol. 46, p. 701 (1985)

[38] M. B. Johnson, T. C. McGill and N. G. Paulter, "Carrier lifetimes in ion-damaged GaAs", Appl. Phys. Lett., vol. 54, p. 2424 (1989)

[39] M. Lambsdorff, J. Kuhl, J. Rosenzweig, A. Axmann and J. Schneider, "Subpicosecond carrier lifetimes in radiation-damaged GaAs", Appl. Phys. Lett., vol. 58, p. 1881 (1991)

[40] B. Lin J.-F., D. P. Kocot, D. E. Mars and R. Jaeger, IEEE Trans. Elec. Dev., vol. ED-37, p. 46 (1990)

[41] M. R. Melloch, D. C. Miller and B. Das, Appl. Phys. Lett., vol. 54, p. 943 (1989)

[42] M. Kaminska, Z. Lilionthal-Weber, E. R. Weber, T. George, B. Kotright, F. W. Smith, B. Y. Tsaur and A. R. Calawa, Appl. Phys. Lett., vol. 54, p. 1881 (1989)

[43] D. C. Look, D. C. Walters, M. O. Mansereh, J. R. Sizalove, C. E. Stutz and K. R. Evans, "Anomalous Hall-effect results in low-temperature molecular-beam-epitaxial GaAs: Hopping in a dense EL2-like band", Phys. Rev. B, vol. 42, p. 3578 (1990)

[44] D. C. Look, D. C. Walters, G. D. Robinson, J. R. Sizelove, M. G. Mier and C. E. Stutz, "Annealing dynamics of molecular-beam epitaxial GaAs grown at 200 ^{o}C", J. Appl. Phys., vol. 74, p. 306 (1993)

[45] M. R. Melloch, N. Otsuka, J. M. Woodall, A. C. Warren and J. L. Freeouf, "Formation of arsenic precipitates in GaAs buffer layers grown by molecular beam epitaxy at low substrate temperatures", Appl. Phys. Lett., vol. 57, p. 1531 (1990)

[46] A. C. Warren, J. M. Woodall, J. L. Freeouf, D. Grischowsky, D. T. McInturff, M. R. Melloch and N. Otsuka, "Arsenic precipitates and the semi-insulating properties of GaAs buffer layers grown by low-temperature molecular beam epitaxy", Appl. Phys. Lett., vol. 57, p. 1331 (1990)

[47] A. C. Warren, J. M. Woodall, P. D. Kirchner, X. Yin, F. Pollak and M. R. Melloch, "Role of excess As in low-temperature-growth GaAs", Phys. Rev. B, vol. 46, p. 4617 (1992)

[48] K. Mahalingam, N. Otsuka, M. R. Melloch, J. M. Woodall and A. C. Warren, "Substrate temperature dependence of arsenic precipitate formation in AlGaAs and GaAs", J. Vac. Sci. Technol. B, vol. 9, p. 2328 (1991)

[49] M. R. Melloch, N. Otsuka, K. Mahalingam, C. L. Chang, J. M. Woodall, G. D. Pettit, P. D. Kirchner, F. Cardone, A. C. Warren and D. D. Nolte, "Arsenic cluster dynamics in doped GaAs", J. Appl. Phys., vol. 72, p. 3509 (1992)

[50] K. Mahalingam, N. Otsuka, M. R. Melloch, J. M. Woodall and A. C. Warren, "Arsenic precipitate accumulation and depletion zones at AlGaAs/GaAs

heterojunctions grown at low substrate temperatures by molecular beam epitaxy", J. Vac. Sci and Technol., vol. B10, p. 812 (1992)

[51] M. R. Melloch, J. M. Woodall, N. Otsuka, K. Mahalingham, C. L. Chang and D. D. Nolte, "GaAs, AlGaAs and InGaAs epilayers containing As clusters: semimetal/semiconductor composites", Mat. Sci. and Eng., vol. B22, p. 31 (1993)

[52] F. W. Smith, H. Q. Le, V. Diadiuk, M. A. Hollis, A. R. Calawa, S. Gupta, M. Frankel, D. R. Dykaar, G. A. Mourou and T. Y. Hsiang, "Picosecond GaAs-based photoconductive optoelectronic detectors", Appl. Phys. Lett., vol. 54, p. 890 (1989)

[53] T. Motet, J. Nees, S. Williamson and G. Morou, "1.4 psec rise-time high-voltage photoconductive switching", Appl. Phys. Lett., vol. 59, p. 1455 (1991)

[54] A. C. Warren, N. Katzenellenbogen, D. Grischkowsky, J. M. Woodall, M. R. Melloch and N. Otsuka, "Subpicosecond, freely propagating electromagnetic pulse generation and detection using GaAs:As epilayers", Appl. Phys. Lett., vol. 58, p. 1512 (1991)

[55] E. S. Harmon, M. R. Melloch, J. M. Woodall, D. D. Nolte, N. Otsuka and C. L. Chang, "Carrier Lifetime vs. Anneal in Low Growth Temperature GaAs", Appl. Phys. Lett., vol. 63, p. 2248-2250 (1993)

[56] D. D. Nolte, R. M. Brubaker, M. R. Melloch, J. M. Woodall and S. E. Ralph, "High-Density Optical Storage based on Nano-Size Arsenic Clusters in Low-Temperature-Growth GaAs", Appl. Phys. Lett, vol. 61, p. 3098 (1992)

[57] G. Montemezzani, P. Rogin, M. Zgonik and P. Gunter, "Interband photorefractive effects: Theory and experiments in $KNbO_3$", Phys. Rev. B, vol. 49, p. 2484 (1994)

[58] Q. N. Wang, R. M. Brubaker, D. D. Nolte and M. R. Melloch, "Photorefractive Quantum Wells: Transverse Franz-Keldysh Geometry", J. Opt. Soc. Am. B, vol. 9, p. 1626 (1992)

[59] E. Yablonovitch, T. Gmitter, J. P. Harbison and R. Bhat, "Extreme selectivity in the lift-off of epitaxial GaAs films", Appl. Phys. Lett., vol. 51, p. 2222 (1987)

[60] E. Yablonovitch, D. M. Hwang, T. J. Gmitter, L. T. Florez and J. P. Harbison, "Van der Waals bonding of GaAs epitaxial liftoff films onto arbitrary substrates", Appl. Phys. Lett., vol. 56, p. 2419 (1990)

[61] D. A. B. Miller, D. S. Chemla, D. J. Eilenberger, P. W. Smith, A. C. Gossard and W. T. Tsang, Appl. Phys. Lett., vol. 41, p. 679 (1982)

[62] G. D. Boyd, A. M. Fox, D. A. B. Miller, L. M. F. Chirovsky, L. A. D'Asaro, J. M. Kuo, R. F. Kopf and A. L. Lentine, Appl. Phys. Lett., vol. 57, p. 1843 (1990)

[63] D. D. Nolte, D. H. Olson, G. E. Doran, W. H. Knox and A. M. Glass, "Resonant photodiffractive effect in semi-insulating multiple quantum wells", J. Opt. Soc. Am., vol. B7, p. 2217 (1990)

[64] A. M. Fox, D. A. B. Miller, G. Livescu, J. E. Cunningham and W. Y. Jan, IEEE J. Quant. Electron., vol. 27, p. 2281 (1991)

[65] A. Partovi, A. M. Glass, G. J. Zydzik, H. M. O'Bryan, T. H. Chiu and W. H. Knox, "Effect of carrier excape time on the performance of semi-insulating photorefractive self-electro-optic-effect devices", Appl. Phys. Lett., vol. 62, p. 3088 (1993)

[66] Q. N. Wang, R. M. Brubaker and D. D. Nolte, "Photorefractive phase shift induced by hot electron transport: multiple quantum well structures", J. Opt.Soc. Am. B, vol. 11, p. 1773 (1994)

[67] Q. N. Wang, R. M. Brubaker and D. D. Nolte, "Photorefractive phase shift induced by nonlinear electronic transport", Opt. Lett., vol. 19, p. 822 (1994)

[68] D. D. Nolte, Q. N. Wang and M. R. Melloch, "Robust infrared gratings in photorefractive quantum wells generated by an above-band-gap laser", Appl. Phys. Lett., vol. 58, p. 2067 (1991)

[69] R. H. Yan, R. J. Simes and L. A. Coldren, "Electroabsorptive Fabry-Perot reflection modulators with asymmetric mirrors", IEEE Phot. Tech. Lett., vol. 1, p. 273 (1989)

[70] R. Yan, R. J. Simes and L. A. Coldren, "Analysis and design of surface-normal Fabry-Perot electrooptic modulators", IEEE J. Quant. Electron., vol. 25, p. 2272 (1989)

[71] B. Pezeshki, D. Thomas and J. S. J. Harris, "Optimization of modulation ratio and insertion loss in reflective electroabsorption modulators", Appl. Phys. Lett., vol. 57, p. 1491 (1990)

[72] J. F. Heffernan, M. H. Moloney, J. Hegarty, J. S. Roberts and M. Whitehead, "All optical high contrast absorptive modulation in an asymmetric Fabry-Perot etalon", Appl. Phys. Lett., vol. 58, p. 2877 (1991)

[73] R. Jin, G. Khitrova, H. M. Gibbs and N. Peyghambarian, "High contrast, submilliwatt power InGaAs/GaAs strained-layer multiple-quantum-well asymmetric reflection modulator", Appl. Phys. Lett., vol. 59, p. 3216 (1991)

[74] G. Livescu, G. D. Boyd, R. A. Morgan, L. M. F. Chirovsky, A. M. Fox, R. E. Leibenguth, M. T. Asom and M. W. Focht, "Role of electrorefraction in quantum-well Fabry-Perot modulators", Appl. Phys. Lett., vol. 60, p. 1418 (1992)

[75] D. D. Nolte, "Dynamic Holographic Phase Gratings in Multiple Quantum Well Asymmetric Reflection Fabry-Perot Modulators", Opt. Lett., vol. 19, p. 819 (1994)

[76] K. M. Kwolek, M. R. Melloch and D. D. Nolte, "Dynamic holography in a reflection/transmission photorefractive quantum-well asymmetric Fabry-Perot", Appl. Phys. Lett., vol. 65, p. 385 (1994)

[77] K. M. Kwolek, M. R. Melloch and D. D. Nolte, "Diffractive Quantum-Well Asymmetric Fabry-Perot: Transverse-Field Photorefractive Geometry", submitted to Appl. Phys. Lett., (1995)

[78] A. Partovi, A. M. Glass, D. H. Olson, G. J. Zydzik, K. T. Short, R. D. Feldman and R. F. Austin, "High-speed photodiffractive effect in semi-insulating CdZnTe/ZnTe multiple quantum wells", Opt. Lett., vol. 17, p. 655 (1992)

[79] D. D. Nolte, "Resolution of electro-optic spatial light modulators: the role of lateral transport", Opt. Commun., vol. 92, p. 199 (1992)

[80] S. L. Smith and L. Hesselink, "Analytical modeling of grating dynamics in Stark geometry quantum-well photorefractive devices", J. Opt. Soc. Am. B, vol. 11, p. 1878 (1994)

[81] W. R. Roach, "Resolution of Electrooptic Light Valves", IEEE Trans. Electron. Dev., vol. ED-21, p. 453 (1974)

[82] A. Partovi, A. M. Glass, D. H. Olson, G. J. Zydzik, K. T. Short, R. D. Feldman and R. F. Austin, "High Sensitivity Optical Image Processing Device based on CdZnTe/ZnTe MQW", Appl. Phys. Lett., vol. 59, p. 1832 (1991)

[83] A. Partovi, A. M. Glass, D. H. Olson, G. J. Zydzik, H. M. O'Bryan, T. H. Chiu and W. H. Knox, "Cr-doped GaAs/AlGaAs semi-insualting multiple quantum well photorefractive devices", Appl. Phys. Lett., vol. 62, p. 464 (1993)

[84] C. S. Kyono, K. Ikossi-Anastasiou, W. S. Rabinovich, S. R. Bowman and D. S. Katzer, "GaAs/AlGaAs multiquantum well resonant photorefractive devices fabricated using epitaxial lift-off", Appl. Phys. Lett., vol. 64, p. 2244 (1994)

[85] A. Partovi, A. M. Glass, T. H. Chiu and D. T. H. Liu, "High-speed joint-transform optical image correlator using GaAs/AlGaAs semi-insulating multiple quantum wells and laser diodes", Opt. Lett., vol. 18, p. 906 (1993)

[86] R. M. Brubaker, Q. N. Wang. , D. D. Nolte, E. S. Harmon and M. R. Melloch, "Steady-State four-wave mixing in AlGaAs/GaAs multiple quantum wells with femtosecond pulses", J. Opt. Soc. Am., vol. 11, p. 1038 (1994)

[87] M. C. Nuss, M. Li, T. H. Chiu, A. M. Weiner and A. Partovi, "Time-to-Space Mapping of Femtosecond Pulses", Opt. Lett., vol. 19, p. 664 (1994)

[88] A. M. Weiner, D. E. Leaird, D. H. Reitze and E. G. Paek, "Femtosecond Spectral Holography", IEEE J. Quant. Electron., vol. 28, p. 2251 (1992)

8

PHOTOREFRACTIVE FIBERS

Lambertus Hesselink
Department of Electrical Engineering
Room 353 Durand
Stanford University
Stanford, Ca 94305-4035

1. INTRODUCTION

Photorefractive processes in bulk materials have been considered in great detail during the last decade [1]. The study of photorefractivity in fibers or waveguides, however, is relatively recent. The body of literature is rather small, even when taking the broader view that photorefractivity encompasses phenomena in which a change in the index of refraction is induced by the presence of optical beams in the medium. In this approach, Stimulated Brillouin Scattering (SBS) and Stimulated Rayleigh Scattering (SRS) are viewed as nondegenerate photorefractive two-wave mixing processes. Early work on phase conjugation in fibers was initiated by Hellwarth and colleagues [2] using SBS, contained in multimode capillary tubes. In the late eighties, at Stanford University, the first fibers of Strontium Barium Niobate (SBN) were grown [3] and their photorefractive properties for optical data storage applications [4] [5] were studied. These studies involve multimode fibers. In single mode fibers, Hill et al [6] have studied the development of gratings giving rise to a reflected signal that increases as a function of time until steady state is reached. The origin of this phase conjugate signal is not fully understood, but it is speculated that the photorefractive effect is responsible for the reflected signal. At approximately the same time Yariv and coworkers were investigating aberrations introduced by image propagation through fibers [7] [8] [9]. They suggested that phase conjugation could be used to correct for modal distortions. In the conventional approach, bulk crystals of a photorefractive media could be used to accomplish compensation for these distortions, but fibers may be more suitable in this particular application [8]. More recently, extensive research on applications of fibers for optical storage [5], signal processing [10], and neural networks has been carried out.

Fibers of photorefractive material exhibit certain advantages over bulk media. In fibers, electromagnetic waves are confined to small cross sections over long distances, thereby giving rise to high intensities. As a result, nonlinear effects are substantial using only modest laser powers. In fibers where the pump beams are confined to the lowest order mode, image fidelity can be achieved with high efficiency, whereas in multimode fibers good fidelity can only be maintained at reduced efficiency. The efficiency of phase conjugation using guided modes can, however, still be higher than for phase conjugation with unconstrained waves under similar power conditions. As an added advantage, the acceptance angle of the waves can be large—as large as the acceptance angle of the guide—and pump beam alignment is not as critical.

In this chapter we discuss three phase conjugation configurations. In the first, three beams are incident on the waveguide: two pump beams P1 and P2, and an image bearing beam S. As a result of the interaction, a phase conjugate signal beam S^* is generated. This arrangement, shown in Figure 1, is analogous to four wave mixing in extended media. In this case, however, there are terms that spoil the phase conjugation process. These terms tend to cancel more completely when the guide length increases, as the modes in the signal beams become more numerous, and when imperfections of the wave guide become more pronounced. In any event, for phase conjugation to occur, one of the pump or signal beams must be well guided. Under these conditions the efficiency of the phase conjugation process is proportional to the pump beam intensities, to the square of the effective waveguide length, to the fraction of one of the pump beams that is phase conjugated to the other pump, to the square of the nonlinear susceptibility coefficient and inversely proportional to the square of the cross sectional area.

The theory of phase conjugation with only one pump beam (P2) involves a slight modification of the previous case. The power of the phase conjugated beam is now proportional to the square of the power in the signal beam, as opposed to the power directly. The phase conjugate signal in either case may be different in frequency from the incident beams. The counter propagating beam may also be generated by Fresnel reflection from the back surface of the waveguide, or by amplification of scattered waves. In that case we deal with self-phase-conjugation. This approach is of particular interest in pixel processing of images. A bundle of fibers could be placed in the Fourier plane of an optical processor allowing individual spatial frequencies incident on it to be phase conjugated with an amplitude that can be modified by controlling the back reflection coefficient of each fiber. Such an approach could also be implemented using integrated optics.

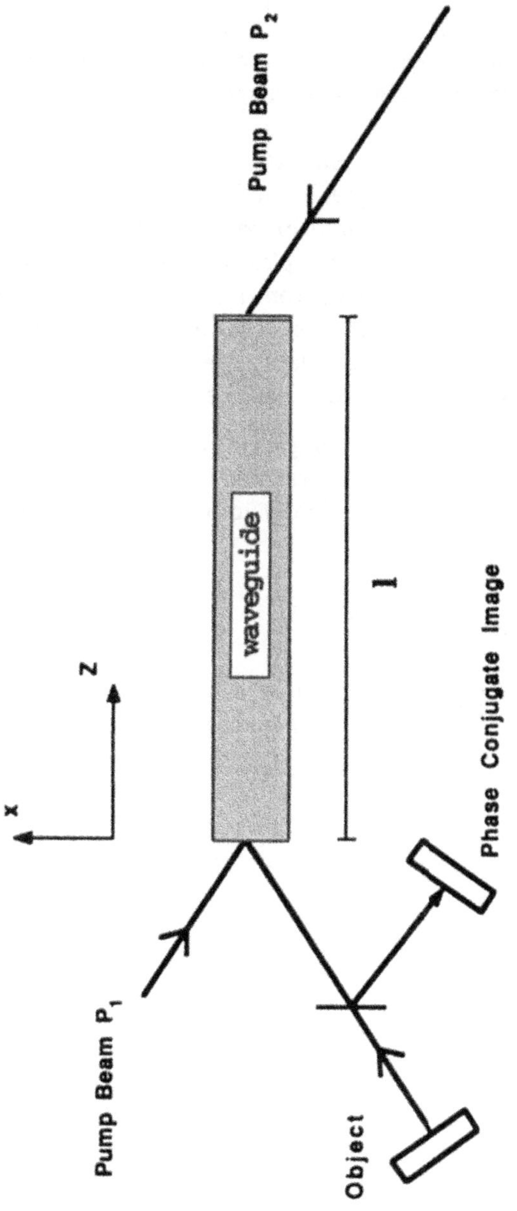

Figure 1: Phase conjugation in fibers.

In this context, a bundle of fibers may be thought of as a synthetic material with interesting properties. For example, in data storage applications, multiple images could be stored in each fiber by a combination of angular, or wavelength and spatial multiplexing. A stack of images, referred to as a page, are isolated from pages stored in adjacent fibers, thereby limiting cross talk between stacks. As an added advantage, the angular sensitivity of multiple images recorded in fibers is three to four times higher than for recording in bulk crystals of equal length. As a result, the angular multiplexing task becomes considerably simpler to implement. For wavelength multiplexing the in-line configuration of object and reference beam are attractive from a system layout point of view. Fiber bundles of this kind allow applications that are not feasible with bulk crystals. In addition, fibers of good optical quality are generally easier to grow than bulk crystals and often exhibit reduced striations.

In the remainder of this chapter a general overview of the analysis for phase conjugation in waveguides is first presented. Then three different cases of phase conjugation are considered, and finally important applications involving nonlinear fibers are discussed.

2. GENERAL THEORY

With reference to Figure 1, consider two beams incident on a nonlinear medium. The two pump beams P1 and P2 confined to the lowest order mode are expressed as:

$$E_{p1} = E_{p1}(x, y) \exp[i(\beta_p z - \omega t)] \tag{1}$$

and

$$E_{p2} = E_{p2}(x, y) \exp[i(-\beta_p z - \omega t)] \tag{2}$$

The signal beam and its phase conjugate replica are expressed in terms of the propagating eigenmodes, neglecting absorption,

$$E_S = \frac{1}{2} \sum_a B_a^S(z) \epsilon_a(x, y) \exp\left[i(\beta_a z - \omega t)\right] + c.c. \tag{3}$$

$$E_R = \frac{1}{2} \sum_a B_a^R(z) \epsilon_a(x, y) \exp\left[i(-\beta_a z - \omega t)\right] + c.c. \tag{4}$$

where ϵ_a satisfy the wave equation with the forcing term equal to zero:

$$\left(\frac{\partial^2}{\partial x^2} + \frac{\partial^2}{\partial y^2} - \beta_a^2 \right) \epsilon_a(x, y) + \omega^2 \mu \epsilon(\mathbf{r}) \epsilon_a(x, y) = 0 \tag{5}$$

The wave equation is given by

$$\nabla^2 \mathbf{E} - \mu \epsilon(\mathbf{r}) \frac{\partial^2 \mathbf{E}}{\partial t^2} = \mu \frac{\partial^2}{\partial t^2} P_{NL}(\mathbf{r}, t) \tag{6}$$

The eigenmodes are orthogonal, satisfying:

$$\int_{-\infty}^{\infty} \epsilon_a(x,y)\epsilon_b(x,y)dx\,dy = \delta_{ab} \tag{7}$$

Upon substitution of (3) and (4) into (6), using (5) and assuming the slowly varying amplitude approximation we get:

$$\sum_a -i\beta_a \frac{dB_a^R}{dz}\epsilon_a \exp[i(-\beta_a z - \omega t)] + \text{c.c.} = \mu\frac{\partial^2}{\partial t^2}P_{NL}(\mathbf{r},t) \tag{8}$$

$$\sum_a +i\beta_a \frac{dB_a^S}{dz}\epsilon_a \exp[i(\beta_a z - \omega t)] + \text{c.c.} = \mu\frac{\partial^2}{\partial t^2}P_{NL}(\mathbf{r},t) \tag{9}$$

Or, by slightly rewriting (8) and (9) using $P_{NL}^{(w_4=w_1+w_2-w_3)} = \chi^{(3)}E_{p1}E_{p2}E_S^*$ we get:

$$\sum_a -\beta_a \frac{dB_a^R}{dz}\epsilon_a(x,y)\exp(-i\beta_a z) = \frac{i\omega^2\mu\chi^{(3)}}{2}E_{p1}E_{p2}\cdot\sum_a B_a^{S*}(z)\epsilon_a(x,y) \exp(-i\beta_a z) \tag{10}$$

Similarly from (9) using conjugation (assuming E_{p1}, E_{p2}, $\chi^{(3)}$ to be real)

$$\sum_a -\beta_a \frac{dB_a^{S*}}{dz}\epsilon_a(x,y)\exp(-i\beta_a z) = -\frac{i\omega^2\mu\chi^{(3)}}{2}\cdot E_{p1}E_{p2}\sum_a B_a^{R*}(z)\epsilon_a(x,y) \exp(i\beta_a z) \tag{11}$$

The coupling equations are then found by multiplying (10) by ϵ_s and integrating over the cross section using (7) to obtain:

$$\frac{dB_s^S}{dz} = -\frac{i\omega^2\mu\chi^{(3)}}{2\beta_s}\sum_a B_a^{S*}(z)\exp[-i(\beta_a-\beta_s)\,z] \iint E_{p1}E_{p2}\epsilon_a\epsilon_s\,dx\,dy \tag{12}$$

$$\frac{dB_s^{R*}}{dz} = -\frac{i\omega^2\mu\chi^{(3)}}{2\beta_s}\sum_a B_s^s(z)\exp[-i(\beta_a-\beta_s)\,z] \iint E_{p1}E_{p2}\epsilon_a\epsilon_s\,dx\,dy \tag{13}$$

From these equations we note that a given mode B_s^S from the phase conjugate wave is coupled to all modes of the incoming signal beam. Substantial coupling is avoided when the pump beam E_{p1} and E_{p2} are approximately constant over the cross section, thereby reducing the overlap integral in (12) and (13) to zero, except when $a = s$. Alternatively, if $l \gg 2\pi(\beta_a - \beta_s)^{-1}$, the two modes are largely mismatched giving rise to little beam coupling. Under these conditions (12) and (13) reduce to

$$\frac{dB_s^R}{dz} = -\frac{i\omega^2\mu\chi^{(3)}}{2\beta_s}E_{p1}E_{p2}B_s^{S*} \tag{14}$$

$$\frac{dB_s^{S*}}{dz} = -\frac{i\omega^2\mu\chi^{(3)}}{2\beta_s}E_{p1}E_{p2}B_s^{R} \tag{15}$$

These equations are similar to those describing four wave mixing with plane waves. Using the boundary conditions that $B_s^S(z=0)$ are given and $B_s^R(l) = 0$ at the end of the fiber we find as solutions:

$$\begin{aligned} B_s^R(z) &= iB_s^{S*}(0)(\sin\kappa_s z - \tan\kappa_s l\cos\kappa_s z) \\ B_s^S(z) &= B_s^S(0)(\tan\kappa_s l\sin\kappa_s z + \cos\kappa_s z) \end{aligned} \tag{16}$$

where

$$\kappa_s = -\frac{\omega^2\mu\chi^{(3)}}{2\beta_s}E_{p1}E_{p2} \tag{17}$$

The reflected wave can now be expanded as

$$B_s^R(0) = -iB_s^{S*}(0)\tan\kappa_s l \tag{18}$$

or by (1) and (2) assuming $\kappa_s = \kappa \neq \kappa_s(s)$

$$E_R(x,y,z=0) = -i\tan\kappa l\sum_a \frac{1}{2}B_a^{S*}(0)\epsilon_a(x,y)\exp(-i\omega t) + \text{c.c.} \tag{19}$$

and since

$$E_S(x,y,z=0) = \sum_a \frac{1}{2}B_a^S(0)\epsilon_a(x,y)\exp(-i\omega t) + \text{c.c.} \tag{20}$$

We get

$$E_{pc} = E_R(x,y,z=0) = -i\tan(\kappa l)E_S^*(x,y,z=0) \tag{21}$$

From this expression we note that the reflected wave is the phase conjugate of the incident beam whose amplitude may be magnified for $\kappa l > \frac{\pi}{4}$.

These results are similar to those obtained in bulk media, except that by guiding the pump beams substantially lower intensities are needed over the interaction region.

In a practical arrangement fiber losses need to be taken into account. An additional term $\exp(-i\alpha z)$ needs to be included into eqn (3) and (4) to get:

$$E_S = \frac{1}{2}\sum_a B_a^S(z)\epsilon_a(x,y)\exp[-\frac{1}{2}\alpha z + i(\beta_a z - \omega t)] + c.c. \tag{22}$$

$$E_R = \frac{1}{2}\sum_a B_a^R(z)\epsilon_a(x,y)\exp[\frac{1}{2}\alpha(z-l) + i(-\beta_a z - wt)] + c.c. \tag{23}$$

where α is assumed to be mode independent.

Upon substitution of these expressions into the wave equation we derive as the coupled wave equations for E_S and E_R:

$$\frac{dB_a^R}{dz} = i\kappa^* \exp(-\alpha z) B_a^{S*} \tag{24}$$

and

$$\frac{dB_a^{S*}}{dz} = i\kappa \exp[\alpha(z-l)] B_a^R \tag{25}$$

where $\kappa^* = \frac{2\pi\omega}{cn}\chi^{(3)}E_{p1}E_{p2}$ where c is the speed of light and n the index of refraction of the medium. The solutions to these equations using the same boundary conditions as before are:

$$E_R = E_{pc}(x,y,o) = \frac{-2i\kappa^* \tan(K_s l)\exp(-\frac{1}{2}\alpha l)}{\alpha\tan(K_s l) + 2K_s} E_S^*(x,y,0) \tag{26}$$

and

$$K_s = [|\kappa|^2 \, exp(-\alpha L) - (\frac{1}{2}\alpha)^2]^{1/2} \tag{27}$$

Under experimental conditions similar to those assumed for this analysis, Au-Yeung et al. [9] observed continuous phase conjugation in a 4 mm hollow capilary tube filled with CS_2 and using plane waves. The two pump beams are counterpropagating and the signal beam was slightly off-axis injected into one end of the fiber. A weak phase conjugate signal was observed (0.45% reflectivity), but the arrangement was not suitable for evaluation of image quality using a signal beam having a large number of resolution spots. In that case the analysis presented here must be modified [2] , because the restriction that the pump beams are plane waves is then not satisfied.

3. FOUR-WAVE MIXING IN OPTICAL FIBERS USING MULTI-MODE PUMP BEAMS

The special case studied in the preceding section requires that the pump beams are coupled to the lowest-order mode in the waveguide. In practice this condition is not usually fulfilled. Here we explore the circumstances under which these conditions can be relaxed, and phase conjugation may still be achieved.

Assume again the form of the solution for the reflected wave and incident beams to be:

$$E_i^R = \sum_a B_a^R(z)\epsilon_{ai}(x,y)\exp\left[i(-\beta_a z - \omega t) + \frac{1}{2}\alpha z\right] \tag{28}$$

$$E_i^S = \sum_a B_a^S(z)\epsilon_{ai}(x,y)\exp\left[i(\beta_a z - \omega t) - \frac{1}{2}\alpha z\right] \tag{29}$$

$$E_i^{p1} = \sum_a B_a^{p1}(z)\epsilon_{ai}(x,y)\exp\left[i(\beta_a z - \omega t) - \frac{1}{2}\alpha z\right] \tag{30}$$

$$E_i^{p2} = \sum_a B_a^{p2}(z)\epsilon_{ai}(x,y)\exp\left[i(-\beta_a z - \omega t) - \frac{1}{2}\alpha z\right] \tag{31}$$

and the coefficients $B_a(z)$ are computed by integrating Maxwell's equations under the assumption that the reflected wave does not appreciably affect the other waves in the guide:

$$B_a^R(0) = (\frac{2\pi\omega}{icn})B_b^{S*}B_c^{p1}B_d^{p2}\int_0^l \exp[i\Delta kz - \frac{1}{2}\alpha z]dz \tag{32}$$
$$\int\int \chi^{(3)}_{pqrs}\epsilon^*_{ap}\epsilon^*_{bq}\epsilon_{cr}\epsilon_{ds}\,dx\,dy$$

and

$$\Delta k \equiv k_a + k_c - k_b - k_d \cong (u_a^2 + u_c^2 - u_b^2 - u_d^2)/2k \tag{33}$$

where $k \equiv n\omega/c$ and u_a are the eigenvalues of the eigenmodes ϵ_{ai}.

To achieve phase conjugation of the signal wave E_S, all terms in (32) must be small, except those for which either

$$a = d, b = c \tag{34}$$

or

$$a = d, c = d \tag{35}$$

Consider the latter case first and note that contributions from condition (35) to $B_a^R(0)$ are:

$$\hat{B}_a^R = \sum_\ell \frac{2\pi\omega Z}{icn}\gamma_{a\ell}B_a^{s*}B_\ell^{p1}B_\ell^{p2}, \tag{36}$$

$$Z = \int_0^l \exp(-\frac{1}{2}\alpha z)dz \tag{37}$$

and

$$\gamma_{a\ell} \equiv \sum_{ijkl}\int \chi^{(3)}_{ijkl}\epsilon^*_{ai}\epsilon^*_{aj}\epsilon_{ck}\epsilon_{cl} \tag{38}$$

In order to assure that the reflected wave is indeed a phase conjugate beam of good fidelity, the term $\sum_\ell \gamma_{a\ell} B_\ell^{p1} B_\ell^{p2}$ should not vary appreciably with index mode a and $\hat{B}_a$ provided that the terms associated with (34) and (35) are dominant. The first condition has been examined in detail by Hellwarth and appears generally satisfied. The second condition can be achieved by assuring that the waveguide has strong enough birefringence or inhomogeneities such that mode degeneracy is restricted to only a few or no modes at all. This is true if the following condition is satisfied:

$$\Delta k Z >> 1 \tag{39}$$

In a practical environment high fidelity phase conjugation can be achieved by using a long fiber in which a large number of high modes are excited, while satisfying condition (36). Inhomogeneities in the guide also tend to enhance the fidelity of the phase conjugate signal.

The efficiency with which a phase conjugate signal can be generated is computed by considering the fraction of energy in the back scattered wave to the energy in the forward signal:

$$R = \frac{| \hat{B}_\ell^S B_\ell^R |^2}{(B_\ell^S B_\ell^{S*})^2} \tag{40}$$

This expression may be evaluated using:

$$\sum_a | B_a^R |^2 \gamma_{ac} \cong \bar{\gamma} \sum_a | B_a^R |^2 = c_{xxxx}/A \tag{41}$$

assuming that $\chi_{ijkl}^{(3)} = 6c_{ijkl}$, and A is the area of a rectangular guide in which mode indices a,c are not equal.

Upon substitution of (41) into (40) we get:

$$R \approx \left| \frac{12\pi\omega Z c_{xxxx} B_a^{p1} B_a^{p2}}{ncA} \right|^2 \tag{42}$$

This expression can be rewritten in terms of the energy fraction

$$E^{p1,p2} = \frac{| B_a^{p1} B_a^{p2} |^2}{B^{p1} B_a^{p1^*} B_a^{p2} B_a^{p2^*}}, \tag{43}$$

$$P^{p1} = \frac{nc}{8\pi} B_a^{p1} B_a^{p1^*} \tag{44}$$

$$P^{p2} = \frac{nc}{8\pi} B_a^{p2} B_a^{p2^*} \tag{45}$$

and

$$\Delta \equiv \frac{| 96\pi^2 w^2 c_{xxxx} |}{n^2 c^2} \tag{46}$$

$$R \approx \frac{\Delta^2}{A^2} Z^2 E^{p1,p2} P^{p1} P^{p2} \tag{47}$$

This expression is valid for $R \leq 0.5$ and must be modified to include higher order terms in B_a^S for $R > 0.5$. The result (47) is similar to that obtained for phase conjugation in bulk media with the modification that the plane-wave pump beam intensities are replaced by P^{p1}/A and P^{p2}/A and the reduction factor $E^{p1,p2}$ accounts for an overall reduction in the phase conjugate signal strength. The main advantage of phase conjugation in fibers as compared with bulk media stems from the lower power requirements of the interacting beams. In particular, for long guides where $Z \gg A/\lambda$, confinement assures high intensities over long interaction lengths.

In examining equation (47), we note that for high R, the overlap factor $E^{p1,p2}$ should be large. This can be achieved in guides that are sufficiently perfect that little mode degeneracy occurs. Under these conditions the terms in the integral corresponding to (34) and (35) may no longer dominate, and the fidelity of the phase conjugation process would be jeopardized. In practice small inhomogeneities usually produce sufficient scattering that degeneracies are virtually eliminated. The efficiency of the phase conjugated signal, however, is then usually small ($R < 0.5$).

4. PHASE CONJUGATION WITH ONE EXTERNAL PUMP BEAM

From a practical point of view, it is desirable to eliminate the forward propagating pump beam, because it makes the setup simpler. In an imperfect guide the forward pump beam E^{p1} and the signal beam may be the same single forward propagating beam E^F. The expression for $\chi^{(3)}$ is then replaced by:

$$P_i^{NL} = 6c_{ijkl}(-\omega, -\omega, \omega, \omega) E_j^{F^*} E_k^F E_l^{p2} \tag{48}$$

As a result, the backward propagating beam then is found by replacing $B_b^{S^*}$ by $B_c^{p1^*}$. The remainder of the argument presented in section 3 is then still valid. Experimentally, this configuration has been investigated by Jensen and Hellwarth [11]. The phase conjugate signal power is now proportional to the square of the power of the input signal as opposed to the first power.

5. PHOTOREFRACTIVE ISSUES SPECIFIC TO FIBERS

In contrast to photographic film, no chemical processing is needed to produce a phase hologram, which makes photorefractive media attractive for real-time optical signal processing applications such as matched filters, holographic data storage systems and optical image processors, requiring an erasable, reusable, read-write medium. In addition, as a consequence of the spatial offset between the light interference pattern and the index of refraction perturbation in the medium, energy can be transferred from one beam to the other. This property of photorefractive materials gives rise to interesting processes including image amplification and beam combining, as discussed elsewhere in this book.

To produce a strong photorefractive phase hologram, the grating vector should be oriented so as to use the largest electro-optic tensor coefficients characterizing the medium. An applied field parallel to the grating vector usually further enhances the index perturbation. This implies that the recording and readout arrangements of the hologram are subject to geometrical constraints. For example, in SBN a grating vector parallel to the c-axis of the crystal produces the largest index perturbation through the r_{33} electro-optic coefficient, using extraordinaryly polarized light. For transmission holograms in SBN this requires that the c-axis is perpendicular to the surface normal, as shown in Figure 2, or parallel to the surface normal for a reflection hologram. In general, transmission holograms are stronger than reflection holograms for the same light intensities of the interfering beams, because the space-charge field saturates at a lower value for smaller grating spacings (large grating k-vectors), with few exceptions most noticeably $LiNbO_3$, in which reflection gratings can be strong due to a large photovoltaic field.

The anisotropic crystal properties also imply preferred directions of energy transfer between interfering beams. For example, in SBN energy is coupled in the direction of the c-axis, which is usually chosen to be parallel to the grating vector. A weak probe beam making an acute angle with the c-axis can be amplified by a strong pump beam symmetrically incident with respect to the surface normal as shown in Figure 2. Directional energy coupling, unfortunately, also has disadvantages. Small compositional variations in the crystal or other index perturbations may scatter light which can be amplified by beam coupling causing beam fanning. A well collimated Gaussian laser beam, for example, may become distorted in its intensity distribution and beam direction while propagating through such a medium. The precise cause of this type of scattering is not well understood for most photorefractive media, but generally depends strongly on crystal quality and uniformity. These effects may be minimized through improved materials.

5.1. Materials Issues

As an added complication, the photorefractive space charge field is erased upon readout. Uniform illumination redistributes the trapped charges evenly, thereby erasing recorded information. A process similar to fixing in photographic film is needed to preserve stored holograms. One of the most effective methods discovered so far involves writing a complementary space charge field at elevated

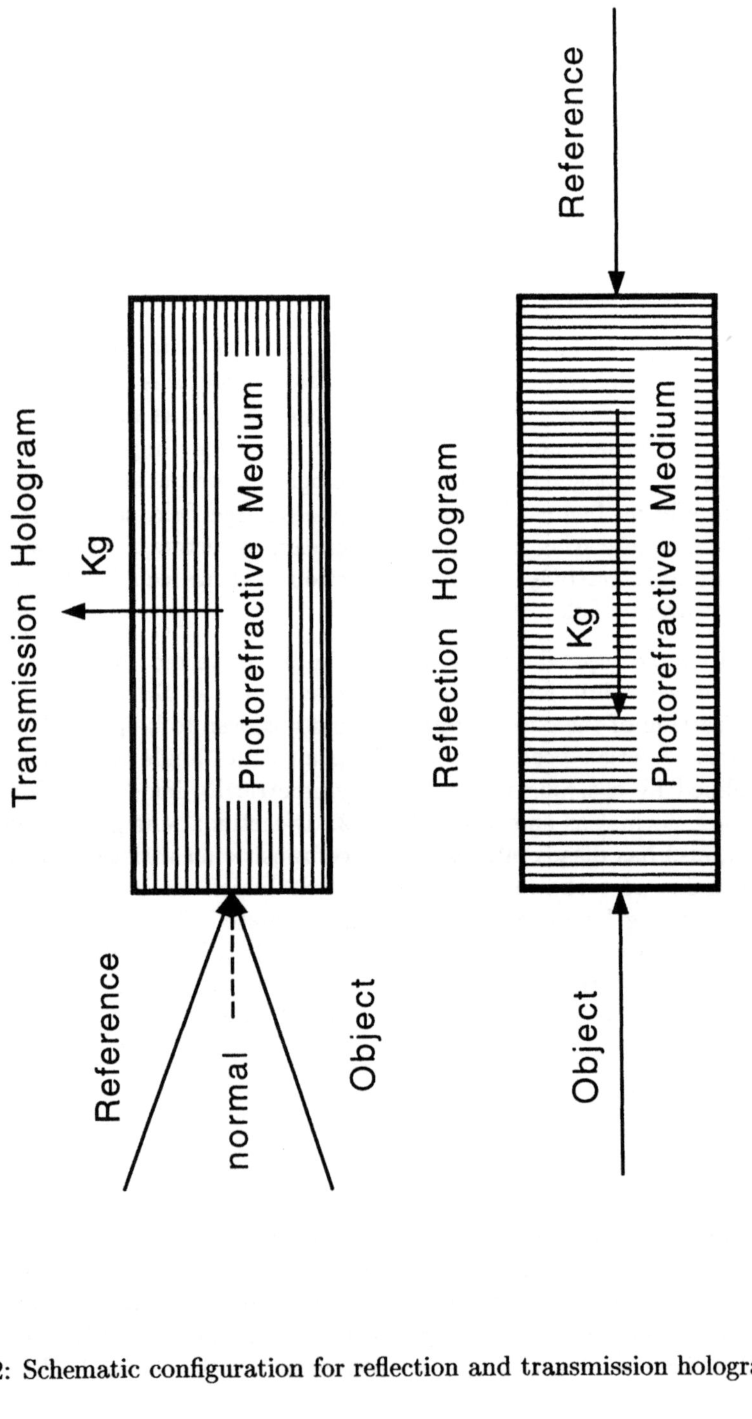

Figure 2: Schematic configuration for reflection and transmission holograms in fibers

temperatures using ionic charge carriers and reading it out at room temperature. The photogenerated space charge field gives rise to a complementary ionic space charge field which partially screens it. At elevated temperatures (approximately 200° C in $LiNbO_3$) the mobility of the ionic charge carriers is sufficiently high to produce a screening grating. Upon readout at room temperature, where ion mobility is rather low, the photo-generated grating is erased revealing the ionic space charge field, usually in the presence of an applied field. This procedure has been most thoroughly studied in specially prepared $LiNbO_3$ crystals, where protons are identified as the ionic charge carriers. Fixing in other ferroelectrics is much less understood and a topic of intense research at present. The most promising approach to overcoming these problems is a combined materials development effort guided by practical architectures for signal processing and holographic data storage applications. Through an integrated, multi-disciplinary effort, considerable progress has been made in recent years in improving the optical quality and photorefractive properties of the most promising ferroelectric media such as SBN, $BaTiO_3$, KTN and $LiNbO_3$. Of considerable significance has been the development of novel crystal growth techniques.

Important photorefractives, including SBN and BSO, are grown in bulk form by the Czochralski technique. The constituent chemicals are melted inside a (platinum) crucible by induction heating, and crystals are grown by dipping a small seed rod into the melt and withdrawing it at a rate commensurate with the growth dynamics. The temperature in the center of the crucible is generally lower than at the edges and the bottom, with the melting isotherm determining the crystal diameter. The effects of random temperature fluctuations in the liquid are reduced by rotating the grown crystal about its axis, while it is being pulled from the melt. Although this stabilizes crystal growth, fluctuations in chemical composition as well as in temperature of the melt cause rotation-induced growth striations. These striations manifest themselves as approximately helical surfaces of varying index of refraction, their pitch (or spacing) being determined by the rotation and pull rate of the boule. The use of extremely pure starting materials and temperature control tends to reduce growth striations and improve diameter control, but crystals of large size, good optical quality, and economical price are difficult to grow consistently by the Czochralski method.

$Sr_xBa_{1-x}Nb_2O_6$ (SBN), for example, is a tungsten bronze ferroelectric material possessing a large electrooptic effect, and as a photorefractive it has good optical sensitivity, a large steady-state response and a relatively short recording time response that can be varied over a wide range by changing dopant concentration and species as well as lattice composition (for x=0.25 to x=0.75). For SBN, the radial temperature gradients imposed by the small difference in melting points between SBN and the crucible are small. The shallow gradients make crystal growth unstable, producing random fluctuations in crystal diameter for all but the smallest size boules, while crystal rotation generally stabilizes growth, but induces striations.

Experimentally it has been found that SBN grows preferentially along the c-axis (i.e., the [001] direction), implying that the striations appear essentially perpendicular to this axis, with a spacing and index of refraction variation similar to those of the desired photorefractive holograms. Unfortunately, this implies that the noise gratings are largely aligned with the desired photorefractive gratings, because in SBN the largest holographic response occurs when the grating vector is parallel to the c-axis. As a result, the noise gratings cause considerable scatter in the direction of the diffracted signal beam. Attempts to grow SBN along other directions have generally been unsuccessful. In other photorefractive materials the effects of striations may not be so detrimental, because growth-induced striations are oriented in other directions than the desired holographic gratings, or are not present. $LiNbO_3$, for example, is available in good optical quality, in large sizes and at relatively low cost, but it has a time constant orders of magnitude slower than SBN for a similar photorefractive response. Through materials research, photorefractive properties and optical quality may be improved.

5.2. Fiber Synthesis

For example, to overcome some of the difficulties of bulk SBN growth, a novel crystal growth technique has been described by Feigelson in 1986 [12], and has been analyzed by Fejer [13]. The laser-heated pedestal growth (LHPG) technique as it is commonly referred to, may be used to produce single-crystal multimode fibers or crystallites of diameters ranging from 100 μm to 800 μm. Referring to Figure 3, the tip of a small source rod is uniformly illuminated and heated by a high intensity CO_2 laser beam using a reflaxicon followed by a parabolic mirror.

A seed crystal is then dipped into the molten zone, and growth proceeds in a miniature float-zone fashion by pulling the fiber crystal (at rates of a few millimeter per minute) away from the melt while feeding the source rod at a commensurately steady-state rate. The first SBN c-axis fibers were grown at Stanford University at the Center for Materials Research [14],[3]. Initial compositions included x=0.46 and x=0.75, which were noted to be essentially striation free.

As noted earlier, it is highly desirable for holographic applications to grow a-axis (i.e., [100]) fibers for obtaining optimal photorefractive response. For SBN:60 at room temperature, for example, $r_{33} \approx 235$ pm/V and $r_{13} \approx 47$ pm/V. This is significant, because compared with reflection holograms in SBN, the lower

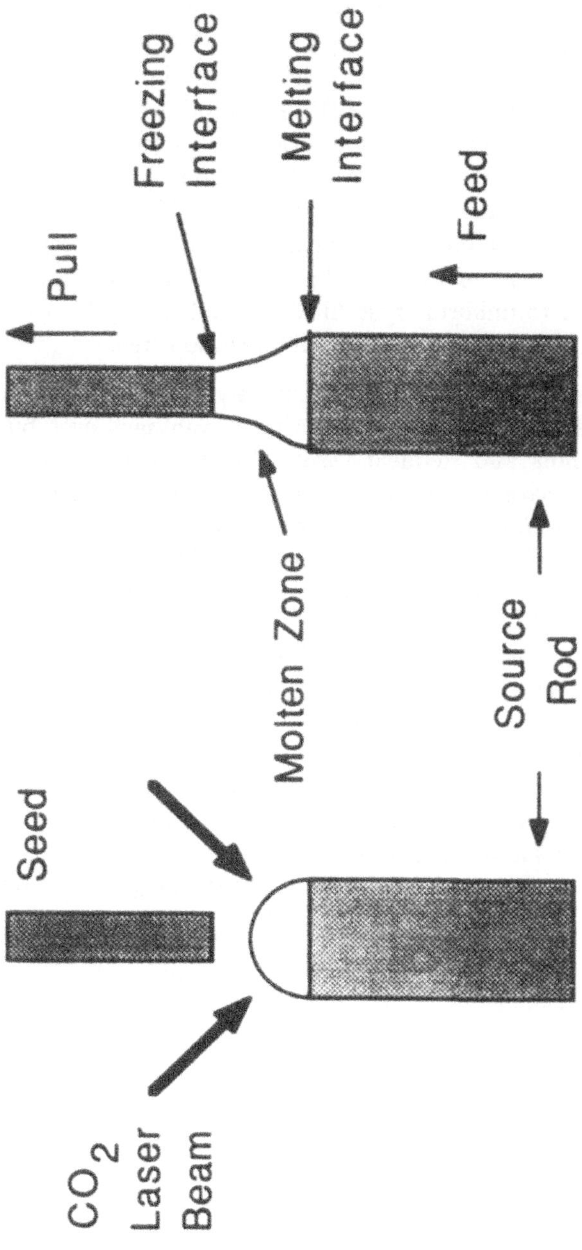

Figure 3: Schematic arrangement for the laser heated pedestal growth technique.

spatial frequency components of the grating allow a much larger space charge field to develop before saturation limits arise. Despite significant anisotropy in the growth characteristics of SBN, good quality, but short, a-axis ([100]) and [110] fibers can be grown with the LHPG method overcoming the limitations of the bulk Czochralski technique. Both orientations produce essentially the same results in terms of growth behavior and fiber morphology. Stable growth of a-axis SBN fibers was independently reported by Wilde et al. [5], and Sugiyama et al. [10] in 1991. The stability is a consequence of steep temperature gradients and tight confinement of the melt by the LHPG method. Since the LHPG method can lead to uniform crystallization, and no crystal rotation is needed, fibers grown in this manner are typically striation free.

Multimode fibers (or perhaps, more accurately described as crystallites) of photorefractive material exhibit certain other advantages over bulk media. For example, Hesselink and Redfield (1988) [3] used c-axis, Ce doped fibers for photorefractive characterization. They found that the angular sensitivity of multiplexed holograms in these fibers is substantially increased when compared with similar recording in bulk crystals. In addition, they noted that fibers may be bundled into an array so that multiple holograms (around 40) may be recorded in each fiber with little or no crosstalk between fibers. Such an array may prove to be a valuable recording medium as a replacement of bulk crystals in a holographic data storage device. In addition, in fibers, confinement of electromagnetic waves to small cross-sections over long interaction lengths gives rise to substantial nonlinear effects at modest laser powers. And photorefractive fibers could possibly be integrated with glass fibers into compact and efficient systems. Another real bonus of the LHPG technique is the efficient manner in which sample fibers can be grown in a few hours, thereby making dopant studies much more practical than when using the CZ technique.

Of course, there are potential disadvantages to the use of fibers for signal processing and storage applications as well. For imaging applications for example, waveguiding and mode coupling induce distortions that need to be corrected. Several such techniques are described below.

6. APPLICATIONS

6.1. Holographic Data Storage

Holographic data storage devices appear extremely attractive for applications requiring large capacity (exceeding Tbytes) and extremely high data transfer rates, on the order of Gbits/sec or more. A generic configuration using an array of SBN fibers or small crystallites that has been investigated at Stanford University since 1985 is shown in Figure 4. An electronic data stream representing information to be stored is transformed into an optical signal by use of

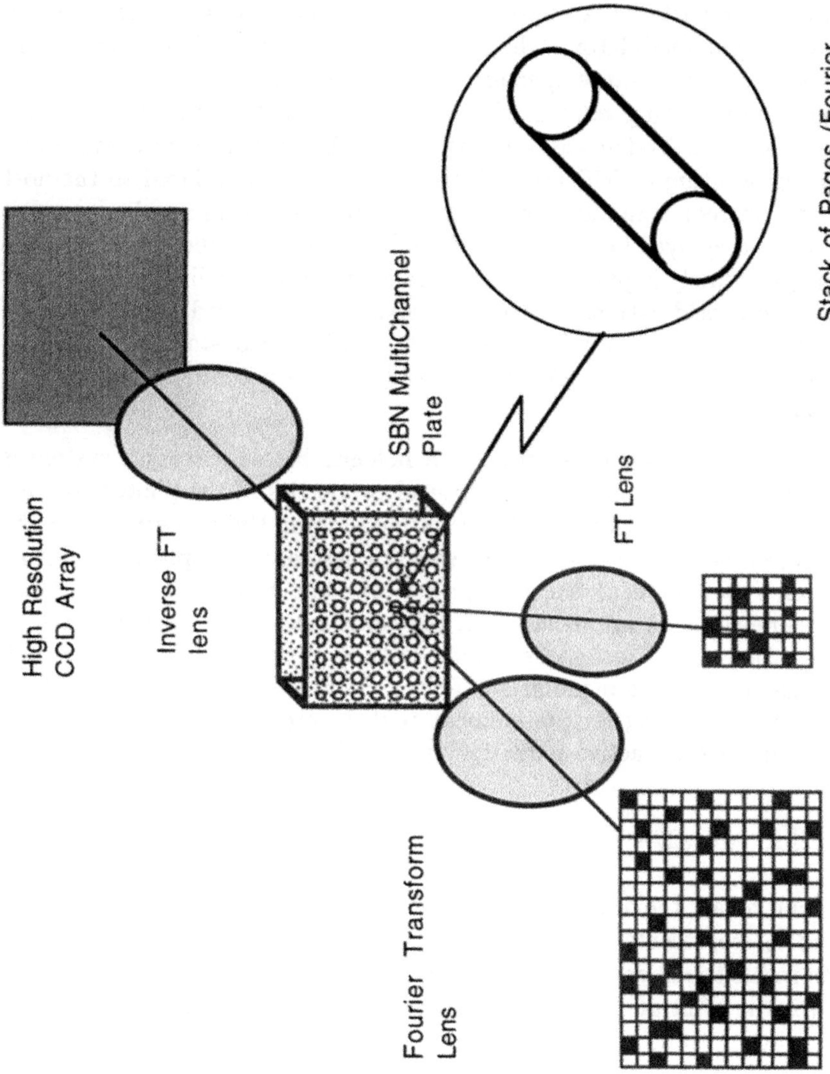

Figure 4: Schematic of a holographic data storage system implemented with fibers (after Hesselink et al, Ref. 14).

a spatial light modulator illuminated by a coherent light source. The Fourier transform of this signal is incident on a recording medium where a reference beam is superimposed. Multiple holograms are stored by either angular, phase encoding or wavelength multiplexing. For phase encoding a set of orthogonal binary phase patterns is imposed on the reference beam using a binary phase spatial light modulator. The superimposed signal and reference beam are scanned over the entrance aperture of the recording medium for utilization of the complete recording volume. Upon readout of the hologram with the appropriate phase-coded reference beam, the stored image is transformed into an electronic signal using a CCD array. This signal is then interfaced to the host processor for further manipulations using electronic signal processing, such as error decoding employed by Heanue et al. [15]. By using parallel readout of the CCD array, extremely high transfer rates are achieved within limitations set by signal-to-noise considerations, because all pixels of the stored bit pattern are reconstructed simultaneously. By using electro-optic or acousto-optic scanning, no moving parts are needed, making this architecture potentially very rugged and reliable.

The specific practical implementation of a holographic data storage system is to a large extent determined by the physical properties of the photorefractive crystal, as noted above. Here we discuss in more detail the properties of SBN, which, in recent years, have been intensively studied for this application. In the transmission geometry depicted in Figure 4, the image-carrying object beam and the reference beam are incident on the same side of the fiber array. The resulting grating vector inside the crystal is then approximately aligned with the c-axis giving rise to a modulation of the refractive index through the large r_{33} coefficient. Assuming for the moment that we are dealing with two plane waves, the effective refractive index for extra-ordinary light can be expressed as [16]:

$$r_{eff} = 1/2[n_o^4 r_{13}(cos2\theta - cos2\phi) + 4n_e^2 n_o^2 r_{42} sin^2\phi + n_e^4 r_{33}(cos2\theta + cos2\phi)]cos\phi \tag{49}$$

where n_o and n_e are the ordinary and extraordinary index of refraction, respectively, and r_{ij} represents the appropriate element of the electrooptic tensor. The angle inside the fiber between the object and reference beam is denoted by 2θ, and ϕ is the slant angle between the grating vector and the c-axis. For an a-axis fiber and a transmission hologram equation (49) simplifies to:

$$r_{eff} = (2n_e^2 n_o^2 r_{42} sin^2\theta + n_e^4 r_{33} cos2\theta)cos\theta \tag{50}$$

For c-axis fibers in a reflection geometry we get:

$$r_{eff} = (2n_e^2 n_o^2 r_{42} cos^2\theta + n_e^4 r_{13} cos2\theta)sin\theta \tag{51}$$

The two configurations may be compared on the basis of a figure of merit, Q expressed as: $Q = \frac{r_{eff}}{n\epsilon}$ where $n = n_e n_o/(n_o^2 cos^2\theta + n_e^2 sin^2\theta)^{1/2}$, and ϵ denotes the dielectric constant. For experiments conducted by Sugiyama et al ([10]), $2\theta = 6.3°$ for transmission holograms, and $2\theta = 173°$ for reflection holograms, implying that $Q = 3.1$ and $Q = 1.5$, respectively, using independently published values for the electrooptic and dielectric constants. A larger figure of merit implies a larger diffraction efficiency, as shown in Figure 5. The grating formation time for a given write power is inversely proportional to the figure of merit, favoring a-axis over c-axis fibers for data storage applications. The growth of long a-axis SBN fibers of good optical quality, however, presents a special challenge. Although striations are much reduced or absent compared with Czochralski grown crystals, in the core region of the fiber the index of refraction varies slightly, giving rise to light scattering. The cause and potential remedies of this perturbation are presently being investigated at several laboratories in the US and Japan.

The first photorefractive fibers of SBN of good optical quality were grown at Stanford University in the Center for Materials Research [3]. Fibers of length exceeding 10 cm and diameters varying from 160 μm to 800 μm have been demonstrated and cerium doping is used to increase the photorefractive response. These fibers are embedded in a host material (bakelite) to provide optical and electrical isolation. Since the fibers preferentially grow along the c-axis, the fiber and c-axis coincide. The fibers are poled along the principal direction. As a result, counter propagating beams are used in storage applications to take advantage of the large electrooptic coefficient.

With reference to Figure 4, two beams of Argon light ($\lambda = 514$ nm) are incident at opposite ends of a 4 mm long, 1 mm diameter fiber. Measurements are taken both for two wave mixing recordings of approximately 0.75 mm diameter collimated Gaussian beams and for Fourier transform hologram recordings of an Air Force RES 1 test pattern. The two wave mixing configuration uses collimating lenses, a beam splitter and mirrors for adjusting beam separation or the spatial frequency of the recording. The transform recording requires a spatial filter to expand the beam, the test pattern, and a transform lens in the object beam path. To adjust the entrance angle of the reference beam path the fiber is mounted on the shaft of a stepper motor in such a fashion that it rotates about the location of the entrance spot of the beam on the fiber. No voltage is applied across the fiber in these initial experiments. The reconstructed beam is detected or captured with either a photodiode or CCD array detector. To rotate the polarization of the beams a 1/2 wave plate is used and is inserted in front of the beam splitter. A high intensity incoherent light source is optionally used for purging the fiber of any remnants from previous recording.

The angular sensitivity of holograms recorded in both a co-propagating and counter-propagating configuration is measured as a function of the reference beam angle with respect to the Bragg angle. The reference beam angle is also varied during recording, resulting in different Bragg angles. These measure-

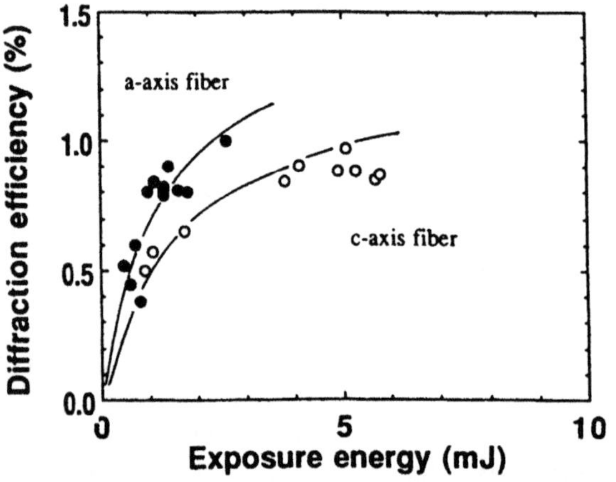

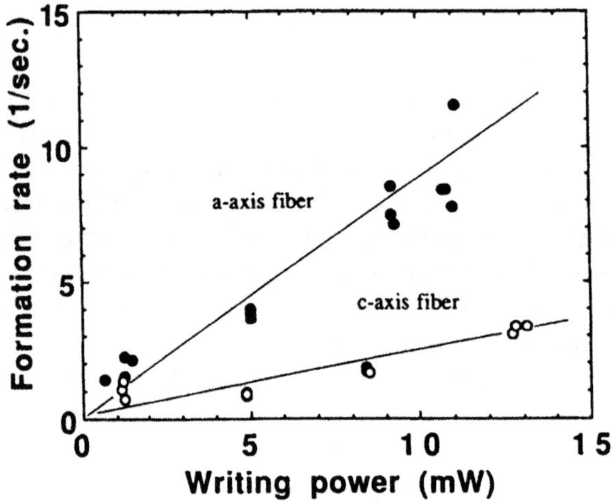

Figure 5: Comparison between a- and c-axis performance; a) diffraction efficiency versus exposure; b) grating formation versus writing power (after Sugiyama, Ref. 10).

ments are carried out for both two wave mixing of collimated beams and for recording of the Fourier transform of an Airforce RES 1 test pattern. Figure 6 shows the angular selectivity results.

It can be seen that the angular sensitivity when the reference beam is almost parallel to the axis is approximately 0.7 degree, which is similar to recording in bulk material of equal length. For larger off-axis reference beam angles, however, the angular sensitivity increases substantially and at a rate faster than would be the case for similar recording in bulk material. This improvement is significant for data storage applications, as explained above, as well as optical interconnect or crossbar switch applications where the number of connections to be made depends on the number of holograms stored in the photorefractive crystal.

The observed angular sensitivity of the reflective holograms recorded in the fiber can be explained by studying in detail internal wave propagation using coupled wave analysis. The rays inside the medium form caustics. As a result, the angles between the object and reference beams are no longer constant, but depend on the geometry of the caustic. From raytracing calculations we estimate that the extreme angles between object and reference beam rays have been increased sevenfold for the 20 degree case (angles measured outside the fiber). As a result the angular sensitivity, which is proportional to $\Delta\theta \approx \lambda/(D * \sin\theta)$ increases for the larger off-axis angles. Schematically, we may represent the reference beam as a filled cone which interferes with the object beam (directed along the axis) to give rise to a cone of grating K_g vectors, as shown in Figure 7. Here, one example of a reference ray lying inside the cone and in the $k_1 = 0$ plane as well as the object ray is depicted. The angular sensitivity is smallest on axis and decreases for the larger angles near the outer surface of the shaded region, indicated by the thick elliptical ring at the perimeter of this region and the larger cone surrounding the reference ray along the K_2 axis. The observed increased angular sensitivity for recording in fibers as compared with bulk materials is qualitatively in accordance with this model. Multiple images recorded in these fibers with an angular offset corresponding to the 10° curve shown in Figure 6 could be reconstructed independently with an angular variation in the reference beam of approximately 0.2 degrees. In the experiments, cross coupling between adjacent images with that much angular separation is of a magnitude similar to the background noise level.

For weak perturbations a phase conjugate readout (where the readout beam is the phase conjugate of the reference recording beam) can compensate for perturbations introduced by the core, as long as most of the scattered light is collected and reversed through the fiber. This approach has additional advantages of being able to (at least largely) compensate for distortions caused by modal phase dispersion, linear mode coupling as a result of fiber bending or inhomogeneities and surface irregularities, and intra-signal photorefractive

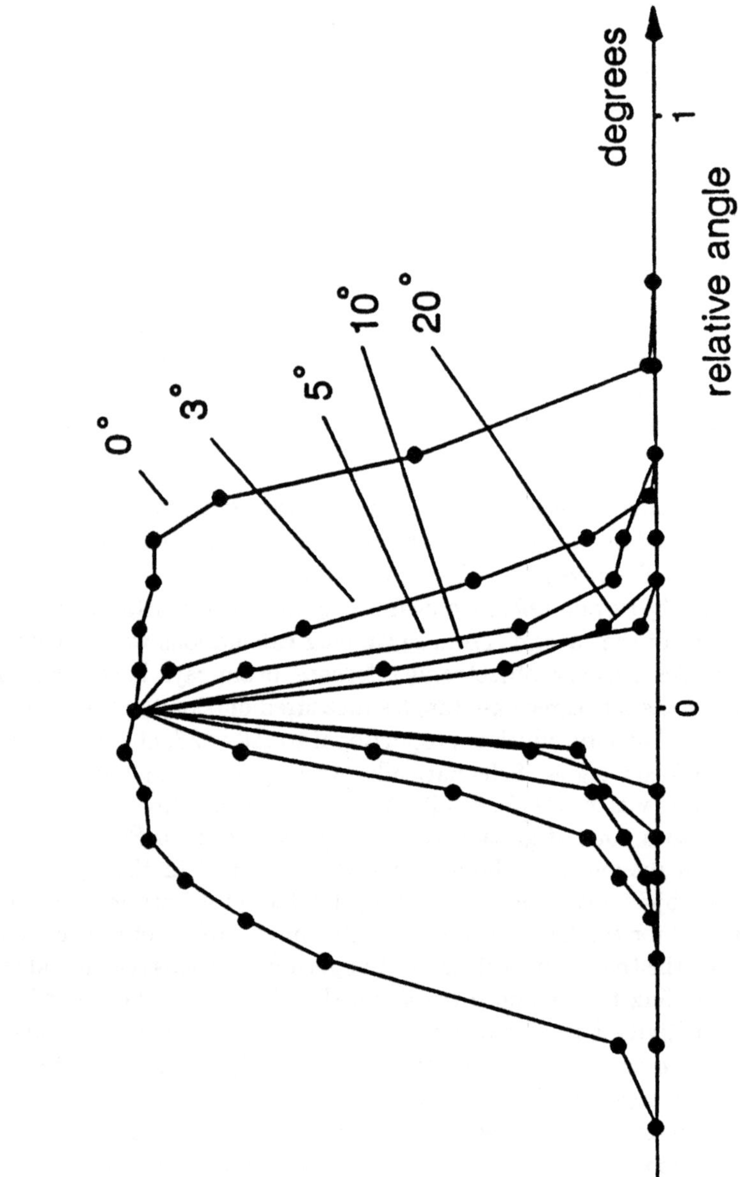

Figure 6: Angular sensitivity results for fiber recording (after Hesselink, Ref. 3).

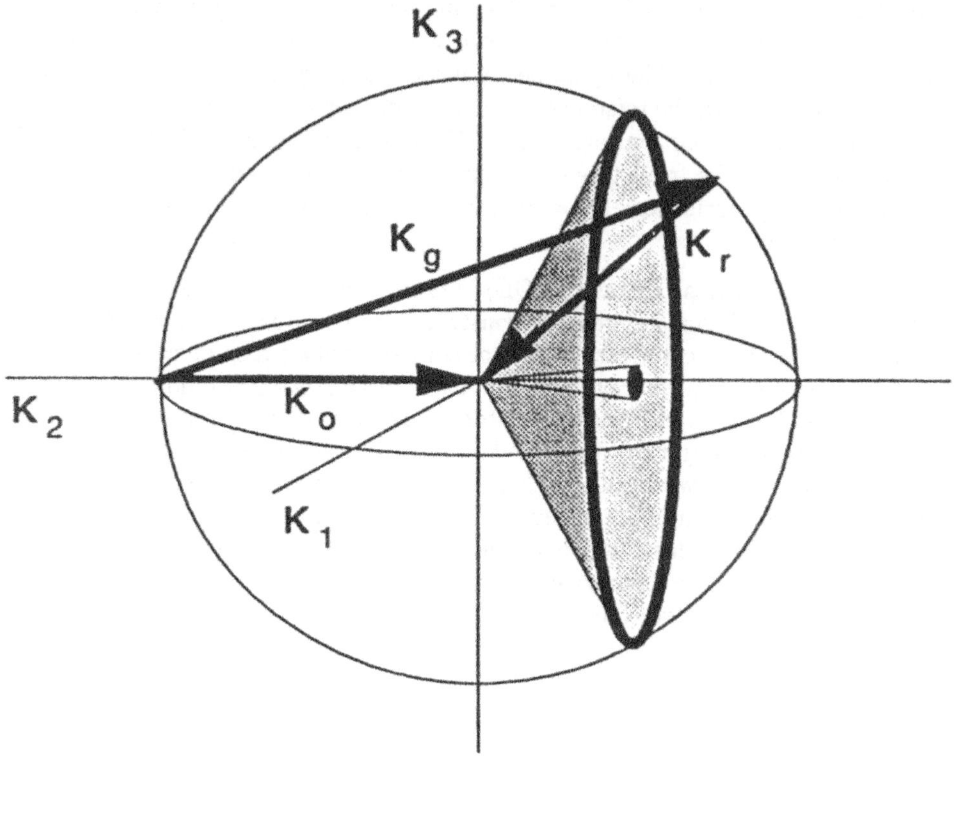

Figure 7: Angular sensitivity diagram for fiber recording (after Hesselink, Ref. 3).

coupling. This latter effect is peculiar to photorefractives, in which plane wave components of the object beam can couple to each other. These distortions are particularly troublesome when using fibers with large numerical apertures and for object waves guided over long distances.

Recently, Bashaw *et al.* [17] have observed phase-conjugate readout in a-axis fibers of SBN . The sample has been doped with 0.02 mole % Ce and is poled at room temperature with 400 V. The crystallite is 450 μm in diameter and 2 mm long. Holograms are written using a transmission arrangement as shown in Figure 8, in which the reference beam is directed down the bore with an incident intensity of $\sim$ 1 W/cm^2. The angle between the signal and reference beam is 12° outside the crystallite, corresponding to an average grating period of 2.4 μm in the fiber. The signal beam, whose intensity is more than an order of magnitude less than the reference beam, consists of an illuminated transparency of the Air Force resolution target and is Fourier transformed such that the back focal plane is at the front face of the crystallite. The resolution is limited to 5 lp/mm by the Fourier lens (f = 100 mm) and the aperture of the front face of the crystallite.

A holographic optical element (HOE) is used to generate the phase-conjugate of the reference beam for image reconstruction. Eventually, the HOE may be replaced by a phase-conjugate resonator consisting of a c-axis fiber, as described below. For hologram storage and playback three steps are followed. First, a hologram of the reference wave as distorted by the fiber is stored in the SBN:61 HOE; shutters (b) and (e) are closed to prevent interference with the forward propagating signal wave and back propagating reference waves, respectively. Second, the hologram of the resolution test target is stored in the fiber; shutters (c), (d), and (e) are closed to isolate the HOE from this interaction. Finally, the HOE is illuminated to reconstruct the phase-conjugate reference wave, which in turn constructs a phase conjugate of the stored signal wave. Shutters (a), (b), and (d) are closed to eliminate scatter from the forward propagating reference beam, interaction with the forward propagating signal beam, and erasure of the HOE by the blocked HOE reference beam.

The results for storage of one image are shown in Figure 9. The resolution of the reconstructed phase-conjugate image is $\sim$ 4 lp/mm. This result has been obtained by using an idling self-pumped $BaTiO_3$ resonator, which has certain systems advantages for storing a single image but which cannot store more than one hologram as needed for multiplexing. By using multiple reference beams R, R_{HOE}, and R^*_{HOE}, more than one image is multiplexed in an a-axis fiber.

This problem has also been investigated in $LiNbO_3$ by a group of researchers [18] at NTT, who showed that for transmission holograms diffraction limited performance can be achieved while using phase conjugate readout, even in the presence of intrasignal beam coupling. Their experimental setup is reproduced

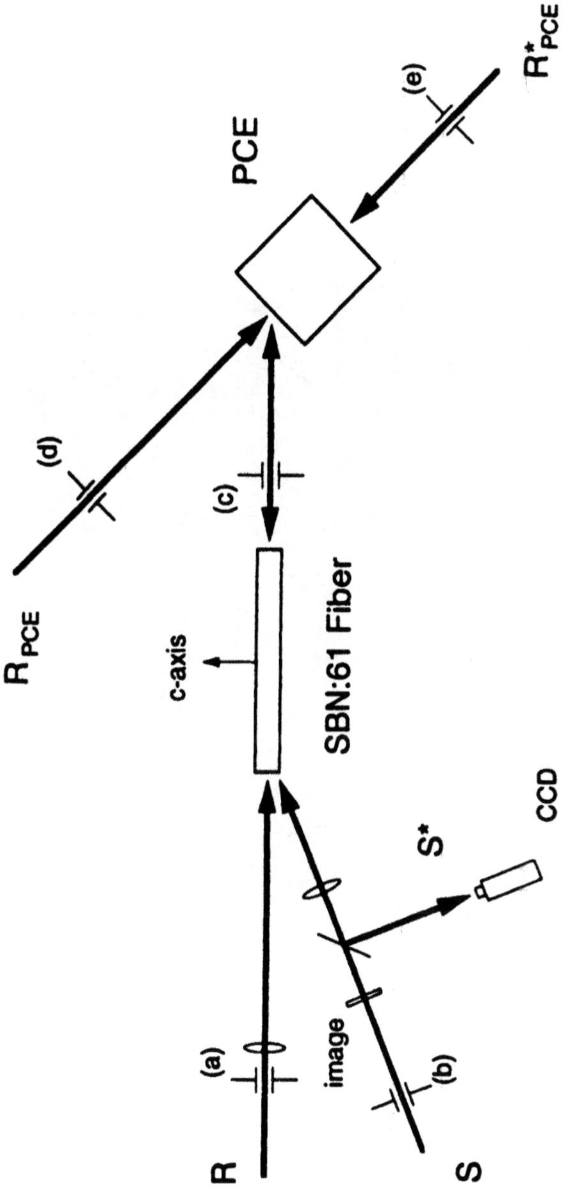

Figure 8: Experimental arrangement for storage of holograms in a-axis fibers and readout with the phase-conjugate reference beam.

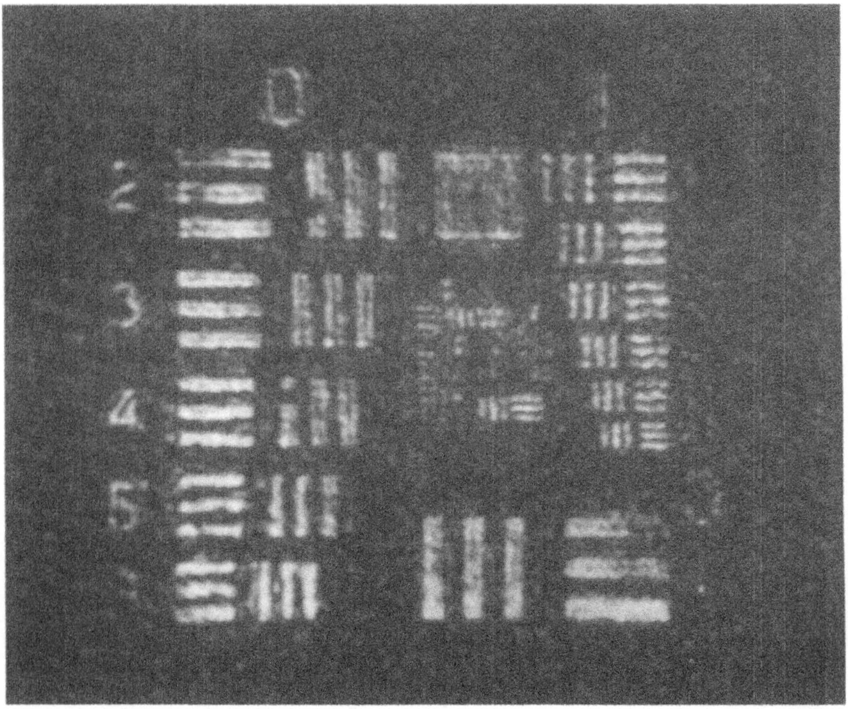

Figure 9: Reconstruction of a hologram stored in an a-axis SBN fiber using phase conjugate readout. The resolution limit imposed by the optics is - 5 lp/mm, and the resolution of the reconstructed image is - 4 lp/mm. The phase conjugate of the reference beam here is generated by an idling self-pumped $BaTiO_3$ resonator.

in Figure 10. Two plane waves and a reference beam are incident on a $LiNbO_3$ fiber (0.5 mm in diameter and 3 mm long, polished around its perimeter to reduce surface loss) to produce a transmission grating. The reference beam is phase conjugated by the the PCM. The coherence length of the laser is short compared with the separation between the PCM and the back end of the fiber to assure that no reflection grating is formed. Analysis shows that intrasignal distortion associated with reflection gratings cannot be compensated for in this manner, favoring again the use of a-axis over c-axis fibers. An attractive idea suggesting itself is to combine an a-axis SBN fiber for data storage and a c-axis SBN fiber for phase conjugate readout. Hesselink et al. have shown that c-axis SBN fibers can be used to produce a self-seeded, relatively high reflectivity (50%) phase conjugate signal by placing a retroreflecting material (3M Scottsguard) closely behind the fiber. This is potentially a viable technique, requiring detailed investigations of conjugation properties to evaluate it. In particular, a- and c-axis SBN fibers have not been coupled in a low-loss fashion, but appear to be at least an intriguing possibility. Alternatively, holograms could be recorded with the phase conjugate of the reference beam, allowing readout with an undistorted beam.

6.2. Optical interconnects and associative memory

Multichannel optical interconnects could possibly be implemented with photorefractive fiber arrays, which appear attractive for similar reasons as mentioned in conjunction with data storage applications. In this case, monochromatic light is usually used, so that wavelength encoding is frequently not a viable option. With reference to Figure 11, a transmission hologram is recorded with the phase conjugate of the reference beam using a polarization preserving phase conjugate mirror. By making the reference beam incoherent with the object beam, the reflection hologram is avoided during recording. Upon readout with the reference beam, distortions caused by modal dispersion, scattering and intra-signal coupling are compensated for. For this application a-axis fibers are again preferred. Selective enhancement of interconnect patterns for switching networks may be achieved by applying distinct electric fields along the fiber axis commensurate with those used during recording [19].

In another application of photorefractive fibers where large storage capacity is needed, Kitayama et al. [20] used two fibers for heteroassociative memory operations. In a proof-of-concept experiment, they used two fibers to recall a stored letter from an imperfect or partial input image.

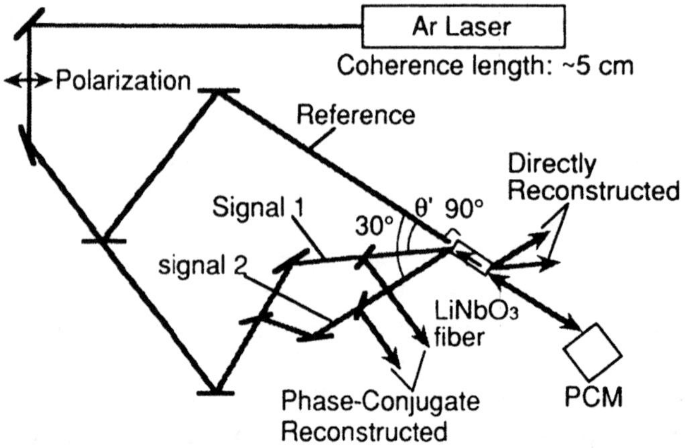

Figure 10: Experimental set-up for compensation of intra- modal distortion using phase conjugate readout of holograms stored in $LiNbO_3$ fibers (After Ito et al., Ref. 22).

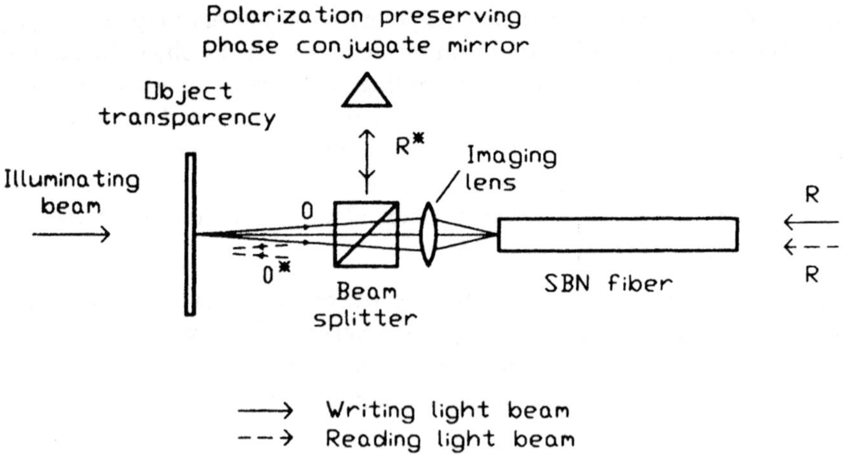

Figure 11: Optical interconnect using phase conjugate reference recording beam (after Wu et al., Ref. 20).

6.3. Two-beam coupling

In photorefractive media, two interfering beams can exchange energy between themselves, as a result of the phase shift between the interference pattern and the resulting index-of-refraction modulation, as described above. In bulk media, where the interacting beams propagate in two distinct directions, energy flow is unidirectional. In fibers, on the other hand, pump and signal beams undergo periodic variations in direction caused by guiding. This results in oscillatory energy exchange between two beams in a-axis fibers. Considering, for example, the geometry of Figure 12, a signal bearing beam propagates along the fiber axis, and the pump or reference beam is incident at an oblique angle, resulting in unidirectional energy transfer from the pump to the signal beam. For the desirable a-axis configuration, where the angle β between the c-axis and the fiber axis is larger than the propagation angle of the pump beam θ, periodic energy transfer between the two beams results, giving rise to a small net effect. This problem can be overcome by periodically poling the crystal as shown in figure 12. Selective enhancement of the signal beam occurs when the propagation angle $\theta_0 = \tan(a/l)$, where a is the fiber diameter, and l denotes the period of the reversal structure. Consequently, intra-signal coupling is reduced and amplification of the signal beam is enhanced as shown in Figure 12b. In an experimental demonstration, Ito et al. [21] used periodically reversed single crystal $BaTiO_3$ crystals, but periodically poled fibers should be suitable as well.

Beam coupling also produces mode conversion in waveguides, in which higher order modes are funneled into lower orders if the signal beam propagates in the direction of the c-axis. This is useful for combining multiple laser beams into one single beam, as is frequently required to produce high powered diode lasers. Vice-versa, antifunneling (scattering into higher order modes) is induced when the signal waves propagate counter to the c-axis [22].

6.4. Double phase conjugation

Double phase conjugation has been demonstrated in bulk materials like $BaTiO_3$ and $Bi_{12}TiO_{20}$ where two incoherent beams are incident on a photorefractive medium giving rise to two phase conjugate beams. The incoherent beams couple in a single 4-wave mixing process, allowing frequency locking, beam profile improvement and signal amplification. Beam fanning is fundamentally responsible for double phase conjugation. Since this process depends on noise amplification by scattering, beam intensity plays a crucial role in the time response and signal strength of the phase conjugate signals.

The fiber configuration lends itself well for this experiment, as long interaction lengths and beam confinement allow very efficient phase holograms to be

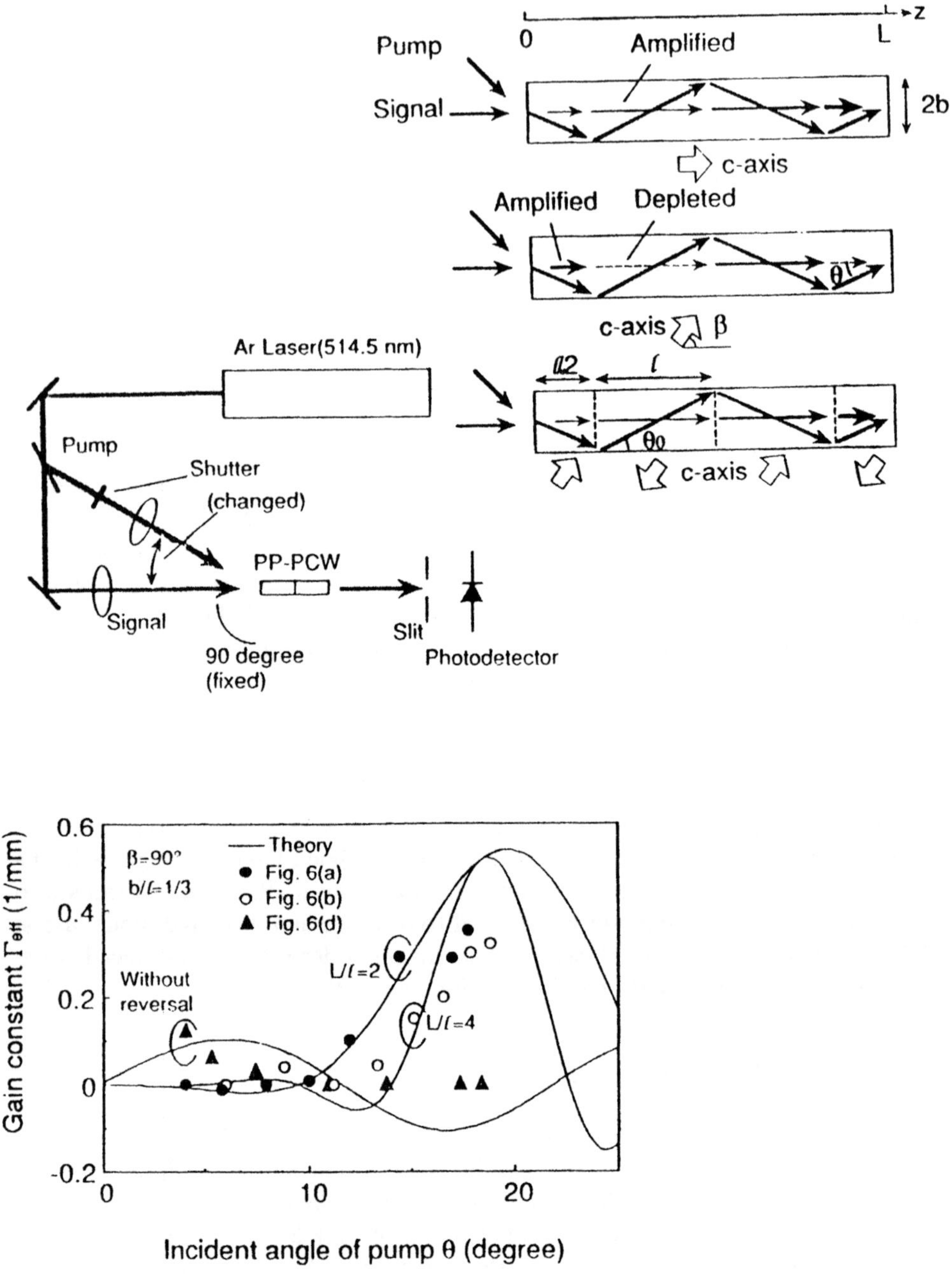

Figure 12: Energy transfer via two-beam coupling in waveguides. Measured gain versus angle of incidence for periodically and non-periodically poled $BaTiO_3$ waveguides (after Ito et al., Ref. 18).

written. Kamshilin et al., have recently demonstrated double phase conjugation in $Bi_{12}TiO_{20}$ (BTO) fibers (actually small rods of crossectional size 0.9 x 1.3 mm). Since the lateral dimension are relatively small compared with bulk crystals, high field strengths are obtainable, exceeding 50 kV/cm. As a result of this geometry, while applying an alternating electric field, phase conjugate reflectivities of 7% have been obtained (not counting reflection losses) using a He-Ne laser of 24 mW power. Phase conjugate signals with a spatial bandwidth of 160 lp/mm have also been phase conjugated with good fidelity in a 10 mm long fiber.

6.5 Summary

Photorefractive fibers appear attractive for a variety of optical signal processing and data storage applications for which traditionally bulk crystals are used. The LHPG technique offers the advantage of short run times and an efficient growth process allowing dopant and materials studies to be carried out conveniently. Since bulk crystals of large size are difficult to grow economically, bundles of fibers may be an attractive alternative for applications such as holographic data storage requiring large storage volumes. Since the optical quality of ferroelectric photorefractive fibers is generally quite good as a result of steep temperature gradients in the melt during fabrication, scatter caused by bulk growth striations is substantially reduced. There are, however, still problems to overcome. A-axis SBN fibers exhibit a small index of refraction variation in the core, a problem that is being studied intensively at several laboratories with good prospects for finding a solution. Arrays of photorefractive fibers or crystallites should be constructed to demonstrate feasibility in data storage systems and to show that practical fabrication issues such as applying fields to a-axis crystallites can be worked out. But if these issues can be dealt with in a satisfactory manner, then there are interesting possibilities for new applications using photorefractive fibers coupled to glass fibers for sensors and optical image processing. Additionally, the advantages of long interaction lengths, and energy confinement in fibers should be exploited for nonlinear optical experiments, providing a powerful stimulus for further research and development on photorefractive materials and devices.

References

[1] P. Gunther and J-P Huignard. *Photorefractive materials and their applications*, pages 1–367. Springer, Berlin, 1988.

[2] R.W. Hellwarth. Theory of phase-conjugation by four-wave mixing in a waveguide. *IEEE, Journal of Quantum Electronics*, 15(2):101–109, 1979.

[3] L. Hesselink and Steve Redfield. Photorefractive holographic recording in strontium barium niobate fibers. *Opt. Lett.*, 13(10):877–879, 1988.

[4] Steve Redfield and Lambertus Hesselink. Enhanced nondestructive holographic readout in SBN. *Opt. Lett.*, 13(10):880–882, 1988.

[5] Jeffrey P. Wilde, Dieter H. Jundt, Ludwig Galambos, and Lambertus Hesselink. Growth of SBN fibers: new results regarding orientation. *J. Cryst. Growth*, 114:500–506, 1991.

[6] K.O. Hill, Y. Fujii, D.C. Johnson, and B.S. Kawasaki. Photosensitivity in optical fiber waveguides:applications to reflection filter fabrication. *Appl. Phys. Lett.*, 32(10):647–649, 1978.

[7] Amnon Yariv. On transmission and recovery of three-dimensional image formation in optical waveguides. *J. Opt. Soc. Am.*, 66(4):301–306, 1976.

[8] Yariv A., J. AuYeung, D. Fekete, and D.M. Pepper. Image phase compensation and real-time holography by four-wave mixing in optical fibers. *Appl. Phys. Lett.*, 32(10):635–637, 1978.

[9] AuYeung, D. Fekete J., D.M. Pepper, and A. Yariv. Continuous backward-wave generation by degenerate four-wave mixing in optical fibers. *Opt. Lett.*, 4(1):42–44, 1979.

[10] Yasayuki Sugiyama, Itaru Yokohama, Ken ichi Kubodera, and Shogo Yagi. Growth and photorefractive properties of a- and c-axis cerium-doped strontium barium niobate single crystal fibers. *IEEEP Photonics Technology Lett.*, 3(8):744–746, 1991.

[11] Jensen S.M. and R.W. Hellwarth. Of the time-reversed replica of a monochromatic optical wave. *Appl. Phys. Lett.*, 32(3), 1978.

[12] R. S. Feigelson. Pulling optical fibers. *J. Crystal Growth*, 79:669–680, 1986.

[13] M.J. Fejer. *Single crystal fibers: Growth dynamics and non-linear optical interactions.* PhD thesis, Stanford University, 1986.

[14] Lambertus Hesselink. Photorefractive fibers for optical data storage and processing. *Int. J. Optoelectron.*, 5:103–124, 1990.

[15] John F. Heanue, Matthew C. Basahw, and Lambertus Hesselink. Volume holographic storage and retrieval of digital data. *Science*, 265:749–752, 1994.

[16] J. Feinberg. Asymmetric self-defocusing of an optical beam from the photorefractive effect. *J. Opt. Soc. Am.*, 72:46–51, 1982.

[17] M. C. Bashaw, A. Aharoni, and L. Hesselink. A-axis strontium barium niobate single-crystal fibers for holographic storage. *to be published*, 1993.

[18] Fumihiko Ito, Ken ichi Kitayama, and Osaka Nakao. Enhanced two-wave mixing in a photorefractive waveguide having periodically reversed c-axis by electrical poling technique. *Appl. Phys. Lett.*, 60(7):793–795, 1992.

[19] Shudong Wu, Andy Mayers, Sumati Rajan, and Francis T.S. Yu. Use of photorefractive fiber in optical interconnections and switching. *Appl. Opt.*, 29(8):1059–1061, 1990.

[20] Ken-Ichi Kitayama and Fumihiko Ito. Optical signal processing using photorefractive effect. *Multidimensional systems and signal processing*, 2:401–419, 1991.

[21] Fumihiko Ito, Ken ichi Kitayama, and Hitoshi Oguri. Compensation of fiber holographic image distortion caused by intrasignal photorefractive coupling by using a phase-conjugate mirror', periodically reversed c-axis by electrical poling technique. *Opt. Lett.*, 17(3):215–217, 1992.

[22] Baruch Fisher and Mordechai Segev. Photorefractive waveguides and nonlinear mode coupling effects. *Appl. Phys. Lett.*, 54(8):684–686, 1989.

Index:

GPSR Compliance

The European Union's (EU) General Product Safety Regulation (GPSR) is a set of rules that requires consumer products to be safe and our obligations to ensure this.

If you have any concerns about our products, you can contact us on ProductSafety@springernature.com

In case Publisher is established outside the EU, the EU authorized representative is:

Springer Nature Customer Service Center GmbH
Europaplatz 3
69115 Heidelberg, Germany

Batch number: 10755284

Printed by Printforce, the Netherlands